AF514296

Mathematics of Planet Earth

Volume 14

This series provides a variety of well-written books of a variety of levels and styles, highlighting the fundamental role played by mathematics in a huge range of planetary contexts on a global scale. Climate, ecology, sustainability, public health, diseases and epidemics, management of resources and risk analysis are important elements. The mathematical sciences play a key role in these and many other processes relevant to Planet Earth, both as a fundamental discipline and as a key component of cross-disciplinary research. This creates the need, both in education and research, for books that are introductory to and abreast of these developments.

Springer's MoPE series will provide a variety of such books, including monographs, textbooks, contributed volumes and briefs suitable for users of mathematics, mathematicians doing research in related applications, and students interested in how mathematics interacts with the world around us. The series welcomes submissions on any topic of current relevance to the international Mathematics of Planet Earth effort, and particularly encourages surveys, tutorials and shorter communications in a lively tutorial style, offering a clear exposition of broad appeal.

Mathilde B. Sørensen · Jörn Behrens ·
Fatemeh Jalayer · Finn Løvholt · Stefano Lorito ·
Irina Rafliana · Mario A. Salgado-Gálvez ·
Jacopo Selva
Editors

Probabilistic Tsunami Hazard and Risk Analysis

A Cookbook

Editors
Mathilde B. Sørensen
Department of Earth Science
University of Bergen
Bergen, Norway

Jörn Behrens
Department of Mathematics
Universität Hamburg
Hamburg, Germany

Fatemeh Jalayer
Department of Disaster Risk Reduction
University College London
London, UK

Finn Løvholt
Offshore Energy
Norwegian Geotechnical Institute
Oslo, Norway

Stefano Lorito
Istituto Nationale di Geofisica e Vulcanologia
Roma, Italy

Irina Rafliana
Research Center for Area Studies
Badan Riset dan Inovasi Nasional (BRIN) Indonesia
Jakarta, Indonesia

Center for Life Ethics
Universität Bonn
Bonn, Germany

Mario A. Salgado-Gálvez
CIMNE
Barcelona, Spain

ERN International
Mexico City, Mexico

Jacopo Selva
Dipartimento di Scienze della Terra, dell' Ambiente e delle Risorse (DiSTAR)
Università di Napoli Federico II
Naples, Italy

ISSN 2524-4264 ISSN 2524-4272 (electronic)
Mathematics of Planet Earth
ISBN 978-3-031-98114-2 ISBN 978-3-031-98115-9 (eBook)
https://doi.org/10.1007/978-3-031-98115-9

This Springer imprint is published by the registered company Springer Nature Switzerland AG
The registered company address is: Gewerbestrasse 11, 6330 Cham, Switzerland

Acknowledgements

This publication is based upon work from COST Action AGITHAR, CA18109, supported by COST (European Cooperation in Science and Technology). We appreciate discussions with and input from a large number of colleagues during AGITHAR meetings and other meetings and conferences. In particular, we thank Anawat Suppasri, Gareth Davies, Eric Geist, Katsuishiro Goda, Tomaso Esposti Ongaro, Vasily Titov, Stuart Fraser, and Denis Chang Seng for reviewing individual chapters of the book, as well as Jörn Lauterjung for reviewing the final draft of the complete book.

Jörn Behrens acknowledges for the ASCETE Earthquake-Tsunami Coupling recipe (R7) support through Volkswagen-Foundation through grant AZ 85434. The computing and archive facilities used in recipe R10 were provided by the European Centre for Medium-range Weather Forecasts (ECMWF) through national quota and the ECMWF Special Projects. Mario A. Salgado-Gálvez acknowledges the support through the Severo Ochoa Centers of Excellence Program (CEX 2018-000797-S) funded by MCIN/AEI/10.13039/501100011033. Jörn Lauterjung, Harald Spahn, and Jörn Behrens acknowledge support from the German Federal Ministry for Education and Research (BMBF), Grant 03TSU01, for recipe R21. Natalia Zamora acknowledges the support through the ChEESE-2P—Center of Excellence in Solid Earth Second Phase, grant agreement ID: 101093038 https://doi.org/10.3030/101093038. Miquel Canals and Galderic Lastras acknowledge the financial support by a Catalan Government Grups de Recerca Consolidats grant (2021 SGR 01195).

The authors would like to acknowledge partial support from Horizon Europe Project Geo-INQUIRE. Geo- INQUIRE is funded by the European Commission under project no. 101058518 within the HORIZON-INFRA-2021-SERV-01 call.

Contents

Part I Ingredients

1 **Introduction** 3
Mathilde B. Sørensen, Jörn Behrens, Stefano Lorito, Finn Løvholt, Mario A. Salgado-Gálvez, Fatemeh Jalayer, Jacopo Selva, and Irina Rafliana

2 **Frameworks for Assessing Tsunami Hazard and Risk** 13
Natalia Zamora, Anita Grezio, Maria Papathoma-Köhle, Fatemeh Jalayer, Dimitra Salmanidou, Tom Parsons, Eric L. Geist, Jacopo Selva, Mathilde B. Sørensen, and Irina Rafliana

3 **Source Selection and Their Activity Rate** 87
Finn Løvholt, Roberto Basili, Andrea Rovida, Stefano Lorito, Fabrizio Romano, Mathilde B. Sørensen, Carl B. Harbitz, Jacopo Selva, Raphaël Paris, Anita Grezio, and Cléa Denamiel

4 **Source Modelling** 157
Finn Løvholt, Shane Murphy, Stefano Lorito, Fabrizio Romano, Carl B. Harbitz, Jacopo Selva, Raphaël Paris, Anita Grezio, and Cléa Denamiel

5 **Hydrodynamic Modelling** 223
Íñigo Aniel-Quiroga, Maria Ana Baptista, Jörn Behrens, Miquel Canals, Cléa Denamiel, Utku Kânoğlu, Galderic Lastras, Finn Løvholt, Jorge Macías, Shane Murphy, and Raphaël Paris

6 **Exposure Modelling** 265
Mario A. Salgado-Gálvez, Naveen Ragu Ramalingam, Fatemeh Jalayer, Danijel Schorlemmer, Helen Crowley, and Rade Rajkovchevski

7 **Vulnerability Modelling** 307
Marco Baiguera, Ignacio Aguirre Ayerbe, Raffaele De Risi, Marta Del Zoppo, Ufuk Hancilar, Rozana Himaz, Maria Papathoma-Köhle, Tiziana Rossetto, and Mislav Stepinac

8 **Communicating Tsunami Risk and Uncertainties** 341
Nikos Kalligeris, Irina Rafliana, Lorenzo Cugliari, Harald Spahn, Andrea Cerase, Jörn Lauterjung, Dimitra Salmanidou, Musavver Didem Cambaz, František Babič, Ignacio Aguirre Ayerbe, and Alessandro Amato

9 **Final Remarks** 383
Jörn Behrens, Fatemeh Jalayer, Shane Murphy, Jacopo Selva, Mathilde B. Sørensen, Mario A. Salgado-Gálvez, Stefano Lorito, Finn Løvholt, and Irina Rafliana

Part II Recipes

10 **R1–The NEAM Tsunami Hazard Model 2018 (NEAMTHM18)–Region-Wide Probabilistic Tsunami Hazard Based on a Large Number of Tsunami Scenarios** 395
Roberto Basili and Stefano Lorito

11 **R2–Local Probabilistic Tsunami Hazard Analysis** 405
Manuela Volpe, Steven J. Gibbons, and Enrico Baglione

12 **R3–Probabilistic Analysis for Tsunami Hazard in Macro Tidal Coasts** 413
Íñigo Aniel-Quiroga, Ernesto Mauricio Gonzalez, Maria Ana Baptista, and Jorge Macías

13 **R4–Event-Tree PTHA of Coastal Inundation with Tidal Uncertainty Treatment** 421
Rachid Omira, Maria Ana Baptista, and Luis Manuel Matias

14 **R5–Combination of Multiple Tsunami Rate Estimates Weighted by Uncertainty with Likelihood Functions: An Example Runup Probability Map for the Caribbean Region** ... 429
Tom Parsons and Eric L. Geist

15 **R6–Worst-Case Tsunami Approach Applied to Catania (Eastern Sicily)** 437
Filippo Zaniboni and Alberto Armigliato

16 **R7—ASCETE Coupled Earthquake-Tsunami Modelling** 447
Jörn Behrens

17 **R8—Probabilistic Tsunami Hazard Analysis for Landslide Sources—The Lyngen Case** 453
Finn Løvholt and Sylfest Glimsdal

18 **R9—Probabilistic Hazard Analysis for Tunamis Generated by Subaqueous Volcanic Explosions in the Campi Flegrei Caldera, Italy** ... 459
Anita Grezio, Raphaël Paris, and Jacopo Selva

19 **R10—Adriatic APTHA** ... 467
Cléa Denamiel

20 **R11—Surrogate-Based PTHA** ... 475
Cléa Lumina Denamiel and Dimitra Salmanidou

21 **R12—Multi-Source (Bayesian) PTHA in the Gulf of Naples, Italy** ... 485
Anita Grezio and Jacopo Selva

22 **R13—Global Risk Model 2015—Tsunami Component—Global Risk and Loss Computation** ... 493
Finn Løvholt, Mario A. Salgado-Gálvez, and Gareth Davies

23 **R14—Event-Based Probabilistic Earthquake and Tsunami Risk Assessment—Application in Acapulco (Mexico) and Callao (Peru)** ... 501
Mario A. Salgado-Gálvez, Mario Ordaz, and Benjamín Huerta

24 **R15—Multi-Hazard Risk Assessment for Earthquakes and Tsunamis** ... 507
Katsuichiro Goda and Raffaele De Risi

25 **R16—Bayesian Empirical Fragility Assessment for Hierarchical Damage Levels: 2018 Sulawesi-Palu Tsunami** ... 515
Fatemeh Jalayer and Hossein Ebrahimian

26 **R17—Nonlinear Pushover Analysis for Tsunami** ... 523
Marco Baiguera, Marta Del Zoppo, and Tiziana Rossetto

27 **R18—Analytical Vulnerability Assessment for Italian Buildings Portfolio** ... 531
Marta Del Zoppo, Marco Di Ludovico, and Andrea Prota

28 **R19—An Indicator-Based Method for the Assessment of the Physical Vulnerability of Buildings to Tsunami** ... 539
Maria Papathoma-Köhle

29 **R20—Monte Carlo Approach for Tsunami Economic Impact Estimation in a Pseudo-Probabilistic Framework** ... 545
Alejandro González del Pino, Jorge Macías, Miguel Llorente Isidro, and Carlos Paredes Bartolomé

30 **R21—GITEWS End-To-End Early Warning** ... 555
Jörn Lauterjung, Harald Spahn, and Jörn Behrens

31 R22—Probabilistic Tsunami Forecasting for Tsunami Warning 565
Stefano Lorito and Jacopo Selva

32 R23—Use of Neural Networks for Tsunami Arrival Time and Maximum Height Prediction 575
Juan Francisco Rodríguez Gálvez and Jorge Macías

33 R24—PTHA-Based Tsunami Inundation Mapping in Italy 583
Roberto Tonini, Daniela Di Bucci, and Pio Di Manna

34 R25—Methodology for Tsunami Hazard Microzoning Applied to Evacuation Decision Making 593
Natalia Zamora, Patricio A. Catalan, and Alejandra Gubler

Contributors

Ignacio Aguirre Ayerbe Instituto de Hidráulica Ambiental (IH Cantabria), Universidad de Cantabria, Santander, Spain

Alessandro Amato Centro Allerta Tsunami, Istituto Nationale di Geofisica e Vulcanologia, Roma, Italy

Íñigo Aniel-Quiroga Instituto de Hidráulica Ambiental (IH Cantabria), Universidad de Cantabria, Santander, Spain

Alberto Armigliato Dipartimento Di Fisica e Astronomia "A. Righi", Alma Mater Studiorum—Università Di Bologna, Bologna, Italy

František Babič Faculty of Electrical Engineering and Informatics, Technical University of Kosice, Kosice-Sever, Slovakia

Enrico Baglione Istituto Nazionale di Geofisica e Vulcanologia, Bologna, Italy

Marco Baiguera Risk Team, Global Earthquake Model (GEM) Foundation, Pavia, Italy

Maria Ana Baptista Instituto Superior de Engenharia de Lisboa, Instituto Politécnico de Lisboa, Lisboa, Portugal

Roberto Basili Istituto Nazionale di Geofisica e Vulcanologia, Roma, Italy

Jörn Behrens Department of Mathematics, Universität Hamburg, Hamburg, Germany

Musavver Didem Cambaz Kandilli Observatory and Earthquake Research Institute, Boğaziçi University, Istanbul, Türkiye

Miquel Canals Department of Earth and Ocean Dynamics, CRG Marine Geosciences, Universitat de Barcelona, Barcelona, Spain

Patricio A. Catalan Departamento de Obras Civiles, Universidad Tecnica Federico Santa Maria, Valparaiso, Chile;
Research Center for Integrated Disaster Risk Management (CIGIDEN), Santiago, Chile

Andrea Cerase Department of Communication and Social Research (CoRiS), Sapienza Università di Roma, Roma, Italy;
National Earthquake Observatory, Istituto Nationale di Geofisica e Vulcanologia, Roma, Italy

Helen Crowley Global Earthquake Model (GEM) Foundation, Pavia, Italy

Lorenzo Cugliari Istituto Nationale di Geofisica e Vulcanologia, Roma, Italy; Department of Communication and Social Research (CoRiS), Sapienza Università di Roma, Roma, Italy

Gareth Davies Geoscience Australia, Symonston, Australia

Marta Del Zoppo Università di Napoli Federico II, Naples, Italy

Cléa L. Denamiel Division of Marine and Environmental Research, Ruđer Bošković Institute, Zagreb, Croatia

Raffaele De Risi School of Civil, Aerospace and Design Engineering, University of Bristol, Bristol, UK

Daniela Di Bucci Italian Civil Protection Department, Presidency of the Council of Ministers, Roma, Italy

Marco Di Ludovico Department of Structures for Engineering and Architecture, Università di Napoli Federico II, Naples, Italy

Pio Di Manna ISPRA-Italian Institute for Environmental Protection and Research, Geological Survey of Italy, Roma, Italy

Hossein Ebrahimian Department of Structures for Engineering and Architecture, Università di Napoli Federico II, Naples, Italy

Eric L. Geist US Geological Survey, Moffett Field, USA

Steven J. Gibbons Natural Hazards, Norwegian Geotechnical Institute, Oslo, Norway

Sylfest Glimsdal Natural Hazards, Norwegian Geotechnical Institute, Oslo, Norway

Katsuichiro Goda Earth Sciences, Western University, London, ON, Canada

Alejandro González del Pino Departamento de Análisis Matemático, Estadística e Investigación Operativa y Matemática Aplicada, Universidad de Málaga, Málaga, Spain

Ernesto Mauricio Gonzalez Instituto de Hidráulica Ambiental (IH Cantabria), Universidad de Cantabria, Santander, Spain

Anita Grezio Istituto Nazionale di Geofisica e Vulcanologia, Bologna, Italy

Alejandra Gubler Research Center for Integrated Disaster Risk Management (CIGIDEN), Santiago, Chile

Ufuk Hancilar Department of Earthquake Engineering, Kandilli Observatory and Earthquake Res. Inst., Boğaziçi University, İstanbul, Türkiye

Carl B. Harbitz Natural Hazards, Norwegian Geotechnical Institute, Oslo, Norway

Rozana Himaz Department of Risk and Disaster Reduction, University College London, London, UK

Benjamín Huerta ERN International, Mexico City, Mexico

Fatemeh Jalayer Department of Disaster Risk Reduction, University College London, London, UK

Nikos Kalligeris Institute of Geodynamics, National Observatory of Athens, Athens, Greece

Utku Kânoğlu Department of Aerospace Engineering, Dokuz Eylül University, Izmir, Türkiye

Galderic Lastras Department of Earth and Ocean Dynamics, CRG Marine Geosciences, Universitat de Barcelona, Barcelona, Spain

Jörn Lauterjung Geophysics, Helmholtz Centre Potsdam GFZ German Research Centre for Geociences, Potsdam, Germany

Miguel Llorente Isidro Geological Risks and Climate Change, IGME (CSIC), Madrid, Spain

Stefano Lorito Istituto Nazionale di Geofisica e Vulcanologia, Roma, Italy

Finn Løvholt Offshore Energy, Norwegian Geotechnical Institute, Oslo, Norway

Jorge Macías Departamento de Análisis Matemático, Estadística e Investigación Operativa y Mathemática Aplicada, Universidad de Málaga, Málaga, Spain

Luis Manuel Matias Instituto Dom Luiz, Faculdade de Ciências da Universidade de Lisboa, Lisboa, Portugal

Shane Murphy Geo-Ocean, Ifremer, Plouzané, France

Rachid Omira Instituto Português do Mar e da Atmosfera–IPMA, Lisboa, Portugal

Mario Ordaz Instituto de Ingeniería, Universidad Nacional Autónoma de México (UNAM), Mexico City, Mexico

Maria Papathoma-Köhle Institute for Mountain Risk Engineering (IAN), BOKU University, Vienna, Austria

Carlos Paredes Bartolomé Geological and Mining Engineering, Escuela de Minas y Energía, Universidad Politécnica de Madrid, Madrid, Spain

Raphaël Paris CNRS, IRD, OPGC, Laboratoire Magmas et Volcans, Université Clermont Auvergne, Aubière, France

Tom Parsons US Geological Survey, Moffett Field, USA

Andrea Prota Università di Napoli Federico II, Naples, Italy

Irina Rafliana Research Center for Area Studies, Badan Riset Dan Inovasi Nasional (BRIN) Indonesia, Jakarta, Indonesia;
Center for Life Ethics, Universität Bonn, Bonn, Germany

Rade Rajkovchevski Faculty of Security–Skopje, St. Kliment Ohridski University, Skopje, North Macedonia

Naveen Ragu Ramalingam University School for Advanced Studies–IUSS Pavia, Pavia, Italy

Juan Francisco Rodríguez Gálvez Departamento de Análisis Matemático, Estadística e Investigación Operativa y Matemática Aplicada, Universidad de Málaga, Málaga, Spain

Fabrizio Romano Istituto Nazionale di Geofisica e Vulcanologia, Roma, Italy

Tiziana Rossetto Faculty of Civil Engineering and Geosciences (CEG), Delft University of Technology (TU Delft), Delft, The Netherlands

Andrea Rovida Istituto Nazionale di Geofisica e Vulcanologia, Roma, Italy

Mario A. Salgado-Gálvez CIMNE, Barcelona, Spain;
ERN International, Mexico City, Mexico

Dimitra Salmanidou Advanced Research Computing Centre, University College London, London, UK

Danijel Schorlemmer Seismic Hazard and Risk Dynamics, Helmholtz Centre Potsdam GFZ German Research Centre for Geosciences, Potsdam, Germany

Jacopo Selva Dipartimento di Scienze della Terra, dell' Ambiente e delle Risorse (DiSTAR), Università di Napoli Federico II, Naples, Italy

Mathilde B. Sørensen Department of Earth Science, University of Bergen, Bergen, Norway

Harald Spahn Independent Consultant, Apen, Germany

Mislav Stepinac Faculty of Civil Engineering, University of Zagreb, Zagreb, Croatia

Roberto Tonini Istituto Nazionale di Geofisica e Vulcanologia, Roma, Italy

Manuela Volpe Osservatorio Nazionale Terremoti, Istituto Nazionale di Geofisica e Vulcanologia, Roma, Italy

Natalia Zamora Wave Phenomena Group, Computer Applications in Science and Engineering, Barcelona Supercomputing Center, Barcelona, Spain

Filippo Zaniboni Dipartimento Di Fisica e Astronomia "A. Righi", Alma Mater Studiorum—Università Di Bologna, Bologna, Italy

Abbreviations

AAL	Average Annual Loss
ABM	Agent-Based Modeling
ADB	Asian Development Bank
ADCIRC	Advanced CIRCulation
ADER-DG	Arbitrary high order DERivative Discontinuous Galerkin
AdriSC	Adriatic Sea and Coast
AGITHAR	Accelerating Global Science in Tsunami Hazard and Risk Analysis
AHP	Analytic Hierarchy Process
AI	Artificial Intelligence
AIC	Akaike's Information Criterion
ANN	Artificial Neural Network
ANOVA	Analysis of Variance
ANUGA	Australian National University and Geoscience Australia tsunami propagation and inundation code
APTHA	Atmospheric Probabilistic Tsunami Hazard Analysis (or Assessment)
AR	Augmented Reality
ARCGIS	Aeronautical Reconnaissance Coverage Geographic Information System
ASCE	American Society of Civil Engineers
ASCE7-22	American Society of Civil Engineers Standard 7-22
ASCETE	Advanced Simulation of Coupled Earthquake Tsunami Events
ASCII	American Standard Code for Information Interchange
ASTARTE	Assessment, STrategy and Risk Reduction for Tsunamis in Europe
ASTER	Advanced Spaceborne Thermal Emission
BBB	Build Back Better
BIC	Bayesian Information Criterion
BIM	Boundary Integral Methods

BMCS	Bayesian Updating and Model Class Selection
BP	Before Present
BPT	Brownian Passage Time
BRIFS	Balearic Rissaga Forecasting System
BS	Background Seismicity
BUI	Bottom-Up Citizen Initiatives
CAM	Continent—Azores—Madeira
CAPRA	Probabilistic Risk Assessment Tool
CAT	Italian Tsunami Alert Centre
CAT-INGV	Centro Allerta Tsunami—Istituto Nazionale di Geofisica e Vulcanologia/Tsunami Alert Center—National Institute of Geophysics and Vulcanology
CDPO	Constant-Depth Pushover
CE	Common or Current Era
CFD	Computational Fluid Dynamics
CFL	Courant-Friederic-Lewy stability criterion
CIA World Factbook	Central Intelligence Agency US World Fact Book
CIAT	Center for Tropical Agriculture, Spain
CIESIN	Center for International Earth Science Information Network
CIMNE	Centre Internacional de Metodes Numerics en Enginyeria
CMeEWS	Croatian Meteotsunami Early Warning System
CMIP6	Coupled Model Intercomparison Project Phase 6
CN IGME-CSIC	Centro Nacional Instituto Geológico y Minero de España—Consejo Superior de Investigaciones Científicas
COAWAST	Coupled Ocean-Atmosphere-Wave-Sediment Transport
COMCOT	Cornell Multi-grid Coupled Tsunami Model
CPT	Cone Penetration Test
CVI	Composite Vulnerability Index
DAS	Distributed Acoustic Sensing
DEM	Digital Elevation Model
DEM	Discrete Element Method
DG	Discontinuous Galerkin
DG-ECHO	The Directorate General for European Civil Protection and Humanitarian Aid Operations
DI	Damage Index
DLR	Deutsches Zentrum für Luft- und Raumfahrt e.V.
DM	Damage Measure
DNB	Day/Night Band
DONET	Dense Oceanfloor Network for Earthquakes and Tsunami, Japan
DPC	Dipartimento della Protezione Civile (Civil Protection Department), Italy
DPCM	Direttiva Del Presidente Del Consiglio Dei Ministri
DS	Damage State

DS1	No damage
DS2	Minor damage
DS3	Moderate damage
DS4	Major damage
DS5	Complete damage
DS6	Collapse
DS7	Wash-away
DTM	Digital Terrain Model
DV	Decision Variable
EAL	Expected Annual Loss
ECMWF	European Centre for Medium-Range Weather Forecast
ECMWF ATOS HPC	European Center for Medium-Range Weather Forecasts—Atos High-Performance Computing
EDP	Engineering Demand Parameters
EEFIT	Earthquake Engineering Field Investigation Team
ELT	Event Loss Table
EMEC	European-Mediterranean Earthquake Catalog
EMODNET	European Marine Observation and Data Network
EMS	European Macroseismic Scale
EPA	Environmental Protection Agency US
EPOS	European Plate Observing System data portal
ERA5	Fifth generation of atmospheric analysis on global climate
ERF	Earthquake Rate Forecast
ESM R2019	European Settlement Map very high resolution for reference year 2015 an resolutions of 2 m
ESRI	Environmental Systems Research Institute US
ESRM20	European Seismic Risk Model 2020
ETAS	Epidemic Type Aftershock Sequence
ETOPO2022	Earth TOPOgraphy
ETRiS	European Tsunami Risk Service
EU	European Union
EV	(Seismic) Event
EV	Explanatory Variables
FAIR	Findable, Accessible, Interoperable and Re-usable
FD or FDM	Finite Difference Method
FE or FEM	Finite Element Method
FEMA	Federal Emergency Management Agency US
FEMAP-646	Federal Emergency Management Agency P-646 Guidelines
FMD	Frequency Magnitude Distributions
FoS	Factor of Safety
FOSM	First-Order Sampling Method
FP	Forecast Point
FP7	EU Framework Program 7
FTRT	Faster-Than-Real-Time

FUNWAVE-TVD	FUNWAVE—Total Variation Diminishing
FV or FVM	Finite Volume Method
GAR15	Global Assessment Report on Disaster Risk Reduction 2015
GAR19	Global Assessment Report on Disaster Risk Reduction 2019
GCM	Global Climate Model
GDP	Gross Domestic Product
GEBCO	General Bathymetric Chart of the Oceans
GED4ALL	Global Exposure Database for ALL
GEM	Global Earthquake Model
GeoClaw	Version of Clawpack for geophysical waves and flows, Clawpack Conservation Laws Package
GHSL	Global Human Settlement Layer
GIS	Geographical Information Systems
GITEWS	German Indonesian Tsunami Early Warning System project
GLM	Generalized Linear Model
Gmsh	Open source 3D finite element mesh generator
GNSS	Global Navigation Satellite System
gPCE	Generalized Polynomial Chaos Expansion
GPU	Graphic Processing Unit
G-R	Gutenberg–Richter magnitude–frequency distribution
GRIB	General Regularly-distributed Information in Binary Form
GTM	Global Tsunami Model
HDF5	Hierarchical Data Format Version 5
HILL	High-Impact Low Likelihood
HPC	High-Performance Computing
HRSC	High-Resolution Stereoscopic Camera
HRSL	High-Resolution Settlement Layer
HTHH	Hunga Tonga–Hunga Ha'apai
HYDE	History Database of the Global Environment
IAEA	International Atomic Energy Agency
IDL	Instituto Dom Luiz
IDR	Interstorey Drift
IFPRI	International Food Policy Research Institute
IFREMER	Institut Français de Recherche pour l'Exploitation de la Mer
IFSAR	Interferometric Synthetic Aperture Radar
IGN	Instituto Geográfico Nacional
IM	Intensity Measure
InaRISK	Web-based dashboard on Risk Assessment in Indonesia
InaTEWS	Indonesian Tsunami Early Warning System
INVG	Istituto Nazionale di Geofisica e Vulcanologia
IOC	Intergovernmental Oceanographic Commission

IOC SLSMF	Intergovernmental Oceanographic Commission Sea Level Station Monitoring Facility
IOC UNESCO	Intergovernmental Oceanographic Commission, United Nations Educational, Scientific and Cultural Organization
IoE	Internet of Everything
IPCC	Intergovernmental Panel on Climate Change
IPMA	Instituto Português do Mar e da Atmosfera
IRGC	International Risk Governance Council
ISEL	Instituto Superior de Engenharia de Lisboa
JRC	European Commission Joint Research Center
KL	Karhunen–Loève expansion
LC1	Load Case 1
LEC	Loss Exceedance Curve
LFRS	Lateral-Force-Resisting System
LiDAR	Light Detection and Ranging
LPTHA	Landslide Probabilistic Tsunami Hazard Analysis (or Assessment)
LSW	Linear Shallow Water models
MaxEqMg	Maximum Earthquake Magnitude
MC	Monte Carlo scheme
MCDEM	Ministry of Civil Defense and Emergency Management, New Zealand
MCMC	Markov Chain Monte Carlo simulation
MDR	Mean Damage Ratio
MECE	Mutually Exclusive and Collectively Exhaustive
MFD	Magnitude–Frequency Distribution
MGWs	Mesoscale Gravity Waves
MHHW	Mean Higher High Water
MICE	Mutual Information for Computer Experiments
MIH	Maximum Tsunami Inundation Height
MinEqMg	Minimum Earthquake Magnitude
MISELA	Minute Sea-Level Analysis
ML	Machine Learning
MLE	Maximum Likelihood Estimate
MLIT	Ministry of Land, Infrastructure and Transportation Japan
MLLW	Mean Lower Low Water
MLP	Multi-Layer Perceptron
MMF	Moment of Momentum Flux
MOGP	Multi-Output Gaussian Process
MOST	Method of Splitting Tsunamis
MPI	Message Passing Interface
MR	Mixed Reality
MSL	Mean Sea Level
Mw	Moment Magnitude
NASA	National Aeronautics and Space Administration

NCEI	National Centers for Environmental Information
NCEP	National Centers for Environmental Prediction
NE	North Eastern
NE Atlantic	North European Atlantic region
NEAM	North-Eastern Atlantic, the Mediterranean and Connected Seas
NEAMTHM18	NEAM Tsunami Hazard Model 2018
NEAMTIC	The Intergovernmental Coordination Group for the Tsunami Early Warning and Mitigation System in the North-Eastern Atlantic, the Mediterranean and Connected Seas—Tsunami Information Centre
NetCDF	Network Common Data Form
NGO	Non-Government Organizations
NGU	Geological Survey of Norway (Norges Geologiske Undersøkelse)
NIR	Near-Infrared
NLP	Natural Language Processing
NLSW	Non-Linear Shallow Water models
NN	Neural Network
NOAA	National Oceanic and Atmospheric Administration
NOAA-20	NOAA satellite
NRMSD	Normalized Root Mean Square Deviation
NSF	National Science Foundation
NSLOTT	Non-seismic Sea-Level Oscillation at Tsunami Timescales
NSWING	Non-linear Shallow Water model with Nested Grids
NTHMP	National Tsunami Hazard Mitigation Program US
OASIS LMF	OASIS Loss Modeling Framework
OCR	Optical Character Recognition
OECD	Organization for Economic Co-operation and Development
OED	Open Exposure Data
OLS	Ordinary Least Squares regression
OpenFOAM	Open Field Operation and Manipulation
OpenMP	Open Multi-processing
ORNL	Oak Ridge National Laboratory
OSM	Open Street Map
PCA	Principal Component Analysis
PCTA	Peak Coastal Tsunami Amplitude
PDC	Pyroclastic Density Currents
PDF	Probability Density Function
PEER	Pacific Earthquake Engineering Center
PGW	Pseudo-global Warning
PIA	Participatory Integrated Assessment
PML	Probable Maximum Loss
PoI	Point of Interest

PROTECTS	PROject for Training, Education and Consulting Tsunami Early Warning System
PS	Predominant Seismicity
PSA	Pseudo-Spectral Approximation
PSHA	Probabilistic Seismic Hazard Analysis (or Assessment)
PTF	Probabilistic Tsunami Forecasting
PTHA	Probabilistic Tsunami Hazard Analysis (or Assessment)
PTRA	Probabilistic Tsunami Risk Analysis (or Assessment)
PTVA	Papathoma Tsunami Vulnerability Assessment
QAFI	Quaternary-Active Faults of Iberia
QGIS	Quantum Geographic Information System
RC	Reinforced Concrete
RCM	Regional Climate Model
RKDG	Runge–Kutta Discontinous Galerkin
ROMS	Regional Ocean Modeling System
RuptGen	Rupture Generator
RV	Response Variables
RVI	Relative physical vulnerability to tsunamis
SAFETI	Structural Assessment to Flow and Extreme Tsunami Inundation
SAT	Simultaneous Approximation Terms
SBP	Summation By Parts
SE or SEM	Spectral Element Method
SFDRR	Sendai Framework for Disaster Risk Reduction
SiAM	Sistema di Allertamento Maremoti
SMART	Science Monitoring and Reliable Telecommunications
SMF	Submarine Mass Failure
S-NET	Seafloor Observation Network for Earthquakes and Tsunamis, Japan
SoVI	Social Vulnerability Index
SPH	Smoothed Particle Hydrodynamics
SPHysics	Smoothed Particle Hydrodynamics Theory
SPTHA	Seismic Probabilistic Tsunami Hazard Analysis (or Assessment)
SRTM	Shuttle Radar Topography Mission
SSHAC	Senior Seismic Hazard Analysis Committee
SSP	Shared Socioeconomic Pathways
SSS	Submarine Seismic Sources
STAR	Study of the Tsunami Aftermath and Recovery
STEP-A	School Tsunami Emergency Response and Preparedness Assessment Apps, Indonesia
StormFlash	An Adaptive Discontinuous Galerkin Method for Geophysical Problems
SUOMI SSP	NASA/NOAA Suomi National Polar-orbiting Partnership
SW	South West

SWE	Shallow Water Equations
SWIM	South West Iberia Margin
SWIT	South West Iberian Transpressive Domain
TCS EPOS	Thematic Core Service, European Plate Observing System
TDY	Time-history Dynamic Analysis
TEWS	Tsunami Early Warning System
TGR	Tapered Gutenberg–Richter
TIGAR	Transient Inertia–Gravity and Rossby wave dynamics model
TRANSFER	Tsunami risk and strategies for the European region
TRRI	Tsunami Relative Risk Index
TSUMAP NEAM	Probabilistic Tsunami Hazard Maps for NEAM region
TsunaFlash	Predecessor of TAM: Tsunami simulation model on Adaptive Meshes
Tsunami HySEA	Hyperbolic System and Efficient Algorithms
TsunAWI	Tsunami modeling tool developed by AWI (Alfred Wegener Institut)
TUNAMI-N2	Tohoku University's Numerical Analysis Model for Investigation of Near field tsunamis with Leap-Frog Scheme
TWS	Tsunami Warning System
UNDRO	United Nations Disaster Relief Organization
UNDRR	United Nations Office for Disaster Risk Reduction
UNEP	United Nations Environment Programme
UNISDR	United Nations International Strategy for Disaster Reduction. Now UNDRR
UNPD	United Nations Population Division
UQ	Uncertainty Quantification
VDPO	Variable Depth Pushover
VIIRS	Visible Infrared Imaging Radiometer Suite
VPTHA	Volcano Probabilistic Tsunami Hazard Assessment
VR	Virtual Reality
WAPMERR	World Agency of Planetary Monitoring and Earthquake Risk Reduction
WCTSA	Worst-Case Credible Tsunami Scenario Analysis
WorldPop	Open Spatial Demographic Data and Research
WRF	Weather Research and Forecasting Model
WV	Water Vulnerability
XR	EXtended Reality

Part I
Ingredients

Chapter 1
Introduction

Mathilde B. Sørensen, Jörn Behrens, Stefano Lorito, Finn Løvholt, Mario A. Salgado-Gálvez, Fatemeh Jalayer, Jacopo Selva, and Irina Rafliana

Probabilistic tsunami hazard and risk assessment methods have evolved quickly over the past 10 to 15 years (e.g., Geist and Parsons 2006; Grezio et al. 2017; Mori et al. 2018; Behrens et al. 2021). These developments have been largely driven by an increasing demand for considering tsunamis in disaster risk management efforts, and for a complete identification and quantification of uncertainties in this type of assessments. Related probabilistic methods have been in use for earthquake hazard since the 1960s (Cornell 1968, McGuire 2008), and for performance-based earthquake engineering and seismic risk since the 1990s (Cornell and Krawinkler 2000; Grossi 2008).

M. B. Sørensen
Department of Earth Science, University of Bergen, Bergen, Norway
e-mail: mathilde.sorensen@uib.no

J. Behrens (✉)
Department of Mathematics, Universität Hamburg, Hamburg, Germany
e-mail: joern.behrens@uni-hamburg.de

S. Lorito
Istituto Nationale di Geofisica e Vulcanologia, Roma, Italy
e-mail: stefano.lorito@ingv.it

F. Løvholt
Offshore Energy, Norwegian Geotechnical Institute, Oslo, Norway
e-mail: finn.lovholt@ngi.no

M. A. Salgado-Gálvez
CIMNE, Barcelona, Spain
e-mail: mario.sal.gal@gmail.com

ERN International, Mexico City, Mexico

F. Jalayer
Department of Disaster Risk Reduction, University College London, London, UK
e-mail: f.jalayer@ucl.ac.uk

M. B. Sørensen et al. (eds.), *Probabilistic Tsunami Hazard and Risk Analysis*, Mathematics of Planet Earth 14, https://doi.org/10.1007/978-3-031-98115-9_1

While earthquake risk can be represented by a single workflow, tsunami risk is characterized by several different and cascading workflows, and it is much more dependent on numerical simulations. One key difference between probabilistic hazard assessment for tsunamis and earthquakes is in fact that the typical lack of tsunami observations, attributable to the relative rarity of events, and the complex influence of bathymetry and topography on wave propagation, imply a strong dependency on numerical simulations, which in turn requires much larger computational resources.

Given the rapidly evolving research landscape, there is a need to establish best practices for probabilistic tsunami hazard and risk assessment (abbreviated PTHA and PTRA, respectively) to improve reliability, comparability and reproducibility of studies applying such methods.

This book has been developed within the framework of Cost Action CA18109 AGITHAR (2019–2023) that intended to improve the scientific foundation for PTHA and PTRA. A group of more than 150 experts in the field have participated in this networking initiative. To materialize the networking activities into guidelines and best practices, more than 50 tsunami scientists have joined forces to develop this so-called *cookbook* providing recommendations and workflows for both PTHA and PTRA.

1.1 Idea

The aim of this cookbook is to provide an overview of existing methods, to unify the descriptions of named workflows, to make best practices examples available to a wider community, and to provide background information to various stakeholder groups. The main audience is scientists or practitioners working on implementing PTHA or PTRA, who will find a large number of references to state-of-the-art methods and examples of applications at different scales and in different regions of the world. But the book should also be useful for users or decision-makers who are interested in understanding the benefits and limitations of the key methods in use today. These groups may improve their knowledge and ability to interpret the results of the assessments, which sometimes introduce unfamiliar notations of uncertainty and probability that intuitively contradict the public expectation of a strict and unique scientific truth.

J. Selva
Dipartimento Di Scienze Della Terra, DellAmbiente E Delle Risorse (DiSTAR), Università di Napoli Federico II, Naples, Italy
e-mail: jacopo.selva@unina.it

I. Rafliana
Research Center for Area Studies, Badan Riset dan Inovasi Nasional (BRIN), Jakarta, Indonesia
e-mail: irin002@brin.go.id

Center for Life Ethics, Universität Bonn, Bonn, Germany

We employ the analogy of a cookbook, because successful PTHA and PTRA workflows can be described by essential building blocks (*ingredients*) combined in specific ways (*recipes*) to serve the purpose of the analysis of actual application fields. Furthermore, when using a cookbook and following recipes a certain quality of the prepared "dish" can be achieved, without limiting the freedom of the "chef" to select different methods. Thus, with some experience, ingredients can be combined in different ways to refine the result. Similar to a cookbook, examples of applications are presented as basic recipes or workflows. Hence, for tsunami scientists, this cookbook may help to achieve the above-mentioned improvements, while allowing further scientific development.

1.2 Structure and Workflow

Given the focus of the cookbook on probabilistic methods and uncertainty treatment, we dedicated special attention to the frameworks for tsunami hazard and risk assessment. The frameworks set the basis for mixing different "ingredients" to achieve a certain desired outcome (the dish). The main framework presented is the fully simulation-based workflow for probabilistic hazard and risk analysis (FW1) based on a comprehensive treatment of uncertainties—a nod to the central role of numerical physics-based simulations of tsunamis. This is a *forward* probabilistic framework which propagates the uncertainties from the genesis of the tsunami all the way towards its consequences on the communities and the built environment. We also dedicate attention to more *contextual* frameworks such as indicators of vulnerability and risk (FW2). This latter category of frameworks lends itself quite well to participation and engagement of the societal actors such as the communities, the authorities, practitioners, and the different sectors in the risk assessment process. These two main categories of frameworks can be seen as complementary (O'Brien et al. 2007) since both are needed for a comprehensive tsunami risk governance, due to its inherent complexity, the stochastic nature, and its transboundary consequences (Renn 2021).

A PTHA is started from tsunami source characterization and modelling, employing tsunami propagation and inundation modelling, from where hazard metrics (e.g., various tsunami intensity measures) and hazard data products can be derived (e.g., hazard curves, hazard maps). A PTRA aims to provide a long-term relationship of the losses and their occurrence frequencies, considering not only the hazard levels but the exposure and vulnerability characteristics of the region of interest. Both approaches aim to identify, quantify, and propagate the different uncertainties throughout the assessments.

The cookbook deals with four major and physically different sources (triggers) of tsunamis: (i) earthquakes; (ii) landslides, possibly triggered by earthquakes; (iii) volcanic events (flank collapses, eruptions, explosions); and (iv) atmospheric (air pressure induced) events.

The four tsunami sources are very distinct in their character and require specific characterization, modelling, and uncertainty quantification methods, resulting in different hazard and risk assessment workflows.

Tsunami intensity measures (IM) such as flow depth, velocity, run-up height, or momentum flux are adopted most frequently for the hazard analyses, and can be communicated through hazard curves, or with either probability, hazard, or intensity maps (the maps can be obtained from the curves). The hazard curves express the probability of exceedance for different values of a given hazard intensity measure in a given time frame. The so-called probability maps express the probability that a fixed intensity value is exceeded in a given time frame. The so-called hazard (or intensity) maps express the intensity which is exceeded with a fixed probability (or equivalently with a fixed return period). Sometimes, annual exceedance rates are used instead of probabilities in a given time frame.

In a forward probabilistic framework for tsunami hazard and risk analysis (FW1), the same concepts described above for PTHA can be generalized to PTRA. Metrics (a.k.a. decision variables) such as the expected number of fatalities, expected annual loss (EAL), and the physical damage are usually adopted for risk assessment. Loss curves and risk maps are typical risk data products. The loss curves express the frequency of exceedance of different loss levels, whereas the risk maps usually depict the spatial footprint of fatalities, losses, and physical damage predicted for prescribed return periods.

Tsunami hazard and risk assessment can be performed at a wide range of spatial scales, from global to regional or national down to local or site-specific scale, in support of different activities, ranging from policy development, land use planning, tsunami-resistant design of buildings and infrastructure at risk, early warning systems, evacuation planning, and local disaster risk management.

A generalized workflow of steps toward the quantification of PTHA and PTRA is defined, as outlined in Fig. 1.1 along with the main considerations for setting up each step. Based on that, we have identified seven key *ingredients* that can be combined into PTHA or PTRA. They are:

1. Framework
2. Source characterization
3. Source modelling
4. Hydrodynamic modelling
5. Exposure modelling
6. Vulnerability modelling
7. Risk communication

Each of these ingredients will be described in a separate chapter and is briefly introduced below. Note that the choice of a specific PTHA or PTRA workflow will depend on several factors, such as the intended use of the results, needs of the funding agency and the spatial scale of the analysis, as well as on the available data, computational, time, monetary, and human resources.

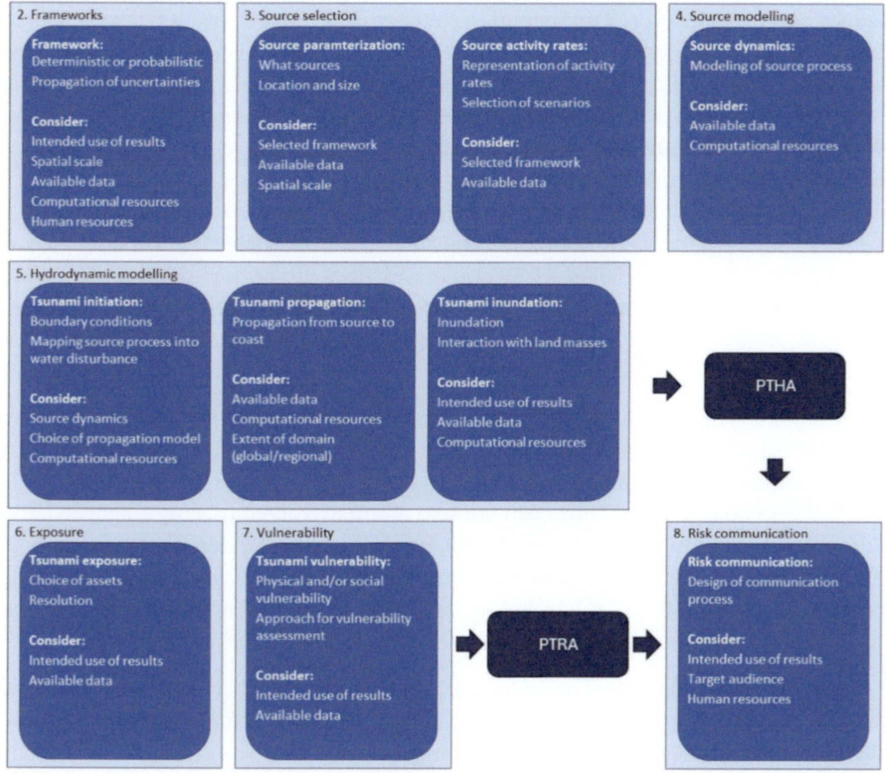

Fig. 1.1 Main steps (dark blue boxes) and associated key considerations in a generalized PTHA/PTRA workflow. Light blue boxes indicate the key ingredients and their relations to the main steps

Framework. The first key ingredient which needs to be considered is the framework used for the hazard or risk assessment. We describe the main frameworks commonly used for tsunami hazard and risk assessments, based on diverse criteria such as the number of scenarios used, the treatment of uncertainties, the statistical methods and the treatment of time dependencies. As mentioned before, two main frameworks are discussed (FW1 and FW2). The fully probabilistic framework (FW1) and different variations of it (the sub frameworks) are employed for scenario-based approaches, catastrophe modelling, considering time-dependent phenomena such as the tidal variations and the sea-level rise, and data-driven workflows. The indicator-based framework (FW2) relies on various vulnerability and risk indicators usually mapped using geographical information systems (GIS). Each framework is described with a focus on potential applications, data and technical requirements, and treatment of uncertainties. Analysts need to consider what framework to use. This may range from simplified to a very detailed fully probabilistic assessment with comprehensive treatment of uncertainties.

Source characterization. The second ingredient in the workflow is tsunami source characterization, which in this book is described separately for each of the four main tsunami source-types (i.e., earthquakes, landslides, volcanic, and atmospheric sources). For each type, the selection is described in three main steps: (i) selection of source areas and typology of sources; (ii) definition of the range of variability for source parameters; (iii) definition of individual scenarios to be modelled and quantification of their occurrence rate. Alternative methods are presented for each source and step. Which type(s) of source to consider and how (densely) to parameterize the individual sources in terms of locations and sizes are critical choices which are intertwined with the choice of the framework. It depends on the available resources, data, and on the time scale of interest since, at specific locations, the frequency of two source types may differ significantly.

Source modelling. The third ingredient regards the tsunami source modelling procedures, simulating various source processes that are the input to tsunami generation modelling. Source modelling is described in separate sections for each of the four tsunami source types. For each source, alternative methods are identified and presented. As for any of the next numerical/statistical modelling steps, a high level of complexity of the source modelling strategy may lead to high accuracy of the results. On the other hand, it would limit to some extent the uncertainty exploration capability, because fewer scenarios would be feasible with the same resources.

Hydrodynamic modelling. The natural next step in a PTHA is the hydrodynamic modelling of initiation, propagation, inundation, and interaction with land masses. The typical sequence of tsunami modelling in a PTHA workflow includes data preparation, interfaces to sources and boundary conditions, propagation, inundation, and testing. Propagation and inundation are often separate steps but may also be performed in one uniform model. Only modelling software that has been thoroughly tested should be used for this step. For each stage in the above sequence, available methods are described with information on data and technical requirements and treatment of uncertainties. A PTHA will typically terminate with the combination of the hydrodynamic modelling, providing IMs, with the source probabilities.

Exposure modelling. Exposure modelling, which encompasses identifying the population and assets at risk, is a fundamental step (ingredient) for any risk assessment. In the context of PTRA, capturing the human exposure and its inherently dynamic features is key and poses critical challenges. Therefore, to assess tsunami losses in a PTRA, there is a need for data on tsunami exposure. Exposure models aim to provide a comprehensive description of the population and assets in the study area, so that combining such information with the PTHA results will give information on what specific assets are exposed to tsunami inundation. For the most important exposure types (e.g., population and built environment), different methods for exposure modelling are presented with information on data and technical requirements, as well as some of the approaches useful for the treatment of uncertainties.

Vulnerability modelling. While vulnerability is also an essential ingredient for any risk assessment, there is an additional twist related to this ingredient: *ambiguity* in terminology. The term vulnerability has different definitions in different frameworks for risk assessment. In a probabilistic tsunami risk framework (FW1), a tsunami vulnerability model represents the link between the tsunami hazard IM (hazard metrics) and risk metrics such as fatalities, damages, and losses to exposed assets. In an indicator-based framework for risk assessment (FW2), vulnerability has a more holistic definition, being a multi-faceted concept which can encompass different dimensions (e.g., physical, functional, economical). For instance, it can capture contextual vulnerabilities (e.g., inequalities, poverty, lack of access to amenities and education) in a society that can both aggravate the consequences of a tsunami and be exacerbated by its occurrence. In this chapter, three main approaches to vulnerability assessment (i.e., empirical, analytical, and indicator-based) are presented with information about data and technical requirements, and treatment of uncertainties. The first two approaches (empirical and analytical) are associated with FW1, whereas the third approach (indicators of vulnerability) is discussed in the context of FW2 and with specific focus on the physical dimension of vulnerability.

The combination of hazard (PTHA), exposure, and vulnerability allows to finalize a PTRA in terms of loss curves and risk maps for specific risk metrics.

Risk communication. The communication of tsunami hazard or risk results (e.g., hazard and risk metrics, hazard and risk data products) with institutional users (such as disaster managers, fire brigades, police units, etc.) and populations not trained to understand and interpret scientific results is an important step and poses many challenges. For example, communication of probabilistic statements that inevitably include uncertainties needs specific attention and considerations (Rafliana et al. 2022). To support the implementation of new tsunami hazard and risk models within disaster risk management activities, the last key ingredient corresponds to tsunami risk communication. While a full elaboration on communicating probabilities and uncertainties goes well beyond the scope of this cookbook, we still strive to raise awareness among the scientific community and practitioners to keep this important aspect of PTHA and PTRA in mind. We aim at providing scientists with guidance and background information on how to appropriately engage in, co-design, prepare, and deliver risk communication, with and to different audiences.

1.3 Recipes

The second part of the book includes 25 "recipes" describing example implementations of PTHA or PTRA. Each recipe provides a schematic description of a specific workflow and is based on already published material. The recipes focus on the ingredients chosen for the application and how uncertainties are treated. The recipes have been selected to represent a wide variety of applications of PTHA and PTRA.

Recipes elaborate on the required resources for implementation so practitioners and scholars may select a recipe based on the scope of the analysis, on the available resources, or based on similarity of their own situation to a successful case. Recipes describe the necessary prerequisites, such as the kind of data required or the type of modelling necessary.

1.4 Further Considerations

While it is often not explicitly mentioned, successful PTHA and PTRA applications require adherence to full transparency with respect to the methods applied, the data used, the assumptions made, and the treatment of uncertainties. This implies to rigorously document each step and make data and methods available for reproducibility. This is not only important for scientific purposes, but also for credibility towards stakeholders, decision-makers, and end users. To this end, we endorse FAIR (for Findable, Accessible, Interoperable, and Re-usable) data policies.

Throughout the cookbook we describe not only the features and promises of the diverse ingredients and recipes, but also their respective limitations. Authors were committed to provide their expertise and point out those pieces of information necessary for reproducing the recipes with the best possible results honestly and openly. However, as in every cookbook, alternatives to certain ingredients are sometimes offered, and the consequences are discussed, when not following the prescribed use cases. And as it is the case with good "cuisine", the "chef" makes the true difference. So, of course, it is necessary to apply the offered methods and workflows with skill and care, being aware of the situation to which the analysis is to be applied, and selecting the ingredients based on available prerequisites, resources, technical, and scientific skills.

1.5 How to Use This Cookbook

This cookbook can be used and read in different ways. On the one hand, and again in analogy to a usual cookbook, readers may browse through recipes, and access the ingredients chapters following the corresponding list of ingredients. The recipes all follow a similar organizational structure, so they can be easily accessed. On the other hand, the book can be read consecutively, starting with the study of ingredients, following the general workflow of PTHA and PTRA. By doing this, scholars will learn in a structured way how to build corresponding hazard and risk assessments. Finally, for more experienced readers, the book may serve as a reference to the current state-of-the-art in this multidisciplinary research area. Individual chapters may serve for filling knowledge gaps and extensive reference lists may guide such readers to more in-depth studies. Tables in each of the ingredient chapters summarize the essential elements comprising the method, the benefits and caveats, the necessary resources and possible restrictions, and allow for a quick overview and accessibility of specific contents.

In the end it needs to be stressed that probabilistic hazard and risk assessments require a large body of disciplinary, interdisciplinary, and transdisciplinary expertise, only available in a dedicated group of experts. In the same spirit this book was prepared by many scientists and practitioners with diverse backgrounds, each providing their knowledge for a description of state-of-the-art methods. It is the wish of us, editors and authors, that this cookbook may be used with care and responsibility for the safety and resilience of coastal communities.

References

Behrens J, Løvholt F, Jalayer F, Lorito S, Salgado-Gálvez M, Sørensen M et al (2021) Probabilistic Tsunami hazard and risk analysis: a review of research gaps. Front Earth Sci 9:p628772. https://doi.org/10.3389/feart.2021.628772

Cornell CA (1968) Engineering seismic risk analysis. Bull Seismol Soc Amer 58:1583–1606

Cornell CA, Krawinkler H (2000) Progress and challenges in seismic performance assessment. PEER Center News 3:1–3

Geist EL, Parsons T (2006) Probabilistic analysis of Tsunami hazards. Nat Hazards 37:277–314. https://doi.org/10.1007/s11069-005-4646-z

Grezio A, Babeyko A, Baptista MA, Behrens J, Costa A, Davies G et al (2017) Probabilistic Tsunami hazard analysis: multiple sources and global applications. Rev Geophys 55:1158–1198. https://doi.org/10.1002/2017RG000579

Grossi P (2008) Modeling seismic mitigation strategies. In: Bostrom A, French S, Gottlieb S (eds) Risk assessment, modeling and decision support. Risk, governance and society, vol 14. Springer, Berlin, Heidelberg. https://doi.org/10.1007/978-3-540-71158-2_10

McGuire RK (2008) Probabilistic seismic hazard analysis: early history. Earthquake Eng Struct Dynam 37:329–338. https://doi.org/10.1002/eqe.765

Mori N, Goda K, Cox D (2018) Recent process in probabilistic Tsunami hazard analysis (PTHA) for mega thrust subduction earthquakes. In: Santiago-Fandiño V, Sato S, Maki N, Iuchi K (eds) The 2011 Japan earthquake and Tsunami: reconstruction and restoration. Advances in natural and technological hazards research 47. Springer, Cham. https://doi.org/10.1007/978-3-319-58691-5_27

O'Brien K, Eriksen S, Nygaard LP, Schjolden ANE (2007) Why different interpretations of vulnerability matter in climate change discourses. Clim Policy 7(1):73–88. https://doi.org/10.1080/14693062.2007.9685639

Renn O (2021) New challenges for risk analysis: systemic risks. J Risk Res 24(1):127–133. https://doi.org/10.1080/13669877.2020.1779787

Rafliana I, Jalayer F, Cerase A, Cugliari L, Baiguera M, Salmanidou D et al (2022) Tsunami risk communication and management: contemporary gaps and challenges. Int J Disaster Risk Reduct 70:102771. https://doi.org/10.1016/j.ijdrr.2021.102771

BY

Chapter 2
Frameworks for Assessing Tsunami Hazard and Risk

Natalia Zamora, Anita Grezio, Maria Papathoma-Köhle, Fatemeh Jalayer, Dimitra Salmanidou, Tom Parsons, Eric L. Geist, Jacopo Selva, Mathilde B. Sørensen, and Irina Rafliana

2.1 Introduction

Tsunami hazard and risk assessment methods have evolved considerably over the last two decades, as observations of large tsunamis provided information that challenged the models used for risk assessment. In contrast to other natural sciences, tsunami science has developed in a unique way owing to recent instrumentation and advances in numerical modelling techniques, as described by Synolakis and Bernard (2006). Tsunamis are multiscale phenomena resulting from a water column displacement that may be induced by multiple sources, and range from local scale inundation processes to ocean-wide scale wave propagation. Different strategies may be required to model tsunami evolution at different scales and to characterize various intensity measures. Research in tsunami hazard and risk has focused mostly on the tsunami effects such as the wave heights or flow depths. To a lesser extent, it has also considered flow, velocity and forces that may incur different levels of damage in the built environment (Charvet et al. 2017; Yeh et al. 2013, 2014).

N. Zamora (✉)
Wave Phenomena Group, Computer Applications in Science and Engineering, Barcelona Supercomputing Center, Barcelona, Spain
e-mail: natalia.zamora@bsc.es

A. Grezio
Istituto Nationale di Geofisica e Vulcanologia, Bologna, Italy
e-mail: anita.grezio@ingv.it

M. Papathoma-Köhle
Institute for Mountain Risk Engineering (IAN), BOKU University, Vienna, Austria
e-mail: maria.papathoma-koehle@boku.ac.at

F. Jalayer
Department of Disaster Risk Reduction, University College London, London, UK
e-mail: f.jalayer@ucl.ac.uk

M. B. Sørensen et al. (eds.), *Probabilistic Tsunami Hazard and Risk Analysis*, Mathematics of Planet Earth 14, https://doi.org/10.1007/978-3-031-98115-9_2

Tsunami risk should be taken into consideration from a multi-hazard and multi-risk perspective, covering the various potential sources of tsunamis, their direct consequences and the cascading phenomena. By definition, tsunami risk assessment frameworks integrate the tsunami hazard (e.g., rate of exceedance of flow depths and velocities) with vulnerability models (e.g. models predicting the impact on the exposed population and assets as a function of tsunami intensity parameters such as flow depth and velocity). This is done using a probabilistic approach for hazard and risk quantification. The spatial scale of interest and the data availability may determine whether hazard is quantified with estimations of flow depth onshore or only with wave heights along the shoreline (e.g. peak coastal tsunami amplitude, PCTA).

2.2 Background: Definition of Frameworks for Tsunami Hazard and Risk Assessment

Two main approaches are followed for the hazard and risk analysis: deterministic and probabilistic. These two approaches have been implemented and described in seminal studies (Geist and Parsons 2006; Grezio et al. 2017; De Risi and Goda 2017; Mori et al. 2018). The reference frame to quantify hazard and the related risk has been derived mainly from seismic hazard and risk analyses (e.g. Cornell 1968), leading to the development of probabilistic tsunami hazard and risk analysis (abbreviated as PTHA and PTRA, respectively). However, differently from seismic hazard and risk, the need to explicitly simulate single tsunami scenarios (due to the lack of accurate

D. Salmanidou
Advanced Research Computing Centre, University College London, London, UK
e-mail: d.salmanidou.12@ucl.ac.uk

T. Parsons · E. L. Geist
US Geological Survey, Moffett Field, USA
e-mail: tparsons@usgs.gov

E. L. Geist
e-mail: egeist@usgs.gov

J. Selva
Dipartimento Di Scienze Della Terra, Dell Ambiente E Delle Risorse (DiSTAR), Università di Napoli Federico II, Naples, Italy
e-mail: jacopo.selva@unina.it

M. B. Sørensen
Department of Earth Science, University of Bergen, Bergen, Norway
e-mail: mathilde.sorensen@uib.no

I. Rafliana
Research Center for Area Studies, Badan Riset dan Inovasi Nasional (BRIN), Jakarta, Indonesia
e-mail: irin002@brin.go.id

Center for Life Ethics, Universität Bonn, Bonn, Germany

empirical techniques to characterize tsunami propagation) has led to concentrate on simulation-based schemes (Grezio et al. 2017; Selva et al. 2016).

Frameworks for conducting computational probabilistic hazard and risk analysis are commonly used for natural phenomena such as hurricanes (e.g., Barbato et al. 2013, Vander Linden et al. 2009), floods (De Risi et al. 2013; Jalayer et al. 2016), volcanoes (Marzocchi et al. 2021), and tsunamis (Attary et al. 2017; Chock et al. 2011, 2016). For tsunamis, at local, regional, and global scales, probabilistic frameworks are increasingly used to inform policy decisions and to improve risk reduction measures such as land-use planning, tsunami building codes, and early warning systems. For local scale studies they are used to help identify and prioritize hazard and risk hotspots (Rafliana et al. 2022). However, decision-making for tsunami risk management is still often connected to deterministic methods, which may be related to common existing limitations, e.g. the assumptions of certain input parameters and the treatment of uncertainty that can be overly computationally expensive. This has been the case for alert systems and evacuation procedures (e.g. Catalán et al. 2020, Kosaka et al. 2023) although lately probabilistic methods have been designed (Selva et al. 2021).

Deterministic approaches have commonly relied on one single scenario or few scenarios that aim to define the so-called "worst-case" scenario, or a range of representative worst-expected scenarios, whose set of conditions would produce the maximum possible negative outcome. Worst-case scenarios may represent the most unfavourable or extreme results without necessarily being associated with a return period. In contrast, the probabilistic approaches strive to cover the entire range of possible scenarios, considering recurrence rates and several sources of uncertainty. Therefore, probabilistic methods rely on a larger number of scenarios and employ different statistical techniques to account for existing *aleatory* and *epistemic* uncertainties. The aleatory uncertainties describe natural randomness in the possible scenarios and the epistemic uncertainties account for potential deficit in knowledge (e.g., Annaka et al. 2007; Behrens et al. 2021).

Different frameworks and workflows can be selected for PTHA and PTRA, depending on the project aims. Use of some frameworks may be led by the data availability (e.g. local topo-bathymetry, historical data, various scales of the built environment fragility curves) and specific requirements, so different levels of complexity can be employed depending on the purpose of the project or study. In this chapter, we review diverse methodologies for the propagation of uncertainties, consideration of time dependency, future climate projections, and physical vulnerability assessment based on qualitative indicators. It is the endeavour of this chapter to provide a description of the commonly used frameworks for PTHA and PTRA, with specific focus on probabilistic frameworks. The aim is to describe their workflows and report a range of studies that have helped defining them. These frameworks will be further exemplified in the various recipes that describe different implementations of workflows to PTHA and PTRA.

In this chapter, we propose the following frameworks (FWs) for PTHA and/or PTRA:

- Framework 1—FW1: quantitative PTHA/PTRA
  (a) FW1A. Fully simulation-based hazard/risk assessment
  (b) FW1B. Simulation-based hazard convolved with vulnerability and loss using numerical integration
  (c) FW1C. Incorporation of dynamic processes with time dependency: tidal and sea level variations
  (d) FW1D. Working on a limited number of scenarios
- Framework 2—FW2: Tsunami risk assessment with special focus on indicator-based vulnerability assessment
  (a) FW2A. Indicator-based physical vulnerability assessment
  (b) FW2B. Indicator-based social (multi-dimensional) vulnerability assessment

The first general framework, FW1, stands for the comprehensive PTHA and PTRA, and relies on a full probabilistic modelling described in terms of four standalone frameworks. FW1A relies on a simulation-based workflow for propagation of uncertainties, whereas FW1B relies on a hybrid workflow and employs numerical integration to propagate uncertainties from hazard to risk. It should be noted that both FW1A and FW2B are simulation-based workflows, and the two FWs differ in the way they convolute the PTHA results with the vulnerability models. FW1C incorporates certain time-dependent processes such as variation in the sea-level due to tides and climate change. The PTHA which includes sea-level variations may not be fully probabilistic. In fact, IPCC scenarios for sea-level rise caused by climate change are limited in number, so different approaches must be considered to carry out the PTHA. Finally, FW1D describes the scenario-based tsunami risk assessment for a single or a limited number of scenarios, typical of deterministic hazard. It should be noted that this FW1D does only a partial characterization of uncertainties since it does not consider all the plausible tsunamigenic sources, and it can be seen as a first step toward PTHA.

The second general framework, FW2, provides aggregated indicator-based estimates of physical vulnerability for the buildings and applies multidimensional approaches to assess tsunami risk. This framework does not strictly follow a consequential forward propagation of uncertainties from hazard to risk. Finally, an approach for tsunami hazard assessment based directly on observed tsunami events is briefly discussed.

Generally speaking, PTHA and PTRA involve evaluating the probability of exceedance of tsunami metrics or impacts in a particular coastal area in a given interval time (Geist and Parsons 2006; Grezio et al. 2017; Goda et al. 2017; Goda 2020, 2021). To move from tsunami hazard towards risk assessment, information about exposure and vulnerability needs to be incorporated (Sect. 2.1.1). Tsunami exposure models define the people and assets at risk. They can entail human exposure and asset exposure (e.g., buildings, infrastructure, crops, and forests). Dynamic exposure models can consider the time-dependence (e.g. summer vs. winter for tourist places, day vs. night) affecting exposure to tsunamis. Tsunami vulnerability is a multidimensional and multi-faceted concept that describes "the conditions determined by physical, social, economic and environmental factors or processes which increase the susceptibility of an individual, a community, assets or systems to the impacts of hazards" (UNISDR 2009). It is a general concept that can have different interpretations in different contexts. In the context of PTRA, vulnerability models (or vulnerability functions) describe the various risk metrics (e.g., fatalities, losses) as a function of a tsunami intensity measure (e.g., flow depth, current velocity, momentum flux). Alternatively, indicator-based approaches are based on spatial datasets that combine various factors that affect the vulnerability. They make use of specific inundation scenarios (deterministic) to determine the relative vulnerability of assets (e.g. buildings) as will be described later in this Chapter.

The main aspects of this introductory section are summarized in Table 2.1. Further examples and methods will be described as part of the descriptions of the different ingredients and in the recipes of this Cookbook.

2.2.1 General Probabilistic Formulation

This section provides an overview of the main probabilistic formulations in PTHA and PTRA. Tsunami hazard is usually calculated through a scenario- and simulation-based probabilistic scheme. Assume we have defined a finite collection of hypothetical tsunamigenic sources which include the sources of all potential tsunami events that could affect a point of interest. Usually, events are assumed to happen at random and independently in time, allowing the assessment to be drawn as a Poisson process. Equation 2.1 then provides the mean annual rate with which intensity measure *IM* exceeds a prescribed threshold *im* at a specific location:

$$\lambda(IM > im) = \sum_{i}^{N} \sum_{e \in Ei} P(M > im|e)\lambda(e) \tag{2.1}$$

where N is the number of tsunamigenic sources, Ei the set of possible tsunami scenarios generated from source i, and e denotes a specific scenario from this set. $P(IM > im|e)$ is the probability of exceeding im given specific event e. IM is the selected intensity measure and represents a hazard metric, for example maximum tsunami wave height or flow depth at the site of interest. (e) represents the rate of occurrence of the given scenario e. It should be noted that the hazard ($IM > im$) is a rate and not a probability, implying that it is not constrained to be less than one.

Figure 2.1 shows a general workflow for PTHA and PTRA, describing their main 'ingredients'. The notation in Fig. 2.1 is explained in the rest of this subsection below.

Table 2.1 Summary of Sect. 2.1 defining frameworks for PTHA and PTRA

What does the ingredient describe?	The main frameworks used for tsunami hazard and risk assessment
Definitions of the ingredient	A framework describes a strategy, or generalized workflow, that can be followed to estimate tsunami hazard or risk
What is/are the metric(s) used for measuring/evaluating the ingredient?	Tsunami hazard is represented by the mean annual rate of exceeding prescribed tsunami intensity measures (e.g., flow depth, momentum flux) Tsunami risk is represented by the mean annual rate of exceeding certain values of the decision variable representing risk (e.g., number of casualties, economic loss) Annual rates may be transformed in probabilities of exceedance Alternatively, vulnerability indicators can be used to provide qualitative measures of hazard and risk in a compound manner
Possible uses/applications of the ingredient	Setting up an appropriate workflow for tsunami hazard or risk assessment
Possible multi-hazard dimensions/applications?	Tsunami hazard and risk assessment is inherently a multi-hazard and multi-risk problem. The multi-hazard aspect is related to a tsunami being a process that can be triggered or influenced by different phenomena. Thus, a multi-hazard analysis is needed for thorough hazard assessment. Since interactions occur also at the level of consequences (for example, a bridge damaged by an earthquake may complicate evacuation for the tsunami triggered by the earthquake), the multi-risk aspect needs to be considered as well
Possible time-dependent/ dynamic aspects	The underlying probabilistic framework is usually defined for stationary phenomena. However, the framework can be extended also to consider non-stationarity (e.g., Sepúlveda et al. 2021)
Short outline of the available methods to be described below	Two general frameworks, with a total of six sub-frameworks, are described following diverse criteria such as the propagation of uncertainties, number of scenarios, the statistical approach regarding time dependency and future projection to climate change, and the uni- and multi-dimensional vulnerability and risk assessment based only on indicators

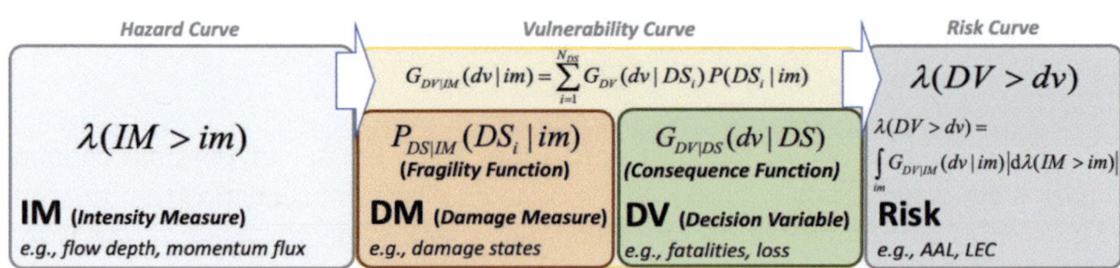

Fig. 2.1 General workflow for PTRA (adapted from Behrens et al. 2021). AAL: average annual loss and LEC: loss exceedance curves

Risk is defined as the mean annual rate of a specific risk metric, called here generically the decision variable *DV* (e.g., number of casualties, direct economic losses, damage), exceeding a threshold *dv*:

$$\lambda(DV > dv) = \int_{IM} P(DV > dv|IM = im)|d\lambda(IM > im)| \tag{2.2}$$

Here, $d\lambda$ $(IM > im)$ is non cumulative hazard, that is the increment of mean annual rate of exceedance of an intensity measure in a specific location of the study region, that is, the increment of hazard, as calculated differentiating Eq. 2.1. $P(DV > dv|IM)$ is the probability of exceeding a given level of risk metric *DV* for a given *IM* value. It is referred to as the vulnerability function or the vulnerability curve (when plotted as a function of *IM*).

Equation 2.2 divides risk assessment into two principal components: the hazard and the vulnerability. The former expresses the mean annual rate of exceedance of a hazard metric (intensity measure *IM*) and is represented by the hazard curve (when plotted versus *IM*), the latter is the exceedance probability for a given risk metric and is often visualized with the use of vulnerability curves. In this equation, λ $(DV > dv)$ represents the mean annual rate of that risk metric expressed by *DV* exceeding a given threshold level *dv*.

The rate of exceedance in Eq. 2.1 can be transformed into the probability that the IM threshold *im* is exceeded at least once in a given time window ΔT:

$$P(IM > im, \Delta T) = 1 - exp(-\lambda(IM > im)\Delta T) \tag{2.3}$$

Equation 2.3 assumes that exceeding specific *IM* levels is expressed by a homogeneous Poisson Process with rate λ $(IM > im)$ as explained in several publications (e.g., Geist and Parsons 2006; Grezio et al. 2017, 2020; Mori et al. 2018). Note that Eq. 2.3 also holds for transforming risk λ $(DV > dv)$ into probability of exceeding the risk metric threshold *dv* at least once in time interval ΔT. This can be done by substituting λ $(IM > im)$ by $\lambda(DV > dv)$ in Eq. 2.3 to get $P(DV > dv, \Delta T)$.

The vulnerability function $P(DV > dv|IM)$ can also be denoted as the complementary probability distribution function *G*, calculated by integrating fragility and the consequence functions as given in Behrens et al. (2021). Hence, the vulnerability function can be calculated as follows:

$$P(DV \geq dv|IM = im) = G_{DV|IM}(dv|im) = \sum_{i=1}^{N} G(dv|DS_i)P(DS_i|IM = im) \tag{2.4}$$

Here, $P(DS_i|IM = im)$ is the probability that the exposed asset of interest (eg, a building, an infrastructure) is in (discrete) damage state *i*, given $IM = im$, calculated as the difference between tsunami fragility curves for consecutive damage levels. The term $G_{DV|IM}$ $(dv|im)$ stands for the tsunami consequence function (the damage-loss function), expressed as the complementary cumulative distribution function of *DV* given *DM*, as given in Behrens et al. (2021).

2.2.2 Overall Uncertainty Propagation

Uncertainties in geosciences take their origin in the (i) limited number of observations and scarce access to direct observations (e.g. of deep seismic sources), (ii) the complexity and the non-linearity of natural processes that make them particularly difficult to forecast, and (iii) limitations in the available capability of modelling (Hetenyi et al. 2022). PTHA and PTRA present relevant gaps and open questions related to source identification and representations, scaling and constraints, which add to the uncertainty characterizing the estimation of recurrence rates and the modelling of the different tsunami sources and coastal processes (Behrens et al. 2021). Additional uncertainties are connected to the time-dependent components in the hazard and risk analyses (DeRisi et al. 2022). Uncertainties may be classified as epistemic or aleatory. The aleatory uncertainties are related to inherent randomness of the occurrences of a set of natural events and may be described as frequencies of a set of events (Marzocchi et al. 2021). Such frequencies are objective probabilities, and they can be potentially measured through a well-defined experimental concept that defines collections of data. Hypotheses about aleatory uncertainty can be tested against observations with frequentist error-statistical methods (e.g. Marzocchi et al. 2014, 2015). Epistemic uncertainties measure the lack of knowledge in the estimation of objective probabilities. The models assessing epistemic uncertainty are often based on expert opinions, thus epistemic uncertainty is described by subjective probabilities (Marzocchi et al. 2015). Epistemic uncertainty measures the convergence/divergence of the scientific community toward a strong estimation of aleatory uncertainty, and this convergence may be favoured by new data and research (Marzocchi et al. 2015). In a broad sense, uncertainties are characterized as epistemic, if the modeller sees a possibility to reduce them by gathering more data or by refining models, and aleatory if the modeller does not foresee the possibility of reducing them (Der Kiureghian and Ditlevsen 2009).

PTHA and PTRA deal, by definition, with aleatory uncertainty, as they are designed to quantify the aleatory uncertainty of the tsunami process through the total probability theorem. The hazard integral is usually evaluated by discretization of the source space or by Monte Carlo sampling sources from source probability distributions. Different techniques have been also suggested to improve the treatment of uncertainties in the numerical integration approach, such as bootstrap resampling or adaptive Markov chain Monte Carlo simulations (De Risi et al. 2017; Jalayer et al. 2023).

The quantification of epistemic uncertainty is also fundamental, to communicate to decision-makers the degree of agreement and sureness in hazard and risk assessment (SSHAC 1997). This can also be considered as the extent to which we think the assessment could possibly change in future, as new information becomes available, representing a fundamental input to define meaningful policies for risk reduction actions. Different techniques or methods exist for the quantification of the epistemic uncertainties. For instance, logic trees are based on the definition of a set of alternative probabilistic approaches. This set consists of a discrete number of possibilities, assumed to be mutually exclusive and collectively exhaustive (MECE), and, as a consequence of the MECE postulation, one path of the tree must represent the true outcome and the probabilities of the logic tree can be combined in a total probability.

Ensemble models (Marzocchi et al. 2015; Selva et al. 2016) are also based on a set of alternative probabilistic models that individually only include aleatory uncertainty, but the alternatives are considered as a sample of an unknown parent distribution of possible models. In this case, the sample is not assumed to be MECE and can derive from one or more logic trees, or from a collection of models with associated weights accounting for the confidence ranges based on expert opinion and/or quantitative evaluations of the forecasting performance. The parent distribution covers the gaps among the models and adds tails to the range defined by the alternatives (Marzocchi et al. 2015).

Bayesian models may also be adopted to model the uncertainty on probabilistic quantifications and reduce this uncertainty by integrating it with local data (Gelman et al. 1995). The prior model can be set up based on the alternative models, similarly to logic tree and ensemble modelling, and then local data can be used to update it through the Bayes' rule (e.g., Grezio et al. 2012a, b; Selva and Sandri 2013).

Epistemic uncertainties can be quite significant for hazard and risk estimations. For example, data is not always available at the desired scales, which leads to oversimplifications that could lead to biased estimations of PTRA and subsequent application of risk prevention measures that may fail (Oetjen et al. 2022; Uchida et al. 2021). Neglecting this source of uncertainty may lead to erroneous hazard and risk estimates. In several studies where sensitivity analysis has been conducted, the hazard values may vary up to 50% depending on the recurrence rates of events (Selva et al. 2016; Sepulveda et al. 2021; Zamora and Babeyko 2020). The selection of the seismic zone or the source characterization (Davies et al. 2015; Müller et al. 2015) could influence the inundation extent. It has also been noted that the selection of IMs, model set up related to Manning values or topo-bathymetric uncertainty can strongly affect the tsunami inundation (Williamson et al. 2020; Harig et al. 2022; Gibbons et al. 2022). Furthermore, associated hazard uncertainties could impact the development of fragility curves inducing further significant biases in the PTRA (De Risi et al. 2017; Straub and Der Kiureghian 2008), and from the standpoint of risk quantification, Gómez-Zapata et al. (2021) draw attention to the uncertainties surrounding the choice of spatial aggregation and fragility models for exposure modelling and loss estimates (e.g. Suppasri et al. 2018).

2.3 Frameworks Used in PTHA and PTRA

2.3.1 Framework 1: Probabilistic Tsunami Risk Assessment

This framework represents the most comprehensive probabilistic tsunami hazard and risk assessment (PTHA and PTRA) workflows. Here, a fully probabilistic method is chosen, where all sources of uncertainty that can be identified and constrained are considered, from the source characterization (Chap. 3) to source modelling (Chap. 4), hydrodynamic modelling (Chap. 5), exposure modelling (Chap. 6), and vulnerability modelling (Chap. 7). Under this framework, four different workflows are described: FW1A is a fully simulation-based workflow for PTHA and PTRA. FW1B follows a simulation-based workflow for PTHA and uses numerical integration to convolute hazard and vulnerability to evaluate the tsunami risk. A third workflow (FW1C) is described here for incorporating non-stationary phenomena such as tidal variations and sea level rise. Finally, a fourth framework (FW1D) is described, using a limited number of scenarios as the first steps to PTHA.

Tsunami hazard analysis normally considers scenarios that cover different geographical scales and spatial resolutions, for example by combining near-field and far-field events. PTHA conducted at different scales could link regional models to local models, based on available computational grids and the computational capabilities (Davies et al. 2022; Lorito et al. 2015; Selva et al. 2016; Volpe et al. 2019; Williamson et al. 2020).

Multiple tsunamigenic source types have not been commonly considered as part of PTRA, with most studies considering one source-type, either for earthquakes, landslides, volcanic, or atmospheric events. In reality, however, there are areas where two or more sources need to be considered. For example, some large earthquakes may not generate tsunami, but in the case of short-term tsunami hazard assessment, the occurrence of such an event may still require updating each single source probability (Grezio et al. 2015, Selva et al. 2022), which may increase the probability of future tsunamigenic earthquakes (Lombardi et al. 2007). Volcanic unrest/eruptions or seismic events can trigger landslides. In those cases, the tsunami hazard evaluations must also be updated and re-considered in the short-term. In this regard, Grezio et al. (2020) show multi-source probabilistic hazard analysis for each individual source, computed on the basis of a large number of tsunamigenic events of each different source type, such as volcanic (e.g. Paris et al. 2019, Zorn et al., 2022), seismic (e.g. Lorito et al. 2008, Parsons and Geist 2009, Tinti et al. 2011, Harbitz et al. 2012, Geist et al. 2014, Davies et al. 2018) and landslides (e.g. Ten Brink et al. 2014). More details about this approach can be found in recipe R12.

The outputs of PTHA or PTRA can be represented in various forms. The most common ways of expressing the risk are through probabilistic risk maps or loss exceedance curves. The curves represent the probability of the decision variable chosen as risk metric (generically referred to as loss) exceeding a specific value over a certain time window in the region of interest. The probabilistic risk maps are also of critical importance for the decision-making process and may depict hazard intensity, damage ratios, or loss ratios for different return periods or probabilities of exceedance (Goda et al. 2021).

The four main sub-frameworks falling under FW1 are described in the following sections.

FW1A. Fully simulation-based risk assessment

Short description of the method

One of the major challenges for tsunami risk assessment is to account for uncertainties and their propagation through the entire workflow, from source characterization to estimation of consequences (e.g., Cremen et al. 2022). The simulation-based framework lends itself very well to a comprehensive uncertainty propagation from source to consequences. In very general terms, the workflow of FW1A, which is usually defined as a Monte Carlo simulation scheme, entails: (1) characterizing the potential tsunamigenic sources and the relative recurrence rates of tsunamigenic events; (2) generating plausible tsunami scenarios from each source by simulating source parameters; (3) hydrodynamic modelling of each tsunami scenario. After this step, hazard curves and their uncertainty intervals can be derived for each point of interest (defined to be compatible with the exposure model) within the tsunami inundation extent. Then, *IM*-values are integrated for each point. (4) Simulating the fatalities and/or economic losses from the vulnerability curve and its confidence interval, based on the information provided by the exposure model and the related uncertainties, and characterizing the loss curve and its uncertainty interval for each point/area of interest (depending on how the exposure is defined). FW1A can be described through the following general formulation:

$$\lambda(DV > dv) = \sum_{I}^{N} \sum_{e \in Ei} P(DV > dv|e)\lambda(e) \tag{2.5}$$

Here, *N* is the number of tsunamigenic sources considered and *Ei* is the set of all plausible tsunamigenic event scenarios considered for source *i*. *P*(*DV* > *dv*|*e*) is the probability that the decision variable (risk metric) exceeds a specific value *dv* given scenario *e*. This can also be written as following, by expanding *P*(*DV* > *dv*|*e*) with respect to *IM* using the total probability theorem:

$$\lambda(DV > dv) = \sum_{i}^{N} \sum_{e \in Ei} \int_{IM} P(DV > dv|IM = im) f(IM = im|e)\, dim\, \lambda(e) \tag{2.6}$$

Here, *P*(*DV* > *dv*|*IM*), as described before, is the vulnerability function. The vulnerability function can further be expanded with respect to discrete levels of damage using Eq. 2.7:

$$\lambda(DV > dv) = \sum_{i}^{N} \sum_{e \in Ei} \int_{IM} \sum_{DS} P(DV > dv|DSi)P(DSi|IM = im)f(IM = im|e)\, dim\, \lambda(e) \quad (2.7)$$

The above equation assumes that the discrete damage states are mutually exclusive and collectively exhaustive (MECE, Jalayer et al. 2023). This means that the damage levels (i.e., the thresholds marking the damage states) are going to be hierarchical. The set of possible damage states is usually referred to as a *damage scale* and the number of distinct damage levels (excluding damage level 0) is called *tiers* (e.g., the European Macroseismic Scale (EMS98; Grunthal 1998) is a 5-tier damage scale). For damage states, we assumed here a discrete probability distribution. Therefore, the set of possible discrete damage states will need to be mutually exclusive and collectively exhaustive in order for them to be an acceptable domain.

Equation 2.7 depicts the PTRA workflow (Fig. 2.1). In FW1A, this equation can be defined as a Monte Carlo Simulation scheme and can be solved starting from scenario generation up to simulation of damage and losses. Examples of applications of FW1A in literature are, for example, Goda and De Risi (2017) and Goda (2023).

Applications of the FW1A

Using many scenarios has been a common approach to obtain *IMs* shown as inundation maps for evacuation purposes (Wood et al. 2020) and probabilistic tsunami forecasting (Selva et al. 2021), or instead to conduct full probabilistic hazard assessment to estimate the probability of exceeding a given threshold at a specific site and in a specific time window, accounting for large variability in the sources (e.g. Davies et al. 2018; Volpe et al. 2019, Sepúlveda et al. 2017). The probabilistic approach to assess the hazard usually relies on numerical simulations or empirical relations to obtain preselected IMs. Next, the hazard aggregation is performed, and hazard curves or hazard maps are generated. FW1A is commonly used in the scientific community and for engineering applications where full uncertainty propagation is required but also may be applied for catastrophe modelling with focus on measuring the monetary losses of the structural component (Fukutani et al. 2019) or at a household level (Salmanidou et al. 2021). Table 2.2 summarizes the main components needed for the inputs, and the expected general outcomes. Good examples of applications that go all the way towards risk assessment include Goda and De Risi (2017), Goda (2020), and Goda and De Risi (2023). It should be noted that the simulation-based PTHA is much more documented in the literature than simulation-based PTRA.

Simulation-based PTRA, known in the catastrophe modelling community as event-based risk assessment, lends itself particularly well to spatial risk assessment (see also recipes R14 and R15).

Data and technical requirements in FW1A

Fully simulation-based PTRA workflows benefit from the advances of computational capabilities. Owing to resources like high performance computers, tsunami numerical models and statistical techniques derived from machine learning and artificial intelligence, more complex and faster risk analysis can be conducted (e.g. Giles et al.

Table 2.2 Purpose, input, and output at different scales in FW1A

Spatial scale	Purpose	Important inputs	Outputs	References to examples
Global	International-DRR prioritization	Source model recurrence intervals, global exposure models	Global hazard and risk maps	Løvholt et al. (2014, 2015); Davies et al. (2018); Davies and Griffin (2019)
Regional	Early warning systems, baseline models	Regional bathymetry	Offshore hazard maps, curves and risk index	Sørensen et al. (2012); Linli et al. (2016), Sepúlveda et al. (2017); Basili et al. (2021)
National	Building codes, evacuation	Digital elevation model and exposure maps, auxiliary data	Tsunami inundation extents, hazard and risk zoning maps	Davies et al. (2018); Power et al. (2018)
Community	Evacuation, land use zoning, built-defences	exposure maps	Onshore hazard, evacuation maps, hazard and risk zoning	Wood et al. (2020); Tonini et al. (2021)
Site specific	Safety of critical infrastructure	Fragility functions	Risk curve	Grezio et al. (2012b); Volpe et al. (2019); DeRisi and Goda (2017); Baba et al. (2022)

2020, Macias et al. 2017, Gibbons et al. 2020, Zaniboni et al. 2021). Some of these advances are described in the recipes of this book.

A full and comprehensive probabilistic tsunami risk assessment conducted with numerical modelling relies on access to and good understanding of the tsunami numerical codes, a digital elevation model, damage models (fragility models), cost models (also known as consequence models), vulnerability/loss models, and exposure models at a resolution that is compatible with the inundation maps.

In general, FW1A is largely dependent on computational resources, on the compromise between resolution and number of scenarios, and on the data availability (e.g., topography and bathymetry resolution). Historical data and open access to high quality datasets of the built environment and high-resolution demographic datasets are limited in most regions of the world. Moreover, unconstrained and/or large uncertainties in the recurrence rates of the triggering events (e.g., earthquakes, submarine landslides, etc.) can contribute significantly to uncertainties in both PTHA and PTRA results.

For a fully probabilistic PTRA, HPC resources may be necessary. The estimates of the tsunami hazard and risk rely mostly on numerical simulations that may require advanced simulation schemes (Harig et al. 2020; Macias et al. 2017). There are of course exceptions, for example, PTRA relying on data-driven runup-catalogues or empirical relations may require less computational resources than the models based on numerical modelling (e.g. Grezio et al. 2012a, b).

Machine learning or surrogate models may be useful in reducing computational burdens, but in an initial phase they require a large number of scenarios for training and validation which may not be feasible if high computational capabilities are not available. However, once the training has been completed, these methods can be used for generating hazard maps or early warning decision making products (e.g. Liu et al. 2021, Makinoshima et al. 2021, Núñez et al. 2022).

The validation and testing of probabilistic hazard and risk maps is not commonly addressed, owing to the lack of historical and geological tsunami records (Baba et al. 2022). Horspool et al. (2014) propose integrating data over large geographical areas in order to have enough samples of observations for testing purposes in regional or global studies. However, other limitations related to resolution may appear. Thus, rigorously testing is still a challenge for probabilistic risk assessment.

Treatment of uncertainties in FW1A

Recent studies on tsunami hazard assessment have demonstrated the caveats of not accounting for the variability of source characteristics or different estimates of recurrence rates of e.g. seismic sources (Davies et al. 2018; Selva et al. 2016), as well as of the assumptions regarding uniform versus variable slip distribution (Davies et al. 2015; Melgar et al. 2016). The standard Monte Carlo simulation scheme is a statistical technique used frequently for solving the PTHA (Eq. 2.1) and PTRA (Eq. 2.2) integrals numerically. In fact, Eqs. 2.5, 2.6 and 2.7 provide the simulation-based versions of PTRA with respect to the possible tsunami scenarios. The Monte Carlo simulation can include treatment of both aleatory and epistemic uncertainties depending on the context (e.g., Davies et al. 2022). While the Monte Carlo scheme is quite straightforward to use, it may not be the most computationally efficient. This shortcoming can be overcome through the use of HPC and use of efficient simulation schemes such as importance sampling (Davies et al. 2022; see also Chap. 3) and the use of data-driven meta-modelling approaches (Rohmer et al. 2018).

The aleatory uncertainties are commonly propagated in the recurrence rates or physical characterizations of the tsunamigenic source and in simulation of damage and loss from mean (median) fragility and vulnerability models. The epistemic uncertainties related to source recurrence rates are usually treated using techniques such as logic trees or ensemble modelling. Epistemic uncertainties for fragility and vulnerability models are usually modelled through confidence intervals defined around the mean or the median fragility and vulnerability curves (e.g. Selva et al. 2013, De Risi and Goda 2017, Lahcene et al. 2021, Jalayer et al. 2023).

For the exposure, the uncertainty can be measured by the completeness of the exposure model and its spatial coverage. This can be done through statistical inference on aggregated models (Pittore et al. 2017). The uncertainties in the losses or fatalities can be described either directly through confidence intervals in the vulnerability curve or through uncertainties in the adopted consequence model. However, not much consolidated literature is available about such approaches in the tsunami risk context.

Damage observations, collected after a disastrous event (e.g. Muhari et al. 2015, Williams et al. 2020), offer a direct link between the tsunami intensity and the incurred damage. Such observations can be used to build representative fragility functions giving the probability that the structure will exceed a threshold of damage. The logic tree approach can be implemented to construct the fragility curves for combinations of building characteristics under the framework of a flood risk assessment to account for epistemic uncertainties (DeRisi et al. 2013). A Bayesian approach has been instead adopted by Selva et al. (2013) to quantify the epistemic uncertainty with alternative fragility models and propagate it to risk quantification.

The general modular Bayesian scheme is an alternative method to account for the associated uncertainties (Grezio et al. 2012a; Smit et al. 2019). The effects of propagating the input data uncertainty and sample size on tsunami fragility functions have been investigated by adopting Bayesian regression methods (DeRisi and Goda 2017; Jalayer et. al. 2023), considering that the lack of observed data may be a limitation for the empirical assessment of input inundation data uncertainty. The magnitude of the input data uncertainty will propagate into the tsunami fragility modelling as well as to tsunami loss estimation. It is important to underline that when developing empirical tsunami fragility curves, accounting for the input data uncertainty makes their use in damage and loss assessment more reliable.

Finally, other sources of uncertainties may affect the inundation mapping to some extent. For example, the use of inappropriate Manning values (Harig et al. 2022; Gibbons et al. 2022), neglecting dispersion in the far field (e.g., Sugawara 2021) may affect the results. Chap. 5 provides a broader description of the numerical models.

A summary of main aspects of FW1A is given in Table 2.3.

Table 2.3 Summary of the main characteristics of the frameworks described in this chapter

Short description of the method	FW1A	FW1B	FW1C	FW1D	FW2A	FW2B	Tsunami data and observations
Theoretical/mechanical background for the method	Workflows in FW1A involve analysis based on many tsunamigenic scenarios, allowing a comprehensive treatment of uncertainties in each step of the tsunami hazard and risk assessment. Applications are wide, but PTRA applications often focus on engineering and urban planning,	Workflows in FW1B are based on several tsunamigenic scenarios to account for a broad range of uncertainties. Convolution of hazard and vulnerability relies on numerical integration. Often used by insurance or reinsurance companies that develop financial risk transfer tools for estimating losses	FW1C can be considered an extension of the other frameworks under FW1, integrating tides and projected sea level variations to the PTHA. Applications are as for the other frameworks under FW1	Workflows in FW1D encompass analysis based on a few tsunamigenic scenarios. The framework allows to assess hazard and consequences from very complex tsunamigenic scenarios. Applied traditionally to generate evacuation maps and to design tsunami countermeasures	Workflows in FW2A consider indicators in physical vulnerability. Indicators are derived based on recommendations for constructing composite indicators (OECD 2008)	Workflows in FW2B consider social or multi-dimentional vulnerability based on indicators (OECD 2008) following a multidimensional approach	This approach outlines methods that can be used for a data-driven PTRA. It takes into account curation and statistical analysis of data from various sources of tsunami information. Different methods are developed to assess potential recurrence distributions and determine which best fits the observed data

(continued)

Table 2.3 (continued)

Short description of the method	FW1A	FW1B	FW1C	FW1D	FW2A	FW2B	Tsunami data and observations
Short outline of how the method works	This method uses a fully simulation-based workflow to propagate the uncertainties related to source characterization and modelling, tsunami propagation and inundation, exposure modelling, and fragility and vulnerability modelling through the PTHA or PTRA	Different modules are considered such as hazard, exposure, vulnerability, and policies. The different modules interact with each other through conditional probabilities	Conditional hazard curves or hazard maps are estimated assuming a given tidal level or sea-level rise. Time dependent changes of the position of the shoreline are combined with time independent tsunami hazard. The non-stationary Poisson process model incorporates sea level rise as a non-stationary process in the PTHA	This method is similar to FW1A or FW1B workflows with the difference that uncertainties are only partially characterized. Therefore, the method does not qualify as PTHA/PTRA	The framework is based on an inundation scenario and the selection of physical vulnerability indicators for individual buildings, followed by aggregation of such indicators into a vulnerability index	This approach is can use one scenario or aggregated tsunami hazard to assess vulnerability, exposure and losses based on a multi-dimensional indicator matrix	Synthetic records can be generated by drawing at random from recurrence distributions and the most frequently successful matches are taken to be the most likely fit to the data. Alternatively, recurrence distribution parameters can be estimated using a modification of the maximum log-likelihood method, accounting for open time intervals and uncertain age dates
How is/are the output(s) of the method represented?	Hazard curves, probabilistic hazard maps, loss curves, probabilistic risk maps	Hazard curves, loss curves, hazard maps (e.g. hazard footprints) and risk maps	Hazard curves and probabilistic hazard and risk maps taking into account the tidal effects and/or projected sea level changes	Inundation maps, scenario-based damage/loss maps, microzoning, evacuation maps	Vulnerability index per building visualized in vulnerability map	Vulnerability index for different dimensions, vulnerability and risk map	Tallies of fits that can be shown as contours, histograms, mean rate curves, Hazard functions
In what way does the method take into account time-dependent/ dynamic aspects?	This framework is used primarily for stationary phenomena. However, it could incorporate time-dependent aspects related to source characterization or dynamic aspects of the sea level changes (see FW1C)	Often implemented in a stationary form but can incorporate time-dependent components	Simulations take into account the time-varying reference water level and the time-varying current which are modulated by tides, as well as long-term changes in sea level	Often implemented in a stationary form but can incorporate time-dependent components (e.g., aftershocks, dynamic aspects like sea level rise or tides (FW1C))	The method relies on dynamic aspects. It is possible to update the inundation scenario and indicators database	It may rely on hazard values that consider dynamic aspects, and may consider, e.g. the social dynamics	Both time dependent and time independent distributions can be applied

(continued)

Table 2.3 (continued)

Short description of the method	FW1A	FW1B	FW1C	FW1D	FW2A	FW2B	Tsunami data and observations
Does the method use parallelisation, machine learning, surrogate models, advanced simulation schemes, other computational reduction strategies (e.g. cluster analysis, etc.)?	When building a comprehensive PTHA or PTRA, code parallelisation or GPUs will aid fast computations. Clustering and scenario-reduction can be used to optimize the use of computational resources. Machine learning or surrogate models are less explored but can be used to generate PTHA. Also, statistical techniques such as data augmentation in areas with scarce information can aid PTHA and PTRA	As for FW1A, it can be used, subject to availability and model set-up	As for FW1A, it can be used, subject to availability and model set-up	The framework mostly relies on numerical simulations that may require advanced numerical schemes. Machine learning may require great amounts of simulated tsunami scenarios that will need large computational resources. Such methods are thus unlikely to be applied in this context unless using frugal learning techniques	No	No	The method uses parallelisation to speed up the process

(continued)

Table 2.3 (continued)

Short description of the method	FW1A	FW1B	FW1C	FW1D	FW2A	FW2B	Tsunami data and observations
What are the pros and cons and peculiarities of this specific method? Are there known shortcomings/benefits of the method? In what cases is this method a good/poor choice?	Pros: The most appropriate and comprehensive way of doing PTHA and PTRA Cons: High computational demands Dependency on computational resources may compromise resolution and number of scenarios Dependency on data availability (e.g., historical catalogues, bathymetry resolution, high-resolution exposure datasets, suitable damage and loss datasets) Overall treatment of uncertainties may not be well addressed. This means that it cannot be guaranteed that uncertainties are properly characterized by the modelling approach	Pros: The workflow is fully probabilistic, relying on a comprehensive treatment of uncertainties Its modular structure facilitates the use of ready products such as outputs of PTHA (i.e., hazard curves, hazard footprints), fragility, and vulnerability curves and the respective epistemic uncertainties Cons: It suffers the same shortcoming as in FW1A, related to lack of data and inadequate characterization of uncertainties	Pros: Adding time dependent processes to hazard assessment reduces uncertainties in coastal tsunami impact Cons: Simulating more scenarios involving the coastal changes is computationally demanding There are uncertainties related to the sea level projections	Pros: More attention can be given to more complex source characterization and modelling of cascading events Cons: Owing to the limited number of scenarios, the aleatory uncertainty is not considered	Pros: Results can be easily updated. There is a direct connection to vulnerability reduction activities Cons: Large data requirements, monetary damage is not assessed, there is no treatment of uncertainties, no quantified decision variables such as loss	Pros: The approach is community oriented and can be extended to damage and property tax vulnerabilities. Uses elicitation processes Cons: Thresholds are subjective. As in FW2A, there is no treatment of uncertainties	Pros: Results are directly linked to past observations. Uncertainties can be considered in the analysis Cons: Results can be surprising when they are drawn from a long-tailed distribution like a log-normal

(continued)

Table 2.3 (continued)

Data and technical requirements for method	FW1A	FW1B	FW1C	FW1D	FW2A	FW2B	Data driven approach
What input data does the method use?	Source characterization: geological data, geodetic measurements, historical trigger event and tsunami catalogues, paleotsunami data, potential tsunamigenic source databases, and uncertainties at the sources Numerical modelling: Topo-bathymetry, digital elevation model Exposure: land-cover and land-use datasets Fragility curves for physical elements exposed to risk Consequence functions for physical elements exposed to risk Vulnerability or fatality functions Datasets of population (human exposure) and built environment (physical exposure)	Similar to FW1A. Hazard footprints, vulnerability curves, exposure datasets	Similar to FW1A, in addition to tide observations and sea-level projections from IPCC scenarios	Similar to FW1A	Wave height, flow depths Topography of the area, data on buildings, georeferencing of the built environment	Wave height, flow depths Topography-bathymetry, buildings, census, georeferenced data, social indicators, community knowledge, Outcomes from surveys and elicitation	Earthquake and tsunami catalogues A time series from observations and a wide range of recurrence distribution parameters

(continued)

Table 2.3 (continued)

Data and technical requirements for method	FW1A	FW1B	FW1C	FW1D	FW2A	FW2B	Data driven approach
What are the technical requirements?	Computational resources: numerical codes for tsunami modelling High-resolution digital elevation models along the coast Atmospheric model for meteotsunamis	Computational resources: a platform or stand-alone codes for propagation of uncertainties and risk analysis Access to the input models and datasets	Access to model and data	Computational resources: numerical code/simulations Digital elevation model. For inundation modelling a high resolution model is required along the coast Atmospheric model for meteotsunamis	GIS technology	Statistical techniques for index integration. GIS technology	Any computational resources
Are HPC resources necessary, would absence of these compromise the use?	Comprehensive PTHA and PTRA generally need HPC resources. However, there are ways to reduce the computational cost (e.g., using advanced simulation schemes such as importance sampling, surrogate models such as kriging, pre-calculated green functions, empirical amplitude functions, etc.)	Required especially for high-resolution hazard modelling	Depends on the main framework used for the analysis	Not always necessary but may be needed for very computationally demanding numerical codes	Not necessary	Not necessary	Not necessary
Is the method/ workflow available as a git/software/jupyter notebook?	No	No	No, except these resources: https://tidesandcurrents.noaa.gov/sltrends/sltrends_global.html https://psmsl.org/gloss/ https://www.ipcc.ch/srocc/	No	Yes, supplementary material is available together with the publication by (DallOsso et al. 2016)	No, except the SoVI framework	No

(continued)

Table 2.3 (continued)

Treatment of uncertainties for the method	FW1A	FW1B	FW1C	FW1D	FW2A	FW2B	Data driven approach
Is the method itself deterministic or uncertainty based?	The method includes comprehensive treatment of uncertainties (probabilistic), although it is difficult to validate realistic characterization, since it may involve subjectivity or be sensitive to the methods used	Uncertainty based, similar to FW1A	Can follow uncertainty propagation as in FW1A. However, the sea-level rise projections are done in a scenario-based manner and at best they are associated with some likelihood values	It can range from deterministic to partially probabilistic	Mostly deterministic (scenario-based)	Mostly deterministic (scenario-based)	The method is intended to treat the complete range of uncertainties associated with paleo and empirical catalogues
Which are potential sources of uncertainties? Which parameters are used to characterize them?	Each module of the risk assessment is associated with uncertainties. This will be covered in the next chapters	The same sources of uncertainties are present as in FW1A	In the sea level projections, anthropogenic causes have high levels of uncertainties related to human and economical choices. Tides are generally well constrained	The same sources of uncertainties are present as in FW1A. However, there is limited or no treatment of uncertainties	Uncertainties are related to the inundation scenario, the building data quality and the expert judgment required for the weighting of the indicators	Uncertainties are similar to FW2A	Open intervals before the first and last events in an observed time series, dating uncertainty associated with paleo events, assumed form of parametric model(s) for inter-event times
How are uncertainties quantified for the identified parameters? Are there significant correlations between different uncertain parameters?	Uncertainties are generally treated using probabilistic modelling based on observed data, logic trees, ensemble approaches, and Bayesian inference. There are significant correlations between different source parameters (e.g., earthquake magnitude and slip)	The same as in FW1A	Tsunami-tide interactions and sea-level variations that are nonlinear in shallow waters, require time-dependent tsunami hazard analysis methods that account for a range of variations (offsets) with respect to mean sea level e.g. in deep waters	The same as in FW1A	The quantification of uncertainties is not included in the framework	The quantification of uncertainties is not included in the framework	Events within open intervals are allowed to happen at any time before the first observation and at any time after the last, up to the present

(continued)

Table 2.3 (continued)

Treatment of uncertainties for the method	FW1A	FW1B	FW1C	FW1D	FW2A	FW2B	Data driven approach
Is the method able to digest and propagate uncertainty in input (previous ingredient) and propagate it to the output (next ingredients)? Does the assessment of this ingredient require uncertainty propagation? If yes, how are uncertainties propagated?	FW1A relies on forward uncertainty propagation from one module to the next (e.g., from hazard to damage to losses) assuming uncertainty propagation in each step of the assessment. Commonly uses logic trees to account for epistemic uncertainties	FW1B follows a modular structure, and the uncertainties are propagated from one module to the next through conditional probabilities. These conditional probabilities work in a "Markovian" way, assuming that each module is only dependent on the module that immediately precedes it	Uncertainties propagate in the simulations of the processes	Not fully. This FW can follow a forward propagation of uncertainties starting from selected scenarios	Not applicable, the hazard input transfers uncertainties to the final outcome	Not applicable, but some cases can use sensitivity analysis of indicators and similar methods exposed in Chapter 8	The method identifies most likely recurrence distribution parameters. The recurrence distribution is itself a representation of uncertainty, as is the probability calculated from that distribution
What are the consequences of lack of data in input or for boundary conditions, or for characterizing certain parameters?	Inaccurate hazard and risk assessments, inaccurate probability/ return periods associated to the intensity measure and the decision variable, inaccurate hazard and risk maps, inadequate tsunami countermeasures	The same as in FW1A, lack of data may lead to inaccuracies in the inputs of each module, affecting the risk assessment outcomes	Underestimation or overestimation of the predicted flow depth and run-up values, and thus, erroneous risk assessment outcomes	Most likely, misestimation of the hazard and risk estimates	Misestimations caused by insufficient data due to the high costs of datasets and fieldwork	Misestimations caused by insufficient data due to the high costs of datasets and fieldwork	Less data implies a wider range of possible solutions and their attendant uncertainty

(continued)

Table 2.3 (continued)

Treatment of uncertainties for the method	FW1A	FW1B	FW1C	FW1D	FW2A	FW2B	Data driven approach
What are the consequences of ignoring or poorly treating uncertainties?	Partial characterization and under-estimations of: tsunami intensity, number of casualties, physical damages, direct and indirect economic losses and their corresponding probabilities/return periods. Overall bias and underestimated/overestimated hazard/risk. This may lead to poor risk management or inadequate design of mitigation measures	The same as in FW1A	The same as in FW1A	The same as in FW1A	Poor risk management or inadequate design of mitigation measures	Poor risk management or inadequate design of mitigation measures. Inadequate urban and economic planning	The method is intended to prevent this

FW1B. Simulation-based hazard convolved with vulnerability and loss using numerical integration

Short description of method

FW1B is distinguished from FW1A because it relies on numerical integration for convolution of hazard and vulnerability (Eq. 2.2). Otherwise, this framework is the same as FW1A. The framework has a modular workflow as shown in Fig. 2.1 and is originally inspired from the Pacific Earthquake Engineering Center (PEER) probabilistic performance-based framework for earthquake engineering (e.g., Krawinkler 1999; Moehle and Deierlein 2004). The approach has its roots in engineering-design applications and has been used in diverse hazard settings (for earthquakes, flooding, hurricanes, etc.). For example, besides earthquakes, the PEER-like framework has been applied to risk assessment for natural hazards such as flooding (De Risi et al. 2013), hurricanes (Barbato et al. 2013), tsunami (Attary et al. 2019) and is considered one of the main inspirations for a modular PTRA framework (Behrens et al. 2021). The main advantage of FW1B with respect to FW1A is its efficiency since the risk calculation is performed independently of the PTHA. The hazard assessment in FW1B can be done in a simulation-based manner, in the same way as described for FW1A. The risk calculation is relatively less computationally demanding as the FW1B relies on pre-calculated hazard curves and their uncertainty. In fact, HPC resources are not always necessary in FW1B, though they are required for large-scale and comprehensive hazard and risk estimates. The Hazus tsunami model developed by the Federal Emergency Management Agency (FEMA), which uses engineering knowledge to describe relationships between hazard and structure types (Frucht et al. 2021), adopts such a modular framework.

Catastrophe models, which form the tool of the (re-)insurance industry to measure peril losses and inform policies, often employ a modular approach (e.g., Woessner et al. 2018). Goda et al. (2021) have used a catastrophe model to calculate the multi-hazard and single-hazard loss distributions for earthquake rupture scenarios in Kuroshio Town in Japan. The models calculate loss by integrating the hazard, exposure, intensity/damage relations and various policies for insurance portfolios. The OASIS loss modelling framework (LMF) is an open-source example that has been employed in the context of tsunami hazard to quantify future economic losses in Indonesia (Salmanidou et al. 2021). Figure 2.2 shows the modular probabilistic approach adopted by OASIS LMF.

Data and technical requirements in FW1B

FW1B integrates different components or modules to assess the risk posed to communities (Fig. 2.2). The different modules focus on the hazard and vulnerability with respect to exposure but may also evaluate the effect of various financial policies. The modules interact with each other through intermediate variables and their conditional probabilities. The approach links the disastrous event (in this case the occurrence and the hazard modules) to the direct and indirect socio-economic impact that it causes

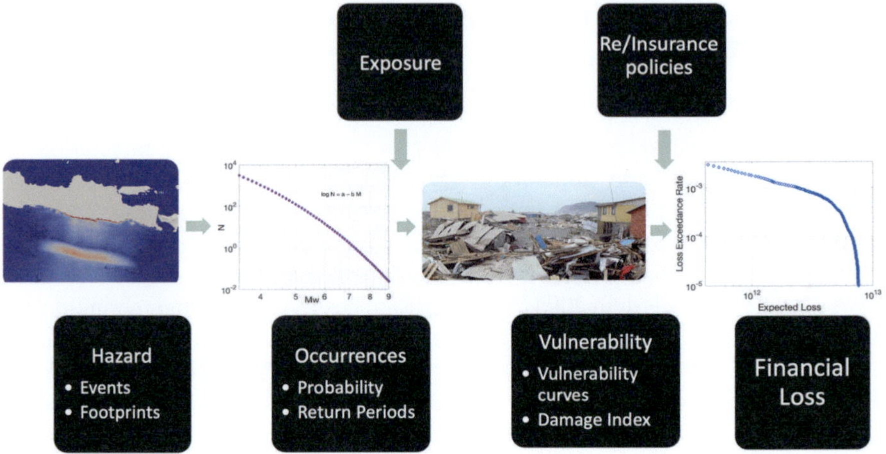

Fig. 2.2 Anatomy of a modular catastrophe model. Adapted from https://oasislmf.github.io/ based on data from Salmanidou et al. (2021). Damage photo: Patricio Winckler

(expressed through vulnerability and the financial loss module). OASIS LMF envisions developing financial risk transfer tools for estimating losses through financial loss and re-insurance policy modules.

In FW1B (and also FW1A), the hazard maps (commonly referred to as footprints) can be formally defined as the maps of *IM* values corresponding to given probabilities of exceedance or return periods. This definition lends itself well to spatial risk analysis applications (e.g., risk analysis for a portfolio of buildings, local or regional risk analysis). The footprints of the hazard may correspond to different IMs, some of the most commonly used are flow depth and inundation extent. Tsunami velocities, important due to development of disruptive currents in ports that pose a risk for marine infrastructure and shipping, also form an important intensity metric.

As in FW1A, this framework is typically implemented in a stationary form but can incorporate recurrence rates based on alternatives such as Poisson or Brownian Passage time models (e.g., Matthews et al. 2002). The framework may also consider the dynamic changes in exposure of a region over time and incorporate climate change effects. Input and output data related to application of this workflow in catastrophe modelling may not be open-source. For example, access to built environment/exposure datasets is often limited, since many datasets used by the insurance industry can be proprietary and thereby not open access.

Multi-event catastrophe modelling frameworks, combining for example earthquake ground shaking and tsunami inundation, have recently started gaining ground for risk assessment studies due to their importance in practical applications (Goda and De Risi 2023).

Similar to other frameworks, FW1B benefits from computational advances (Gibbons et al. 2020), yet in some cases the absence of accessible computational resources or high-quality data, such as bathymetry or exposure data, have led to simplified and poor uncertainty propagation. Thus, a great limitation is the dependency on computational resources that compromises resolution versus the number of scenarios. The lack of data (e.g., for damage estimation, fragility/vulnerability curves) or data of poor quality can make FW1B hard to implement and prone to errors.

Input data for the tsunami hazard in FW1B are similarly treated as in FW1A (e.g. Goda et al. 2023). The recurrence rates and hazard are inputs to the vulnerability and exposure modules (e.g. Salmanidou et al. 2021) but are treated separately. Uncertainties are comprehensively treated a priori by the risk analyst and represented as an input to the model. The model requires information about the hazard intensity for the exposed assets with damage thresholds, which allows accounting for uncertainty in different ways. The total insured value of the assets under consideration, for example at district scale, can also be used to quantify how different financial values may be applied. As in other modules, the risk assessment outputs are compromised in cases where uncertainties in the exposure are not adequately addressed.

Treatment of uncertainties in FW1B

The uncertainties in FW1B are quantified by incorporating different event outcomes in the process and allowing for their probability of occurrence to be integrated in the modelling. However, it must be noted that aleatory and epistemic uncertainty are inherent in each module/component of the process (hazard, vulnerability, exposure) and thus difficult to treat.

As in FW1A, there is epistemic uncertainty in the evaluation process for each module. The hazard modelling is a major source of uncertainty in the tsunami risk assessment workflow (Song and Goda 2019). Many of the various modules have uncertainty that relates to the source distributions, the lack of data and the unknowns (assumptions in the model that are not well quantified due to lack of knowledge or bias). Therefore, large uncertainties can be propagated from the hazard intensity estimates, the exposure data, and damage and cost/consequence models (Song and Goda 2019). As described for FW1A in Sect. 2.1.2, there are a number of methods that can be used to quantify the uncertainties of the identified parameters.

As all FWs, also FW1B is heavily affected by the potential lack of data. A poorly defined source characterisation may lead to reduced accuracy in the loss calculations. This is especially evident when modelling tsunami sources (e.g. landslide tsunamis) for which the magnitude-occurrence rate relationships are not yet fully understood. In a similar way, coarse spatial resolution of data or outdated information (e.g. in the exposure datasets) may lead to an overestimation or underestimation of risk. For example, data from field surveys after major tsunami events can be used to assess damage levels and associated vulnerability, but changes in socio-economic factors over time are not easy to capture and may result in increased costs for similar types of events in the future. In other cases, such surveys are inaccessible or non-existent as similar events have not occurred in the past.

The future projections of tsunami impacts are complex to characterize. Ignoring or inappropriate treatment of uncertainties can lead to inaccurate predictions. For example, variations in future population exposure may lead to underestimation or overestimation of the risk when not incorporated in the model. In a similar way, inaccurate damage assessments and lack of structural information may lead to an underestimation of the costs and fatality rates.

A summary of the main aspects of FW1B is given in Table 2.3.

FW1C. Incorporation of dynamic processes with time dependency: tidal and sea level variations

Short description of method

This framework considers the time dependency and dynamic aspects of tides and projected sea-level rise, mainly due to future climate change. It is conceived as an 'extension case' that could be part of any other workflow falling under FW1, in order to integrate the cumulative effects of tsunami hazard and sea level variations for long-term coastal development planning (Koyano et al. 2022; Grezio et al. 2024).

FW1C is characterized by combining dynamic time dependent processes, such as climate change, sea-level variations, land subsidence, glacial cycles (Lambeck and Chapell 2001), and tides, with tsunami waves produced by the different tsunamigenic sources. The time interval between tsunami generation and inundation is of the order of seconds to hours with varying intensities of the devastating effects. Other phenomena, such as relative sea-level and shoreline variations caused by geological factors (e.g. subsidence) and mean sea-level rise due to global climate change, may occur at time scales of years to centuries and may severely impact coastal areas and the long-term tsunami hazard and risk assessments (Li et al. 2018).

Tides interact with tsunamis by modifying wave propagation, amplification, and dissipation. Tidal amplitudes change the sea depth and consequently the tsunami wave elevation, velocity, momentum advection, bottom friction, and water flux (Shelby et al. 2016). These processes are strongly dependent on local bathymetry, coastal geometry, tidal phase and amplitude (Kowalik et al. 2010), and on tsunami dynamic effects and tsunami wave directivity (Ayca et al. 2016).

Tides and tsunamis are long waves, and their propagation can be modelled using appropriate wave theory including nonlinear and dispersive effects (typically some variant of the nonlinear shallow water equations). Simulations can take into account the time-varying reference water level and the time-varying currents which are modulated by tides. Results of each dynamic tsunami–tide simulation are compared to those of the corresponding simulation done over a static water level, to identify the extent to which tsunami–tide interactions may increase flooding and currents in some areas depending on tide phase (Ayca and Lynett 2016; Mofjeld et al. 2007).

PTHA methods considering tides follow different procedures:

- Linear superposition of tsunami and tides, which are added sequentially using a year-long record (Gaussian Method; Adams et al. 2015) or added to the water depth at hourly intervals in the tsunami simulation before solving the nonlinear equations of mass and momentum conservation (Latcharote et al. 2018)
- Incorporation of a probability density function of the difference between initial tsunami amplitude and tides (Mofjeld et al. 2007)
- Numerical simulations including tides in idealized channels (Kowalik et al. 2010), realistic sites with estuaries and rivers (Zhang et al. 2011), and shallow water environments (Lee et al. 2015)
- Incorporating synthetic seismic catalogs and tidal conditions based on the probabilistic distribution of the involved variables (García et al. 2019).

Similarly to FW1A, FW1B and FW1D (working on a limited number of scenarios), hazard curves and probabilistic hazard and risk maps are obtained, however in FW1C they take into account the tidal effects on coastal tsunami impact.

One of the major effects of climate change is the rising sea level which is estimated to occur at a rate of 4–9 mm yr^{-1} under the RCP2.6 IPPC scenario and 10–20 mm yr^{-1} under the RCP8.5 IPCC scenario on global average (Oppenheimer et al. 2019). Including projected sea-level rise in the PTHA is a challenge because sea-level rise could be site-dependent and is mainly related to increased inundation areas and changes in the propagation depths, which both modify the tsunami impacts. Moreover, sea-level rise is not homogeneous globally on the coasts and further local bathymetry changes are expected due to vertical land movements (e.g., due to local subsidence or human activities).

Methods to implement the IPCC sea-level rise scenarios in a time dependent PTHA or PTRA framework typically consider the tsunami hazard of seismic sources. The following approaches have been described:

- Conditional hazard curves in Japan (Alhamid et al. 2022), or conditional hazard maps in South China Sea (Li et al. 2018) estimated by a given sea level rise as indicated by the IPCC scenarios;
- Time dependent changes of the position of the shoreline due to different sea-level rise values which are combined with time-independent deterministic earthquake modelling in the Alaska-Aleutian Zone (Dura et al. 2021) or probabilistic tsunami hazard (Grezio et al. 2024);
- A non-stationary Poisson process model which takes advantage of the independence of the arrivals on the sea-level rise, included as a non-stationary process in the PTHA in the South China Sea (Sepúlveda et al. 2021).

Data and technical requirements in FW1C

FW1C can be seen as a special case of FW1A, FW1B, or FW1D, as a similar treatment is expected in each step of this workflow. Both tides and relative sea-level changes due to natural or anthropogenic actions can be included in the PTHA while using the same methods to propagate uncertainties as explained in FW1A and FW1B. A variety of numerical schemes and cluster analysis are used in general circulation models with and without nesting, and they are parallelized to carry out probabilistic studies and tsunami modelling, as described in more detail in Chap. 5.

Including tide-tsunami interaction is crucial in some areas. For example, large differences can be observed in maximum current speeds due to nonlinear interaction between tsunamis and tides inside harbours and bays (Garcia-Aristizabal et al. 2019, Shelby et al. 2016). Tsunami penetration in rivers, inlets, and estuaries is strongly affected by tidal conditions because the changes in the depth alter the tsunami propagation conditions (e.g., Ayca et al. 2016, Kowalik et al. 2006, Kowalik et al. 2010, Lee et al. 2015, Mofjeld et al. 2007, Shelby et al. 2016, Tolkova et al. 2015, Tolkova et al. 2016, Zhang et al. 2011). Several studies have been carried out to understand whether tide-tsunami interactions affect tsunami elevation, flow speed, and arrival times. Adams et al. (2014) overcome the assumption of linear superposition of the tide and tsunami waves by using multiple tsunami model runs at different tidal stages and introducing nonlinearities in the inundation process. Zang et al. (2011) compare simulations conducted with and without tide-tsunami interactions and indicate that large site-specific differences are possible, in some cases tsunami-tide interactions can account for as much as 50% of the run-up and up to 100% of the inundation (e.g., flow depths). Thus, the flow characteristics of the tsunami are very sensitive to the tidal phase and the choice of modelling approach is crucial for the hazard assessment.

Climate scenarios for the present century indicate a large variation of the sea-level which is important to include in long-term PTHA. While the average rate of sea-level rise between the 1901 and 1990 has been 0.81–1.95 mm yr^{-1} (Oppenheimer et al. 2019), the IPCC Business-as Usual Scenario (RCP8.5 IPCC scenario) predicts a global-mean sea level at year 2030 that is 8–29 cm higher than today, with a best-estimate of 18 cm. In 2070, the sea-level rise is predicted to be 21–71 cm, with a best estimate of 44 cm (Warrick et al. 1990, https://www.ipcc.ch/report/ar1/wg1/sea-level-rise/).

Applying FW1C requires access to data that quantify the tides and/or projected sea level changes. Elevation and phase of tides are generally provided by tide gauges close to the site of interest or adopted from the NOAA Tides and Currents database. Also, a DEM and local records of tidal variations must be available, as well as high-resolution coastal bathymetry data. For sea-level variations, regional assessments are required, and local sea-level projections should indicate the relation between geological processes such as coastal deformation and sea-level rise effects due to climate change. Comprehensive source-based hazard assessment should include sea-level projections which are not on the same scale, however, this is not yet typically done for PTHA or PTRA.

Treatment of uncertainties in FW1C

Most of the probabilistic approaches that explore aleatory and epistemic uncertainties neglect the tide-tsunami interactions. The methods to implement the IPCC sea-level rise scenarios in a time dependent framework for tsunami hazard are mostly following a "hybrid" approach. In this approach, the assessments take into account both the probabilistic tsunami hazard analysis with the associated uncertainties and the IPCC projections with their relative likelihoods based on model outputs. Tsunami hazard is estimated considering e.g. the uncertainties associated with the source parameters of the triggering event (for example, magnitude and slip rates in the case of earthquake source). Those uncertainties can be integrated with those related to the tides and sea-level projections, although it is commonly assumed that uncertainties in tides are smaller than in source parameters and phenomena associated with sea-level.

The aleatory uncertainties related to the source and to the tidal variations have been considered in several studies (e.g. Omira et al. 2016). Gonzalez et al. (2021) proposed a PTHA methodology that treats aleatory uncertainties both in the seismic source and the tidal stage through a Monte Carlo type framework deriving probabilistic coastal hazard maps (wave heights, current speed, and total force) for different return periods. In this framework, the tides were included as an aleatory variable in the PTHA scheme.

It is straightforward to forecast potential tsunami impact, given the tsunami wave height in the region of interest, at both high tide and low tide, which may lead to considerably different effects. With this assumption, there is no epistemic uncertainty in the tide, as tides are relatively well known and modelled, and we have only the aleatoric uncertainty associated with the fact that the triggering event could happen at any time during the tidal cycle (Adams et al. 2015; Lee et al. 2015).

Ignoring tide-related issues may be problematic in certain areas. For example, in the NE Atlantic region, the peak-to-peak amplitude of tidal variation can be as high as 4–5 m, while the tide at the time of tsunami occurrence cannot be predicted. In order to include tidal stage uncertainty in PTHA, one must calculate the probability that a tsunami impact (wave height, flow depth) exceeds a particular threshold for an event realization that includes a specific tide level.

Shelby et al. (2016) simulated the maximum inundation in the Hudson River Estuary resulting from dynamic interactions between the significant maximum tsunamis and a tide. They note that dynamic simulations predict a slight increase in inundation, which may be comparable to the uncertainties related to modelling of tsunami sources (such as their initial water elevation), and in bottom friction and bathymetry used in tsunami grids.

To avoid poorly treating uncertainties for tide-tsunami interaction, the tsunami models need to be of the appropriate resolution. In fact, due to coarse grids, Myers et al. (2001) showed only modest influence of tides on tsunami runup in comparison to other model uncertainties such as those related to bottom friction and grid resolution.

The epistemic uncertainties related to both climate models and the IPCC scenarios are considered the main uncertainties in identifying sea-level rise projections for tsunami hazard (Alhamid et al. 2022). Sterodynamic, glacier, and ice sheet sea-level rise are considered components largely affected by climate change with epistemic uncertainties related to parameterizations, monthly data measurements and global mean surface temperature. Many authors disregard land water storage from hydrological processes and inland water utilization, with the assumption that those components are negligible, even if locally their contributions to the sea-level rise may be significant. Common sea-level rise projections are conditionally dependent on plausible future climate change emission scenarios, which are known as the representative concentration pathways and defined as anthropogenic emissions from energy and land-use trends.

Since there is no consensus regarding the likelihood of each future anthropogenic emission scenario, the simplest assumption possible is that the occurrence probabilities associated with the individual climate change scenarios are identical. IPCC experts provide two probability ranges for each scenario. The sea-level rise is assessed to have a 66% probability of being within the inner range (17th to 83rd percentiles), corresponding to the IPCC's definition of a 'likely' range. There is an assessed probability of 90% that the sea-level rise will be within the outer range (5th to 95th percentiles), corresponding to the IPCC's definition of a 'very likely' range.

Alhamid et al. (2022) point out that tsunamis with intensified impacts due to climate change can deteriorate the functionality of infrastructure systems and generate a more significant amount of disaster-generated-waste, which implicitly extends to economic losses. It is possible to more precisely quantify the risks to structures and infrastructure systems when taking sea-level rise into account while evaluating tsunami hazards. For instance, the effects of sea level rise are expected to cause the tsunami hazard in the Hokusei region in Japan to double in intensity. Hence it is shown that neglecting these effects when assessing the tsunami hazard can result in inadequate structural safety assessments and improper prioritizing of disaster countermeasures in the area (e.g., Alhamid et al. 2022).

Tsunami hazard may be underestimated when future sea-level projections are not considered in PTHA. In some cases, earthquake and tsunami modelling combined with local probabilistic relative sea-level rise projections show an increased potential for more frequent, lower magnitude earthquakes to produce far-field tsunamis that exceed historically observed maximum nearshore tsunami heights. Dura et al. 2021 demonstrate such an example for the Ports of Los Angeles and Long Beach. Sepúlveda et al. (2021) carried out sensitivity analyses of a non-stationary PTHA in Hong Kong and Kao Hsiung demonstrating that the sea-level rise produces an impact comparable to the earthquake recurrence model uncertainty when bathymetry measurement accuracy and resolution are taken into account along with other relevant sources of uncertainty. This is discussed further in Chap. 5.

A summary of the main aspects of FW1C is given in Table 2.3.

FW1D. Working on a limited number of scenarios

Short description of the method

Development of hazard studies and inundation maps based on a few scenarios has been a common approach, due to restricted capabilities for numerical simulations. This framework provides an overview of a traditional approach applied in early warning systems, evacuation decision making, and tsunami inundation mapping. Contrary to the full probabilistic treatment, in this framework, risk analysis is performed considering only selected (presumably representative) scenarios or scenarios pertaining to a certain range of tsunami magnitudes. The analysis is based on individual (or a few) events, either for illustrative reasons, or to provide an estimate of the order of magnitude of the hazard. By considering one or few representative large tsunami scenarios, decision-makers would get an idea of the most severe outcomes and implement strategies to mitigate risks or handle adverse events effectively. However, their application for decision making could be disputable and has been justified only in absence of probabilistic estimates. A major problem associated with the partial characterization of uncertainties or adopting "worst case" scenarios, is that we do not have any guarantee that we have effectively defined and constrained the dominant or worst possible cases. A probabilistic approach, on the other hand, strives to provide and aggregate a systematic range of all plausible events and their consequences conditioned on the data we have, without claiming to constrain the "worst possible" event. That said, even in PTHA we cannot ensure that "all plausible events" have been correctly considered, since we cannot exclude the possibility that events larger than those considered in the PTHA will happen in the future (e.g., Mulargia et al. 2017, Stirling 2014, Zanetti et al. 2023).

A common approach to assess tsunami hazard in FW1D is to aggregate of a few scenarios (e.g., Álvarez et al. 2013, Chacón Barrantes and Arozarena-Llopis 2021, Harbitz et al. 2012, Løvholt et al. 2012a, b, Wronna et al. 2015). Alternatively, the assessment may be based on an "expected worst scenario' or historical records (Martínez et al. 2020) without necessarily associating scenarios with a return period or probability of exceedance. In some cases, return periods can be estimated based on historical data or other background knowledge.

Restricting the analysis to few scenarios allows focusing on other details, like the assessment of cascading processes from an individual event (e.g., Wang et al. 2023, Terry et al. 2022). More robust characterization of tsunamigenic sources is possible using physics-based approaches. In this regard, tsunami hazard assessment with a limited number of scenarios may allow scenarios to be represented with fewer simplifying assumptions. Selecting a few representative sources allows characterizing the tsunamigenic sources of a complex system, and more robust methods may replace the typically used scaling relations. Such methods include physics-based source modelling that accounts for stress and friction parameters (Kaneko et al. 2008; Ulrich et al. 2022), seismic rupture dynamics (Murphy et al. 2016; Ramos et al. 2022; Wirp et al. 2021), coupled-cascading effects or complex ruptures (Ulrich et al. 2019;

Abrahams et al. 2023) and kinematics (Fuentes et al. 2022; Williamson et al. 2020), as well as high resolution bathymetry leveraged by computational resources that are lately available for such tasks (Wirp et al. 2021).

Risk and impact assessment in FW1D are also going to be conditioned on the selected scenarios, expressed in terms of conditional probabilities of exceeding certain risk metric thresholds for the locations of interest. In that regard, the workflow for risk assessment is very similar to those of FW1A and FW1B. In cases where it is possible to estimate the rates of occurrence of the selected scenarios, these rates can be multiplied by the above-mentioned conditional exceedance probabilities in order to calculate the rates of exceeding certain risk metric thresholds. However, these rates are going to represent partial contributions to the overall rates of exceedance (the risk estimated from a comprehensive PTRA procedure as in FW1B and FW1A).

Hazard estimates based on one or few scenarios have been applied to generate mostly evacuation maps by estimating "worst-case" scenarios defined based on geological constraints, historical maximum event or expert judgment (Mori et al. 2018). When using more than one historical event or maximum expected scenario, the potential tsunamigenic sources have been treated mainly by selecting the maximum wave heights from all the sources or aggregating them. Hazard assessment based on a selection of this tsunamigenic sources can be treated by means of the deterministic or pseudo probabilistic schemes (Gonzalez et al. 2022) or deterministic approaches (Al-Habsi et al. 2022; Berryman et al. 2005; Fruct et al. 2021; Sato et al. 2003; Wijetunge et al. 2014).

Previous studies considering few scenarios have used historical data or possible large future scenarios (e.g., Harbitz et al. 2012; Lorito et al. 2008; Tinti et al. 2008; Tonini et al. 2011) defined within the geological context of the study area (Basili et al. 2013).Strong limitations (or bias) in the hazard estimates may emerge if only historical events or paleo-tsunami data are used, or, on the contrary, if tsunami observations are not included in the source recurrence modelling (e.g. Stein and Okal, 2007; Kagan and Jackson, 2013).

When adopting FW1D for risk assessment, the analyst should be aware that the characterization of uncertainties is partial. This means that, even if probabilistic risk estimates (in terms of rates of exceedance of certain risk metric thresholds or maps of inundation and loss corresponding to given return periods) are developed, these results are representing partial (desirably significant) contributions to the total risk that, by definition, needs to take into account all plausible tsunamigenic scenarios.

Scenario-based tsunami inundation and risk (e.g., damage, loss) maps are key outcomes from numerical simulations, where flow depths, arrival times, damages and losses are shown. Gomez-Zapata et al. (2021) provide a comprehensive example of the application of this FW in Lima, Peru. FW1D lends itself well to building scenarios of complex and cascading interactions between events such as Tonga-like volcanic explosions (Ramírez-Herrera et al. 2022) or Anak Krakatau-like volcanic cascades (Walter et al. 2019), and of interaction of tsunamis with slow-onset events and the societal-ecological context. As such, it provides a framework for participatory scenario-building with the objective of engaging local communities (e.g., Johnson et al. 2012) or devising counterfactual scenarios (Woo 2019).

Data and technical requirements in FW1D

The resources required for FW1D are similar to those needed in the other sub-frameworks of FW1. Physics-based models for a limited number of scenarios may not always require HPC resources. Thus, the main requirements are the numerical codes (Chap. 5) and a digital elevation model. The latter may vary in the spatial resolution which may limit the hazard analysis and exposure assessment, and thus the overall risk estimate. As an alternative to numerical modelling, the hazard and risk estimation could also rely on empirical relations where the tsunami wave heights or runups are estimated from values offshore with the aid of amplification factors or empirical runup relations.

Treatment of uncertainties in FW1D

The tsunami IMs are calculated based on numerical simulations to generate scenario-based tsunami inundation maps. The quantification of uncertainties in this FW is limited. Since the framework is limited to one or few scenarios, it is clear that a comprehensive treatment of uncertainties related to all the possible tsunami source mechanisms cannot be carried out. However, conditioned on a given tsunami scenario uncertainty propagation can be done. This means that the uncertainties in the exposure, fragility and vulnerability models can be taken into account even if a limited number of tsunami scenarios are considered. In many cases, it is possible to associate the selected scenarios with a return period or rate of occurrence, though caution must be taken since single scenarios usually give limited information about the return period of tsunami waves at onshore sites of interest. For instance, we might have a reasonable estimate of the recurrence rate of a given earthquake magnitude, but due to the presence of other seismic sources or the possibility that other magnitude scenarios may be tsunamigenic, this may not correspond to the rate of "tsunami inundation depth" exceedance at any given site of interest (e.g. Davies et al. 2022).

The potential sources of uncertainties in FW1D can include magnitude scaling relations, slip distribution, uncertainty in parameters related to tsunami offshore and onshore propagation, uncertainty in the exposed assets attributes, uncertainty in the fragility and the vulnerability models and their parameters. Different to FW1, more attention can be given to more complex source characterization and modelling of cascading events but owing to the limited number of scenarios, the aleatory uncertainty is not considered.

A summary of the main aspects of FW1D is given in Table 2.3.

2.3.2 *Framework 2: Tsunami Risk Assessment with Special Focus on Indicator-Based Vulnerability Assessment (FW2)*

This framework represents approaches to tsunami risk assessment where vulnerability assessment relies on indicators rather than numerical modelling. According to Birkmann et al. (2006), a vulnerability indicator is a "variable which is an operational representation of a characteristic or a quality of a system able to provide information regarding its susceptibility, coping capacity and resilience of a system to an impact of an albeit ill-defined event linked with a hazard of natural origin". It is worth noting that although physical vulnerability has often been defined as the degree of loss of an element subject to the impact of a natural process (UNDRO 1984), the use of indicators is more related to the general definition of vulnerability by UNISDR (2009). Here, vulnerability is described as "the conditions determined by physical, social, economic and environmental factors or processes which increase the susceptibility of an individual, a community, assets or systems to the impacts of hazards" suggesting that vulnerability is a pre-existing condition.

The required data collection and analysis (weighting, classification) usually include the participation of different stakeholders. Technologies such as Geographic Information Systems (GIS) and remote sensing are essential for the organization of the acquired data, as well as the visualization and updating of the results. This kind of approach enhances accuracy, supports inclusiveness through an elicited process, encourages a stronger sense of ownership of risk evaluation up to community level, and provides a good basis for effective risk communication applying techniques described in Chap. 8. GIS plays a central role in incorporating participatory-based outcomes in the tsunami risk assessment, allowing for the collection, storage, analysis, and visualization of spatial data related to tsunami hazard and risk components (Fagerholm et al. 2021; Iles and Kedzior 2023). GIS technology offers resources for gathering, storing, processing, analysing, and visualizing spatial data on the risk factors and tsunami threat, supporting the assessment and decisions with georeferenced information (González-Riancho et al. 2014). GIS databases may be updated easily compared to other static or analogue maps and may provide the basis for risk communication and collaboration among interdisciplinary teams as well as among authorities and the public.

The approaches presented for this framework have a special focus on the vulnerability of elements at risk (e.g., population, buildings, infrastructure) and they use indicators to assess it and provide the basis for tsunami risk assessment. This assessment is commonly done using deterministic approaches, but outcomes from PTHA can be adapted and used for different return periods. The concept of these approaches follows the schema of Fig. 2.3, showing the steps of the hazard and vulnerability mapping and the applications of such products by different end users. Regional or local stakeholders include the education sector, local tourism industries and business, among others.

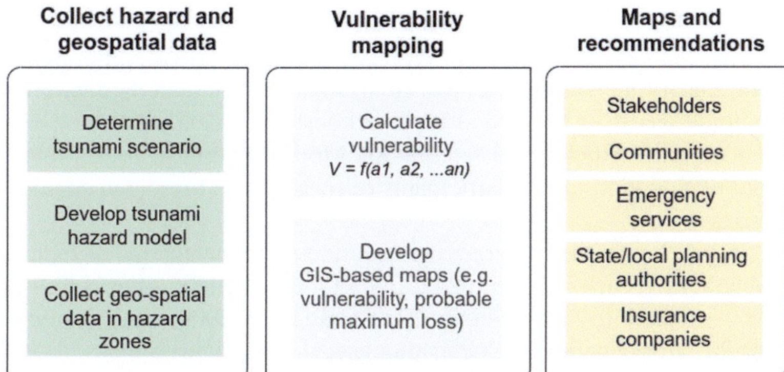

Fig. 2.3 The concept of a typical indicator-based tsunami vulnerability assessment method (modified from Tarbotton et al. 2012)

The first part requires identification of the inundation zone based on historical data, numerical modelling, or even expert judgment. The second part involves identification of all the characteristics of the element at risk under investigation that control the robustness or sensitivity of the inundation zone to the impact of the tsunami. These characteristics (the so-called indicators) can be found directly in the literature or determined following participatory methods (focus groups, interviews, questionnaires, etc.). Following the selection of the indicators, the weighting can also be developed through participatory methods, often involving expert judgment (e.g. the Analytic Hierarchy Process; AHP). The AHP approach (Saaty 1987) is used to solve multi-criteria decision problems and is based on a pairwise comparison of different factors to order them hierarchically. Moreover, to assess vulnerability and overall tsunami risk, methods such as the Participatory Integrated Assessment (PIA) can be used as a multidisciplinary field-based approach (Salter et al. 2010). Such methods may support the integration of local data and knowledge and incorporate risk perception via elicitation methods where subjective probabilities can result from stakeholders, experts, and the overall engaged communities. Elicitation, in the search for expert consensus, can facilitate the quantitative expression of subjective judgment (Dias et al. 2018). This approach has also been used for risk assessment and mitigation purposes (Grezio et al. 2017).

In more detail, participatory methods involve engaging a diverse group of stakeholders, including scientists, engineers, local communities, policymakers, household owners, and the general public, in the risk assessment process. Participatory approaches are essential because they recognize that tsunami risk assessment is not just a technical endeavour, but also a social one. Moreover, they make use of local knowledge, experience, and expert knowledge, which is important when other types of data are not available, as they provide complimentary information. Finally, participatory methods can ensure the availability and accessibility of data in cases where empirical data are absent or of bad quality and modelling is not possible.

There are many advantages but also limitations to using participatory methods. According to Mercer et al. (2008), advantages include opportunities for knowledge transfer, learning and involvement of all stakeholders including hard-to-reach groups. Limitations may be related to time consuming data collection, language barriers and lack of experienced facilitators. Moreover, ethical issues related e.g. to potential exploitation of the participants or ownership of research should not be underestimated.

Many approaches in the literature demonstrate implementations of indicators to assess vulnerability, each one having a different focus. Some studies concentrate on the physical vulnerability of buildings (Dall'Osso et al. 2009, 2016; Voulgaris and Murayama 2014; Sathiparan 2020; Papathoma et al. 2003; Papathoma and Dominey-Howes 2003; Alberico et al. 2015; Izquierdo et al. 2018), some focus on the vulnerability of infrastructure (Aguire-Ayerbe et al. 2018), whereas some use indicators that express aspects of social (Wood et al. 2010), environmental or economic vulnerability in combination with physical vulnerability to develop composite or multidimensional indices (González-Riancho et al. 2014). Geographically, there are many case studies from different parts of the world including the Mediterranean (Turchi et al. 2022; Papathoma-Köhle et al. 2019a, b), Australia (Dall'Osso et al. 2009), USA (Dominey-Howes et al 2010), and many Asian countries such as Indonesia (Nisaa et al. 2021), Turkey (Tufekci et al. 2018), Japan (Voulgaris and Murayama 2014), Sri Lanka (Hasalanka et al. 2021), and Oman (Aguirre-Ayerbe et al. 2018). Furthermore, many studies can be found in Central and South America like Chile (Izquierdo et al. 2018), Ecuador (San Martín et al. 2018), El Salvador (González-Riancho et al. 2014), as well as in Africa (Omira et al. 2009).

Some of the approaches that focus on physical vulnerability are based on characteristics of buildings and their immediate surroundings, whereas others develop indices that relate to the geomorphology and other characteristics of the surrounding area, combining them (Tavares et al. 2017) or not (Cankaya et al. 2016; Tufekci et al. 2018) with building characteristics. Moreover, some of the approaches propose or demonstrate practical uses of the results of the vulnerability assessment for the implementation of risk management measures and disaster risk reduction strategies. The assessment based on FW2 and its visual representation can be directly used as a knowledge basis to inform evacuation planning, land-use planning, early warning system development and to locate structural mitigation measures. Indicator-based vulnerability assessment methods may indirectly enhance disaster risk reduction by promoting participation of different groups.

Dall'Osso et al. (2010) tested indicator-based tsunami vulnerability assessment in the Aeolian Islands in Italy and concluded that the results can be used for future urban planning and management of movable objects. Santos et al. (2014) assessed physical vulnerability in Portugal and suggested that there is a need for inclusion of social indicators due to extensive commuting in the area, so that the results can be used for urban planning. Moreover, in the same study, it was clear that critical infrastructure (port facilities) was particularly vulnerable, showing that there is a need for intervention in this respect. In Japan, the results of a physical vulnerability assessment of buildings were combined with population data showing that there are many buildings with only one resident and that there is a high concentration of aged population, a finding that could guide evacuation planning and targeted awareness programs (Voulgaris and Murayama 2014). Very often, the results of physical vulnerability assessments have been used to identify shelters for vertical evacuation (Alberico et al. 2015; Sathiparan 2020; León et al. 2018), to support evacuation planning (e.g., Sathiparan 2020; Mück et al. 2013; Ismail et al. 2012), for focused building reinforcement (Mück et al. 2013), for resource allocation for reducing vulnerability (Alberico et al. 2015) or for introducing building codes through legislation and regulations (Alberico et al. 2015). Moreover, Ismail et al. (2012) suggested integrating the Physical Vulnerability Index (PVI) in a Coastal Zone Management Decision Support System to guide development, structural protection, coastal defence plans, and agricultural areas. Nevertheless, proposing risk management measures based on vulnerability and risk assessments does not guarantee their implementation. For example, Gonzalez-Riancho et al. (2014) assessed tsunami vulnerability in El Salvador based on indicators associated with human, environmental, socioeconomic and infrastructure vulnerability. Based on the results, they proposed three risk reduction measures related to raising awareness, reforestation of mangroves and relocation/reinforcement of specific buildings. However, they also pointed out that the measures would be economically and politically difficult to implement. In another case study in Oman (Aguirre-Ayerbe et al. 2018), a Risk Reduction Measures (RRM) catalogue was developed and connected to the results of a tsunami risk assessment aiming at preparedness and prevention. As far as preparedness is concerned, activities related to raising awareness, capacity building and education were proposed together with risk mapping and emergency planning (e.g., early warning and evacuation). Activities related to preparedness included engineered-based measures (e.g., seawalls), nature-based measures (e.g., beach nourishment), and coastal planning and architectural measures (e.g., building standards). Prioritization and localisation of the measures were based on the results of the risk assessment that indicated specific hotspots for their implementation.

Indicator-based methods can be used directly for risk management planning and they can contribute indirectly to enhanced participation in the risk management process. For example, the choice and weighting of the indicators are often decided at stakeholder workshops (Gonzalez-Riancho et al. 2014). An alternative strategy for data acquisition is to use applications such as Open Street Map, that engages voluntary digitizing of maps and provides open access to data. The key challenges in such an approach are, among others, related to data validation and reliance on non-experts (Nurmaya et al. 2023). Indicator-based methods require a large amount of local information on buildings, surroundings or population, that are often not available. Local population and stakeholders can support the collection of such data and their weighting according to their importance. Nisaa et al. (2021) stress the importance of participatory methods in tsunami risk assessment, however they point out that the quality of the results depends on the knowledge and experience of people, making the outcome very area-specific. Moreover, since tsunamis are relatively rare phenomena, lack of collective memory and experience with past events affects the risk perception of the population, making tsunami risk communication particularly

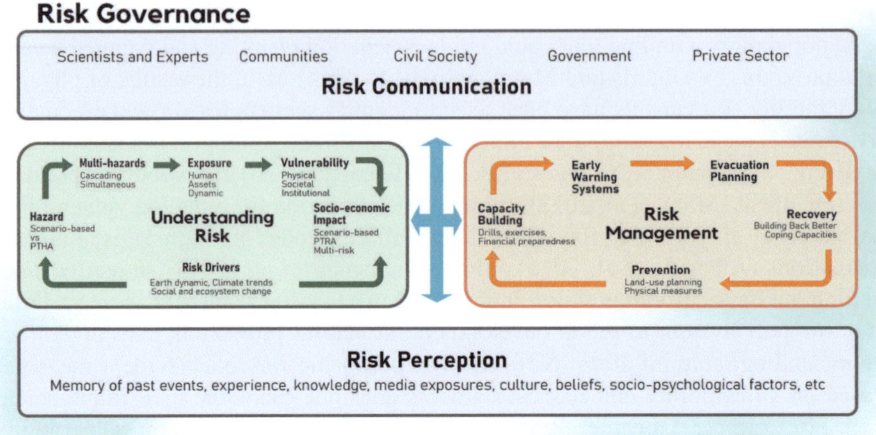

Fig. 2.4 The applications of tsunami risk: from hazard assessment to risk management. *Source* Rafliana et al. (2022)

challenging (Rafliana et al. 2022). By looking at historical and recent events, such risk communication challenges and gaps connected to lack of a tsunami risk assessment approach have been identified (Rafliana et al. 2022).

The increased number of studies on risk perception (see for example Cugliari et al. 2022) and its relations to how risks are understood, communicated, and managed has demonstrated the importance of the topic. It is, however, challenging to quantify perception, since perceptions are embedded social constructs and continuously changing. While risk perception is argued to contribute significantly to vulnerability, it is still unclear how it can be incorporated in the aggregated risk assessment. The schema presented in Fig. 2.4 hence addresses risk perceptions as external to the risk assessment (understanding risk) processes.

Risk managers may struggle to make a clear link between the findings of risk assessment and management, making it difficult to draw conclusions about how to minimize risk at a given location (Gonzaléz-Riancho et al. 2014). In this regard, Rafliana et al. (2022) emphasize that risk assessment should inform land use planning, decisions related to nature-based and engineering-based solutions as well as building codes and adapted design.

Within the context of FW2, two approaches are presented in the following paragraph. The first (FW2A) has a focus on physical vulnerability. It uses participatory methods to weigh the indicators and GIS to display the outcome of the analysis (Papathoma et al. 2003; Papathoma and Dominey-Howes 2003; Dall'Osso et al. 2009, 2016). The second approach (FW2B) represents a multi-dimensional vulnerability assessment that focuses on population, environment, socio-economic setting and infrastructure and uses participatory methods (stakeholder workshops) for the collection and analysis of data (González-Riancho et al. 2014). Some qualitative approaches for assessing vulnerability are also briefly mentioned under FW2B.

FW2A. Indicator-based physical vulnerability assessment

Short description of method

FW2A does not rely on the general PTHA and PTRA probabilistic framework in the strict sense and may use qualitative scoring criteria for aggregating various indicators of physical vulnerability into a vulnerability index. It is useful for capturing certain aspects of vulnerability that might not be captured properly through other frameworks that have a consequential structure to them (i.e., through forward propagation of uncertainties). In other words, the vulnerability indicators can have an inherent nature and might not necessarily be a direct causal consequence of a tsunami. They may identify critical situations where existing vulnerabilities will be exacerbated and accentuated by the occurrence of a tsunami. This approach may rely on hazard estimates following FW1D for the identification of inundation zones for certain tsunami scenarios, or it may consider IMs with a specific return period from any of the other sub-frameworks of FW1.

FW2A focuses mainly on the physical vulnerability of buildings using their characteristics and main features of their surroundings, represented by vulnerability indicators. The framework is based on the Papathoma Tsunami Vulnerability Assessment (PTVA) method that was developed in Greece in 2003 (Papathoma et al. 2003; Papathoma and Dominey-Howes 2003) and was further developed by Dall'Osso et al. (2009, 2016). The method makes use of deterministic tsunami scenarios and focuses on the built environment within the specified inundation zone at the local scale.

The approach considers building characteristics and characteristics of their surroundings (indicators) to develop an overall vulnerability index for buildings. The different characteristics are scored, normalized, standardized, weighted, and aggregated into a single vulnerability index. The index is assigned to each building in the inundation area.

The first step of the method is to determine the potential inundation zone as the geographical basis of the study. In general, the inundation zone can be determined through modelling, empirical data (event documentation), or historical data. If modelling is not possible, an investigation of existing catalogues, archives and historical data is required to identify a possible scenario and its characteristics. If the scenario is based on a recent event, the wave height or the inundation inland may have been recorded, or it can be estimated by photographic documentation. For example, during the development of the PTVA, the inundation zone was identified by historical data suggesting that a tsunami in 1650 AD overtopped a wall in Heraklion, Crete, (Fig. 2.5) such that the inundation area was identified as the zone between the coastline and the 5 m contour (Papathoma et al. 2003).

Based on the chosen tsunami scenario, the inundation area is given, and the buildings located within this area are considered in the further analysis. For each building, the indicators are collected and entered into a GIS database. The detailed selection of indicators and their scoring is described in Chap. 7. The indicator identification is based on existing descriptions of past events. The indicator selection, scoring, normalization, and weighting have been based on OECD (2008). The indicators

Fig. 2.5 The stone wall of Heraklion (Crete, Greece). Photo: M. Papathoma-Köhle

used for developing the vulnerability index for the latest version of PTVA (Dall' Osso et al. 2009) are presented in Fig. 2.6.

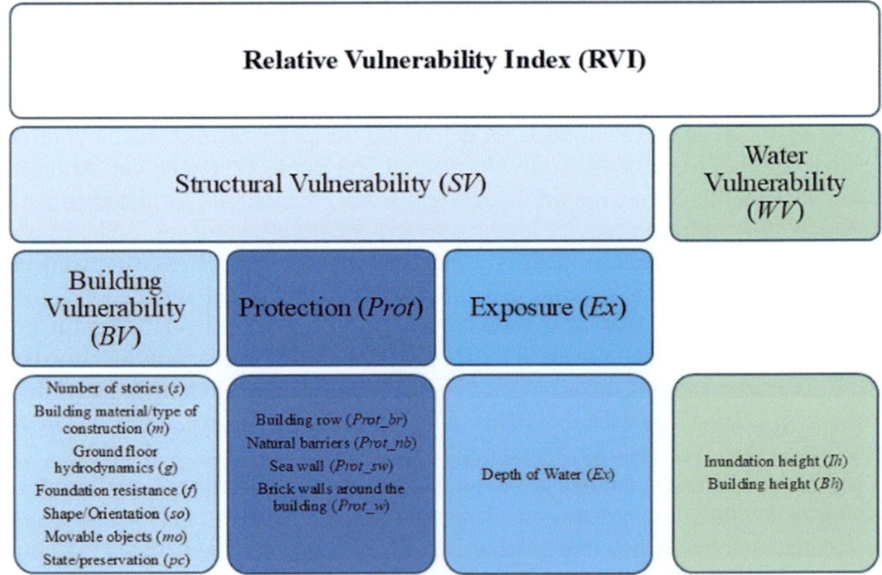

Fig. 2.6 The variables and the attributes of the vulnerability index (own figure based on Dall'Osso et al. (2009))

One of the most important parts of developing a vulnerability index is the weighting of the individual indicators. Different weighting methods can lead to different vulnerability indices. These differences may affect decision-making and plans for the reduction of tsunami vulnerability and risk. Nevertheless, most indicator-based risk assessment methods use equal weights, meaning that they consider all the indicators to be equally important (Beccari et al. 2016). Alternative weighting methods include participatory and statistical methods. Participatory methods involve expert judgment (e.g. Analytic Hierarchy Process) and statistical methods make use of existing empirical data that are not often available.

The outputs of the assessment are indices assigned to each building indicating its relative physical vulnerability to tsunamis (RVI). The method for classifying buildings into vulnerability categories depends on the aim of the study. By using an "equal interval" approach, the vulnerability assessment results are easy to interpret for non-experts (e.g. authorities or decision-makers). However, if the results are to be used for setting priorities a "quantile" classification scheme may be more appropriate. The "quantile" classification is based on the principle that each class contains an equal number of features, making it easy to identify e.g. the top 10% of vulnerable buildings that should be reinforced. Finally, a "standard deviation" scheme defines classes based on how much a feature's attribute varies from the mean. In this way, authorities and decision-makers can decide which vulnerability index values are acceptable for them and which are not (Papathoma-Köhle et al. 2019b).

The results of the assessment can be visualized in maps that can form the basis of decision-making and planning by several end-users. Indicator-based methods may be reproduced in areas with similar building characteristics, or they can be adapted for areas with special building architecture, materials, and design.

The geodatabase containing the indicators for each building and the overall vulnerability index is a dynamic tool. The database may be easily updated and the visualization of the new spatial distribution of the results can be used to evaluate the performance of a Build Back Better (BBB) strategy or as a basis for planning reinforcement of buildings or developing local adaptation measures. Some additional advantages associated with the method include:

- The characteristics of buildings and their surroundings are taken into consideration. This can lead to more focused recommendations for vulnerability reduction.
- Communities and settlements can be assessed in order to identify hotspots of vulnerability and guide plans and decisions.
- Indicator-based approaches have a strong predictive power. They can be used to determine vulnerability where no empirical data are available.

Nevertheless, before applying the method, challenges have to be considered including the difficulty in acquiring the data needed, the subjectivity of the experts in weighting and classification as well as transferability issues (e.g., relevance of indicators or additional indicators given local architectural peculiarities).

Data and technical requirements

Implementation of FW2A relies on data such as information on past events to identify a specific scenario and its potential inundation zone, or data (topographic, bathymetric, information on potential tsunamigenic sources, etc.) for numerical modelling of the tsunami run-up and inundation. The framework also relies on detailed data on buildings (i.e., building features, materials, condition, height) and their surroundings (natural or artificial barriers, vegetation, etc.). Ideally, the building data should be geo-referenced to enable the development of a GIS database. Online tools (e.g., Google Street View, Google Earth, etc.) or remote sensing data can be used for collecting some of the data.

Since the approach is participatory, an expert panel is needed to decide the weighting of the indicators and the classification of vulnerability categories after the assessment has been done. Should enough empirical data exist, the weighting can be done by using statistical rather than participatory methods.

Along with various vulnerability indicators, the severity of the event, expressed as water height, is taken into account when determining the inundation area. Additionally, the water vulnerability WV indicates the percentage of a building that is impacted by the water, and the exposure Ex stands for the height of the water at that particular site. However, only one parameter, the wave height, is used to represent the wave. Other aspects of the process, such as flow direction, impact velocity, and impact time, are not taken into account.

Treatment of uncertainties

It must be stressed that the hazard assessment in FW2 uses a simplified approach compared to FW1, and this is one of the main sources of uncertainty associated with the framework. The main sources of uncertainties associated with FW2A are listed below:

- Inundation scenario and potential inundation zone: the potential inundation zone is often identified from historical records of varying quality and reliability.
- Data quality/availability: The availability of detailed data on the scale needed is always limited and most of the data needs to be collected through fieldwork. Limited visibility through dense vegetation or city plan, restricted access to the vicinity of some buildings, and lack of georeferenced building footprints may be only some of the difficulties in obtaining a reliable indicator database.
- Expert judgment (weighting of indicators): the weighting of indicators through expert judgment may vary in different geographical areas since it dependends on the experts' experience, interests and education (OECD 2008). OECD (2008) suggests that "alternatively, participatory methods that incorporate various stakeholders—experts, citizens and politicians—can be used to assign weights. This approach is feasible when there is a well-defined basis for a national policy (Munda 2005, 2007). For international comparisons, such references are often not available, or deliver contradictory results".

The uncertainties inherent to this approach can be reduced by increasing the quality of the required data. This can be achieved through better data collection methods and by developing detailed damage reports following large events. Not recognizing or ignoring uncertainties may lead to underestimation or overestimation of physical vulnerability, which will have significant consequences for decision-making regarding disaster risk reduction strategies. Nevertheless, to date, no treatment of uncertainties has been implemented for this specific method. Applications of the method in disaster risk reduction and risk management are described in Chap. 7.

A summary of the main aspects of FW2A is given in Table 2.3.

FW2B. Indicator-based social (or multi-dimensional) vulnerability assessment

Short description of method

As an alternative to physical vulnerability, social or multi-dimensional factors may be considered to describe the degree of vulnerability and exposure in a more comprehensive manner. To account for the complexity and variability of coastal areas, tsunami vulnerability and risk assessment can be approached from a multi-dimensional perspective. In this regard, vulnerability varies across physical space and between and within social groups and in terms of scale-dependence in time, space, and units of analysis. The vulnerability assessment is considered a dynamic process, as the characteristics and drivers of vulnerability change over time (Vogel and O'Brien 2004; Raduszynski and Numada 2023).

Despite the paucity of information on applications of various theoretical and conceptual frameworks, several authors have proposed a tsunami risk assessment based on an integral approach combining hazard, vulnerability and risk assessment, and incorporating risk management (González-Riancho et al. 2014; Tavares et al. 2017). This comprehensive approach combines and aggregates data on human and natural systems while also considering the dynamic and scale-dependent processes. The multi-dimensional assessment is an extension of FW2A as it covers the exposed elements (i.e., human, environmental, socioeconomic and infrastructure) and adds the effects resulting from feedback loops such as the impacts on incomes or institutions, or effects on critical infrastructure. It is crucial that the elements are separated into different indicators, to avoid losing information while aggregating the outcomes (González-Riancho et al. 2014).

FW2B has its roots in multi-risk approaches using various parameters that compose the multidimensionality (Tinti et al. 2011; Birkmann et al. 2013, among others). According to Tavares et al. (2017), such dimensions are described as morphological, structural, social, and tributary components of vulnerability. The approach can be extended to different geographic regions with focus on regional or local contexts. Local applications related to morphology, land use, and occupation can be conducted, enabling stakeholders or authorities to discuss and implement measures to prevent, reduce, and mitigate tsunami risk using a straightforward methodology (González-Riancho et al. 2014).

One of the great challenges in risk analysis is accounting for complexities. Social science scholars have consistently raised concerns about social complexities and emphasised that simplifications through metrics or linearity of time may be reductive, and hence also excluding (Mol and Law 2002). Users, including policy makers, are robustly responding to such simplifications, and therefore the exclusion of multi-layered social complexities is multiplying. It is argued that qualitative analysis is the best approach to mitigate such dilemmas, although at the moment there are few tangible examples for the case of PTRA. Triyanti et al. (2023) analysed the tsunami in Palu in 2018, and argued how the event should be seen from a systemic risk framework perspective, allowing to unfold social inequalities, historical conflicts and war, absence of local knowledge in the urbanized coasts, and many others important factors that created severe impacts once the areas were exposed by cascading hazard events. The Palu case, as well as many other tsunami events worldwide, has demonstrated how the entanglement of these social complexities also can lead to further social problems post disaster, such as sexual abuse, discrimination, and poor governance. While empathizing with the challenges of disaster managers and stakeholders on the ground to operationalize complex risks, it is still important to continue the attempt to incorporate reflective vulnerability and risk analysis through co-production of knowledge and reciprocal dialogue among diverse scientists, government authorities, and societal stakeholders.

Data and technical requirements

The definition of multidimensional indicators is based on elicitation processes or expert opinions. As an alternative, González-Riancho et al. (2014) follow criteria of e.g. 'analytical soundness' and 'independence' based on predetermined indicators of official institutions. The approaches have been established in the Handbook on Constructing Composite Indicators (OECD 2008). Because of the indicator system's resilience, sensitivity, and transparency, data can be managed at the index level, divided into distinct indicators, and work directly with base data. This is crucial for preventing data loss during result aggregation and for developing suitable risk mitigation strategies (González-Riancho et al. 2014). The multidimensional aggregated indicator-based risk assessment can consider the exposure (human, environmental, socioeconomic, infrastructure), the human sensitivity indicators (age, illiteracy, extreme poverty, disability, isolation, critical evacuation), environmental sensitivity (protection, singularity, threat, degradation), socioeconomic sensitivity (job generation, contribution to GDP and to foreign trade), infrastructure sensitivity (critical infrastructure, critical buildings, vertical evacuation, building materials), and resilience (coping capacity, recovery capacity). There is also a need to incorporate qualitative vulnerability assessment (e.g., tsunami eye-witness accounts, local knowledge, videos, storylines, and ethnographic approaches) in the vulnerability/risk analysis to make an integral and inclusive assessment (Strunz et al. 2010). Specific examples of matrix construction can be found in González-Riancho et al. (2014; Table 2.2) and Tavares et al. (2017; Table 2.3).

Different components (or classes) of vulnerability can be described (e.g., González-Riancho et al. 2014; Tavares et al. 2017; Birkmann et al. 2013):

(a) Morphological component: the flooded area can be evaluated using a matrix based on field surveys, remote sensing such as satellite imagery, integrated in a GIS software. Parameters like average slope, distance to coastline, land use, and land cover are considered. The matrix provides intrinsic attributes with values ranging from $+1$ to -1, representing increased or decreased vulnerability.
(b) Structural building assessment: potentially affected buildings can be assessed using numerical modelling to determine tsunami inundation parameters, focusing on structural, architectural, occupational, and functional aspects. Specifically, parameters related to building occupation, functionality, and the structures' age, underground floors, occupation form, number of usable units, average daily visitors per unit, and floating occupation, strengthen the analysis and introduce new information. The structural vulnerability is calculated using a multicriteria analysis to define weights assigned to each parameter and attributes with values between 0 and 1 (Tavares et al. 2017).
(c) Social vulnerability: this component can be evaluated using factor analysis, which involves standardizing variables to z-scores, calculating Pearson correlation matrix, implementing factor analysis with Varimax rotation, analysing variance rate parameters, adjusting the Kaiser-Meyer-Olkin sample measurement or implementing principal component analysis (PCA). A social vulnerability index (Sv) results by summing all factors using e.g. equal weighting. The vulnerability may be varied according to the standard deviation, following the recommended classification SoVI that was originally adopted in developing the social vulnerability index (e.g., Cutter et al. 2003) and applied in recent studies (e.g., Tavares et al. 2017; Mah et al. 2023). Mengal et al. (2021) also used the SoVI framework. The social vulnerability index can be obtained by combining principal component analysis with group census variables (e.g., Martínez et al. 2020).
(d) Property tax or tributary: The analysis of inferred property value involves four allocation coefficients (housing, commerce, services, and industry) for calculating municipal property tax. These coefficients help classify areas based on activities and identify exposed elements. The vulnerability is calculated using the coefficients, resulting in five vulnerability levels based on standard deviation classes (Tavares et al. 2017).
(e) Institutional vulnerability: Institutional vulnerability is related to the possibility of affecting governance systems with legal or customary rules. Such factors may be compelled to alter in response to a disaster, exposing weaknesses (Birkmann et al. 2013). This class is more related to risk management yet could influence the levels of vulnerability during a disaster.

Considering the above indicators, aggregation is suggested for each dimension in accordance with expert opinion. Taking into consideration the distribution of the data, indices are converted into groups (i.e., categories) whose thresholds are shown with a colour code to reflect the information geographically (see e.g. Damm 2010). The result can be integrated into a GIS system and designed to provide the best possible value distribution across multiple classes. Different risk matrices are proposed in literature that include the classes identified for the hazard and vulnerability indices or, in the case of partial data, using the hazard and sensitivity indices to assess the risk (González-Riancho et al. 2014, Greiving et al. 2006, Jelínek et al. 2009). In such studies, the exposure has been incorporated into the matrix since the exposed elements form the ground for assessing sensitivity and vulnerability.

The Composite Vulnerability Index (CVI) represents a technique used to define values from morphological, structural, social, and taxable property vulnerabilities, ranging each in accordance with the standard deviation and defined categories. The weights assigned to each parameter can vary between 1–100 (Birkmann et al. 2013; González-Riancho et al. 2014; Tavares et al. 2017).

Treatment of uncertainties

The most appropriate way of treating uncertainties in FW2B is by means of sensitivity analyses, based on elicitation techniques. The main sources of uncertainty come from the application of the worst case scenario or the inundation scenario derived from the 'envelope' of maximum expected inundation. As for FW2A, the analysis usually refers to worst case scenarios following a deterministic approach, hence excluding the IMs for different return periods.

González-Riancho et al. (2014) discuss the potential errors that may result from aggregating the different dimensions. The OECD (2008) minimum–maximum (Min–Max) method can be applied to normalize the indicators in order to correct the imbalance caused by the different variable units and enable their comparison and combination, as proposed by Damm (2010).

An additional form of uncertainty that is further discussed in Chap. 8 is social uncertainty. Social uncertainty is an important factor that heavily influences the effectiveness of risk communication. This uncertainty, related to human and social dimensions, affects how communities (individuals, social groups, and institutions) perceive risk and behave in different situations, and in this context upon threat, which adds another dimension to risk assessment and management.

A summary of the main aspects of FW2B is given in Table 2.3.

2.4 Tsunami Data and Observations in PTHA

The probability of various tsunami scenarios occurring during a specific time period in a specific place is assessed using the comprehensive PTHA methodology, mostly using synthetic scenarios as described in previous sections. However, tsunami observations play a relevant role as they provide significant information for understanding the features of previous tsunamis and their potential effects on coastal communities and are invaluable for testing models (e.g., Davies 2019). Historical earthquake and tsunami records in recent catalogues, paleorecords, or oceanographic observations provide information regarding frequency, recurrence intervals and patterns (Satake

2015; Cisternas et al. 2005), as well as the size of past events, all which can be used to forecast tsunami likelihood or the probability of tsunamis aiding forecasting of catastrophes (Satake and Atwater 2007). Hazard and risk analysts can more accurately forecast potential future tsunami sources and their associated probability by including past events (e.g., Acuña et al. 2021; Parsons 2012; Zöller 2018).

Tsunamis can leave distinctive sediment deposits and geological evidence, hence paleoseismic studies (Atakan 2015) and paleotsunami records such as sand layers and erosion patterns (e.g., McCalpin 2009; Masuda et al. 2022) can be an extended source of information about the frequency, clustering, and magnitude of previous events (Engel et al. 2020; Griffin et al. 2020). Particularly, observations recorded by ocean bottom seismometers, tide gauges, and ocean buoys provide real-time tsunami measurements that are essential for tsunami warning systems and for enhancing accuracy and reliability of inversion methods (e.g. Cienfuegos et al. 2018).

Besides, tsunami observations can be key for validation and calibration of numerical tsunami codes (Horspool et al. 2014). Such numerical models can be validated and enhanced using observations of tsunami wave heights and periods from tide gauges and deep-ocean buoys. Accurate tsunami wave amplitude and duration measurements aid in determining the possible effects on certain coastal areas. Lately, such observations can be augmented or approximated using assimilation or machine learning techniques. These data can be trained and used in the context of early warning and PTHA (e.g. Makinoshima et al. 2021, Liu et al. 2021, Núñez et al. 2022, Rodríguez et al. 2022, see also recipe R23).

Uncertainty quantification is crucial (Selva et al. 2016; Sepúlveda et al. 2017), and assumptions made to forecast tsunami hazard may lead to biased hazard results. In this regard, tsunami observations provide critical data for the development, validation, and improvement of probabilistic tsunami hazard assessments by incorporating historical and real-time observations into sophisticated numerical modelling approaches.

The following sections consider the hazard assessment based on a combination of tsunami data using empirical and numerical modelling. Data-driven tsunami hazard assessment involves using available observational data and reliable historical data (e.g., NCEI/WDS 2024) to analyse and understand the potential hazards posed by tsunamis.

Workflows are presented for giving weight to empirical data in a tsunami hazard assessment that can include historical tsunami records, and any other available information related to tsunamis. Such workflows for developing data-driven hazard models may involve preprocessing that includes merging different datasets, data curation, filling data gaps, using data augmentation techniques and standardization of data formats (e.g., Wilkinson et al. 2016) to conduct statistical analyses.

These datasets can be used in tsunami hazard assessment to simulate past earthquake or tsunami measurements and for data augmentation using various statistical methods (e.g., Ogata 2017; Geist et al. 2009). These methods are covered in further detail in Chap. 3, which describes common applications for various tsunami sources.

2.4.1 Dealing with Empirical and Paleo Data

Here, focus is given to empirical and paleo data that are used to develop runup rate models for different probability density functions with treatment of radiocarbon dating and open interval uncertainties. Data-driven tsunami hazard assessments are an ongoing process that enables and requires updates and refinement as new data become available, which can be addressed using Bayesian statistical modelling (e.g., Grezio et al. 2010; Boumis et al. 2023). Methods to determine recurrence statistics from empirical runup observations in historical catalogues, as well as paleo tsunami and paleo earthquake observations with uncertainty are described.

This approach can also be used to determine whether a time dependent or time independent probability density function (PDF) is best fit to the data. The Maximum Likelihood and Monte Carlo fitting methods can be useful on short event sequences and are sensitive to the histogram of the data (Parsons 2008; Geist et al. 2013). Results reflect epistemic uncertainties by showing the range of parameter distributions that are consistent with observations and their uncertainties. Estimation of the event rate (inverse of the recurrence interval) for a small number of observations is subject to considerable uncertainty. Some of these techniques are exemplified in recipe R5, showing an example on the application of a Monte Carlo technique to tsunami runup rate determination for the Caribbean region. Additionally, an example of using the Monte Carlo method is given to determine whether a time dependent or time independent recurrence distribution is most applicable to a paleo series. A formulation for maximum likelihood methods and a comparison of the methods is also provided.

2.4.2 Treatment of Uncertainties

In the Monte Carlo approach, the range of possible rate parameters that fit empirical observations can be achieved by finding those that most commonly reproduce observed tsunami frequencies. The first step is to construct a series of distributions that covers all reasonable rates (i.e. 1/10 to 1/3000). Rates can be randomly drawn, for example 100000 times from each series, and assembled into tsunami event sequences. In the Caribbean case, Parsons and Geist (2009) apply Monte Carlo sequences that begin with events given freedom to occur any time before a given date. In this case, the extra events are not counted into the frequencies but are used to establish starting points for such Monte Carlo sequences.

Tsunamis affect broad areas as they propagate away from their sources, and a region as complex as the Caribbean has widely distributed tsunamigenic sources. Since tsunamis resulting from multiple seismic sources operating at different recurrence intervals can affect a given coast, the best probability model is one that is time-independent, such as a Poisson process built around an exponential distribution (e.g., Geist and Parsons 2006; Parsons and Geist 2009). Observed tsunami runups can thus be used to calculate rate parameters (λ) in the exponential probability density distribution as

$$f(t) = \lambda e^{-\lambda t}, t > 0 \tag{2.8}$$

Each event sequence that matches observed frequencies is tallied. A distribution of matches to the observed record is produced, and the mode, median, or mean of that distribution can be taken to represent the rate parameter. This approach simultaneously estimates uncertainty in the rate parameter resulting from small sample-size, and the influence of the open intervals.

An example distribution of possible probability values for sites of different frequencies is shown in Fig. 2.7. A central question surrounding exclusive use of empirical tsunami runup rates for future hazard calculations in this example is how complete the empirical record is, as shown by Acuña et al. (2021) for central Chile, and specifically how representative the past ~500 yr period has been in the case study of the Caribbean (Parsons and Geist 2009). In the latter, there was a much higher frequency of observations in the 19th century as compared with the previous centuries, implying that increased population density increased the number of observations. Providing that a seismic catalogue is dominated by local-earthquake generated tsunamis, a moment balance test can be conducted. Here, two assumptions can be made, either that the region has released all the expected seismic moment, or instead, that there is the potential to generate more large earthquakes to satisfy a deficit. This can be achieved by modelling expected slip rates from relative plate motions to assess if there has been a moment balance.

Given the wide range of credible probability results shown in Fig. 2.7, it is difficult to know what single value might best represent tsunami hazard. The mean value is an obvious choice, although the distributions of possible probability values are not symmetrical (e.g. Gaussian).

2.4.3 Time Dependent Versus Time Independent Rates

Tests can be conducted to prove whether there is enough data to justify a time dependent forecast. This can be achieved by comparing the relative success of time dependent and time independent distributions at reproducing an earthquake/tsunami event series with open intervals and dating uncertainty. In this example, the test becomes an assessment of whether an observed paleoevent series with dating uncertainty is most consistent with an exponential time-independent PDF (Eq. 2.8), or alternatively a time-dependent Brownian Passage Time (inverse Gaussian) PDF (Kagan and Knopoff 1987, Matthews et al. 2002) as

$$f(t, \mu, \alpha) = \sqrt{\frac{\mu}{2\pi\alpha^2 t^3}} exp\left(-\frac{(t-\mu)^2}{2\mu\alpha^2 t}\right) \tag{2.9}$$

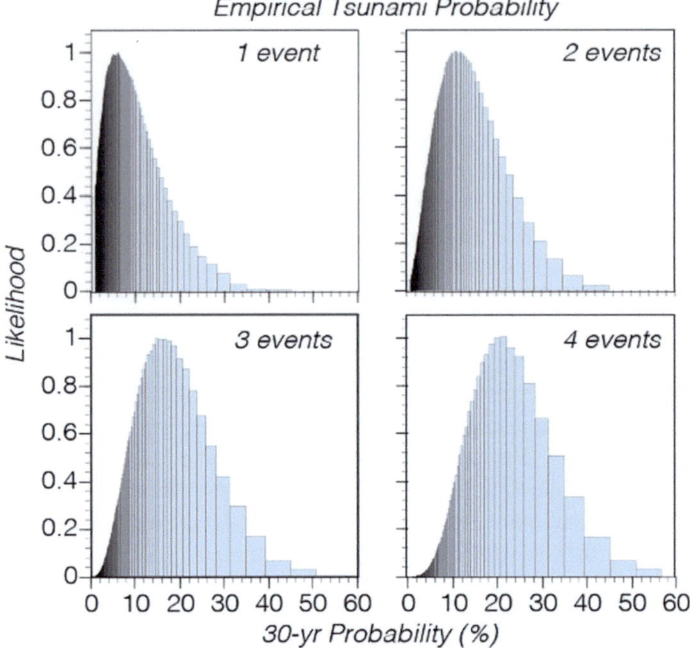

Fig. 2.7 Example of normalized histograms giving the likelihood of a given 30-yr Poisson probability resulting from calculated rate parameters for 20 km by 20 km cells with observations of 1 to 4 events in ~500 yrs (Geist and Parson 2009)

where μ is the recurrence interval and α is the coefficient of variation. The Brownian Passage Time distribution is nearly indistinguishable (in terms of its shape) from the lognormal distribution for earthquake recurrence. Eleven samples, provided with an associated uncertainty value, are utilized for the test. A series of distributions covering all reasonable mean recurrence intervals (such as 10 years to 5000 years) is developed with coefficient of variation values ranging from 0.01 to 0.99 for every mean recurrence interval. Groups of event times are randomly drawn millions of times from each possible recurrence distribution and assembled into sequences.

With this method, sequence means are identified directly from the parameters of parent distributions rather than from taking arithmetic means of sequences. Those sequences that have a single event occurring in order during each observed event window (range of possible event times as constrained by radiocarbon dating), and no earthquakes in the intervals between event windows are tallied, and a ranking of matches vs. mean recurrence and coefficient of variation to the observed record is produced (Fig. 2.8).

To summarize the technique, exponential distributions with mean recurrence intervals from 10 to 5000 yrs are generated in 10-yr increments, and a group of Brownian Passage Time distributions are generated with mean recurrence intervals from

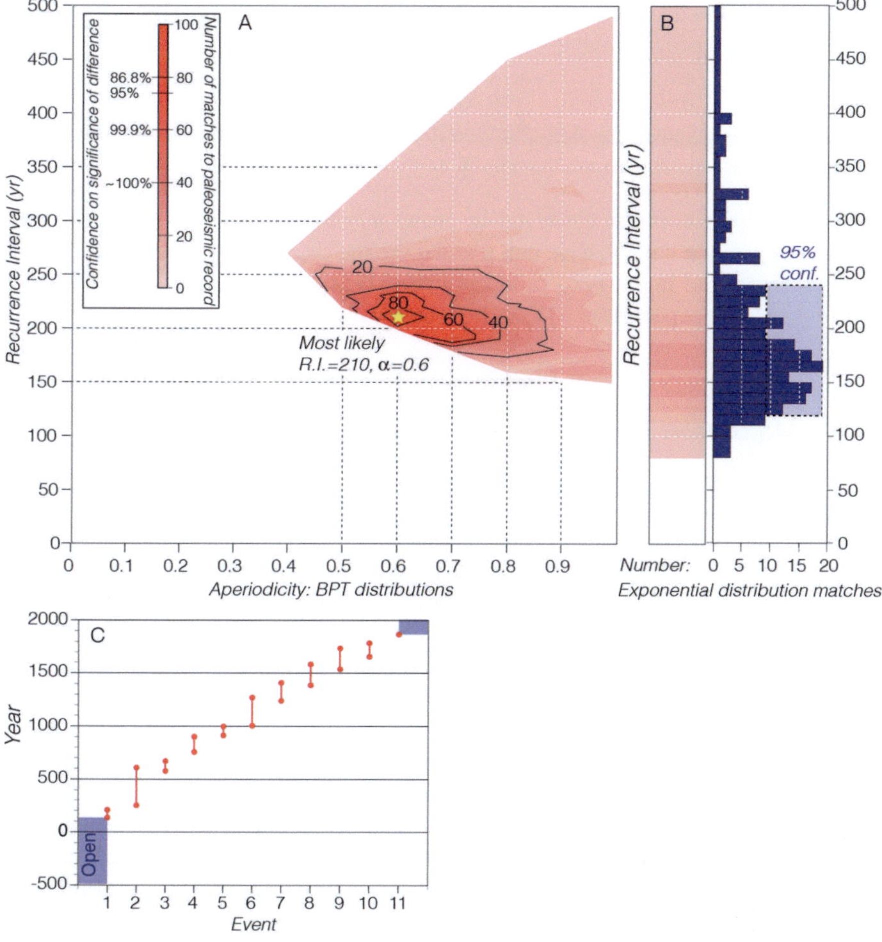

Fig. 2.8 **a** Time dependent (Brownian Passage Time) and **b** time independent (exponential) recurrence distributions fits to a **c** paleoevent series with uncertainties are contoured. The best-fit distributions are time dependent, with recurrence intervals of μ ~210 yr, and a coefficient of variation α ~0.6. Confidence (Z-test) on the significance of relative proportions is keyed to the contour intervals. The best-fit exponential distributions have significantly fewer matches, leading to the conclusion that earthquake recurrence for this series is time dependent. A histogram of exponential distribution matches is shown in (**b**), with 95% confidence of significance shaded, which gives the same information as the adjacent contour mapping, but in more detail

10 to 5000 yrs in 10-yr increments. The time dependent distributions also range through coefficients of variation between 0.01 and 0.99 (0.01 increments) for every mean recurrence interval. Each individual distribution is given 15 million chances to reproduce the entire paleoevent series within 95% confidence intervals on radiocarbon dating uncertainties. The number of successful matches is tallied for every possible distribution. A highly significant difference between the number of hits for

the time-dependent distributions vs time-independent ones is found using a two-tailed Z-test for relative proportions as in the following Eq. 2.10:

$$Z = \frac{P_a - P_b}{\sqrt{\frac{pq}{n_a} - \frac{pq}{n_b}}} \tag{2.10}$$

where p_a is the proportion observed in group a with size n_a, p_b is the proportion observed in group a with size n_b, and p and q are the overall proportions. The procedure is faster if the intent is to produce a time independent probabilistic forecast, requiring only an exponential PDF to be tested.

2.4.4 Maximum Likelihood Methods

The likelihood function represents the probability that a fixed set of observations would be randomly generated by a probability distribution with a given set of parameters (Aitkin 2010). The parameters that maximize the log-likelihood function (θ) are termed the maximum likelihood estimate (MLE). The logarithm of likelihood is used because the associated expressions involve sums of terms, rather than products, and because its gradient is well scaled compared to just using the values of likelihood. Accuracy of the inferred parameters are typically reported using likelihood-based confidence intervals (Pawitan et al. 2001).

Standard or "plug-in" (Nomura et al. 2011) expressions for the log-likelihood function and MLE are known for the probability models described in Chap. 3, assuming that the occurrence times are known exactly and not considering the open intervals. However, these expressions may be significantly biased for a small number of samples. Ogata (1999) developed general expressions of the likelihood function that consider the open intervals and events t_i over $T \leq t_1 \leq t_2 \ldots \leq t_n \leq S$. Then, the log-likelihood function for a stationary renewal process (Daley and Vere-Jones 2003; Ogata 1999) is:

$$\ln L_{[S,T]}(\theta) = -\ln V_\theta + \ln \phi_\theta(t_1 - S) + \sum_{i=2}^{n} \ln f_\theta(t_i - t_{i-1}) + \ln \phi_\theta(T - t_n) \tag{2.11}$$

where v_θ is equivalent to the mean recurrence time over the entire distribution (not the arithmetic mean of the samples) and is given by Eq. 2.12:

$$V_\theta = \int_0^\infty \phi_\theta(t)dt = \int_0^\infty t f_\theta(t)dt \tag{2.12}$$

This expression of the log-likelihood function differs from previous paleoseismic studies that do not include the open interval before the oldest dated event and after the last observation. Different likelihood-based methods, such as the one above, can be assessed using a model selection criterion, such as Akaike's Information Criterion (AIC) (Ogata 1999).

A type of empirical Bayesian method developed by Ogata (1999) accounts for uncertain event times, in addition to the effect of open intervals. Unlike traditional Bayesian inference techniques in which the prior and posterior distributions are a function of the parameter space (θ), the Ogata (1999) method is formulated such that the prior and posterior distributions are the probability densities of each event age distribution. Thus, event age distributions and parameters associated with a specific renewal model are both considered in this method. Different renewal models can be evaluated with AIC or similar metrics. The prior distributions for event ages used by Ogata (1999) include Dirac delta functions (for precisely known ages), and uniform and triangular distributions (for uncertain ages). Bayes' theorem is then formulated as in Eq. 2.13:

$$\varphi(t_1, \ldots, t_n|\theta) = \frac{L_{[S,T]}(\theta) \prod_{i=1}^{n} \psi_i(t_i)}{L_{[S,T]}(\theta)} \tag{2.13}$$

where $\varphi(t_1, \ldots, t_n|\theta)$ is the posterior distribution and is termed the integrated or marginal likelihood (Aitkin 2010; Ogata et al. 2000). $L[S, T(\theta)]$ is a constant that ensures that Eq. 2.13 gives a valid probability density and integrates the likelihood function with the joint distribution of the priors as shown in Eq. 2.14:

$$\mathcal{L}_{S,T}(\theta) = \int_S^T \ldots \int_S^T L_{[S,T]}(\theta; t_1, \ldots t_n) \prod_{i=1}^{n} \psi_i(t_i)\, dt_1, \ldots, dt_n \tag{2.14}$$

Using Dirac delta functions for the priors recovers the likelihood expression given in Eq. 2.13. The model parameters that maximize the integrated likelihood (Eq. 2.14) is termed, for this type of statistical inference, Type II maximum likelihood (Good 1965). The multiple integrations involved in computing are such that maximization techniques may not be successful.

Both the maximum likelihood techniques and the Monte Carlo technique for estimating the occurrence rates include the effect of the open time intervals on parameter estimation. The Monte Carlo technique and the Type II maximum likelihood technique also account for uncertainty in age dating events. Geist et al. (2014) compared the Type II likelihood and Monte Carlo time-dependent techniques for a paleo series and noted that the Type II likelihood technique appeared to be estimating longer return times and a significantly lower value for the shape parameter (coefficient of variation, α) than the Monte Carlo technique. The difference in the results from the two methods may relate to how the open intervals are handled. In the case of the likelihood methods, no event is assumed to occur before the first observation, whereas in the case of the Monte Carlo method an event can occur at any time prior to the first observation. The computational requirement of each technique depends on the number of events and event-age uncertainty relative to the mean return time and the probability model used. Geist et al. (2013) found that for the Monte Carlo results, twenty times the number of runs needed to be made. In contrast, the computational requirement for the Type II likelihood method depends on the number of events, in terms of the number of integrations required to define the likelihood function and the analytical form of the probability model.

A summary of the main aspects of applying the data-driven approach in PTHA is given in Table 2.3.

2.5 Final Remarks

This chapter provides an overview of the main frameworks that can be considered to assess tsunami hazard and risk. Quantification of tsunami risk is commonly defined by convoluting hazard outcomes with vulnerability functions. These results can inform the decision-making process for disaster risk reduction. To address this, two main frameworks are described to provide a structured approach to quantifying the hazard and the levels of risk associated with tsunamis. FW1 relies on a comprehensive characterization of uncertainties. Depending on the technique used for convoluting hazard and vulnerability functions, two workflows are identified: one based on a fully simulation-based workflow (FW1A) and another that uses numerical integration (FW1B). Both FW1A and FW1B could take into account sea-level rise and tidal variations (FW1C). FW1D relies on a partial characterization of uncertainties and/or identification of "worst-case scenarios" and constitutes the traditional workflow used to inform tsunami risk management decision-making (e.g., evacuation planning, design of tsunami countermeasures). FW2 uses participatory vulnerability and risk mapping based on physical vulnerability and multidimensional indicators (FW2). This chapter provides an overall scheme for each framework, whereas specific details of each step of the analysis are provided in the following chapters. Implementations of the different frameworks are exemplified in illustrative recipes in the second part of the book.

Probabilistic schemes for tsunami hazard and risk analysis provide a more comprehensive basis for decision-making for risk reduction and emergency planning compared to the deterministic methods. A comprehensive PTHA or PTRA combines all possible tsunami sources considering their relative importance, but in practice, in many regions the PTHAs consider only one type of tsunami source because it is assumed to be the prevalent one with more data available. This approach is acceptable if the probabilities of occurrence of the non-prevalent sources are negligible and/or if their probabilities of exceeding a selected tsunami intensity parameter are around zero. Common phases of the assessment use data, such as source characterization, including historical and paleo tsunami records, seismological data, synthetic data, population- and built environment databases, to calculate and evaluate e.g. tsunami hazard, vulnerability, and levels of risk.

With recent advances in modelling capabilities and high-performance computing, tsunami hazard and risk models can become more accurate. The integration of data, different statistical techniques and advanced modelling contributes to decision-making, risk reduction measures, and effective emergency planning to minimize the impacts of tsunamis on coastal communities.

Overall, frameworks in tsunami hazard and risk assessment provide a variety of structured approaches, ranging from simplified to more comprehensive, for understanding and evaluating the potential risks associated with tsunamis. The choice of framework will depend on the scope of the study and any limitations in e.g. data availability (e.g., high resolution bathymetry datasets, high resolution exposure models, suitable vulnerability models) and resources (e.g., human, computational). For any method, a poor treatment of uncertainties may lead to inaccurate results.

Finally, one aspect that is not always tackled is how hazard and risk model outcomes can be validated and how results can be improved. Here, historical data, independent datasets, and expert evaluation play a crucial role. The validation process enhances the credibility and reliability of the assessments and identifies areas for refinement and improvement in future iterations.

2.6 Final Note to the Readers

General recommendations on which methods to adopt for particular goals are given in each chapter of the book. Particularly, this chapter provides a description of the general frameworks for assessing tsunami hazard and risk. An effort has been made to offer an overview of the primary applications and examples of applications in various research projects. Depending on the goals of their project and the data at hand, users must carefully select the most appropriate workflow in order to provide a responsible and reliable analysis that incorporates all potential uncertainties.

References

Abrahams LS, Krenz L, Dunham EM, Gabriel AA, Saito T (2023) Comparison of methods for coupled earthquake tsunami modelling. Geophys J Int 234(1):404–426. https://doi.org/10.1093/gji/ggad053

Acuña F, Montalva GA, Melnick D (2021) How good is a paleoseismic record of megathrust earthquakes for probabilistic forecasting? Seismol Res Lett 93(2A):739–748. https://doi.org/10.1785/0220210044

Adams LM, LeVeque RJ, González FI (2015) The pattern method for incorporating tidal uncertainty into probabilistic tsunami hazard assessment (PTHA). Nat Hazards 76:19–39. https://doi.org/10.1007/s11069-014-1482-z

Aguirre-Ayerbe I, Martínez-Sánchez J, Aniel-Quiroga Í, González-Riancho P, Merino M, Al-Yahyai S, González M, Medina R (2018) From tsunami risk assessment to disaster risk reduction—the case of Oman. Nat Hazards Earth Syst Sci 18:2241–2260. https://doi.org/10.5194/nhess-18-2241-2018

Aitkin M (2010) Statistical Inference: an integrated Bayesian/likelihood approach, CRC Press, Boca Raton, FL, pp 236

Alberico I, Di Fiore V, Iavarone R, Petrosino P, Piemontese L, Tarallo D, Punzo M, Marsella E (2015) The tsunami vulnerability assessment of urban environments through freely available datasets: the case study of Napoli city (Southern Italy). J Marine Sci and Eng 3(3):981–1005. https://doi.org/10.3390/jmse3030981

Alhamid AK, Akiyama M, Ishibashi H, Aoki K, Koshimura S, Frangopol DM (2022) Framework for probabilistic tsunami hazard assessment considering the effects of sea-level rise due to climate change. Struct Saf 94:102152. https://doi.org/10.1016/j.strusafe.2021.102152

Álvarez-Gómez JA, Aniel-Quiroga Í, Gutiérrez-Gutiérrez O, Larreynaga Q, González JM, Castro M, Gavidia F, Aguirre-Ayerbe I, González-Riancho P, Carreño E (2013) Tsunami hazard assessment in El Salvador, Central America, from seismic sources through flooding numerical models. Nat Hazards Earth Syst Sci 13(11):2927–2939. https://doi.org/10.5194/nhess-13-2927-2013

Al-Habsi Z, Hereher M, El-Hussain I, Omira R, Baptista MA, Deif A, Al-Awadhi T, Al-Nasiri N (2022) Tsunami hazard and risk zoning for Qurayyat in northeast Oman coast: worst-case credible scenarios along the Makran Subduction zone, Western Asia. J Asian Earth Sci: X 8:100103. https://doi.org/10.1016/j.jaesx.2022.100103

Annaka T, Satake K, Sakakiyama T, Yanagisawa K, Shuto N (2007) Logic-tree approach for probabilistic tsunami hazard analysis and its applications to the Japanese coasts. Pure Appl Geophys 164:577–592. https://doi.org/10.1007/s00024-006-0174-3

Anzidei M, Scicchitano G, Scardino G, Bignami C, Tolomei C, Vecchio A, Serpelloni E, De Santis V, Monaco C, Milella M, Piscitelli A, Mastronuzzi G (2021) Relative sea-level rise scenario for 2100 along the coast of South Eastern Sicily (Italy) by InSAR data, satellite images and high-resolution topography. Rem Sens 13(6):1108. https://doi.org/10.3390/rs13061108

Aristizabal C, Bard P-Y, Beauval C (2022) Site-Specific PSHA: combined effects of single-station-sigma, host-totarget adjustments and nonlinear behaviour: a case study at Euroseistest. Ital J Geosci 141(1):5–34. https://doi.org/10.3301/IJG.2022.02

ASCE (2017) ASCE 7–16 Tsunami design zone maps for selected locations. Am Soc Civil Eng. https://doi.org/10.1061/9780784480748

Atakan K (2015) Paleoseismology: integration with Seismic Hazard. In: Beer M, Kougioumtzoglou IA, Patelli E, Au SK (eds) Encyclopedia of earthquake engineering. Springer, Berlin, Heidelberg. https://doi.org/10.1007/978-3-642-35344-4_39

Attary N, Unnikrishnan VU, Van De Lindt JW, Cox DT, Barbosa AR (2017) Performance-based tsunami engineering methodology for risk assessment of structures. Eng Struct 141:676–686. https://doi.org/10.1016/j.engstruct.2017.03.071

Attary N, Van De Lindt JW, Barbosa AR, Cox DT, Unnikris-Ahnan VU (2019) Performance-based tsunami engineering for risk assessment of structures subjected to multi-hazards: tsunami following earthquake. J Earthq Eng 7:1–20. https://doi.org/10.1080/13632469.2019.1616335

Ayca A, Lynett PJ (2016) Effect of tides and source location on nearshore tsunami-induced currents. J Geophys Res: Oceans 121:8807–8820. https://doi.org/10.1002/2016JC012435

Baba T, Kamiya M, Tanaka N et al. (2022) Probabilistic tsunami hazard assessment based on the Gutenberg–Richter law in eastern Shikoku, Nankai subduction zone, Japan. Earth Planets Space 74(156). https://doi.org/10.1186/s40623-022-01715-1

Barbato M, Petrini F, Unnikrishnan VU, Ciampoli M (2013) Performance-based hurricane engineering (PBHE) framework. Struct Saf 45:24–35. https://doi.org/10.1016/j.strusafe.2013.07.002

Basili R, Tiberti MM, Kastelic V, Romano F, Piatanesi A, Selva J, Lorito S (2013) Integrating geologic fault data into tsunami hazard studies. Nat Hazards Earth Syst Sci 13:1025–1050. https://doi.org/10.5194/nhess-13-1025-2013

Basili R, Brizuela B, Herrero A, Iqbal S, Lorito S, Maesano FE, Murphy S, Perfetti P, Romano F, Scala A, Selva J, Taroni M, Tiberti MM, Thio HK, Tonini R, Volpe M, Glimsdal S, Harbitz CB, Løvholt F, Baptista MA, Carrilho F, Matias LM, Omira R, Babeyko A, Hoechner A, Gürbüz M, Pekcan O, Yalçıner A, Canals M, Lastras G, Agalos A, Papadopoulos G, Triantafyllou I, Benchekroun S, Agrebi Jaouadi H, Ben Abdallah S, Bouallegue A, Hamdi H, Oueslati F, Amato A, Armigliato A, Behrens J, Davies G, Di Bucci D, Dolce M, Geist E, Gonzalez Vida JM, González M, Macías Sánchez J, Meletti C, Ozer Sozdinler C, Pagani M, Parsons T, Polet J, Power W, Zaytsev SM, A, (2021) The making of the NEAM tsunami hazard model 2018 (NEAMTHM18). Front Earth Sci 8:616594. https://doi.org/10.3389/feart.2020.616594

Beauval C, Bard PY, Danciu L (2020) The influence of source- and ground-motion model choices on probabilistic seismic hazard levels at 6 sites in France. Bull Earthquake Eng 18:4551–4580. https://doi.org/10.1007/s10518-020-00879-z

Beccari B (2016) A comparative analysis of disaster risk, vulnerability and resilience composite indicators. PLoS current disasters 14(8). https://doi.org/10.1371/currents.dis.453df025e34b682e9737f95070f9b970

Behrens J, Løvholt F, Jalayer F, Lorito S, Salgado-Gálvez MA, Sørensen M Abadie S, Aguirre-Ayerbe I, Aniel-Quiroga Í, Babeyko AY, Baiguera M, Basili R, Grezio A, Johnson K, Murphy S, Paris R, Rafliana I, DiRisi R, Vyhmeister E (2021) Probabilistic Tsunami hazard and risk analysis: a review of research gaps. Front Earth Sci 9:628772. https://doi.org/10.3389/feart.2021.628772

Berryman K (ed) (2005) Review of Tsunami hazard and risk in New Zealand. Geological and Nuclear Sciences (GNS). Client Report 2005/104, pp. 149. Institute of Geological and Nuclear Sciences, New Zealand

Berryman K, Wallace L, Hayes G, Bird P, Wang K, Basili R, Lay T, Pagani M, Stein R, Sagiya T, Rubin C, Barrientos S, Kreemer C, Litchfield N, Stirling MW, Gledhill K, Haller K, Costa C (2015) The GEM faulted earth subduction characterisation project ver 2.0. GEM Faulted Earth project

Beven KJ, Aspinall WP, Bates PD, Borgomeo E, Goda K, Hall JW et al (2018) Epistemic uncertainties and natural hazard risk assessment—part 2: what should constitute good practice? Nat Hazards Earth Syst Sci 18:2769–2783. https://doi.org/10.5194/nhess-18-2769-2018

Birkmann J (2006) Indicators and criteria for measuring vulnerability: theoretical bases and requirements. In: Birkmann J (eds) Measuring vulnerability to natural hazards: towards disaster resilient societies. Tokyo: UNU Press

Birkmann J, Cardona O, Tibaduiza M, Barbat A, Pelling M, Schneiderbauer S, Kienberger S, Keiler M, Alexander D, Zeil P, Welle T (2013) Framing vulnerability, risk and societal responses: the MOVE framework. Natural Haz 67:193–211. https://doi.org/10.1007/s11069-013-0558-5

Bommer JJ, Scherbaum F (2008) The use and misuse of logic trees in probabilistic seismic hazard analysis. Earthq Spectra 24(4):997–1009. https://doi.org/10.1193/1.2977

Boumis G, Geist EL, Lee D (2023) Bayesian hierarchical modeling for probabilistic estimation of tsunami amplitude from far-field earthquake sources. J Geophys Res: Oceans 128(12):e2023JC020002. https://doi.org/10.1029/2023JC020002

Cankaya ZC, Suzen ML, Yalciner AC et al (2016) A new GIS-based tsunami risk evaluation: MeTHuVA (METU tsunami human vulnerability assessment) at Yenikapı. Istanbul Earth Planet Sp 68:133. https://doi.org/10.1186/s40623-016-0507-0

Catalán PA, Gubler A, Cañas J, Zuñiga C, Zelaya C, Pizarro L, Cienfuegos R (2020) Design and operational implementation of the integrated tsunami forecast and warning system in Chile (SIPAT). Coast Eng J 62(3):373–388. https://doi.org/10.1080/21664250.2020.1727402

Chacón-Barrantes S, Arozarena-Llopis I (2021) A first estimation of Tsunami hazard of the Pacific coast of Costa Rica from local and distant seismogenic sources. Ocean Dyn 71:793–810. https://doi.org/10.1007/s10236-021-01467-8

Charvet I, Macabuag J, Rossetto T (2017) Estimating Tsunami-induced building damage through fragility functions: critical review and research needs. Front in Built Environ 3. https://doi.org/10.3389/fbuil.2017.00036

Chock GY, Robertson I, Riggs HR (2011) Tsunami structural design provisions for a new update of building codes and performance-based engineering, in solutions to coastal disasters 2011. In: Wallendorf LA, Jones C, Ewin L (eds), pp 423–435. https://doi.org/10.1061/41185(417)38

Chock G (2016) Design for Tsunami loads and effects in the ASCE 7–16 standard. J Struct Eng 142(11):04016093. https://doi.org/10.1061/(ASCE)ST.1943-541X.000156

Cienfuegos R, Catalán PA, Urrutia A, Benavente R, Aránguiz R, González G (2018) What can we do to forecast tsunami hazards in the near field given large epistemic uncertainty in rapid seismic source inversions? Geophys Res Lett 45. https://doi.org/10.1029/2018GL076998

Cisternas M, Atwater B, Torrejón F, Sawai Y, Machuca G, Lagos M, Eipert A, Youlton C, Salgado I, Kamataki T, Shishikura M, Rajendran C, Malik J, Rizal Y, Husni M (2005) Predecessors of the giant 1960 Chile earthquake. Nature 437:404–407. https://doi.org/10.1038/nature03943

Cornell CA (1968) Engineering seismic risk analysis. Bull Seismol Soc Amer 58(5):1583–1606. https://doi.org/10.1785/BSSA0580051583

Cremen G, Galasso C, McCloskey J (2022) Modelling and quantifying tomorrow's risks from natural hazards. Sci of Total Envir 817:152552. https://doi.org/10.1016/j.scitotenv.2021.152552

Cooke RM, Goossens LHJ (2004) Expert judgment elicitation for risk assessments of critical infrastructures. J Risk Res 7(6):643–656. https://doi.org/10.1080/1366987042000192237

Cugliari L, Cerase A, Amato A (2022) Tsunami risk perception, a state-of-the-art review with a focus in the NEAM region. Front Earth Sci. https://doi.org/10.3389/feart.2022.995536

Cutter SL, Boruff BJ, Shirley WL (2003) Social vulnerability to environmental hazards. Soc Sci Q 84(2):242–261. https://doi.org/10.1111/1540-6237.8402002

Daley DJ, Vere-Jones D (2003) An introduction to the theory of point processes, 2nd edn. Springer, New York, p 469

Dall'Osso F, Gonella M, Gabbianelli G, Withycombe G, Dominey-Howes D (2009) A revised (PTVA) model for assessing the vulnerability of buildings to tsunami damage. Nat Haz Earth Syst Sci 9:1557–1565. https://doi.org/10.5194/nhess-9-1557-2009

Dall'Osso F, Maramai A, Graziani L, Brizuela B, Cavalletti A, Gonella M, Tinti S (2010) Applying and validating the PTVA-3 Model at the Aeolian Islands, Italy: assessment of the vulnerability of buildings to tsunamis. Nat Hazards Earth Syst Sci 10:1547–1562. https://doi.org/10.5194/nhess-10-1547-2010

Dall'Osso F, Dominey-Howes D, Tarbotton C et al (2016) Revision and improvement of the PTVA-3 model for assessing tsunami building vulnerability using 'international expert judgment': introducing the PTVA-4 model. Nat Haz 83:1229–1256. https://doi.org/10.1007/s11069-016-2387-9

Damm M (2010) Mapping social-ecological vulnerability to flooding. Graduate Research Series, Ph.D. Dissertations, Publication Series of UNU-EHS Vol. 3, ISBN: 978-3-939923-46-6. United Nations University—Institute for Environment and Human Security (UNU-EHS), Bonn.

Davies G, Griffin J (2018) The 2018 Australian probabilistic Tsunami hazard assessment. Technical Report. Record 2018/41. Geoscience Australia, Canberra. https://doi.org/10.11636/Record.2018.041

Davies G, Griffin J, Løvholt F, Glimsdal S, Harbitz C, Thio HK et al (2018) A global probabilistic tsunami hazard assessment from earthquake sources. Geol Soc Lond Spec Publ 456:219–244. https://doi.org/10.1144/SP456.5

Davies G, Griffin J (2020) Sensitivity of probabilistic Tsunami hazard assessment to far-field earthquake slip complexity and rigidity depth-dependence: case study of Australia. Pure Appl Geophys 177:1521–1548. https://doi.org/10.1007/s00024-019-02299-w

Davies G, Horspool N, Miller V (2015) Tsunami inundation from heterogeneous earthquake slip distributions: evaluation of synthetic source models. J Geophys Res Solid Earth 120:6431–6451. https://doi.org/10.1002/2015JB012272

Davies G (2019) Tsunami variability from uncalibrated stochastic earthquake models: tests against deep ocean observations 2006–2016. Geophys J Int 218:1939–1960. https://doi.org/10.1093/gji/ggz260

Davies G, Weber R, Wilson K, Cummins P (2022) From offshore to onshore probabilistic tsunami hazard assessment via efficient Monte Carlo sampling. Geophys J Int 230(3):1630–1651. https://doi.org/10.1093/gji/ggac140

Der Kiureghian A, Ditlevsen O (2009) Aleatory or epistemic? Does it matter? Struct Saf 31(2):105–112. https://doi.org/10.1016/j.strusafe.2008.06.020

De Risi R, Jalayer F, De Paola F, Iervolino I, Giugni M, Topa ME, Mbuya E, Kyessi A, Manfredi G, Gasparini P (2013) Flood risk assessment for informal settlements. Nat Hazards 69(1):1003–1032. https://doi.org/10.1007/s11069-013-0749-0

De Risi R, Goda K (2017) Simulation-based probabilistic Tsunami hazard analysis: empirical and robust hazard predictions. Pure Appl Geophys 174:3083–3106. https://doi.org/10.1007/s00024-017-1588-9

De Risi R, Goda K, Mori N, Yasuda T (2017) Bayesian Tsunami fragility modeling considering input data uncertainty. Stoch Env Res Risk Assess 31(5):1253–1269. https://doi.org/10.1007/s00477-016-1230-x

De Risi R, Muhammad A, De Luca F, Goda K, Mori N (2022) Dynamic risk framework for cascading compounding climate-geological hazards: a perspective on coastal communities in subduction zones. Front Earth Sci 10:1023018

Dias LC, Morton A, Quigley J (2018) Elicitation: state of the art and science. In: Dias L, Morton A, Quigley J (eds) Elicitation, Int Series in Operations Res and Manag Sci 261. Springer, Cham. https://doi.org/10.1007/978-3-319-65052-4_1

Dominey-Howes D, Dunbar P, Varner J et al (2010) Estimating probable maximum loss from a Cascadia Tsunami. Nat Hazards 53:43–61. https://doi.org/10.1007/s11069-009-9409-9

Dura T, Garner AJ, Weiss R, Kopp RE, Engelhart SE, Witter RC, Briggs RW, Mueller CS, Nelson AR, Horton BP (2021) Changing impacts of Alaska-Aleutian subduction zone tsunamis in California under future sea-level rise. Nature Comm 12:7119. https://doi.org/10.1038/s41467-021-27445-8

Dyanati M, Huang Q, Roke D (2017) Sensitivity analysis of seismic performance assessment and consequent impacts on loss analysis. Bull of Earthq Eng 15:1–40. https://doi.org/10.1007/s10518-017-0150-6

El-Hattab MM, Mohamed SA, El Raey M (2018) Potential Tsunami risk assessment to the city of Alexandria, Egypt. Environ Monit Assess 190(9):496. https://doi.org/10.1007/s10661-018-6876-z

Engel M, May S, Pilarczyk J, Brill D, Garrett E (2020) Geological records of Tsunamis and other extreme waves: concepts, applications and a short history of research. https://doi.org/10.1016/B978-0-12-815686-5.00001-8

Fagerholm N, Raymond CM, Olafsson AS, Brown G, Rinne T, Hasanzadeh K, Broberg A, Kyttä M (2021) A methodological framework for analysis of participatory mapping data in research, planning, and management. Int J Geog Inform Sci 35(9):1848–1875. https://doi.org/10.1080/13658816.2020.1869747

FEMA (2017a) Federal emergency management agency. Hazus tsunami model user guidance. FEMA: Fed Emerg Manag Agency, Washington DC. https://www.fema.gov/media-library-data/1511284000292-4f18206fb0c7bab3c5ecbbbdf504b9fd/Hazus_40_Tsunami_User_Manual.pdf

FEMA (2017b) Federal emergency management agency. Hazus tsunami model technical guidance. FEMA, Fed. Emerg. Manag. Agency, Washington DC. https://www.fema.gov/media-library-data/1511284000276-4f18206fb0c7bab3c5ecbbbdf504b9fd/Hazus_40_Tsunami_Tech_Manual.pdf

Frucht E, Salamon A, Rozell J, Levi T, Calvo R, Avirav V, Burns JN, Zuzak C, Gal E, Trapper P, Galanti B, Bausch D (2021) Tsunami loss assessment based on Hazus approach—the Bat Galim, Israel, case study. Eng Geology 289:106175. https://doi.org/10.1016/j.enggeo.2021.106175

Fuentes M, Riquelme S (2022) Stochastic Tsunami modeling induced by kinematic complex sources. Sci Rep 12:16815. https://doi.org/10.1038/s41598-022-21336-8

Fukutani Y, Anawat S, Imamura F (2016) Uncertainty in tsunami wave heights and arrival times caused by the rupture velocity in the strike direction of large earthquakes. Natural Haz 80:1749–1782. https://doi.org/10.1007/s11069-015-2030-1

Fukutani Y, Moriguchi S, Terada K, Kotani T, Otake Y, Kitano T (2019) Tsunami hazard and risk assessment for multiple buildings by considering the spatial correlation of wave height using copulas. Nat Hazards Earth Syst Sci 19:2619–2634. https://doi.org/10.5194/nhess-19-2619-2019

Garcia-Aristizabal A, Kocot J, Russo R et al (2019) A probabilistic tool for multi-hazard risk analysis using a bow-tie approach: application to environmental risk assessments for geo-resource development projects. Acta Geophysica 67:385–410. https://doi.org/10.1007/s11600-018-0201-7

Geist EL, Parsons T (2006) Probabilistic analysis of tsunami hazards. Nat Hazards 37:277–314. https://doi.org/10.1007/s11069-005-4646-z

Geist E, Parsons T, ten Brink U, Lee H (2009) Chapter 4-Tsunami probability. In: Bernard ED, Robinson AR (eds) The Sea 15:93–135. ISBN 978-0-674-03173-9

Geist EL, Chaytor JD, Parsons T, ten Brink U (2013) Estimation of submarine mass failure probability from a sequence of deposits with age dates. Geosphere 9:287–298. https://doi.org/10.1130/GES00829.1

Geist EL, ten Brink US, Gove M (2014) A framework for the probabilistic analysis of meteotsunamis. Nat Hazards 74:123–142. https://doi.org/10.1007/s11069-014-1294-1

Gelman A, Carlin JB, Stern HS, Rubin DB (1995) Bayesian data analysis. Chapman and Hall, London

Gibbons S, Lorito S, Macías J, Løvholt F, Selva J, Volpe M et al (2020) Probabilistic tsunami hazard analysis: high performance computing for massive scale inundation simulations. Front Earth Sci 8:623. https://doi.org/10.3389/feart.2020.591549

Gibbons S, Lorito S, Asunción M, Volpe M, Selva J, Macías J, Sánchez-Linares C, Brizuela B, Vöge M, Tonini R, Lanucara P, Glimsdal S, Romano F, Meyer J, Løvholt F (2022) The sensitivity of Tsunami impact to earthquake source parameters and manning friction in high-resolution inundation simulations. Frontiers Earth Sci 9. https://doi.org/10.3389/fearst.2021.757618

Giles D, Gopinathan D, Guillas S, Dias F (2020) Faster than real time tsunami warning with associated hazard uncertainties. Front Earth Sci 8:66. https://doi.org/10.3389/feart.2020.597865

Goda K, Yasuda T, Mori N, Maruyama T (2016) New scaling relationships of earthquake source parameters for stochastic Tsunami simulation. Coastal Eng J 58(3):1650010-1-1650010-40. https://doi.org/10.1142/S0578563416500108

Goda K, Song J (2016) Uncertainty modeling and visualization for tsunami hazard and risk mapping: a case study for the 2011 Tohoku earthquake. Stoch Environ Res Risk Assess 30:2271–2285. https://doi.org/10.1007/s00477-015-1146-x

Goda K, De Risi R (2017) Probabilistic tsunami loss estimation methodology: stochastic earthquake scenario approach. Earthq Spectra 33:1301–1323. https://doi.org/10.1193/012617eqs019m

Goda K (2020) Multi-hazard portfolio loss estimation for time-dependent shaking and Tsunami hazards. Front Earth Sci 8:592444. https://doi.org/10.3389/feart.2020.592444

Goda K (2021) Multi-hazard parametric catastrophe bond trigger design for subduction earthquakes and tsunamis. Earthq Spectra 37:1827–1848. https://doi.org/10.1177/8755293020981974

Goda K, De Risi R, De Luca F, Muhammad A, Yasuda T, Mori N (2021) Multi-hazard earthquake-tsunami loss estimation of Kuroshio Town, Kochi Prefecture, Japan considering the Nankai-Tonankai megathrust rupture scenarios. Int J Disaster Risk Reduct 54:102050. https://doi.org/10.1016/j.ijdrr.2021.102050

Goda K, De Risi R (2023) Future perspectives of earthquake-tsunami catastrophe modelling: from single-hazards to cascading and compounding multi-hazards. Front Built Env 8. https://doi.org/10.3389/fbuil.2022.1022736

Gomez-Zapata JC, Brinckmann N, Harig S, Zafrir R, Pittore M, Cotton F, Babeyko AY (2021) Variable-resolution building exposure modelling for earthquake and tsunami scenario-based risk assessment: an application case in Lima, Peru. Nat Hazards Earth Syst Sci 21:3599–3628. https://doi.org/10.5194/nhess-21-3599-2021

González M, Álvarez-Gómez JA, Aniel-Quiroga I, Otero L, Olabarrieta M, Omira R, Luceño A, Jelinek R, Krausmann E, Birkman J, Baptista M, Mira M, Aguirre-Ayerbe I (2021) Probabilistic Tsunami hazard assessment in meso and macro tidal areas. Application to the Cádiz Bay, Spain. Frontiers in Earth Sci 9:591383. https://doi.org/10.3389/feart.2021.591383

González A, Fernández M, Llorente M, Macías J, Sánchez-Linares C, García-Mayordomo J, Paredes C (2022) Pseudo-probabilistic design for high-resolution Tsunami simulations in the Southwestern Spanish coast. GeoHazards 3(2):294–322. https://doi.org/10.3390/geohazards3020016

González-Riancho P, Aguirre-Ayerbe I, García-Aguilar O, Medina R, González M, Aniel-Quiroga I, Gutiérrez OQ, Álvarez-Gómez JA, Larreynaga J, Gavidia F (2014) Integrated tsunami vulnerability and risk assessment: application to the coastal area of El Salvador. Nat Hazards Earth Syst Sci 14:1223–1244. https://doi.org/10.5194/nhess-14-1223-2014

Good IJ (1965) The estimation of probabilities: an essay on modern Bayesian methods, The MIT Press, Cambridge, MA, p 109

Greiving S, Fleischhauer M, Lückenkötter J (2006) A methodology for an integrated risk assessment of spatially relevant hazards. J Environ Planning Manage 49(1):1–19

Grezio A, Marzocchi W, Sandri L, Gasparini P (2010) A Bayesian procedure for probabilistic Tsunami hazard assessment. Nat Hazards 53:159–174. https://doi.org/10.1007/s11069-009-9418-8

Grezio A, Sandri L, Marzocchi W, Argnani A, Gasparini P, Selva J (2012a) Probabilistic tsunami hazard assessment for Messina Strait area (Sicily, Italy). Nat Hazards 64:329–358. https://doi.org/10.1007/s11069-012-0246-x

Grezio A, Gasparini P, Marzocchi W, Patera A, Tinti S (2012b) Tsunami risk assessments in Messina, Sicily–Italy. Nat Hazards Earth Syst Sci 12. https://doi.org/10.5194/nhess-12-151-2012

Grezio A, Tonini R, Sandri L, Pierdominici S, Selva J (2015) A methodology for a comprehensive probabilistic tsunami hazard assessment: multiple sources and short-term interactions. J Marine Sci and Eng 3:23–51. https://doi.org/10.3390/jmse3010023

Grezio A, Babeyko A, Baptista MA, Behrens J, Costa A, Davies G et al (2017) Probabilistic tsunami hazard analysis: multiple sources and global applications. Rev of Geophys 55:1158–1198. https://doi.org/10.1002/2017RG000579

Grezio A, Cinti FR, Costa A, Faenza L, Perfetti P, Pierdominici S, Pondrelli S, Sandri L, Tierz P, Tonini R, Selva J (2020) Multisource Bayesian probabilistic Tsunami hazard analysis for the Gulf of Naples (Italy). J Geophys Res 125(2):e2019JC015373. https://doi.org/10.1029/2019JC015373

Grezio A, Anzidei M, Baglione E, Brizuela B, Di Manna P, Selva J, Taroni M, Tonini R, Vecchio A (2024) Including sea-level rise and vertical land movements in probabilistic tsunami hazard assessment for the Mediterranean Sea. Sci Rep 14:28873. https://doi.org/10.1038/s41598-024-79770-9

Griffin J, Stirling M, Wang T (2020) Periodicity and clustering in the long-term earthquake record. Geophy Res Lett 47. https://doi.org/10.1029/2020GL089272

Grünthal G (1998) European Macroseismic Scale. Centre Européen de Géodynamique et de Séismologie, Luxembourg, vol 15

Grünthal G, Thieken AH, Schwarz J, Radtke KS, Smolka A, Merz B (2006) Comparison risk assessments for the city of Cologne-storm, floods, earthquakes. Nat Haz 30:21–44

Hanks T, Beroza G, Toda S (2012) Have recent earthquakes exposed flaws in or misunderstandings of probabilistic seismic hazard analysis? Seismol Res Lett 83:759–764. https://doi.org/10.1785/0220120043

Harbitz C, Glimsdal S, Bazin S, Zamora N, Løvholt F, Bungum H, Smebye H, Gauer P, Kjekstad O (2012) Tsunami hazard in the Caribbean: regional exposure derived from credible worst case scenarios. Continental Shelf Res 38:1–23. https://doi.org/10.1016/j.csr.2012.02.006

Harig S, Immerz A, Weniza W, Griffin J, Weber B, Babeyko A, Rakowsky N, Hartanto D, Nurokhim A, Hayani T, Weber R (2020) The Tsunami scenario database of the Indonesia Tsunami early warning system (InaTEWS): evolution of the coverage and the involved modeling approaches. Pure and Appl Geophys 177. https://doi.org/10.1007/s00024-019-02305-1

Harig S, Zamora N, Gubler A, Rakowsky N (2022) Systematic comparison of Tsunami simulations on the chilean coast based on different numerical approaches. GeoHazards 3. https://doi.org/10.3390/geohazards3020018

Hetényi G, Balázs L, Barcza Z et al (2022) The inherent uncertainty in geosciences. Acta Geod Geophys 57:411–418. https://doi.org/10.1007/s40328-022-00392-6

Horspool N, Pranantyo I, Griffin J, Latief H, Natawidjaja D, Kongko W, Cipta A, Bustaman B, Anugrah S, Thio H (2014) A probabilistic tsunami hazard assessment for Indonesia. Nat Haz Earth Syst Sci 14:3105–3122. https://doi.org/10.5194/nhess-14-3105-2014

Iles K, Kedzior SB (2023) Operationalizing participation: experiences and perspectives of participatory GIS program coordinators. J Geogr Syst 25:539–565. https://doi.org/10.1007/s10109-023-00416-x

Izquierdo I, Fritis E, Abad E (2018) Analysis and validation of the PTVA tsunami building vulnerability model using the 2015 Chile post-tsunami damage data in Coquimbo and La Serena cities. Nat Hazards Earth Syst Sci 18:1703–1716. https://doi.org/10.5194/nhess-18-1703-2018

Hasalanka H, Siriwardana CSA, Kularatne D, Dias WPS (2021) Development of a structural robustness index against tsunamis for hospitals. Civil Eng and Environ Sys 38(2):85–101. https://doi.org/10.1080/10286608.2021.1890045

Ismail H, Abd Wahab AK, Mohd Amin MF, Mohd Yunus MZ, Jaffar Sidek FJ, Esfandier B (2012) A 3-tier tsunami vulnerability assessment technique for the north-west coast of Peninsular Malaysia. Nat Hazards 63:549–573. https://doi.org/10.1007/s11069-012-0166-9

Jalayer F, Carozza S, De Risi R, Manfredi G, Mbuya E (2016) Performance-based flood safety-checking for non-engineered masonry structures. Eng Struct 106:109–123. https://doi.org/10.1016/j.engstruct.2015.10.007

Jalayer F, Ebrahimian H (2020) Seismic reliability assessment and the nonergodicity in the modelling parameter uncertainties. Earthq Eng Struct Dyn 49(5):434–457. https://doi.org/10.1002/eqe.3247

Jalayer F, Ebrahimian H, Trevlopoulos K, Bradley B (2023) Empirical tsunami fragility modelling for hierarchical damage levels. Nat Hazards Earth Syst Sci 23(2):909–931

Jelínek R, Eckert S, Zeug G, Krausmann E (2009) Tsunami vulnerability and risk analysis applied to the city of Alexandria, Egypt. EUR 23967 EN. Office for Official Publications of the European Communities - JRC53316, Luxembourg

Jelínek R, Krausmann E, González M, Álvarez-Gómez JA, Birkmann J, Welle T (2012) Approaches for tsunami risk assessment and application to the city of Cádiz, Spain. Nat Hazards 60:273–293. https://doi.org/10.1007/s11069-011-0009-0

Jenks GF (1967) The data model concept in statistical mapping. Int Yearbook Cartog 7:186–190. Box 4-MS 347. University of Kansas. Kenneth Spencer Research Library

Jonientz-Trisler C, Simmons RS, Yanagi BS, Crawford GL, Darienzo M, Eisner RK, Petty E, Priest GR (2005). Planning for Tsunami-Resilient Communities. In: Bernard EN (eds) Developing Tsunami-Resilient Communities. Springer, Dordrecht. https://doi.org/10.1007/1-4020-3607-8_8

Johnson KA, Dana G, Jordan NR, Draeger KJ, Kapuscinski A, Olabisi LKS, Reich PB (2012) Using participatory scenarios to stimulate social learning for collaborative sustainable development. Ecology and Society 17(2):9. https://www.jstor.org/stable/26269034

Kagan YY, Knopoff L (1987) Random stress and earthquake statistics: time dependence. Geophys Jour Res Astron Soc 88:723–731

Kagan YY, Jackson DD (2013) Tohoku earthquake: a surprise? Bulletin of the Seismol Soc of Amer 103(2B):1181–1194. https://doi.org/10.1785/0120120110

Kaneko Y, Lapusta N, Ampuero J-P (2008) Spectral element modeling of spontaneous earthquake rupture on rate and state faults: effect of velocity-strengthening friction at shallow depths. J Geophys Res 113:B09317. https://doi.org/10.1029/2007JB005553

Kappes M, Papathoma-Köhle M, Keiler M (2012) Assessing physical vulnerability for multi-hazards using an indicator-based methodology. Appl Geog 32(2):577–590. https://doi.org/10.1016/j.apgeog.2011.07.002

Koyano K, Takabatake T, Esteban M, Shibayama T (2022) Magnification of Tsunami risks due to sea level rise along the Eastern Coastline of Japan. J Coastal and Hydr Struct 2. https://doi.org/10.48438/jchs.2022.0012

Kosaka N, Koshimura S, Terada K, Matsubara H (2023) Decision-making support utilizing real-time tsunami inundation and damage forecast. Int J Disaster Risk Reduct 94:103807. https://doi.org/10.1016/j.ijdrr.2023.103807

Kowalik Z, Proshutinsky T, Proshutinsky A (2006) Tide-tsunami interactions. Sci Tsun Haz 24(4):246–256

Kowalik Z, Proshutinsky A (2010) Tsunami–tide interactions: a cook inlet case study. Cont Shelf Res 30:633–642. https://doi.org/10.1016/j.csr.2009.10.004

Krawinkler H (1999) Challenges and progress in performance-based earthquake engineering. In: International seminar on seismic engineering for tomorrow–in honor of Professor Hiroshi Akiyama, vol. 26

Lahcene E, Ioannou I, Suppasri A, Pakoksung K, Paulik R, Syamsidik S, Bouchette F, Imamura F (2021) Characteristics of building fragility curves for seismic and non-seismic tsunamis: case studies of the 2018 Sunda Strait, 2018 Sulawesi-Palu, and 2004 Indian Ocean tsunamis. Nat Hazards Earth Syst Sci 21(8):2313–2344. https://doi.org/10.5194/nhess-21-2313-2021

Latcharote P, Al-Salem K, Suppasri A et al (2018) Tsunami hazard evaluation for Kuwait and Arabian Gulf due to Makran Subduction zone and subaerial landslides. Nat Hazards 93:127–152. https://doi.org/10.1007/s11069-017-3097-7

Lambeck K, Chappell J (2001) Sea level change through the last glacial cycle. Science 292:679–686. https://doi.org/10.1126/science.1059549

Lee HS, Shimoyama T, Popinet S (2015) Impacts of tides on tsunami propagation due to potential Nankai Trough earthquakes in the Seto Inland Sea, Japan. J Geophys Res: Oceans 120(10):6865–6883. https://doi.org/10.1002/2015JC010995

León J, Mokrani C, Catalán P, Cienfuegos R, Femenías C (2018) Examining the role of urban form in supporting rapid and safe tsunami evacuations: a multi-scalar analysis in Viña del Mar, Chile. Procedia Eng 212:629–636. https://doi.org/10.1016/j.proeng.2018.01.081

Li L, Switzer AD, Chan C-H, Wang Y, Weiss R, Qiu Q (2016) How heterogeneous coseismic slip affects regional probabilistic tsunami hazard assessment: a case study in the South China Sea. J Geophys Res 121(8):6250–6272. https://doi.org/10.1002/2016JB013111

Li L, Switzer AD, Wang Y, Chan CH, Qiu Q, Weiss R (2018) A modest 0.5-m rise in sea level will double the tsunami hazard in Macau Sci Adv 4. https://doi.org/10.1126/sciadv.aat1180

Liu Z, Nadim F, Garcia-Aristizabal A, Mignan A, Fleming K, Luna BQ (2015) A three-level framework for multi-risk assessment. Georisk Assess Manage Risk Eng Syst Geohazards 9(2):59–74. https://doi.org/10.1080/17499518.2015.1041989

Liu CM, Rim D, Baraldi R, LeVeque RJ (2021) Comparison of machine learning approaches for tsunami forecasting from sparse observations. Pure Appl Geophys 178:5129–5153. https://doi.org/10.1007/s00024-021-02841-9

Lombardi AM, Marzocchi W (2007) Evidence of clustering and nonstationarity in the time distribution of large worldwide earthquakes. J Geophys Res 112:B02303. https://doi.org/10.1029/2006JB004568

Lorito S, Tiberti MM, Basili R, Piatanesi A, Valensise G (2008) Earthquake-generated tsunamis in the Mediterranean sea: scenarios of potential threats to Southern Italy. J Geophys Res 113:B01301. https://doi.org/10.1029/2007JB004943

Lorito S, Selva J, Basili R, Romano F, Tiberti MM, Piatanesi A (2015) Probabilistic hazard for seismically induced tsunamis: accuracy and feasibility of inundation maps. Geophys J Int 200(1):574–588. https://doi.org/10.1093/gji/ggu408

Løvholt F, Glimsdal S, Harbitz C, Zamora N, Nadim F, Peduzzi P, Dao H, Smebye H (2012a) Tsunami hazard and exposure on the global scale. Earth-Sci Rev 110. https://doi.org/10.1016/j.earscirev.2011.10.002

Løvholt F, Kühn D, Bungum H, Harbitz C, Glimsdal S (2012b) Historical tsunamis and present tsunami hazard in eastern Indonesia and the southern Philippines. J Geophys Res 117:B09310. https://doi.org/10.1029/2012JB009425

Løvholt F, Glimsdal S, Harbitz CB, Smebye H, de Bono A, Syrris V, Gardi A (2014) Global tsunami hazard and exposure due to large co-seismic slip. Int J Disaster Risk Reduct 10:406–418. https://doi.org/10.1016/j.ijdrr.2014.04.003

Løvholt F, Griffin J, Salgado-Gálvez MA (2015) Tsunami hazard and risk assessment on the global scale. In: Meyers RA (ed) Encyclopedia of complexity and systems science. Springer, Berlin & Heidelberg, pp 1–34. https://doi.org/10.1007/978-3-642-27737-5_642-1

Macías J, Castro M J, Ortega S, Escalante C, González-Vida J M (2017) Performance benchmarking of tsunami-HySEA model for NTHMP's inundation mapping activities. Pure Appl Geophys 174:3147–3183. https://doi.org/10.1007/s00024-017-1583-1

Mah JC, Penwarden JL, Pott H, Theou O, Andrew MK (2023) Social vulnerability indices: a scoping review. BMC Public Health 1:1253. https://doi.org/10.1186/s12889-023-16097-6

Makinoshima F, Oishi Y, Yamazaki T, Furumura T, Imamura F (2021) Early forecasting of tsunami inundation from tsunami and geodetic observation data with convolutional neural networks. Nat Commun 12:2253. https://doi.org/10.1038/s41467-021-22348-0

Martínez C, Cienfuegos R, Inzunza S, Urrutia A, Guerrero N (2020) Worst-case tsunami scenario in cartagena bay, central Chile: challenges for coastal risk management. Ocean Coast Manag 185:105060. https://doi.org/10.1016/j.ocecoaman.2019.105060

Marzocchi W, Garcia-Aristizabal A, Gasparini P, Masellone ML, Di Ruocco A, Novelli P (2012) Basic principles of multi-risk assessment: a case study in Italy. Nat Haz 62:551–573. https://doi.org/10.1007/s11069-012-0092-x

Marzocchi W (2013) Seismic hazard and public safety. Eos Trans AGU 94(27):240

Marzocchi W, Jordan TH (2014) Testing for ontological errors in probabilisticforecasting models of natural systems. Proc Nat Acad Sci PNAS 111(3):11973–11978. https://doi.org/10.1073/pnas.1410183111

Marzocchi W, Taroni M, Selva J (2015) Accounting for epistemic uncertainty in PSHA: logic tree and ensemble modeling. Bull Seismol Soc of Amer 105(4):2151–2159. https://doi.org/10.1785/0120140131

Marzocchi W, Selva J, Jordan T (2021) A unified probabilistic framework for volcanic hazard and eruption forecasting. Nat Hazards Earth Syst Sci 21:3509–3517. https://doi.org/10.5194/nhess-21-3509-2021

Masuda H, Sugawara D, Abe T, Goto K (2022) To what extent tsunami source information can be extracted from tsunami deposits? Implications from the 2011 Tohoku-oki tsunami deposits and sediment transport simulations. Progress Earth Planet Sci 9. https://doi.org/10.1186/s40645-022-00527-x

Matthews M (2002) A brownian model for recurrent earthquakes. Bull Seismol Soc Amer 92:2233–2250. https://doi.org/10.1785/0120010267

McCalpin J (2009) Chapter 9 Application of paleoseismic data to seismic hazard assessment and neotectonic research. Int Geophysics 95:1–106. https://doi.org/10.1016/S0074-6142(09)95009-4

Mercer J, Kelman I, Lloyd K, Suchet-Pearson S (2008) Reflections on use of participatory research for disaster risk reduction. Area 40:172–183. https://doi.org/10.1111/j.1475-4762.2008.00797.x

Melgar D, LeVeque RJ, Dreger DS, Allen RM (2016) Kinematic rupture scenarios and synthetic displacement data: an example application to the Cascadia subduction zone. J Geophys Res Solid Earth 121:6658–6674. https://doi.org/10.1002/2016JB013314

Mengal A, Goda K, Ashraf M et al (2021) Social vulnerability to seismic-tsunami hazards in district Gwadar, Balochistan, Pakistan. Nat Hazards 108:1159–1181. https://doi.org/10.1007/s11069-021-04724-7

Mitchell-Wallace K, Jones M, Hillier J, Foote M (2017) Natural catastrophe risk management and modelling: a practitioner's guide. Wiley, p 544

Mignan A, Wiemer S, Giardini D (2014) The quantification of low probability-high consequences events: part I generic multi rish approach. Nat Hazards 73:1999–2022. https://doi.org/10.1007/s11069-014-1178-4

Moehle J, Deierlein GG (2004) A framework methodology for performance-based earthquake engineering. In: Proceedings of the 13th world conference on earthquake engineering, vol. 679, WCEE Vancouver

Mofjeld HO, González FI, Titov VV, Venturato AJ, Newman JC (2007) Effects of tides on maximum tsunami wave heights: probability distributions. J Atmosph and OceanTech 24(1):117–123

Mol AM, Law J (2002) Complexities: social studies of knowledge practices. Duke University Press, Durham

Mori N, Goda K, Cox D (2018) Recent process in probabilistic tsunami hazard analysis (PTHA) for mega-thrust subduction earthquakes. In: Santiago-Fandiño V, Sato S, Maki N, Luchi K (eds) The 2011 Japan earthquake and tsunami: reconstruction and restoration, Springer, Cham 47:469–485

Mueller C, Power WL, Fraser S, Wang X (2015) Effects of rupture complexity on local tsunami inundation: implications for probabilistic tsunami hazard assessment by example. J Geophys Res Solid Earth 120:488–502. https://doi.org/10.1002/2014JB011301

Mück M, Taubenböck H, Post J, Wegscheider JS, Strunz G, Sumaryono S, Ismail FA (2013) Assessing building vulnerability to earthquake and tsunami hazard using remotely sensed data. Nat Hazards 68:97–114. https://doi.org/10.1007/s11069-012-0481-1

Muhari A, Charvet I, Tsuyoshi F, Suppasri A, Imamura F (2015) Assessment of tsunami hazards in ports and their impact on marine vessels derived from tsunami models and the observed damage data. Nat Hazards 78:1309–1328. https://doi.org/10.1007/s11069-015-1772-0

Mulargia F, Stark PB, Geller RJ (2017) Why is probabilistic seismic hazard analysis (PSHA) still used? Phy Earth Planet Inter 264:63–75. https://doi.org/10.1016/j.pepi.2016.12.002

Munda G (2005) Multiple criteria decision analysis and sustainable development. In: Figueira J, Greco S, Ehrgott M (eds) Multiple criteria decision analysis: state of the art surveys. Springer, pp 953–986

Munda G (2007) Social multi-criteria evaluation for a sustainable economy. Springer, Berlin

Murphy S, Scala A, Herrero A et al (2016) Shallow slip amplification and enhanced tsunami hazard unravelled by dynamic simulations of mega-thrust earthquakes. Sci Rep 6:35007. https://doi.org/10.1038/srep35007

Myers EP, Baptista AM (2001) Analysis of factors influencing simulations of the 1993 Hokkaido Nansei-Oki and 1964 Alaska Tsunamis. Nat Hazards 23:1–28. https://doi.org/10.1023/A:1008150210289

NCEI/WDS (2024) Global Historical Tsunami Database, 2100 BC to Present. National Geophysical Data Center (NGDC), World Data Service. Washington, D.C, United States. https://www.ngdc.noaa.gov/hazel/view/hazards/tsunami. Accessed 18 Feb 2024

Nisaa RM, Sartohadi J, Mardiatno D (2021) Participatory GIS approach to assessing building vulnerability to tsunamis in pangandaran regency. Forum Geografi 35(2). https://doi.org/10.23917/forgeo.v35i2.14003

Nomura S, Ogata Y, Komaki F, Toda S (2011) Bayesian forecasting of recurrent earthquakes and predictive performance for a small sample size. J Geophys Res 116. https://doi.org/10.1029/2010JB007917

Núñez J, Catalán PA, Valle C, Zamora N, Valderrama A (2022) Discriminating the occurrence of inundation in tsunami early warning with one-dimensional convolutional neural networks. Sci Rep 12:10321. https://doi.org/10.1038/s41598-022-13788-9

Nurmaya A, Herbanu PS, Nisaa RM, Arifati A, Wardana RA, Sahid S, Firmansyah RA, Yudhanti FY (2023) Tsunami risk assessment on public facilities in Southern Part of Bantul Regency, Yogyakarta. In: E3S web of conferences, vol. 447, p. 01015. EDP Sciences

OECD, European Union, EC-JRC, (2008) Handbook on constructing composite indicators: methodology and user guide. OECD Publishing, Paris,. https://doi.org/10.1787/9789264043466-en

O'Donnell J (2016) Simplifying the papathoma Tsunami vulnerability assessment: towards an open and GIS-based data collection and processing method. Master Thesis, University of Sheffield, p 78

Oetjen J, Sundar V, Venkatachalam S et al (2022) A comprehensive review on structural tsunami countermeasures. Nat Hazards 113:1419–1449. https://doi.org/10.1007/s11069-022-05367-y

Ogata Y (1999) Estimating the hazard of rupture using uncertain occurrence times of paleoearthquakes. J Geophys Res 104(17):995–918,014

Ogata Y (2017) Forecasting of a large earthquake: an outlook of the research. Seismol Res Lett 88(4):1117–1126. https://doi.org/10.1785/0220170006

Ogata Y, Katsura K, Keiding N, Holst C, Green A (2000) Empirical Bayes age-period-cohort analysis of retrospective incidence data. Scand J Stat 27:415–432. http://www.jstor.org/stable/4616614

O'Loughlin KF, Lander JF (2003) Caribbean Tsunamis: A 500-Year history from 1498–1998. Kluwer Academic Publishers, Dordrecht, The Netherlands)

Omira R, Baptista MA, Miranda JM et al (2009) Tsunami vulnerability assessment of Casablanca-Morocco using numerical modelling and GIS tools. Nat Hazards 54:75–95. https://doi.org/10.1007/s11069-009-9454-4

Omira R, Matias L, Baptista MA (2016) Developing an event-tree probabilistic Tsunami inundation model for NE Atlantic coasts: application to a case study. Pure Appl Geophys 173:3775–3794. https://doi.org/10.1007/s00024-016-1367-z

Oppenheimer M, Glavovic BC, Hinkel J, van de Wal R, Magnan AK, Abd-Elgawad A, Cai R, Cifuentes-Jara M, DeConto RM, Ghosh T, Hay J, Isla F, Marzeion B, Meyssignac B, Sebesvari Z (2019) Sea level rise and implications for low-lying islands, coasts and communities. In: Pörtner H-Ö, Roberts DC, Masson-Delmotte V, Zhai P, Tignor M, Poloczanska E, Mintenbeck K, Alegría A, Nicolai M, Okem A, Petzold J, Rama B, Weyer NM (eds) IPCC Special Report on the Ocean and Cryosphere in a Changing Climate. Cambridge University Press, Cambridge, UK and New York, NY, USA, pp. 321–445. https://doi.org/10.1017/9781009157964.006

Papathoma-Köhle M, Cristofari G, Wenk M, Fuchs S (2019a) The importance of indicator weights for vulnerability indices and implications for decision making in disaster management. Int J Dis Ris Red 36:1–12. https://doi.org/10.1016/j.ijdrr.2019.101103

Papathoma-Köhle M, Schlögl M, Fuchs S (2019b) Vulnerability indicators for natural hazards: an innovative selection and weighting approach. Sci Rep 9:15026. https://doi.org/10.1038/s41598-019-50257-2

Papathoma M, Dominey-Howes D (2003) Tsunami vulnerability assessment and its implications for coastal hazard analysis and disaster management. Nat Hazards Earth Syst Sci 3(6):733–744. https://doi.org/10.5194/nhess-3-733-2003

Papathoma M, Dominey-Howes D, Zong Y, Smith D (2003) Assessing Tsunami vulnerability, an example from Herakleio Crete. Nat Hazards Earth Syst Sci 3(5):377–389. https://doi.org/10.5194/nhess-3-377-2003

Paris R, Ulvrova M, Selva J, Brizuela B, Costa A, Grezio A, Lorito S, Tonini R (2019) Probabilistic hazard analysis for tsunamis generated by subaqueous volcanic explosions in the Campi Flegrei caldera, Italy. J Volc Geoth Res 379. https://doi.org/10.1016/j.jvolgeores.2019.05.010

Park H, Cox DI, Alam M, Barbosa A (2017) Probabilistic seismic and Tsunami hazard analysis conditioned on a megathrust rupture of the cascadia subduction zone. Frontiers in Built Environment. 3:32. https://doi.org/10.3389/fbuil.2017.00032

Parsons T (2008) Monte Carlo method for determining earthquake recurrence parameters from short paleoseismic catalogs: example calculations for California. J Geophys Res 112. https://doi.org/10.1029/2007JB004998

Parsons T, Geist EL (2009) Tsunami probability in the caribbean region. Pure Appl Geophys 165:2089–2116. https://doi.org/10.1007/s00024-008-0416-7

Parsons T (2012) Paleoseismic interevent times interpreted for an unsegmented earthquake rupture forecast. Geophys Res Lett 39:L13302. https://doi.org/10.1029/2012GL052275

Pawitan Y (2001) In all likelihood: statistical modelling and inference using likelihood, Oxford University Press, Oxford, p 528

Pittore M, Wieland M, Fleming K (2017) Perspectives on global dynamic exposure modelling for geo-risk assessment. Nat Hazards 86(1):7–30. https://doi.org/10.1007/s11069-016-2437-3

Power W, Wang X, Wallace L, Clark K, Mueller C (2018) The New Zealand probabilistic tsunami hazard model: development and implementation of a methodology for estimating tsunami hazard nationwide. Geological Society, London, Special Publications 456. https://doi.org/10.1144/SP456.6

Raduszynski T, Numada M (2023) Measure and spatial identification of social vulnerability, exposure and risk to natural hazards in Japan using open data. Sci Rep 13:664. https://doi.org/10.1038/s41598-023-27831-w

Rafliana I, Jalayer F, Cerase A, Cugliari L, Baiguera M, Salmanidou D, Necmioglu O, Aguirre I, Lorito S, Fraser S, Løvholt F, Babeyko AY, Salgado-Gálvez MA, Selva J, De Risi R, Sørensen MB, Behrens JO, Aniel-Quiroga I, Del Zoppo M, Belliazzi S, Pranantyo IR, Amato A, Hancilar U (2022) Tsunami risk communication and management: Contemporary gaps and challenges. Int J Disaster Risk Red 70(102771):1–30. https://doi.org/10.1016/j.ijdrr.2021.102771

Ramírez-Herrera MT, Coca O, Vargas-Espinosa V (2022) Tsunami effects on the coast of Mexico by the Hunga Tonga-Hunga Ha'apai Volcano Eruption Tongs. Pure Appl Geophys 179(4):1117–1137. https://doi.org/10.1007/s00024-022-03017-9

Ramos MD, Thakur P, Huang Y, Harris RA, Ryan KJ (2022) Working with dynamic earthquake rupture models: a practical guide. Seismol Res Lett 93(4):2096–2110. https://doi.org/10.1785/0220220022

Rohmer J, Rousseau M, Lemoine A, Pedreros R, Lambert J, Benki A (2018) Source characterisation by mixing long-running tsunami wave numerical simulations and historical observations within a metamodel-aided ABC setting. Stoch Environ Res Risk Assess 32:967–984. https://doi.org/10.1007/s00477-017-1423-y

Rodríguez JF, Macías J, Castro MJ, de la Asunción M, Sánchez-Linares C (2022) Use of neural networks for tsunami maximum height and arrival time predictions. GeoHazards 3(2):323–344. https://doi.org/10.3390/geohazards3020017

Saaty TL (1987) The analytic hierarchy process: what it is and how it is used. Math Model 9(3–5):161–176. https://doi.org/10.1016/0270-0255(87)90473-8

Saaty TL (2008) Decision making with the analytic hierarchy process. Int J Serv Sci 1(1):83–98. https://doi.org/10.1504/IJSSCI.2008.017590

Salmanidou DM, Ehara A, Himaz R, Heidarzadeh M, Guillas S (2021) Impact of future tsunamis from the Java trench on household welfare: merging geophysics and economics through catastrophe modelling. IJDRR 61:102291. https://doi.org/10.1016/j.ijdrr.2021.102291

Salter J, Robinson J, Wiek A (2010) Participatory methods of integrated assessment – a review. Wiley Interdiscip Rev Clim Change 1(5):697–717. https://doi.org/10.1002/wcc.73

San Martín TV, Rodríguez Rosado G, Arreaga Vargas P et al (2018) Population and building vulnerability assessment by possible worst-case tsunami scenarios in Salinas, Ecuador. Nat Hazards 93:275–297. https://doi.org/10.1007/s11069-018-3300-5

Santos A, Tavares AO, Emidio A (2014) Comparative tsunami vulnerability assessment of an urban area: an analysis of Setúbal city, Portugal. Appl Geogr 55:19–29. https://doi.org/10.1016/j.apgeog.2014.08.009

Satake K (2015) Geological and historical evidence of irregular recurrent earthquakes in Japan. Philos Trans A Math Phys Eng Sci 28(373):20140375. https://doi.org/10.1098/rsta.2014.0375

Satake K, Atwater B (2007) Long-term perspectives on giant earthquakes and Tsunamis at subduction zones. Ann Rev Earth and Planet Sci 35:349–374. https://doi.org/10.1146/annurev.earth.35.031306.140302

Satake K, Heidarzadeh M, Quiroz M, Cienfuegos R (2020) History and features of trans-oceanic tsunamis and implications for paleo-tsunami studies. Earth-Sci Rev 202:103112. https://doi.org/10.1016/j.earscirev.2020.103112

Sathiparan N (2020) An assessment of building vulnerability to a tsunami in the Galle coastal area, Sri Lanka. J Build Eng 27:100952. https://doi.org/10.1016/j.jobe.2019.100952

Sato H, Murakami H, Kozuki Y et al (2003) Study on a simplified method of Tsunami risk assessment. Nat Hazards 29:325–340. https://doi.org/10.1023/A:1024732204299

Selva J (2013) Long-term multi-risk assessment: statistical treatment of interaction among risks. Nat Hazards 67:701–722. https://doi.org/10.1007/s11069-013-0599-9

Selva J, Tonini R, Molinari I, Tiberti MM, Romano F, Grezio A et al (2016) Quantification of source uncertainties in seismic probabilistic tsunami hazard analysis (SPTHA). Geophys J Int 205:1780–1803. https://doi.org/10.1093/gji/ggw107

Selva J, Sandri L (2013) Probabilistic seismic hazard assessment: combining Cornell-like approaches and data at sites through Bayesian inference. Bull Seismol Soc Am 103(3):1709–1722. https://doi.org/10.1785/0120120091

Selva J, Argyroudis S, Pitilakis K (2013) Impact on loss/risk assessments of inter-model variability in vulnerability analysis. Nat Hazards 67:723–746

Selva J, Lorito S, Volpe M et al (2021) Probabilistic tsunami forecasting for early warning. Nat Commun 12:5677. https://doi.org/10.1038/s41467-021-25815-0

Sepúlveda I, Liu PL-F, Grigoriu M, Pritchard M (2017) Tsunami hazard assessments with consideration of uncertain earthquake slip distribution and location. J Geophys Res Solid Earth 122:7252–7271. https://doi.org/10.1002/2017JB014430

Sepúlveda I, Haase JS, Liu PL-F, Grigoriu M, Winckler P (2021) Non-stationary probabilistic tsunami hazard assessments incorporating climate-change-driven sea level rise. Earth Future 9:e2021EF002007. https://doi.org/10.1029/2021EF002007

Shelby M, Grilli ST, Grilli AR (2016) Tsunami hazard assessment in the Hudson River Estuary based on dynamic tsunami–tide simulations. In: Geist EL, Fritz HM, Rabinovich AB, Tanioka Y (eds) Global tsunami science: past and future, vol I. Pageoph Topical Volumes. Birkhäuser, Cham. https://doi.org/10.1007/978-3-319-55480-8_17

Smit A, Stein A, Kijko A (2019) Bayesian inference in natural hazard analysis for incomplete and uncertain data. Environmetrics 30:e2566. https://doi.org/10.1002/env.2566

Song J, Goda K (2019) Influence of elevation data resolution on Tsunami loss estimation and insurance rate-making. Front Earth Sci 7:246. https://doi.org/10.3389/feart.2019.00246

Sørensen MB, Spada M, Babeyko AY, Wiemer S, Grünthal G (2012) Probabilistic tsunami hazard in the Mediterranean Sea. J Geophys Res 117:B01305. https://doi.org/10.1029/2010JB008169

SSHAC (1997) Senior seismic hazard analysis committee. Recommendations for probabilistic seismic hazard analysis: guidance on uncertainty and use of experts. U.S. Nuclear Regulatory Commission, U.S. Dept. of Energy, Electric Power Research Institute, NUREG/CR-6372, UCRL-ID-122160, vol. 1–2

Stein S, Okal EA (2007) Ultralong period seismic study of the December 2004 Indian Ocean earthquake and implications for regional tectonics and the subduction process. Bull Seismol Soc Amer 97(1A):S279–S295

Stein S, Geller R, Liu M (2012) What to do given that earthquake hazard maps often fail. Tectonophysics 562–563:1–25. https://doi.org/10.1016/j.tecto.2012.06.047

Stirling MW (2014) The continued utility of probabilistic seismic-hazard assessment. In: Shroder JF, Bobrowsky P (eds) Earthquake hazard, risk and disasters. Academic Press, pp 359–376. https://doi.org/10.1016/B978-0-12-394848-9.00013-4

Straub D, Der Kiureghian A (2008) Improved seismic fragility modeling from empirical data. Struct Saf 30(4):320–336. https://doi.org/10.1016/j.strusafe.2007.05.004

Strunz G, Post J, Zosseder K, Wegscheider S, Mück M, Riedlinger T, Muhari A (2010) Tsunami risk assessment in Indonesia. Nat Hazards Earth Syst Sci 11(1):67–82. https://doi.org/10.5194/nhess-11-67-2011

Sugawara D (2021) Numerical modeling of tsunami: advances and future challenges after the 2011 Tohoku earthquake and tsunami. Earth-Sci Rev 214:103498. https://doi.org/10.1016/j.earscirev.2020.103498

Suppasri A, Imamura F, Koshimura S (2011) Effects of the rupture velocity of fault motion, ocean current and initial sea level on the transoceanic propagation of tsunami. Coastal Eng J 52(2):107–132. https://doi.org/10.1142/S0578563410002142

Suppasri A, Mas E, Charvet I et al (2013) Building damage characteristics based on surveyed data and fragility curves of the 2011 Great East Japan tsunami. Nat Hazards 66:319–341. https://doi.org/10.1007/s11069-012-0487-8

Suppasri A, Fukui K, Yamashita K, Leelawat N, Ohira H, Imamura F (2018) Developing fragility functions for aquaculture rafts and eelgrass in the case of the 2011 Great East Japan tsunami. Nat Hazards Earth Syst Sci 18:145–155. https://doi.org/10.5194/nhess-18-145-2018

Synolakis C, Liu P, Philip HA, Carrier G, Yeh H (1997) Tsunamigenic sea-floor deformations. Science 278:598–600

Synolakis CE, Bernard EN (2006) Tsunami science before and beyond boxing day 2004. Phil Trans Royal Soc A 3642231–2265. https://doi.org/10.1098/rsta.2006.1824

Tavares AO, Barros JL, Santos A (2017) Multidimensional approach for tsunami vulnerability assessment: framing the territorial impacts in two municipalities in Portugal. Risk Anal 37(4):788–811. https://doi.org/10.1111/risa.12678

Tarbotton C, Dominey-Howes D, Goff JR, Papathoma-Köhle M, Dall'Osso F, Turner LL (eds) (2012) GIS-based techniques for assessing the vulnerability of buildings to tsunami: current approaches and future steps. Geological Society, London

ten Brink U, Chaytor JD, Geist E, Brothers D, Andrews B (2014) Assessment of tsunami hazard to the U.S. Atlantic margin. Marine Geology. 353. https://doi.org/10.1016/j.margeo.2014.02.011.

Terry JP, Goff J, Winspear N et al (2022) Tonga volcanic eruption and tsunami, January 2022: globally the most significant opportunity to observe an explosive and tsunamigenic submarine eruption since AD 1883 Krakatau. Geosci Lett 9:24. https://doi.org/10.1186/s40562-022-00232-z

Tierz P, Woodhouse MJ, Phillips JC, Sandri L, Selva J, Marzocchi W, Odbert HM (2017) A framework for probabilistic multi-hazard assessment of rain-triggered lahars using Bayesian belief networks. Front in Earth Sci 5:73. https://doi.org/10.3389/feart.2017.00073/full

Tinti S (1991) Assessment of tsunami hazard in the Italian seas. Nat Hazards 4:267–283. https://doi.org/10.1007/bf00162792

Tinti S, Zaniboni F, Pagnoni G et al (2008) Stromboli Island (Italy): scenarios of tsunamis Generated by Submarine Landslides. Pure Appl Geophys 165:2143–2167. https://doi.org/10.1007/s00024-008-0420-y

Tinti S, Chiocci FL, Zaniboni F, Pagnoni G, de Alteriis G (2011) Numerical simulation of the tsunami generated by a past catastrophic landslide on the volcanic island of Ischia, Italy. Mar Geophys Res 32(1):287–297. https://doi.org/10.1007/s11001-010-9109-6

Thywissen K (2006) Core terminology of disaster reduction: a comparative glossary. In: Birkmann J (ed) Measuring vulnerability to natural hazards. UNU Press, Tokyo, pp 448–496

Tolkova E (2016) Tsunami penetration in tidal rivers, with observations of the Chile 2015 tsunami in rivers in Japan. Pure Appl Geophys 173. https://doi.org/10.1007/s00024-015-1229-0

Tolkova E, Tanaka H, Roh M (2015) Tsunami observations in rivers from a perspective of tsunami interaction with tide and riverine flow. Pure Appl Geophys 172. https://doi.org/10.1007/s00024-014-1017-2

Tonini R, Armigliato A, Pagnoni G, Zaniboni F, Tinti S (2011) Tsunami hazard for the city of Catania, eastern Sicily, Italy, assessed by means of worst-case credible tsunami scenario analysis (WCTSA). Nat Hazards Earth Syst Sci 11:1217–1232. https://doi.org/10.5194/nhess-11-1217-2011

Tonini R, Di Manna P, Lorito S, Selva J, Volpe M, Romano F, Basili R (2021) Testing tsunami inundation maps for evacuation planning in Sicily. Front Earth Sci 9:628061. https://doi.org/10.3389/feart.2021.628061

Triantafyllou I, Novikova T, Charalampakis M et al (2019) Quantitative tsunami risk assessment in terms of building replacement cost based on tsunami modelling and GIS Methods: the case of crete Isl. Hellenic Arc. Pure Appl Geophys 176:3207–3225. https://doi.org/10.1007/s00024-018-1984-9

Triyanti A, Surtiari GAK, Lassa J, Rafliana I, Hanifa NR, Muhidin MI, Djalante R (2023) Governing systemic and cascading disaster risk in Indonesia: where do we stand and future outlook. Disaster Prev Manage Int J 32(1):27–48. https://doi.org/10.1108/DPM-07-2022-0156

Tufekci D, Suzen ML, Yalciner AC et al (2018) Revised MeTHuVA method for assessment of tsunami human vulnerability of Bakirkoy district, Istanbul. Nat Hazards 90:943–974. https://doi.org/10.1007/s11069-017-3082-1

Turchi A, Di Traglia F, Gentile R, Fornaciai A, Zetti I, Fanti R (2022) Relative seismic and tsunami risk assessment: the Stromboli Island (UNESCO site, Italy) case study. Int J of Disaster Risk Red 76:103002. https://doi.org/10.1016/j.ijdrr.2022.103002

Uchida N, Bürgmann R (2021) A decade of lessons learned from the 2011 Tohoku-Oki earthquake. Rev of Geophys 59, e2020RG000713. https://doi.org/10.1029/2020RG000713

Ulrich T, Vater S, Madden EH et al (2019) Coupled, physics-based modeling reveals earthquake displacements are critical to the 2018 Palu Sulawesi Tsunami. Pure Appl Geophys 176:4069–4109. https://doi.org/10.1007/s00024-019-02290-5

Ulrich T, Gabriel A-A, Madden EH (2022) Stress, rigidity and sediment strength control megathrust earthquake and tsunami dynamics. Nat Geosci 15:67–73. https://doi.org/10.1038/s41561-021-00863-5

UNDRO (1984) Disaster Prevention and mitigation-a compendium of current knowledge, New York, UN

UNISDR (2009) UNISDR Terminology on Disaster Risk Reduction, UNISDR-20–2009-Geneva, Switzerland, UN, p 35

Van de Lindt JW, Dao TN (2009) Performance-based wind engineering for wood-frame buildings. J Struct Eng 135(2):169–177. https://doi.org/10.1061/(asce)0733-9445(2009)135:2(169)

Van Zelst I, Rannabauer L, Gabriel A-A, van Dinther Y (2022) Earthquake rupture on multiple splay faults and its effect on tsunamis. J Geophysical Research: Solid Earth 127:e2022JB024300. https://doi.org/10.1029/2022JB024300

Villagra P, Herrmann MG, Quintana C et al (2017) Community resilience to tsunamis along the Southeastern Pacific: a multivariate approach incorporating physical, environmental, and social indicators. Nat Hazards 88:1087–1111. https://doi.org/10.1007/s11069-017-2908-1

Vogel C, O'Brien K (2004) Vulnerability and global environmental change: rhetoric and reality (AVISO No. 13). Global Environmental Change and Human Security Project, Carleton University

Volpe M, Lorito S, Selva J, Tonini R, Romano F, Brizuela B (2019) From regional to local SPTHA: efficient computation of probabilistic tsunami inundation maps addressing near-field sources. Nat Hazards Earth Syst Sci 19(3):455–469

Voulgaris G, Murayama Y (2014) Tsunami vulnerability assessment in the Southern Boso Peninsula, Japan. Int J Disaster Risk Reduct 10A:190–200. https://doi.org/10.1016/j.ijdrr.2014.08.001

Walter TR, Haghshenas MH, Schneider FM, Coppola D, Motagh M, Saul J, Babeyko A, Dahm T, Troll VR, Tilmann F, Heimann S (2019) Complex hazard cascade culminating in the Anak Krakatau sector collapse. Nature Comm 10(1):4339. https://doi.org/10.1038/s41467-019-122 84-5

Wang Y, Imai K, Mulia IE, Ariyoshi K, Takahashi N, Sasaki K et al (2023) Data assimilation using high-frequency radar for tsunami early warning: a case study of the 2022 Tonga volcanic tsunami. J Geophys Res Solid Earth 128:e2022JB025153. https://doi.org/10.1029/2022JB025153

Warrick RA, Oerlemans J, Beaumont P, Braithwaite RJ, Drewry DJ, Gornitz V, Grove JM, Haeberli W, Higashi A, Leiva JC, Lingle CS, Lorius C, Raper SCB, Wold B, Woodworth PL (1990) Sea level rise. In: Climate Change: The IPCC Scientific Assessment, Cambridge University Press, pp 257–282. https://www.ipcc.ch/report/ar1/wg1/sea-level-rise/

Wijetunge LJ (2014) A deterministic analysis of tsunami hazard and risk for the southwest coast of Sri Lanka. Cont Shelf Res 79:23–35

Williams J, Wilson T, Horspool N, Paulik R, Wotherspoon L, Lane E, Hughes M (2020) Assessing transportation vulnerability to tsunamis: utilising post-event field data from the 2011 Tōhoku tsunami, Japan, and the 2015 Illapel tsunami, Chile. Nat Haz Earth Sys Sci 20:451–470. https://doi.org/10.5194/nhess-20-451-2020

Williamson AL, Rim D, Adams LM, LeVeque RJ, Melgar D, González FI (2020) A source clustering approach for efficient inundation modeling and regional scale probabilistic tsunami hazard assessment. Front in Earth Sci 8:591663. https://doi.org/10.3389/feart.2020.591663

Wirp S, Gabriel AA, Schmeller M, Madden E, Zelst I, Krenz L, van Dinther Y, Rannabauer L (2021) 3D linked subduction, dynamic rupture, tsunami, and inundation modeling: dynamic effects of supershear and tsunami earthquakes, hypocenter location and shallow fault slip. Front in Earth Sci 9:626844. https://doi.org/10.3389/feart.2021.626844

Woessner J, Farahani RJ, Williams C, Porto NM, Masuda M, Dollarhide E, Bingi S (2018) Tsunami risk for insurance portfolios in Japan. In: Proceedings of the 16th European Conference on Earthquake Engineering, Thessaloniki, Greece, pp 1–12

Woo G (2019) Downward counterfactual search for extreme events. Front in Earth Sci 7:340. https://doi.org/10.3389/feart.2019.00340

Wood NJ, Burton CG, Cutter SL (2010) Community variations in social vulnerability to Cascadia-related tsunamis in the U.S Pacific Northwest. Nat Hazards 52:369–389. https://doi.org/10.1007/s11069-009-9376-1

Wood NJ, Schmidtlein MC (2013) Community variations in population exposure to near-field tsunami hazards as a function of pedestrian travel time to safety. Nat Haz 65:1603–1628. https://doi.org/10.1007/s11069-012-0434-8

Wood N, Peters J, Wilson R, Sherba J, Henry K (2020) Variations in community evacuation potential related to average return periods in probabilistic tsunami hazard analysis. Int J Disaster Risk Red 50:101871. https://doi.org/10.1016/j.ijdrr.2020.101871

Wronna M, Omira R, Baptista MA (2015) Deterministic approach for multiple-source tsunami hazard assessment for Sines. Portugal Nat Hazards Earth Syst Sci 15:2557–2568. https://doi.org/10.5194/nhess-15-2557-2015

Yeh H, Sato S, Tajima Y (2013) The 11 March 2011 East Japan earthquake and tsunami: tsunami effects on coastal infrastructure and buildings. Pure Appl Geophys 170:1019–1031. https://doi.org/10.1007/s00024-012-0489-1

Yeh H, Barbosa A, Ko H, Cawley J (2014) Tsunami loadings on structures: review and analysis. Coastal Eng Proceed 1(4). https://doi.org/10.9753/icce.v34.currents.4

Xu B, Zhai CH, Xu L (2022) Time-dependent risk assessment of a containment building subjected to mainshock-aftershock seismic sequences. J Structural Eng 148:04022010. https://doi.org/10.1061/(ASCE)ST.1943-541X.0003300

Zamora N, Babeyko AY (2020) Probabilistic tsunami hazard assessment for local and regional seismic sources along the pacific coast of Central America with emphasis on the role of selected uncertainties. Pure and Appl Geophys 177:1471–1495. https://doi.org/10.1007/s00024-019-02372-4

Zamora N, Catalán PA, Gubler A, Carvajal M (2021) Microzoning tsunami hazard by combining flow depths and arrival times. Front in Earth Sci 8:591514. https://doi.org/10.3389/feart.2020.591514

Zanetti L, Chiffi D, Petrini L (2023) Philosophical aspects of probabilistic seismic hazard analysis (PSHA): a critical review. Nat Hazards 117:1193–1212. https://doi.org/10.1007/s11069-023-05901-6

Zaniboni F, Pagnoni G, Paparo MA, Gauchery T, Rovere M, Argnani A, Armigliato A, Tinti S (2021) Tsunamis from submarine collapses along the eastern slope of the gela basin (Strait of Sicily). Front Earth Sci 20(8). https://doi.org/10.3389/feart.2020.602171

Zhang YJ, Witter RC, Priest GR (2011) Tsunami–tide interaction in 1964 Prince William Sound tsunami. Ocean Mod 40(3–4):246–259. https://doi.org/10.1016/j.ocemod.2011.09.005

Zöller G (2018) A statistical model for earthquake recurrence based on the assimilation of paleoseismicity, historic seismicity, and instrumental seismicity. J Geophys Res: Solid Earth 123:4906–4921. https://doi.org/10.1029/2017JB015099

Zorn EU, Orynbaikyzy A, Plank S, Babeyko A, Darmawan H, Robbany IF, Walter TR (2022) Identification and ranking of volcanic tsunami hazard sources in Southeast Asia. Nat Haz Earth Syst Sci 22(9):3083–3104. https://doi.org/10.5194/nhess-22-3083-2022

Chapter 3
Source Selection and Their Activity Rate

Finn Løvholt, Roberto Basili, Andrea Rovida, Stefano Lorito, Fabrizio Romano, Mathilde B. Sørensen, Carl B. Harbitz, Jacopo Selva, Raphaël Paris, Anita Grezio, and Cléa Denamiel

3.1 Introduction

This chapter presents several methods to conduct tsunami source selection procedures (we note that modeling of the mechanics of tsunami sources are included in Chap. 4). The chapter is subdivided into four sections comprising the most frequent tsunami sources, namely Earthquake sources, Landslides sources, Volcanic sources, and Atmospheric sources. Volcanic sources may also be linked to landslides and

F. Løvholt (✉)
Offshore Energy, Norwegian Geotechnical Institute, Oslo, Norway
e-mail: finn.lovholt@ngi.no

R. Basili · A. Rovida · S. Lorito · F. Romano
Istituto Nazionale di Geofisica e Vulcanologia, Roma, Italy
e-mail: roberto.basili@ingv.it

A. Rovida
e-mail: andrea.rovida@ingv.it

S. Lorito
e-mail: stefano.lorito@ingv.it

F. Romano
e-mail: fabrizio.romano@ingv.it

M. B. Sørensen
Department of Earth Science, University of Bergen, Bergen, Norway
e-mail: mathilde.sorensen@uib.no

C. B. Harbitz
Natural Hazards, Norwegian Geotechnical Institute, Oslo, Norway
e-mail: carl.bonnevie.harbitz@ngi.no

J. Selva
Dipartimento di Scienze della Terra, dell' Ambiente e delle Risorse (DiSTAR) , Università di Napoli Federico II, Naples, Italy
e-mail: jacopo.selva@unina.it

M. B. Sørensen et al. (eds.), *Probabilistic Tsunami Hazard and Risk Analysis*, Mathematics of Planet Earth 14, https://doi.org/10.1007/978-3-031-98115-9_3

atmospheric pressure disturbances, and in these cases the source description is given only in the landslide or atmospheric related sections. The main ingredients that are identified for these types of sources are listed in Table 3.1. The selection process follows three main steps, of which the general workflow is displayed graphically in Fig. 3.1. These steps include the following operations:

Step 1 consists of the selection of the potential source areas and the selection of the types of sources that may impact the selected target areas. It includes a qualitative analysis of the source areas relevant to the hazard or risk analysis to be conducted.

Step 2 represents the source parameterisation. It consists of defining the source position, density, parameter selection and variability (e.g., in space and size), and is applied only to the source(s) selected in step 1. This can be used to model the variability of the sources.

Step 3 defines a finite list of individual scenarios and quantifies the probability of occurrence or the mean annual rate of occurrence in the reference time window (the exposure time) for each individual source. It is applicable mostly only to probabilistic methods (see Chap. 2).

This Chapter describes the different sources in separate subsections: Earthquake sources (Sect. 3.2), Landslide sources (Sect. 3.3), Volcano sources (Sect. 3.4), and Atmospheric sources (Sect. 3.5). Each subsection presents the different methods available for each of the steps (1, 2, 3) described above.

For each source type, a synthesis of the methods is presented in a table, which provides a summary of the method descriptions, requirements, and uncertainty treatment. Finally, Sect. 3.6 provides very brief remarks about pros and cons of the various methods in order to help users select the appropriate models.

3.2 Earthquake Sources

Earthquakes represent the main cause of tsunamis globally, and are hence the tsunami source analysed the most, with a large body of literature covering seismic PTHA (SPTHA; e.g., Geist and Parsons 2006; Grezio et al. 2017). Much of the techniques and treatment of earthquake sources used in SPTHA originate from Probabilistic Seismic Hazard Analysis (PSHA, Cornell 1968). Over time these have been adapted

R. Paris
CNRS, IRD, OPGC, Laboratoire Magmas et Volcans, Université Clermont Auvergne, Aubière, France
e-mail: raphael.paris@uca.fr

A. Grezio
Istituto Nazionale di Geofisica e Vulcanologia, Bologna, Italy
e-mail: anita.grezio@ingv.it

C. Denamiel
Division of Marine and Environmental Research, Ruđer Bošković Institute, Zagreb, Croatia
e-mail: cdenami@irb.hr

Table 3.1 Source parametrization and rates

What does the ingredient describe?	The process of defining and selecting tsunami sources and quantifying their future probability of occurrence
Definitions of the ingredient	This process typically consists of: The selection of source types, e.g. earthquakes, landslides, volcanoes, atmospheric sources The selection of the source zones and/or of the volcanoes and, for each of them, the type(s) of volcanic source The definition of source parameterization and variability in each of the source zones The quantification of the probability of occurrence and/or of annual rates of the individual sources
What is/are the metric(s) used for measuring/evaluating the ingredient?	(Initial) source position, geometry, kinematics, material parameters, and size of each source Annual rates or probabilities of occurrence of individual sources in the exposure time (for each combination of the source parameters)
Possible uses/applications of the ingredient	Source model for hazard analysis
Possible multi-hazard dimensions/applications?	Yes, by combining source models for several types of hazards (earthquake hazard, landslide hazard, volcano hazard, atmospheric hazards). It is stressed that most tsunamigenic events are also a hazard per se (e.g. landslides, volcanic explosions) or generate other hazards (e.g. seismic ground shaking, volcanic tephra fall), thus adding additional impacts to the tsunami impact Cascading hazards examples: Landslides triggered by earthquakes or volcanic activity. Climate effects—e.g. enhancing probability of slope failure, climate hazard influencing intensity and probability of atmospheric sources
Possible time-dependent/ dynamic dimensions	Earthquake rates (temporal probability of occurrence) can be significantly time-dependent, for example following a large earthquake Landslide slope failure linked to precipitation and seasonal variation (winter, summer) Life-cycles and states of volcanoes (quiescence, unrest, eruptions phases) Atmospheric sources are fully dynamic and time-dependent (i.e., continuous variations in time and space)

(continued)

Table 3.1 (continued)

List of the available methods to be described below	**Earthquakes:** Seismicity Rates and mechanisms based on past seismicity Seismicity Rates and mechanisms based on tectonics Logic trees and ensemble modelling, non-structured expert judgment or structured expert elicitation methods, to quantify epistemic uncertainty Scenarios compatible with historical events or plausible due to different tectonic settings **Landslides:** Methods based on past landslide data—basing the hazard on analysis of past events. Depending on the data availability, they can be both local and regional Methods based on slope stability models—they are often more heuristic, although they can be mixed with past landslide frequency information, based on expert judgment, and therefore often most appropriate for local hazard analysis **Volcanoes:** Selected scenarios compatible with historical events or plausible due to tectonic and volcanic settings For eruptive tsunamis, all methods for eruption forecasting and volcanic hazards, including event trees, belief networks, expert elicitations, Markov chains, stratified sampling, based on past data and/or on monitoring For non-eruptive tsunamis, there are no consolidated methods; methods strongly vary on the source type and its triggering conditions To quantify epistemic uncertainty, Bayesian methods, logic trees, ensemble modelling, and different flavours of expert elicitation can be used **Atmospheric:** Methods based on the type of atmospheric sources: weather or acoustic Methods based on the spatial scale: local or global Methods based on the type of study: forecast, climate or hazard assessment

for use in SPTHA, for example the generation of empirical scaling relations that focus on subduction zones, or the adaptation of stochastic slip models for use on complex subduction zone geometries. Regardless of the specific approach and techniques used, the first step for an earthquake tsunami hazard or risk assessment is generally to set up a (probabilistic) earthquake model.

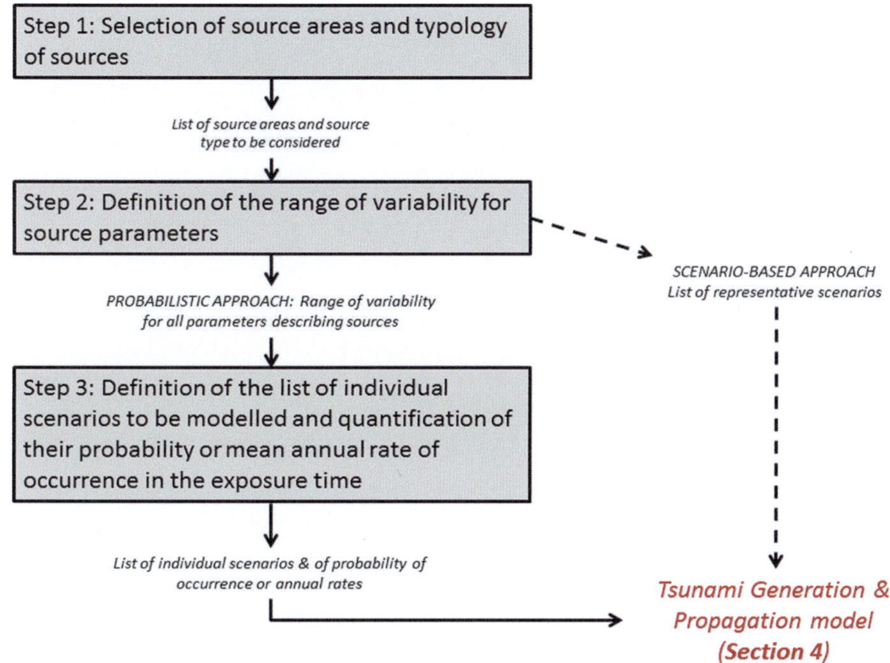

Fig. 3.1 Workflow for the source selection, parameterisation, and rate quantification

3.2.1 Short Description of the Earthquake Source Selection Method

Earthquakes occur due to the reactivation of pre-existing faults (Sibson 1985). By knowing faults, one can predict the locations, mechanisms, and rates of future earthquake ruptures. At the global scale, most earthquakes occur along plate boundaries (Fig. 3.2). On the other hand, oceanic and continental plate interiors account for less than 4% of all global seismicity (Kagan et al. 2010). According to the plate boundary model by Bird (2003), plate boundaries comprise 278,000 km in length, 77% (213,000 km) of which is offshore.

There are three main types of plate boundaries: (i) transform, where two plates slide past each other; (ii) divergent, where plates move away from each other; and (iii) convergent, where plates move towards each other. In the case where one plate sinks below the other, it is referred to as a subduction. Subduction zones cover more than 60,000 km of the total convergent plate boundary length (Bird 2003; Berryman et al. 2015; Hayes et al. 2018; Hasterok et al. 2022).

Earthquake activity rates can be expressed with a magnitude-frequency distribution (MFD). Two methods exist to estimate MFDs: one assumes a constant seismicity rate, and the other constant seismic moment rate (Youngs and Coppersmith 1985).

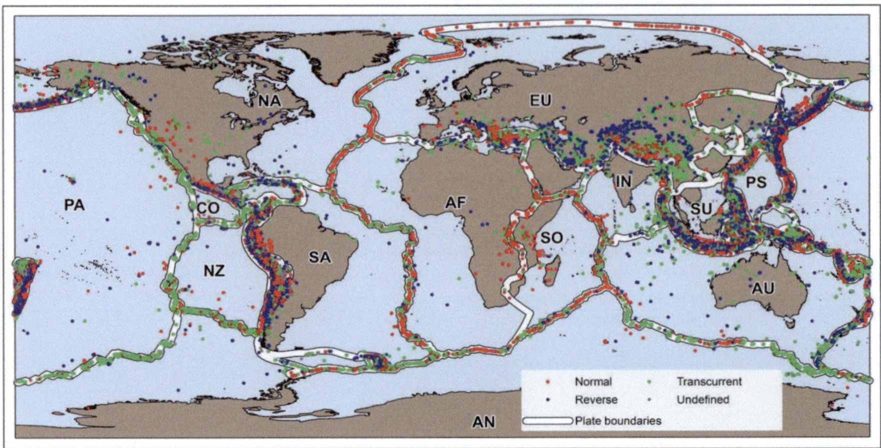

Fig. 3.2 Map of tectonic plate boundaries (data from Bird 2003) overlain by earthquake epicentres from the global centroid moment tensors (data from Ekström et al. 2012), color-coded based on the faulting mechanisms

The first requires earthquake catalogues and fault association strategies; the second needs geologic/geodetic fault slip rates (Fig. 3.3).

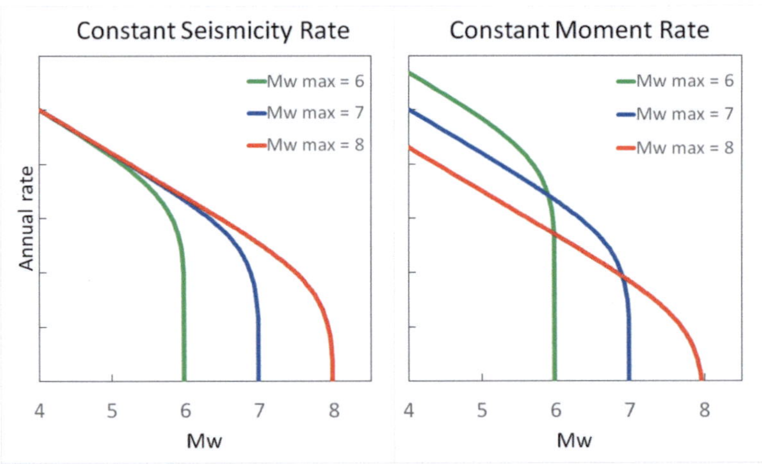

Fig. 3.3 Schematic showing the effect of varying the upper bound magnitude on the magnitude-frequency distribution (i.e. the annual number of earthquakes above a particular magnitude, in this example, magnitude 4) of two alternative approaches: (left) constant seismicity rate, if the number of earthquakes in the historical seismicity record is assumed to be a constraint on the activity rate, and (right) constant seismic moment rate, if the fault slip rate is the principle constraint on the activity rate. (Redrawn from Youngs and Coppersmith 1985). Notice that the tick marks on the Y axis are in base-10 logarithmic scale

For earthquake sources, the main steps in tsunami source characterization follow the three basic steps outlined in Fig. 3.1. Step 1 is related to defining potential earthquake sources, either based on faults, earthquake catalogues or other sources. Step 2 links these sources to specific potential earthquake scenarios, either as fault sources, area sources, or gridded sources. Step 3 defines the occurrence probabilities of the potential scenarios. Available methods for each of those steps will be described in the following sections.

3.2.2 *Step 1: Basis for Source Selection*

Fault sources

Subduction zones are usually considered the main tsunami hazard drivers, and are often the only seismic source type considered in SPTHA, especially at large-scale (González et al. 2009; Davies et al. 2018). SPTHA at the regional scale tends to also consider crustal intraplate and other plate-boundary fault sources (Horspool et al. 2014; Basili et al. 2021a, b).

Subduction zones (Fig. 3.4) include structures that generate different earthquake types (Satake and Tanioka 1999). The slab interface, or megathrust, where the two plates are in contact, generates the largest and thus most relevant earthquakes for SPTHA, but also ruptures originating on the megathrust and propagating in the wedge (Clift and Vannucchi 2004) on splay faults branching upward from the interface may be important (van Zelst et al. 2022). Intraslab earthquakes are of more limited interest to SPTHA, except for relatively shallow earthquakes (Yeh et al. 1995), or outer-rise earthquakes generated where the subducting plate is bent upward (Chapple and Forsyth 1979).

Crustal faults occur on all types of plate boundaries. Although plate boundaries are prominently elongated features, the deforming area around them can be as wide as a few hundred kilometres, as in the case of back-arc basins or collision zones. These deforming areas host the majority of active crustal faults, but not all of them can be defined in sufficient detail for SPTHA (Basili et al. 2013). Additionally, crustal faulting can occur far away from plate boundaries. Depending on the level of knowledge of faults, hazard modelers may resort to different strategies for source parameterization (see Step 2).

Earthquake catalogues

Earthquake catalogues provide crucial information on earthquake sources. They commonly include the occurrence time, hypocentre location, and size of past events. For the size, the moment magnitude (M_w) scale is commonly used for seismic and tsunami hazard analyses due to its direct link to the size of the earthquake rupture. Instrumental recordings from global seismometer networks have been used to systematically assess earthquake parameters since the 1960s (e.g., International Seismological Centre 2023; Storchack et al. 2017; Storchack et al. 2020; Ekström

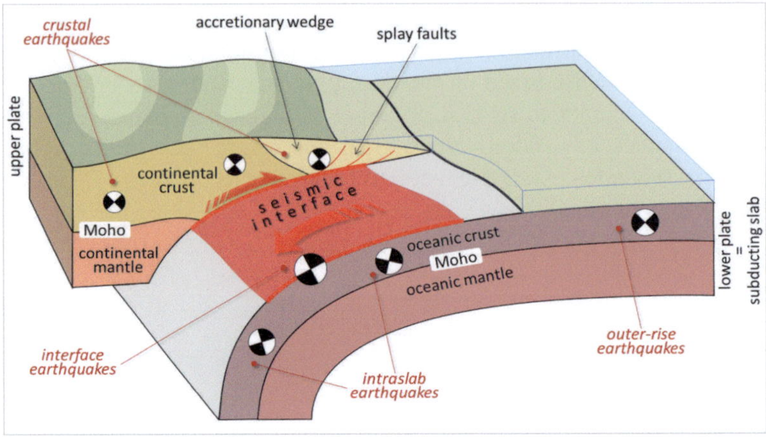

Fig. 3.4 Schematic block model of a subduction zone showing typical locations of different earthquake types and their fault sources

et al. 2012). However, instrumental catalogues have limited temporal coverage, so pre-instrumental earthquake data from historical documents are used to extend earthquake catalogues back in time. A quantitative estimate of the ground shaking at a site is obtained by comparing the effects of an earthquake on humans, buildings, and the environment with typical earthquake effects described in macroseismic scales (see overviews by Musson et al. 2010; Musson and Cecic 2012). Spatial distributions of macroseismic intensity data are then used to estimate earthquake locations and sizes (e.g., Musson and Cecic 2012; Bakun and Wentworth 1997; Gasperini et al. 1999, 2010; Provost and Scotti 2020). Harmonized earthquake catalogues combining instrumental and macroseismic data are common in Europe and worldwide (e.g., Fäh et al. 2011; Grünthal and Wahlström 2012; Manchuel et al. 2018; Rovida et al. 2020; Felzer and Cao 2008; Beauval et al. 2013; Zare et al. 2014; Mikhailova et al. 2015; Brax et al. 2019; Sawires et al. 2019). For SPTHA, large-scale catalogues can be compiled by merging existing regional catalogues and harmonizing magnitudes (Grünthal and Wahlström 2012; Albini et al. 2013, 2014).

3.2.2.1 Other Data Sources

Robust seismicity forecasts for hazard analysis should rely on multiple data sources in addition to fault locations and earthquake catalogues (Gerstenberger et al. 2020; Grezio et al. 2017; Behrens et al. 2021). Including faults and their slip rates could help reduce the uncertainty at long average return periods. Palaeoseismological (hardly available for offshore faults) and paleotsunami observations can help extend the analysis in the past to improve the occurrence rate forecast of future earthquakes and large tsunamis. Geodetic strain rates and geodetically constrained coupling, even if covering only tens of years, with stationarity assumptions, can introduce further constraints (Wang et al. 2012).

3.2.3 *Step 2: Source Parameterisation*

Fault sources

Fault-source modelers use 3D surfaces to represent subduction interfaces or crustal faults and simplify fault surfaces by reducing them to a down-dip planar shape or oriented rectangle. Stochastic distributions of fault positions may be used in cases with limited fault knowledge. Although 3D fault modelling is desirable (Tonini et al. 2020; Serra et al. 2021; Gómez de la Peña et al. 2022), rectangular faults are often used due to computational requirements. The spatial heterogeneity of fault dislocation is modelled by distributing different slip vectors onto the fault surface (González et al. 2009; Davies and Griffin 2018, 2020; Basili et al. 2021a, b). Dutta et al. (2021) showed that models of earthquake ruptures using planar faults can be affected by substantial bias and that simultaneous inversion of fault geometry and slip can solve the problem (LeVeque et al. 2016). Hence, the combination of fault geometry and slip distribution also affects the reconstruction of past earthquake ruptures.

Area sources

Area sources (or seismogenic zones) are 2D polygons with vertical boundaries extending to the bottom of the seismogenic layer. Area sources are regions with spatially uniform seismicity in terms of earthquake occurrence rates, magnitude distribution, maximum magnitude, expected rupture mechanism, and other relevant parameters. Despite criticism of the reliability of the underlying assumptions (Woo 1994), area sources have been conveniently used in PSHA to complement the limited knowledge about seismogenic faults often encountered even in tectonically active regions. The concept of area sources has also been adopted in PTHA (Sørensen et al. 2012; Salazar-Monroy et al. 2021).

Gridded sources

Gridded sources are another strategy to address the possible lack of knowledge of individual seismogenic faults. Differently from area sources, gridded sources are a geometrically regular subdivision of the seismic source domain (squares or rectangles) used to accommodate irregular distributions of earthquake occurrence rates, maximum magnitude, and rupture mechanisms (Necmioglu and Özel 2015; Basili et al. 2021a, b).

3.2.4 Step 3: Probabilistic Treatment

General concepts

In earthquake hazard analyses, seismicity is often assumed to follow a stationary Poisson process, with earthquake activity rates defined by a MFD. The exponential Gutenberg-Richter (G-R) model (Gutenberg and Richter 1944, 1949) is the most common MFD, but other shapes are used such as the maximum magnitude and characteristic earthquake models (Bommer and Stafford 2009). The G-R model gives the annual number of earthquakes of a given magnitude as a function of two parameters, the a-value describing the overall level of seismicity and the b-value describing the relation between small and large magnitudes. The Poisson model assumes constant earthquake occurrence rates and does not account for earthquakes triggering other earthquakes. Time dependent models like the ETAS (Ogata 1998) and Brownian Passage Time (BPT; Ellsworth et al. 1999; Murru et al. 2016) have been developed to address these limitations. Renewal models, such as the BPT, have also been used in SPTHA (Goda 2020).

There is significant uncertainty associated with earthquake sources, their activity rates, and the probabilistic framework, which can be addressed using logic trees, Bayesian inference, or expert elicitation (Geist and Parsons 2006; Annaka et al. 2007; Grezio et al. 2010; Selva et al. 2010; Neri et al. 2008; Selva et al. 2012a, b; Basili et al. 2021a, b).

Seismicity-based earthquake rate forecasting

MFDs related to specific areas can be derived from earthquake catalogues, which contain records of past earthquakes. A complete catalogue, containing all the earthquakes above a given magnitude, is needed for accurate estimates. Observation gaps may exist due to a seismic network's limitations or historical factors, particularly for smaller events or earthquakes at sea. Figure 3.5 shows that global earthquake catalogues contain significantly fewer earthquakes when one goes back in time, demonstrating that also large historical events of interest for SPTHA might be missing from global and regional catalogues (e.g., Kagan 2003; Hough 2013; Albini et al. 2013, 2014). It is thus important to account for catalogue completeness when defining seismicity rates.

Statistical methods to determine catalogue completeness often assume a stationary seismogenic process (Stepp 1972; Grünthal et al. 1998; Mulargia and Tinti 1985; Rotondi et al. 1994; Albarello et al. 2001; Rotondi and Garavaglia 2002; Wössner and Wiemer 2005; Hakimhashemi and Grünthal 2012). Statistical methods are not suitable for estimating the completeness time of the rarest and largest earthquakes. In these cases, historical methods assess completeness by evaluating the likelihood of earthquakes of a given impact at a locality being unreported by historical sources (Ambraseys and Melville 1982; Stucchi et al. 2004; Albini and Rovida 2018; Albini 2020).

To calculate earthquake probabilities, events below the completeness magnitude and older than the completeness time are excluded. Traditional seismicity-based

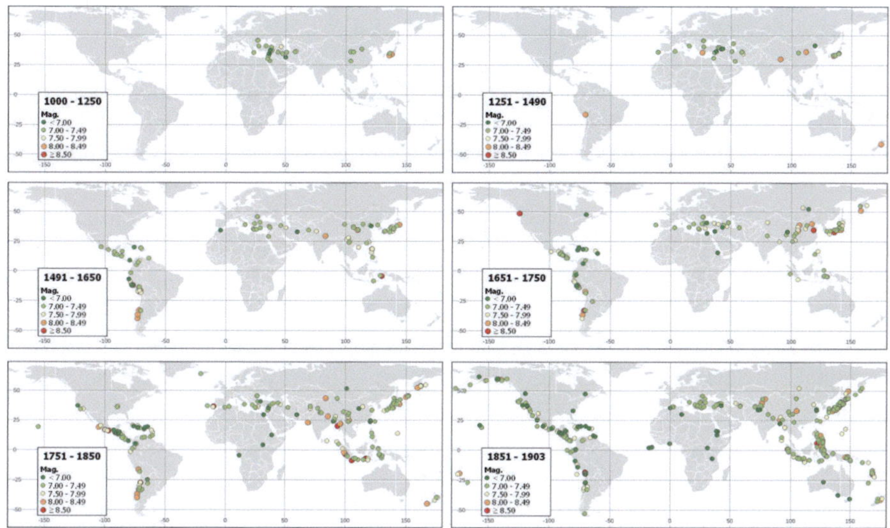

Fig. 3.5 Global distribution of M7 + earthquakes according to the the Global Historical Earthquake Catalogue—GHEC v1.0 (Albini et al. 2013) in different time periods (between 1000 and 1903)

earthquake rate forecasting for PSHA (e.g., Cornell 1968) also requires the removal of dependent events (such as foreshocks, aftershocks and swarm events) to comply with the Poisson process assumption. Various methods exist for identifying dependent events based on time and space windows (Gardner and Knopoff 1974; Reasenberg 1985).

The upper bound (or maximum) magnitude of the MFD can be estimated using statistical methods. The classical G-R distribution assumes an infinite upper bound, with the maximum magnitude obtained by extrapolating the MFD to larger magnitudes until the frequency becomes negligible for a practical application. However, the finiteness of fault systems and the energy available for generating earthquakes requires modifications of the MFD at larger magnitudes. The truncated or tapered MFDs are common modifications of the classical G-R distribution (e.g., Kagan 2002a). In some practical applications, this upper limit has been estimated from the magnitude of the largest observed earthquake in the region under study plus a conservative factor or on a statistical basis with confidence intervals (e.g., Kagan 2002a; Bird and Kagan 2004; Kagan et al. 2010). However, the upper limit is hardly testable (Holschneider et al. 2014). It is important to note that the choice of the maximum magnitude and the functional form of the truncation or tapering of the distribution can affect the hazard estimation.

Seismic-moment-based earthquake rate forecasting

This method relies on the seismic moment conservation principle by equating the geologic or geodetic moment rate to the seismic moment rate, such that

$$\dot{m}_s = \chi \dot{m}_g = \chi \mu L W \dot{S}$$

where $\dot{m}_s$ is seismic moment rate, $\dot{m}_g$ is geologic/geodetic moment rate, χ is seismic efficiency or long-term coupling, μ is rigidity, L and W are fault length and width, respectively, and $\dot{S}$ is slip rate for crustal faults or convergence rate for subduction interfaces.

Different methods use various constraints to convert moment rate estimates into a convenient MFD. The first constraint is the MFD upper bound, which can be obtained in this case also by calculating the magnitude corresponding to the largest earthquake rupture that the fault source can host based on empirical scaling relations, or with methods described in the previous section. The conversion between earthquake magnitude and the seismic moment is readily obtained by the classic relation

$$Mw = \log_{10}(M_0)/1.5 - 6.07$$

where M_w is the magnitude and M_0 is the seismic moment in Nm (Kanamori and Brodsky 2004).

For various MFD forms, one must opt for the decay of the large-size end of the distribution (Utsu 1999) between the truncated and the tapered models (Anderson and Luco 1983; Kagan 2002b; Bird and Kagan 2004).

In most MFDs, the b-value controls the power-law part for small and moderate earthquakes, and the universal estimation of b ~ 0.95 can be used if seismicity data is unavailable (Kagan 1999, 2002a; Bird and Kagan 2004; Kagan et al. 2010). Different formulations can be used to calculate the a-value (Anderson and Luco 1983; Kagan 2002b). The characteristic earthquake model requires an additional parameter that determines the upper end of the exponential part of the distribution (Youngs and Coppersmith 1985). The MFD is transformed into the non-cumulative form and discretized into magnitude bins. Different MFD forms can affect the annual rates of moderate-to-large earthquakes, as shown in Fig. 3.6.

Individual earthquake ruptures

The final step in the source characterization is to define individual earthquake ruptures (with an associated occurrence probability) that can form the basis of tsunami modelling. Each earthquake rupture must be defined with location, fault dimensions and average displacement. Individual earthquakes can be defined by modelling all possible events on a fault, or within an area or a volume, and partitioning among them the total earthquake rate. Examples include systematic sampling of the earthquake parameter space (e.g., Horspool et al. 2014; Davies et al. 2018; Basili et al. 2021a, b), and Monte Carlo methods to create synthetic source catalogues spanning a given amount of time (Sørensen et al. 2012), or even a combination of the two as in a recently proposed stratified importance sampling approach (Davies et al. 2022).

The generalized functional form between the most commonly measured rupture dimensions (length, width, area, and average displacement) and moment magnitude (M_w) is

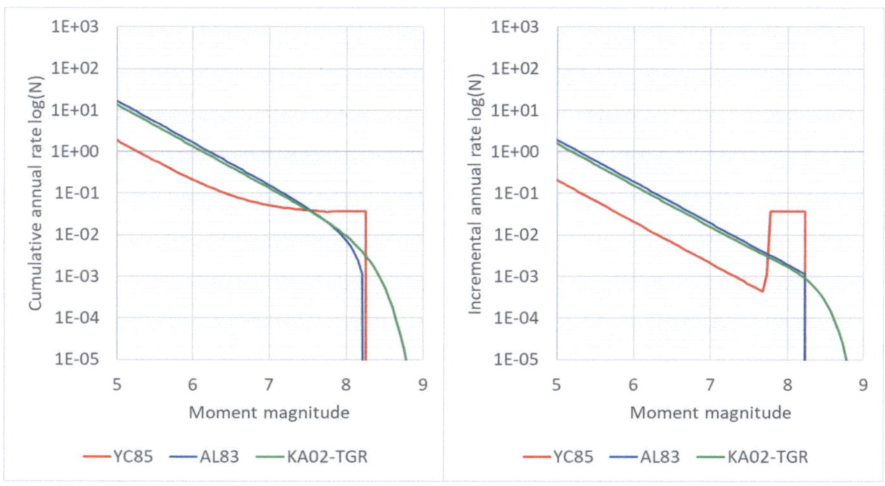

Fig. 3.6 Diagrams showing the effect on the MFD of adopting different typical forms (left panel is cumulative and right panel is incremental) based on the seismic-moment balancing. YC85 (red) is the characteristic earthquake model (Youngs and Coppersmith 1985), AL83 (blue) is the exponential truncated model (Anderson and Luco 1983), and KA02-TGR (green) is the exponential tapered model (Kagan 2002b). The three models balance the same moment rate ($\dot{M}_s = 5.6E + 19\,Nm$) and use the same maximum magnitude in the truncated case and corner magnitude in the tapered case ($M_w = 8 \pm 0.25$) and slope of the power law part ($b = 1$)

$$log_{10}(X) = a + bMw$$

where X is the rupture dimension under consideration, and a and b are coefficients empirically derived from observed ruptures. Such scaling relations for subduction interface, or megathrust, earthquakes significantly differ from those for crustal fault earthquakes (Fig. 3.7).

Comparing several different scaling relations, Skarlatoudis et al. (2016) found that the rupture areas of subduction interface earthquakes are about 1.7 times larger, and their average slip is about 0.4 times smaller than those of crustal earthquakes with the same seismic moment. Different scaling also exists for crustal fault earthquake ruptures of different slip types (Blaser et al. 2010). This highlights the importance of using appropriate scaling relations when modelling individual ruptures, as they can result in different seafloor deformation.

To predict future rupture mechanisms, geologic and geodetic information, or statistics of past earthquakes, can be used. Crustal fault sources are typically reported with a rake value associated with fault geometry (strike and dip) derived from geologic information. For subduction interfaces, the rake is often conservatively assumed to be 90° (pure reverse mechanism), or derived from the local mean direction of plate convergence (relative motion of the upper and lower plates). In areas

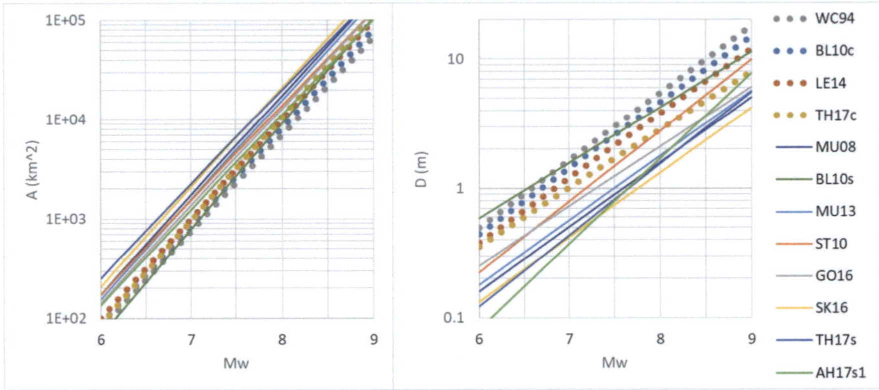

Fig. 3.7 Comparison between subduction interface (solid lines) and reverse crustal fault (dotted lines) earthquake rupture scaling relations between rupture area (A) and earthquake moment magnitude (left panel) and mean slip (D) and earthquake moment magnitude (right panel). WC94 (Wells and Coppersmith 1994), BL10 (Blaser et al. 2010), LE14 (Leonard 2014), TH17 (Thingbaijam et al. 2017), MU08 (Murotani et al. 2008), MU13 (Murotani et al. 2013), ST10 (Strasser et al. 2010), GO16 (Goda et al. 2016), SK16 (Skarlatoudis et al. 2016), AH17 (Allen and Hayes 2017). Only BL10 for subduction interface earthquakes falls within the range of crustal fault relations, though with a different slope. Notice that the average displacement for WC94, BL10, and ST10 was derived from the seismic moment

without known faults, the faulting mechanism can be derived from focal mechanism statistics (e.g. Taroni and Selva 2021), which includes all angles but cannot distinguish between the two auxiliary nodal planes.

A few studies treat slip probability as non-uniform on the fault plane (Davies 2019; Scala et al. 2020), such as in subduction interfaces where depth-dependent features of rigidity and shallow slip amplification effects enhance tsunami hazard (Murphy et al. 2016). Most SPTHA studies model slip distribution as a stochastic process (see Chapter 4), whereas they rarely include additional complexity such as laterally varying seismic coupling in the seismic model (Behrens et al. 2021).

3.2.5 Summary Table

See Table 3.2.

Table 3.2 Summary table: earthquake source selection

Short description of method	
Theoretical/mechanical background for the method(s)	• Earthquakes occur on faults most of which are along active plate boundaries but also within the plate interiors • The earthquake MFD has an exponentially decreasing part controlled by the rate of relatively small earthquakes and a more undefined tail controlled by the behaviour of large earthquakes • The characteristics of individual earthquakes depend on the local structure, pre-stress, and frictional states • Not all the faults can be known, so earthquakes may be initially assumed to occur everywhere; yet large earthquakes are more frequently observed on principal faults and subduction zones
Short outline of how the method(s) works	• The first step of the definition of the source model for a SPTHA is to select both local and distant sources that are potentially relevant to the target area, based on the basin shape, known faults, past seismicity and tsunami • In SPTHA two typologies of faults are considered, that is subducting plate interface and crustal faults • Then, for each typology, different approaches can be used for the parameterisation of the sources (fault, area, gridded sources) • Different approaches can also be used to parameterise the MFD of the earthquakes (seismicity- or seismic moment-based)
How is/are the output(s) of the method(s) represented?	List of potential future earthquake ruptures within a chosen range of magnitude or seismic moment, including their geographic location and depth, orientation (strike and dip), sense of movement (rake), size (length and width, or area and aspect ratio, and slip), frequency (e.g., yearly number of events), and probability. The granularity of these parameters depends on the discretization type and level of each rupture
In what way do(es) the method(s) take into account time-dependent/dynamic aspects?	A time-dependent model is sometimes considered based on recent earthquakes at a specific location
Do(es) the method(s) use parallelisation, machine learning, surrogate models, advanced simulation schemes, other computational reduction strategies (e.g. cluster analysis, etc.)?	We are at the dawn of using AI for preparing input data, for example regarding the construction of earthquake catalogues

(continued)

Table 3.2 (continued)

What are the pros and cons and peculiarities of this specific method(s)? Are there known shortcomings / benefits of the method(s)? In what cases is this method(s) a good/poor choice	• The method overcomes to some extent the scarcity of tsunami data by recurring to tsunamigenic seismic source analysis and numerical tsunami modelling • Nonetheless, the tails of the magnitude-frequency earthquake distributions remain largely unconstrained, as well as the frequency of specific slip distributions at a given locality (e.g. shallow slip probability/tsunami earthquakes) • The stationarity of the seismic process is a convenient working hypothesis in a regime of large uncertainty regarding its time-dependent features; yet it would be in principle better to constrain the modelled future occurrence with the past local seismicity • The seismicity-based and the seismic moment-based models are not necessarily consistent with one another
Data and technical requirements for method	
What input data does the method(s) use?	The source selection uses as input data • Earthquake catalogues including historical events • Geometry of the subduction plate interfaces • Crustal fault sources • Area sources • Gridded sources The source parameterization uses as input data • Rectangular geometry for crustal sources whose sizes depend on the earthquake moment magnitude and empirical scaling relations • Faulting mechanism and slip vector • Geometry of the subduction plate interface opportunely meshed in a discrete number of elementary sub-faults
What are the technical requirements?	• The earthquake catalogues have to be complete and declustered to not contain foreshocks and aftershocks if assuming that the seismicity follows a Poisson process • The distribution of the seismicity in the area sources has to be uniform • The regions containing the seismic events in the gridded sources have to be regular (square, rectangular)
Are HPC computing resources necessary, would absence of these compromise the use?	No
Is the method(s)/workflow(s) available as a git/software/jupyter notebook?	No

(continued)

Table 3.2 (continued)

Treatment of uncertainties for the method	
Is the method(s) itself deterministic or uncertainty based?	Generally (but not always) the fault/source characterization is deterministic while the characterization of its behaviour (mechanism, recurrence) is uncertainty based
Which are potential sources of uncertainties? Which parameters are used to characterize them?	Unknown or never-observed earthquake sources, variability in the realization of earthquakes from known sources
How are uncertainties quantified for the identified parameters? Are there significant correlations between different uncertain parameters?	The uncertainties are characterized based on consistency with observations of past events and geology/seismotectonics, as well as based on the assumed theoretical framework (e.g. functional form of the MFD relation, depth-dependency of the slip occurrence)
Is the method(s) able to digest and propagate uncertainty in input (previous ingredient) and propagate it to the output (next ingredients)? Does the assessment of this ingredient require uncertainty propagation? If yes, how are uncertainties propagated?	Yes, in principle uncertainty in input can be propagated. For example, when assigning different weights to alternative models
What are the consequences of lack of data in input or for boundary conditions, or for characterizing certain parameters?	Potentially missing important sources of hazard, with consequent underestimation of the total hazard, or large uncertainty stemming from unconstrained parameters
What are the consequences of ignoring or poorly treating uncertainties?	Potentially missing important sources of hazard, with consequent underestimation of the total hazard, or a sense of false precision if the natural variability is not explored sufficiently

3.3 Landslide Sources

3.3.1 Short Description of Method

Identification and demarcation of landslide sources in tsunami hazard analysis involve large uncertainty, both with respect to volumes and rates. Landslide source treatment depends on the expected type of landslide from subaerial slides, volcano sector or flank collapses, slump-like submarine landslides, or long run-out giant submarine slides (Fig. 3.8). As landslides are not tied to tectonic structures as earthquakes and volcanoes are, they can in principle happen everywhere where there is enough material to be released on a slope. Another reason for the large uncertainty is that acquiring all data needed for a broad landslide source probability identification is often not feasible. This uncertainty in forecasting landslide volume probability is a direct reason for the Landslide Probabilistic Tsunami Hazard Assessment (LPTHA) being relatively sparsely developed and used, and that consequently past hazard analyses have still often turned to the credible worst case scenario method (e.g. Tinti et al. 2008; Harbitz et al. 2014b; Grilli et al. 2017). Both the amount of data and the understanding of the local or regional slope stability conditions for large landslides are imperative for the source selection methodology. In order to place landslide sources into a probabilistic framework, several simplifying steps are necessary for computational feasibility. Only a few examples of LPTHA exist to date (e.g. Lane et al. 2016; Løvholt et al. 2020). In these examples, the hazard based on averaged rates is aggregated, but the hazard uncertainty is not quantified as for the more sophisticated hazard aggregation procedures used for SPTHA (e.g. Selva et al. 2016).

Landslide source parameterisation for PTHA follows three steps, where Step 1 mostly concerns how the available data can be used to set up a synthetic scenario database, while Steps 2 and 3 concern building the database of scenarios setting source parameters and rates, respectively. Because the first step is mostly a qualitative assessment where the strategy for selecting sources is chosen, the feasibility and strategies for Steps 2 and 3 will be assessed already in this first step.

Broader discussions encompassing the uncertainty and maturity of landslide tsunami hazard analysis methods are found in Behrens et al. (2021), Geist and Lynett (2014a), Harbitz et al. (2014a), and Løvholt et al. (2020). There is presently no single streamlined and complete LPTHA method. While various building blocks exist, all

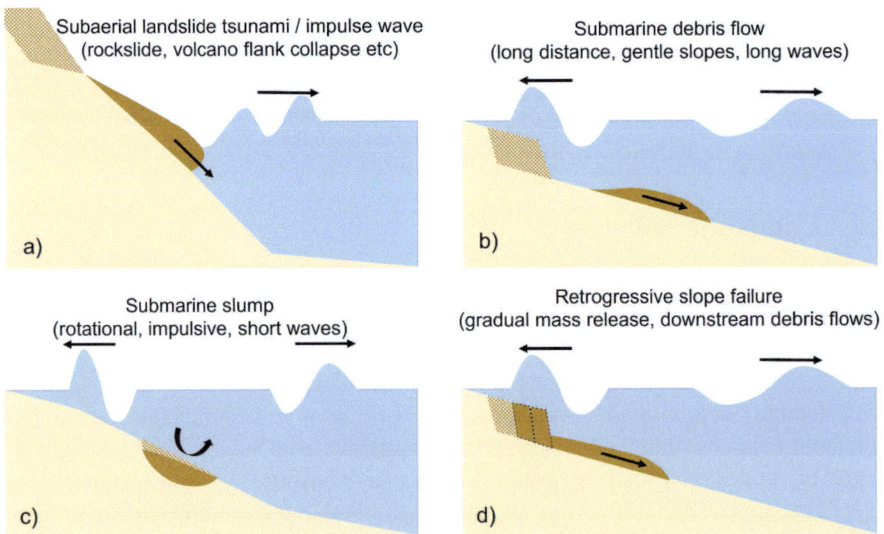

Fig. 3.8 Different types of landslides that may generate different types of waves. **a** Subaerial landslides, including rockslides and volcano flank collapses, that generate impulsive short-crested waves. **b** Submarine debris flows, often involving large volumes that can reach long run-out distances on gentle slopes. **c** Rotational slumps, short run-out and duration, can generate higher and shorter waves than debris flows. **d** Retrogressive landslides, where the failure process takes place in a staged fashion, implying that the landslide volume is not mobilized in the same instance

these blocks have not been tested and combined into a single and complete LPTHA. For the family of methods described here, it is assumed that the sources are meant to be used for simulation-based PTHA, where many different synthetic sources serve as input to numerical tsunami simulations that again provide the aggregated probabilities (frameworks FW1A and FW1B as described in Chap. 2). In cases where sufficient data (of either slope stability evaluation or past relevant landslide data) do not exists, establishing temporal source probabilities may not be feasible. The rigor of the analysis may then be reduced by removing the temporal part of the hazard analysis, and further by reducing the number of scenarios spanning the epistemic uncertainties related to e.g. modelling landslide dynamics. Hence, the same methods may also be used in scenario-based methods not considering probabilities.

3.3.2 Step 1: Basis for Source Selection and Area Demarcation of Sources

LPTHA and related landslide source selection mainly follow two alternative pathways: (i) determining the hazard and sources based on occurrence of past events, and (ii) determining the sources based on slope stability analyses. For method (ii), the slope stability methods may also be mixed with information from past events. Below, these two pathways are discussed in brief.

Landslide sources and rates based on past events

Similar to the approach for earthquakes, statistics of past landslide events can be used to establish source rates. Such methods have the benefit that they are independent of the local trigger. First, an area covering a source region must be defined. This can either be a confined region with homogeneous deposits (e.g. Cook Strait New Zealand, Lane et al. 2016), or broader regions that cover a wider range of source types (e.g. Mediterranean, Urgeles and Camerlenghi 2013). The simplest possible way of establishing source probabilities is to assume that the rates of past events can be used to represent present-day source probabilities.

For this type of source treatment, a comprehensive record of past events is necessary, including assessment of the completeness of the record (e.g. Geist and Parsons 2014b). Acquiring such landslide data requires many years of expensive offshore operations that are normally results of investigations for other purposes, e.g. oil and gas field exploration or seabed mapping carried out by governmental organizations. As such data are available only in a few places (e.g. Mediterranean Sea, USA, New Zealand, Norway), the method based on past data is not applicable for all regions worldwide. MFD curves for landslide sources relate the landslide volumes with their frequency of occurrence. Dating information can be used to bring these MFD curves from relative rates to annual source rates. Alternatively, time periods encompassing geological age ranges can be used (e.g., if the past landslides cannot be dated but interpreted to a given geological time such as the Pleistocene). On the other hand, this would typically imply a broader uncertainty range due to the date being more imprecise. Sedimentation rates and thickness of deposits can often constrain uncertainties to some degree, and most often only the geologically youngest deposits most representative for today's situation (e.g. late Pleistocene, ice age) are used. Available age dating may have influence on the confidence in rate estimation (e.g. Geist et al. 2013).

Landslide sources and rates based on slope stability evaluations

Geotechnical or engineering geology-based source estimation implies knowledge of the triggers and geological structures (either soil thickness distribution or fracture compositions) and may sometimes be tied to assessments of more semi-probabilistic or heuristic nature as the rate in this case cannot be tied to statistical landslide data. As for the statistical source model explained above, slope stability methods require elaborate data, but the investigations are on a more local level. Slope stability evaluation constitutes a large discipline on its own, and we will hence limit the discussion only to methods that have already been used for geohazard evaluations related to landslide tsunamis. These methods are divided into engineering geology methods used for unstable rock slopes, and geotechnical methods for soils. To an extent, they are also validated based on regional landslide statistics and probability parameters are therefore not always transferable from one geographical location to another. These two different branches are described individually below.

Source determination for unstable rock slopes

We here follow the Norwegian risk categorisation method for rock slopes (Hermanns et al. 2012, for a full description see NGU 2012), applicable both for risk related to direct landslide run-out and risk related to secondary impacts such as tsunamis (outputs of the Hermanns et al. 2012 method are used in Løvholt et al. 2020). The classification scheme is related to single rock slopes and translates different observables of the slope to an index-based hazard classification scheme ("very low" to "very high" hazard). The rockslide hazard indexes are in turn translated to annual probabilities as the risk scheme is linked to the legislation (while in reality they are based on a structured expert judgment scheme, which implies that the rate estimation is based on expert judgement). The different indexes are partly tuned to comply with regional statistics of past landslides when a large body of unstable slopes is considered. The following unstable slope parameters are taken into account in the complex scheme that is briefly schematised in the following: (i) development of the back scarp of the slope (degree of opening of the back scarp fracture), (ii) identification and dip angle of potential failure surfaces/initial glide planes, (iii) degree of motion on the slope that has already taken place, (iv) kinematic slope failure feasibility, which takes into account the different faults' orientation and potential of failure, (v) expression/characteristics of basal failure surface (morphologic expression), (vi) rate of present-day displacement, (vii) change in rate of displacement (acceleration), (viii) change of rate in rock fall activity, (ix) presence of other post-glacial (geologically recent) events.

Applying this scheme necessitates close and detailed monitoring of the rock slope, including local field surveys as well as long-term monitoring of displacements and rock falls. It is also beneficial to carry out a geophysical mapping of the substructure. Consequently, the scheme is mainly applicable to a single or a few slopes encompassing a limited sized region, and hence for source characterisation for local LPTHA only. In cases where such information is not available, it might be necessary to carry out the hazard analysis without addressing annual probabilities (i.e. only relative probabilities). This classification scheme estimates both the average return period and the uncertainty of a potential slope failure of the entire volume but does not provide an MFD curve as the assessment for each slope only concerns a single-source volume. On the other hand, information from several slopes can be combined to build a set of synthetic sources (Løvholt et al. 2020).

Source determination for soils and submarine landslides

We here refer to the method outlined by Carlton et al. (2019), also described by Løvholt et al. (2022). The starting point for this method is based on the infinite slope stability analysis, which provides only point values for the static Factor of Safety (FoS), or seismically induced failures linked to the Probabilistic Seismic Hazard Analysis (PSHA). In the former case, the spatial distribution of FoS forms the background for setting up maps for setting both slope areas and thicknesses in a non-probabilistic manner. In the absence of geophysical data, the pure infinite slope FoS can hence be used to make a screening of source areas and volumes prone to landsliding and provide input to non-probabilistic scenario-based methods. Yet, this might often lead to conservative or biased probability estimates as uncertainties related to volumes are large and return periods are unconstrained. On the other hand, if combined with geotechnical and geophysical information, PSHA outputs allow for earthquake-related strain-rate maps to be linked to probability of failure (in the case of earthquake triggered landslides). This implies that the possibility of estimating source probabilities should be considered already in this first step. The method can be supplemented with local slope stability analyses for increased sophistication. This method requires extensive input data including geophysical characterisation of the substructure, typically combined with geotechnical Cone Penetration Tests (CPT) or laboratory tests to quantify the geo-technical strength properties used to establish input parameters to the landslide model in LPTHA. As these methods necessitate different submarine operations for acquiring the field data even for a limited source area, they are often local in nature. They can work in a semi-probabilistic fashion but can further be combined with statistical information of past landslides to form the input to a fully probabilistic source description, unless slope failure can be linked to seismic triggering and PSHA. A preliminary attempt to demonstrate such a method is described in more detail by Løvholt et al. (2022).

3.3.3 Step 2: Source Parameterisation

The slope stability methods described above yield areas for potential slope failure as well as potential failure depth. The rock-slope engineering-geology based method will typically give a single volume as input to the initial condition in a landslide dynamics model, which so far has not been associated with an uncertainty. The infinite slope method returns an area and a depth profile that may be used to form the input for several source volumes but can be supplemented by local slope stability analyses to further constrain the volumes. The actual interpretation of the unstable slopes involves expert judgment, and often the logically connected potentially unstable slopes are used to set up volumes that consist of almost the whole unstable area (see e.g. examples in Carlton et al. 2019).

In most cases, the volumes obtained from slope stability analyses will be direct initial conditions in numerical landslide simulations. Examples are complex worst-case connected polygons (e.g. Carlton et al. 2019) forming source grids that may be used in a similar fashion as unit sources in earthquakes (e.g. Løvholt et al. 2022). Because landslide tsunami generation is time dependent and involves tsunami generation away from the initial failure area, simple unit sources including linear superposition (that are popular for earthquakes) can most often not be employed for landslides, implying that all the source realizations must be modelled separately (see e.g. Løvholt et al. 2022).

Another potentially useful way of setting up initial source geometries that has not been employed so far in PTHA, is using scaling laws that connect the landslide source area to the volume, in order to establish ratios between the landslide area and the slide thickness (Urgeles and Camerlenghi 2013; Zengaffinen-Morris et al. 2022). The scaling relationships also include uncertainty in the initial geometry of the landslide. On the other hand, it is purely statistical of nature, and will not take into account local slope conditions even though site-to-site variability is expected.

Once the geometries of the sources have been set up, the landslide parameters must be associated to each scenario (how to select source models is outlined further in Chap. 4). In Lane et al. (2016), the landslide scenario is modelled as initial conditions, based on a set of overall (non-local) Computational Fluid Dynamics (CFD) simulations that take into account only the water depth, volume, and slope, and therefore do not necessitate further parameterisation. In Løvholt et al. (2020), the landslide kinematics (run-out distance, velocity, and frontal area of the landslide) is subject to epistemic uncertainty, and parameter values for these will need to be built into the model. In Carlton et al. (2019), each scenario must be associated with rheological parameters for each landslide scenario by combining geophysical information with CPT data (see Løvholt et al. 2022 for more details).

3.3.4 Step 3: Probabilistic Treatment

Where a sufficient record over a defined time window of past (large volume) landslides is available, statistical methods can be used to set up MFDs for the landslide sources, relating either the landslide volume or the landslide area (see further discussion below) to the occurrence rate. The modeller then needs to assess how well the past record can be used to represent today's situation. Maximum Likelihood methods (Geist and ten Brink 2019) can be used for setting up the rate functions, both for the median and the mean rates as well as for the uncertainty in the rates. Typically, these rate functions are nonlinearly tapered in the tail, examples of MFD functions include for instance tapered Pareto functions.

For forecasting the source probabilities, Poissonian (stationary) processes are normally assumed. As a first pass, the same MFD can be assumed homogeneous over an entire region to set up the source rates using the landslide volume only as the input. Yet, it has been demonstrated by submarine landslide data statistics (Løvholt et al. 2022) that the relative landslide volume distribution with slope is non-uniform, i.e., there is a tendency that the smaller-volume landslides are more frequent for steeper slopes, and that larger volumes have a higher relative frequency at lower slope angles. In addition, different regions can have different MFD's, for instance it has been observed that landslides are more frequent but have smaller maximum volumes in seismically active regions (Urgeles and Camerlenghi 2013; ten Brink et al. 2016). Hence, it is possible that the rate should use at least a two-parameter function consisting of both the volume and the slope angle. Further complexity such as correlation of rates with faults can be examined, and it is also possible to compartmentalize source subregions within a greater area, to increase the granularity and the realism in the source rates. Yet, it is stressed that in the LPTHA applications carried out to date (e.g. Lane et al. 2016), only regional one-parameter rates have been considered. A principle sketch of such cumulative curve showing the relative frequencies of different landslide volumes is shown in Fig. 3.9. In this example, age constraints (see e.g. Urgeles and Camerlenghi 2013; Geist and Parsons 2014b) from the landslide database are needed to convert these statistics into a MFD relating the volume to the frequency of occurrence.

Once the volume vs rate information is established, it can be embedded into the scenario probability estimation. The most basic steps for setting up the hazard aggregation (summing and ranking probabilities of the tsunami exceeding a given height) are discussed below. In the probabilistic methods applied in present literature, the probabilistic analysis is represented by ensembles of landslide source scenarios. Aggregation procedures such as event trees are used to discretize the source probabilities for each scenario. The main parameter (top branch) is the landslide volume or location. Below, we present some examples of how the probabilities are interpreted.

Volume discretization: Lane et al. (2016) and Løvholt et al. (2022) investigate different synthetic scenario volume probabilities for submarine landslides. For feasibility, only three different volumes are considered. Rough discretisation representing wide volume ranges are necessary for feasibility purposes. It is ensured that the

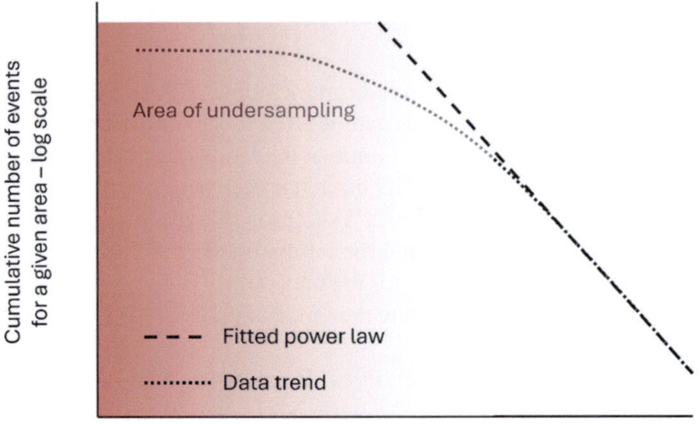

Fig. 3.9 A generic sketch of a cumulative MFD of landslide volumes. The dotted line indicates the distribution of data, and the dashed line a fit to the largest landslide volumes. For the smallest landslide volumes there is typically an area where landslides are under-sampled. Examples of such empirical MFD curves can be found in Zengaffinen-Morris et al. (2022), and Geist and ten Brink (2019)

summed rates over the whole set of ensemble scenarios' discrete realizations are equal to the sum of empirical rates. In this way only landslide volumes in the range of past slope failures are included in the analysis. As noted above, this omits modelling potentially larger and less probable/more consequential events.

Uncertainty treatment of the landslide kinematics: In Løvholt et al. (2020), uncertainties are discretized in an event tree that includes velocity, run-out distance, and frontal areas (see recipe R8). These represent kinematically prescribed landslide dynamics quantities (elaborated further in Chapter 4.3). They are obtained by combining landslide run-out observations, calibrating the model towards past tsunamis, and numerical modelling. A high degree of expert judgment is used to set up parameter values and conditional probabilities in the assessment for each parameter in the event tree branch.

Uncertainty treatment of the landslide soil properties: In Løvholt et al. (2022), a similar uncertainty treatment is set up, but now using soil parameters as input to landslide dynamics models rather than prescribed kinematics. The parameter values are established based on landslide run-out calibrations using in-situ soil properties for the initial yield strength on the average depth of the soil profile as input.

Some potential future improvements: It is noted that the models outlined above all assume time independent rates. On the other hand, it is likely that the tsunami hazard is linked to time-dependent features, such as short-term climatic drivers (e.g. rainfall increasing slope instability) or more slowly time-variable triggers (e.g. earthquake cycles, glacial cycles, volcanic unrest), and these are not embedded into the source models reviewed here. It is further stressed that release of unstable slopes can involve time dependent volume release such as retrogressive failure that can significantly influence the tsunami-genesis reducing the effective volume or even splitting into several independent tsunami events. As such, it has been demonstrated that these effects can influence tsunami-genesis for Holocene slides such as Storegga and Trænadjupet (Løvholt et al. 2017). The time separation between individual volume releases may also imply that individual retrogressive volumes act as separate tsunamigenic events, as observed for e.g. for the ancient Canary Island slope failures (e.g. Masson et al. 2006).

3.3.5 Summary Table

See Table 3.3.

Table 3.3 Summary table: landslide source selection

Short description of methods	
Theoretical/mechanical background for the methods	Evidence and data describing past landslides and their statistics, slope behaviour, geology and geophysics of the substrata, related to possible instabilities that may indicate a potential for future tsunamigenic landslides
Short outline of how the methods work	The method gives lists of synthetic scenarios and rates. There are two main classes of methods, both can be combined with event tree analysis: • Methods based on past landslide data—source selection methods are sparsely available and mostly prototypical • Methods based on slope stability models—they are often more heuristic, where the rates used in the analysis are often based on expert judgment, and often appropriate for local hazard analysis. Regional statistics of past landslides are still used to calibrate model probabilities. Volumes are taken from unstable slope information
How are the output(s) of the methods represented?	A set of landslide sources / synthetic databases. For each landslide source in the database: Initial position, lateral source extent and thickness, landslide parameters (kinematics or material properties), as well as related conditional probabilities

(continued)

Table 3.3 (continued)

In what way do the methods take into account time-dependent/dynamic aspects?	Although subaerial slope failures are linked to precipitation and season (winter, summer), this is not yet accounted for in presently available models
Do the methods use parallelisation, machine learning, surrogate models, advanced simulation schemes, other computational reduction strategies (e.g. cluster analysis, etc.)?	No, but future applications will likely soon include this due to the fast acceleration of AI based methods
What are the pros and cons and peculiarities of these methods? Are there known shortcomings/benefits of the methods? In what cases are these methods good/poor choices?	The methods require large amounts of data in order not to have excessive uncertainty bounds, with less data, uncertainties are pronounced. Not all steps are tested operationally, and the examples of use are so far limited. Severe lack of data may imply inaccurate source representation for probabilistic analysis as well as excessive uncertainties
Data and technical requirements for method	
What input data does the method(s) use?	Depending on the method, one or more of the following is needed to decide on source model selection: • Databases of landslides • Age date constraints on the landslide data • Earthquake soil or rock displacements from PSHA • Soil parameters and seismic information, information about basal layers • Data from slope stability investigations, scheme for slope stability evaluation, or aggregated slope stability data
What are the technical requirements?	Step 1—mostly qualitative, assessing how available data can be used to constrain sources in hazard analysis Step2—high-quality slope stability analysis can sometimes require rigorous processing of geophysical data. Past data for calibration can help constraining the analysis, this is linked to the source modelling (Sect. 4.3) Step 3—is mostly statistically based with few technical requirements
Are HPC computing resources necessary, would absence of these compromise the use?	No

(continued)

Table 3.3 (continued)

Is the method(s)/workflow(s) available as a git/software/jupyter notebook?	No
Treatment of uncertainties for the method	
Are the method(s) itself deterministic or uncertainty based?	The single source representation has deterministic characteristics. On the other hand, each source output can be associated with different probabilities
Which are potential sources of uncertainties? Which parameters are used to characterize them?	The main sources of uncertainties are: • Poor/limited information of past events • Projections of past statistics into today's situation • Sparse material data or spatial representation of the ground properties determining failure probability and volumes • Limited knowledge of the landslide mechanisms Parameters that can describe the uncertainties are: • Landslide volumes/frontal areas, run-out distance, soil properties • Landslide kinematics and run-out from past model hindcasts
How are uncertainties quantified for the identified parameters? Are there significant correlations between different uncertain parameters?	Parameters such as landslide kinematics and run-out are associated with different weights representing the epistemic uncertainty. There are few / no examples with uncertainty treatment for the aggregated hazard. Representations of parameterised landslide kinematics (impact velocity and run-out distance) have been used. There can be internal correlations between these kinematic parameters
Are the methods able to digest and propagate uncertainty in input (previous ingredient) and propagate it to the output (next ingredients)?	The uncertainty, if included, is only aggregated to the mean hazard in the presently available workflows
What are the consequences of lack of data in input or for boundary conditions, or for characterizing certain parameters?	Large uncertainties in the models. Potential under / overestimation of large impact components of the hazard curve. If too little data is present, exhaustive probabilistic analysis may not be feasible
What are the consequences of ignoring or poorly treating uncertainties?	Limited understanding of the hazard

3.4 Volcanic Sources

3.4.1 *Short Description of the Method*

What we commonly define as a "volcanic tsunami" is a tsunami generated by either eruptive processes, rapid ground deformation or flank instability at volcanoes (Day et al. 2015; Paris 2015). Following this definition, volcanic tsunamis represent about 5% of all listed tsunamis during the four last centuries (NCEI global tsunami database 2023). Historical examples and theoretical studies on tsunami generation have made it possible to distinguish different types of volcanic tsunamis (Fig. 3.10).

A large variety of volcanic sources may generate tsunamis (Begét 2000; Day et al. 2015; Latter 1981; Kienle et al. 1987; Paris 2015; Grezio et al. 2017) but they are grouped in eruptive (strictly connected to eruption dynamics) and non-eruptive sources (related to the structure of the volcanic edifice and to the existence of specific triggering phenomena). Eruptive tsunamis are generated directly by eruptions, including underwater explosions, pyroclastic flows, collapse of the eruptive column, caldera collapse, and atmospheric acoustic-gravity waves created by major explosions. Non-eruptive volcanic tsunamis are instead related to subaerial and submarine landslides, debris flows (e.g. lahars), or even volcanic earthquakes. All those source mechanisms have different characteristics in terms of duration, volume, mass flux, and energy, which has consequences on the waves generated and the methods used to simulate them. The volcanic tsunami dynamics clearly highlight two substantial features, both for the eruptive and the non-eruptive tsunamis: the multi-hazard character (as most of the triggering events may represent a potential hazard independently from the tsunami) and the time-dependent elements (as triggering conditions are strongly related to life-cycles of volcanoes, e.g. quiescence,

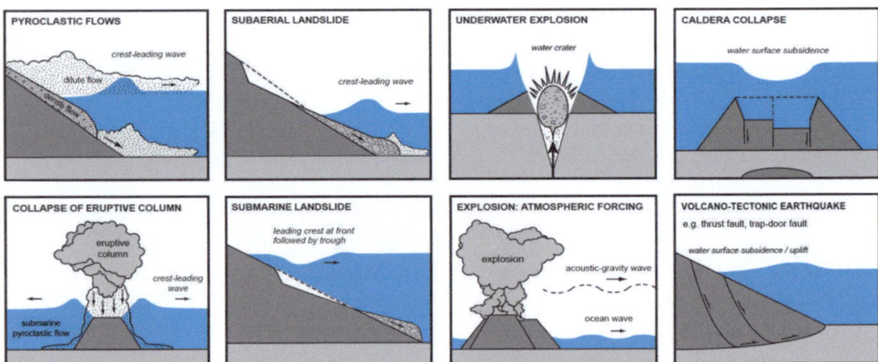

Fig. 3.10 Different types of volcanic tsunamis, i.e. tsunamis generated by volcano activity and instability (updated from Paris et al. 2014)

unrest, eruptions), which are also related to other external and potentially time-varying phenomena, like rainfall, strongly varying in the short-term (e.g., seasons) as well as in the long-term (e.g., climate change).

The volcanic source and estimation of the rate of occurrence is the first step in the hazard assessment, and it consists of the identification and quantification of the annual rates of all the potential volcanic sources of tsunamis that may affect a pre-selected target location. The general workflow graphically reported in Fig. 3.1 fully applies to volcanic sources. For volcanic sources, Step 1 consists of the selection of the potential source volcanoes or volcanic areas, and the selection of the types of sources to be considered for each volcano; Step 2 is applied only to the selected source(s), and encompasses the definition of the source variability (in space and size) to explore the natural variability of the important parameters for tsunami generation, and the kinematics or dynamics of the selected types of sources; and Step 3, applicable only to probabilistic methods, defines a specific list of individual scenarios and quantifies their probabilities or their mean annual rates of occurrence in the exposure time.

The final result of the volcanic source and rate estimation is the definition of a (potentially very long) list of possible sources, described by their characteristics like size, position, geometry, etc., and, for each source, a quantification of the probability of occurrence in the time frame selected for the hazard/risk assessment. This list is typically provided as input to tsunami generation and propagation models (Chap. 4), to complete the hazard quantification. In deterministic scenario-based approaches, Step 3 is omitted, and a list of individual sources is defined directly in Step 2, while a probabilistic quantification of annual rates is omitted.

In general, methods differ significantly for eruptive and non-eruptive tsunamis. The former is connected to the eruption forecasting, both in the characterization of the position of the events (vent opening) and in the constraining of source size (eruption magnitude or size). The latter is instead more related to the specific triggering phenomena, which may or may not be influenced by ongoing volcanic phenomena during unrest or eruptions (e.g. heavy rains, new material available for lahars just after the eruptions, increased seismicity and deformation, or hydrothermal alteration of materials for slope instability).

Most existing volcanic hazard quantifications are focused on single volcanoes (e.g., Neri et al. 2015) rather than on target areas, and they rarely include tsunami hazards. Considering that a few volcanic phenomena may propagate far from the source (typically tephra fall and tsunamis, Selva et al. 2019), a full volcanic tsunami hazard and risk quantification should define firstly the list of volcanoes that may potentially trigger a tsunami in a target area (IAEA 2016; Grezio et al. 2017; Behrens et al. 2021). However, given that each volcano has its own peculiarities, each source requires a volcano-specific analysis to be constrained (see next section) and sometimes specific sources may require computationally expensive procedures to model the tsunami generation and propagation process (see Sect. 4.3). Consequently, the exploration of the sources in volcanic probabilistic applications may be reduced. In the volcanic hazard literature, there are only a few examples of multi-volcano hazard or risk quantifications (Loughlin et al. 2015; Jenkins et al. 2022), also due to the substantial lack of homogeneous methods to quantify annual rates or probabilities for the different volcanic systems (Selva et al. 2022). Other methods range from the selection of one or few fully parameterized reference scenarios to a few examples of full exploration of the parameter space (Paris et al. 2019b; Grezio et al. 2020). The former approach is based on fully parametrizing a short list of scenarios, which allows skipping the quantification of the probability of each individual scenario (discussed in Step 3) but prevents the quantification of source uncertainty. Single scenarios are usually defined through expert opinion, guided by past occurrences or by geological and structural maps to identify the relevant position and size of sources.

3.4.2 *Step 1 Basis for Source Selection and Constraining the Area of Sources*

The first steps for the volcanic source and rate estimation consists of the selection of the volcanoes and, for each volcano, the type(s) of source(s) to be considered.

Once one (or more) volcano is selected, the relevant tsunami-generating phenomena should be identified, among all the potential types of sources that may generate tsunamis at volcanoes. The volcano is often preselected due to its relevance or relatively high potential tsunami generation. This selection is usually based on qualitative considerations regarding the main types of sources, for example the explosivity of eruptions, the main character (dilute vs dense) of pyroclastic flows, the presence of significant slopes or submerged part of the edifice, the depth of potential submarine sources, etc.

To select volcanoes and typologies of sources, we may define the following procedures with increasing level of completeness:

1. Preselected volcano, preselected type of source: a model considering, at a preselected volcano, a specific source typology and exploring its variability (in size and location);
2. Preselected volcano, all types of sources: a model considering, at a preselected volcano, all potential source typologies and exploring, for each of them, their natural variability;
3. All volcanoes and preselected type of source: a model exploring the entire natural variability of one preselected type of source, including all potential volcanoes that may generate tsunamis reaching the target area;
4. All volcanoes and all types of sources: a model exploring the entire natural variability of the sources that may generate a volcano tsunami reaching the target area, including all the volcanoes in the area and all potential source typologies.

In all cases, the output of these procedures is a list of volcanoes and potential types of sources for the subsequent steps toward hazard and risk quantification. The selection is based on databases of active volcanoes or volcanic areas, local records of past eruptions and past tsunamis, as well as the existing theoretical knowledge about potential sources of tsunamis (e.g. tsunami records at global scale). In some cases, the selection may be driven by the present state of the volcano (e.g. during volcanic unrest), which may lead to some time-dependency.

To our knowledge, all existing examples of volcano source tsunami hazard quantification are limited to point 1 above involving qualitative and deterministic procedures, which restrict the exploration of the natural variability of the sources and the potential uncertainties. In fact, considering only an inadequate number of potential sources (in terms of volcanoes, source typologies or limited natural variability) implies that we adopt a strong filter to potential tsunami sources, with possible substantial biases in the volcanic hazard and risk quantifications. While standard Volcano Probabilistic Tsunami Hazard Assessment (VPTHA) procedures may account for the full natural variability of the selected sources (e.g. Paris et al. 2019b; Grezio et al. 2020), we are not aware of formal or quantitative procedures to select which volcanoes should be included in a complete VPTHA, as well as which types of sources should be considered the most important at each volcano. This represents an important gap in VPTHA that should be resolved in the future (Behrens et al. 2021). So far, this selection is qualitative, that is, based either on a specific interest (e.g., needs for the building of a dedicated warning system, like in Stromboli), or on the availability of all the required input information (size variability and probability).

3.4.3 Step 2 Source Parameterisation

The goal of this step is to establish the parameters describing individual sources selected in Step 1, and to define their range of variability to describe the variability of the tsunami generation process. This is often done by defining a list of fully parameterized individual scenarios that are defined by combining the expert judgment with the analysis of past data, both from the specific volcano and from analogous situations. The specific source parameters selected to describe source models are position, size, duration, and geometry of the possible sources and are different for each source typology.

Probabilistic hazard quantification requires the definition of a range of variability for each source parameter. Ranges covering the entire natural variability are typically set from structural considerations and from the analysis of past activity, in the target volcano or from analogous volcanoes (e.g. Tierz et al. 2019), including local records of past eruptions and tsunamis, records from similar volcanoes, geological and geophysical maps used to identify ranges for vent location, extension and volume of potential sources for collapse. For eruptive phenomena, the main parameters to describe the source are the vent location and a measure of the energy of the eruption (magnitude, volcanic explosivity index, etc., e.g. Newhall and Hoblitt 2002) and may be linked to the standard volcanic hazard. Unknown or never-observed sources, limited knowledge on the structure of the volcano, or other limitations of this type may lead to an underestimation of the ranges, in particular of their upper limits, and this should be possibly included in the epistemic uncertainty quantification.

Once ranges are defined for all the specific source parameters, a list of scenarios may be produced sampling each parameter a large number of times. To make the evaluation of the tsunami propagation computationally feasible, the trade-off between full exploration and computational feasibility should be solved (e.g. as for earthquake tsunamis, Lorito et al. 2015; Selva et al. 2016; Davies and Griffin 2020). For this reason, the definition of the list of individual scenarios from the parameter ranges is discussed in Step 3.

In principle, the definition of the range of natural variability of the individual sources is not time-dependent. However, significant structural changes in the volcano may strongly impact the ranges, for example when a large eruption occurs. In this case, new edifices and caldera depressions may be formed, obviously changing the potential range of tsunamigenic sources. Also, variations in the state of the volcano (e.g. quiescence, unrest, eruption) and the occurrence of other phenomena (in the short-term the increasing of flank deformation, rapidly accumulated material on the flanks, heavy rainfall, regional tectonic earthquakes, or in the long-term extreme weather conditions due to climate change, land use changes, deforestation, modifications of soil conditions) may alter the conditions for the occurrence of the different sources. However, in those cases the temporal variability may be treated by decoupling the ranges of parameters from the probabilistic assessment of occurrences (Paris et al. 2019b) or introducing time-dependence in the probability of individual sources (e.g. Selva et al. 2014).

The actual discretization of the parameter space is very different for the different types of sources, which are consequently treated separately in the following subsections.

Landslides and debris flows

Flank instability of volcanoes has been widely documented. Volcano flanks display a broad variety of instabilities, from rock falls, cliff collapses and small landslides (with volumes typically in the order of 10^5–10^6 m^3) to large debris avalanches (with volumes that may range up to order of e.g. 10^8–10^9 m^3). As for landslide sources in general (Sect. 3.3), the main source parameters are the volume of the landslide Froude number (and maximum velocity), relative thickness, mass flux and volume and, for submarine sources, its initial acceleration (Fritz 2004; Viroulet et al. 2013; Yavari-Ramshe and Ataie-Ashtiani 2016; Lee 2022; Ward 2001; Grilli and Watts 2005; Løvholt et al. 2015).

Volcano flanks are particularly unstable, and this instability is the result of both endogenous (structural discontinuities, hydrothermal alteration, and magmatic intrusions inside the edifice, rapid growth by accumulation of tephra and lava flows) and exogenous factors (earthquake, tectonic uplift, weather and climatic event, sea level variations).

Many volcanoes are entirely (volcanic seamounts) or partly (volcanic islands) submarine or sub-lacustrine. As for subaerial landslides, the volumes implied by submarine landslides range from small-scale (10^5–10^6 m^3) events (e.g. collapses of coastal lava deltas, landslides in submarine canyons) to massive collapses on the submarine flanks of ocean islands (up to 10^7–10^9 m^3 or more). The number of tsunamis generated by submarine landslides on volcanoes is probably underestimated due to a lack of observations and data. For this reason, there are few unequivocal historical examples such as Ritter Island 1888 (Johnson 1987; Ward and Day 2003; Karstens et al. 2019), and the more recent examples of Stromboli 2002 (Bonaccorso et al. 2003) and Anak Krakatau 2018 (Muhari et al. 2019; Grilli et al. 2021).

The triggering processes at volcanoes may be particularly complex, making their probabilistic quantification challenging (see Step 3). On the contrary, in terms of tsunami generation, there is no specific difference between landslides observed on the flanks of volcanoes and other landslides, despite each event having its own structural setting, lithology and rheology (see Sect. 3.3).

Pyroclastic flows

Pyroclastic flows are hot mixtures of gas and particles generated by volcanic eruptions, particularly in the case of a dome collapse or a plume (eruptive column) collapse. Pyroclastic flows can generate tsunamis, as demonstrated by recent examples at Montserrat in 2003 and Stromboli in 2019 (Pelinovsky et al. 2004; Herd et al. 2005; Selva et al. 2022), but the conditions required to generate a tsunami and the mechanisms of interaction between the flow and the water are still poorly understood due to lack of observations and the absence of physical models.

The dense basal component of the pyroclastic flow is the main source of tsunami, but other phenomena such as steam explosion, flow pressure and shear, and pressure impulse could theoretically generate small waves (Watts and Waythomas 2003). The important parameters controlling the generation of a tsunami by pyroclastic flows are the flow volume and mass flux, together with the flow density and permeability (ash-rich flows being more tsunamigenic because of their low permeability), the angle of incidence, and the transport distance from the eruptive vent (Watts and Waythomas 2003; Maeno and Imamura 2011; Bougouin et al. 2020; Grezio et al. 2020). High-velocity pyroclastic flows with a bulk density near or even below that of water may generate waves, irrespective of their temperature (Bougouin et al. 2020; Freundt 2003).

Caldera collapse

Large explosive eruptions may result in the collapse of the central part of the edifice, thus forming a caldera. In case of an underwater eruption, the caldera collapse produces subsidence of the water surface that initiates the propagation of a leading trough. The amplitude of the water subsidence depends not only on the volume and geometry of the collapse, but above all on its duration (Gray and Monaghan 2003; Maeno et al. 2006; Ulvrova et al. 2016).

Large collapses lasting just a few minutes are theoretically tsunamigenic but probably unrealistic. Indeed, based on recent examples, the duration of a caldera collapse during an explosive eruption typically lasts more than 30 min (Stix and Kobayashi 2008). There are different types of geometries and collapse mechanisms (Roche et al. 2000; Stix and Kobayashi 2008). Consequently, there is no unequivocal example of a tsunami generated by a caldera collapse. Other phenomena such as underwater explosions, eruptive column collapse and pyroclastic flows may generate tsunamis during an explosive caldera-forming eruption, which makes it difficult to determine the source of the tsunami(s), as illustrated by the near-field tsunami that impacted the Tonga Islands during the 2022 eruption of Hunga Tonga–Hunga Ha'apai volcano (Omira et al. 2022).

Underwater explosion

The theory of water waves generated by underwater explosions is well documented, and has been applied to nuclear, chemical and volcanic explosions (Le Méhauté 1971; Duffy 1992; Le Méhauté and Wang 1996; Egorov 2007). After an underwater volcanic explosion and while different jet flows of fragmented lava and gas are ejected, the development of a water crater might be initiated, depending on water depth and energy of explosion, which are the two main input parameters for initiating wave generation. Subsequent expansion, rise and gravitational collapse of the crater create two successive bores followed by a number of smaller undulations propagating radially from the source. In theory, all volcanic eruptions above 500 m water depth are potentially tsunamigenic, but in reality, only a few such events are significant.

The variability of the sizes and positions of potential underwater volcanic explosions is controlled by the natural variability of eruptive scenarios. Vent position forecasts control the spatial extent to be considered for sources, which in turn is typically controlled by structural considerations like caldera borders or volcanic apron (e.g. Scandone et al. 1991; Orsi et al. 1995; Selva et al. 2012a, b; Connor et al. 2019; Cañón-Tapia 2022). The sizes of the tsunami-generating explosions are instead related to the expected variability of eruption size at the volcano, which strongly varies from volcano to volcano. Explosion energy may be roughly related to eruption size measures like explosivity or magnitude, and even to geometrical observables of past eruptions like crater radius (Torsvik et al. 2010; Ulrova et al. 2016; Paris et al. 2019b).

Collapse of eruptive column

An eruption column or eruption plume is a cloud of volcanic tephra (mostly ash) suspended in gasses emitted during an explosive eruption, potentially up to 40–50 km altitude. When the material (i.e., the tephra) in the column can no longer be supported by convection, it falls under gravity, forming pyroclastic flows on the flanks of the volcano. The intensity of the collapse is variable from one eruption to another and, for stable plumes, it may evolve as a function of the buoyancy ratio (Carazzo et al. 2015).

If the eruption comes from a small island or a shallow-water volcano, the eruptive column may collapse directly into the water. Pyroclastic flows resulting from the collapse are then subaqueous and the collapse itself becomes the main source of tsunami. Although there is abundant literature on eruptive column collapse in volcanology, it is not commonly addressed as a source of volcanic tsunami. However, this neglected mechanism may have played a role in the generation of near-field tsunami during explosive eruptions such as Krakatau 1883 and Hunga Tonga-Hunga Ha'apai 2022. The main input parameters influencing tsunami generation by an eruptive column collapse are the mass flux at the impact, the duration of the collapse, the water depth, and the sea floor geometry (e.g., the presence of a submerged volcanic cone beneath the impact might change the geometry of the initial waves).

Atmospheric forcing following a large explosion

The atmospheric acoustic-gravity waves that are produced during a major volcanic explosion can generate a tsunami. This rare phenomenon was first documented for the 1883 eruption of Krakatau (Press and Harkrider 1966; Pelinovsky et al. 2005) and extremely well recorded for the 2022 eruption of Hunga Tonga-Hunga Ha'apaii volcano in the Tonga Islands. Tsunamis generated by such an atmospheric forcing represent the only type of volcanic tsunamis that may have a global reach.

The largest explosive eruptions generate a broad range of waves in the atmosphere, including acoustic-gravity waves that may reach the ionosphere (Astafyeva et al. 2022). Among these different types of atmospheric waves, compressional surface-guided Lamb waves traveling at the speed of sound can produce long-period waves in the ocean (Kubota et al. 2022; Omira et al. 2022). The tsunami generation and propagation processes are similar to other atmospheric sources, as discussed in Sects. 3.5 and 4.5.

Volcano-tectonic earthquake

Although earthquakes are often mentioned as a source of tsunami preceding or during a volcanic eruption, historical examples are poorly documented because the distinction between tectonic earthquakes, volcanic earthquakes, or other source mechanisms of tsunami (e.g., landslide) is often unclear. Among all kinds of earthquakes related to volcanic and magmatic processes, only volcano-tectonic (high-frequency) earthquakes can involve ground deformation large enough to generate a tsunami. Volcano-tectonic earthquakes result from the accumulation of stress induced by magma ascent. They are characterized by seismic swarms at shallow depth (<10 km), with magnitudes typically lower than $M_w = 6$, thus generating very small-magnitude tsunamis, if any (Paris 2015).

Two special cases are mentioned here. First, earthquakes with magnitude $M_w >$ 6 on large reverse faults at the base of the oceanic shield volcanoes (e.g., Hawaii) can produce tsunamis, as demonstrated by the 1975 Kalapana earthquake at Kilauea volcano, Hawaii (Ando 1979; Ma et al. 1999; Day et al. 2005). Second, earthquakes with magnitude $M_w < 6$ resulting from trapdoor faulting of submarine caldera floor can generate small-amplitude tsunamis, as frequently observed around the Sumisu caldera, Japan (Sandanbata et al. 2022). Tsunami generation and propagation for volcano-tectonic earthquakes is similar to other earthquake sources, as discussed in Sects. 3.3 and 4.3.

3.4.4 Step 3 Probabilistic Treatment

The goal of this step is to evaluate the probabilities of occurrence or the annual rates for a list of individual scenarios that cover the natural variability of the selected tsunami sources. The parameters and the ranges provide the input to the probability evaluation made in Step 3 (Fig. 3.1).

The list of individual scenarios is provided as input to tsunami generation and propagation models (Sect. 4.4), while the probability of occurrence or the annual rate of each scenario is provided as input to hazard aggregation (Grezio et al. 2017; Basili et al. 2021a, 2021b). This step is typically skipped when non-probabilistic scenario-based methods are selected (e.g. FW1D in Sect. 2.2; see Fig. 3.1), with the consequence that the potential impact of some representative scenarios is evaluated without accounting for uncertainty or quantifying the likelihood of such events.

The output of this step is a list of scenarios with the associated probabilities of occurrence or annual rates in the exposure time. The list may be defined to cover the source space and, subsequently, quantify their probability of occurrence (e.g. Selva et al. 2016), or sampling from scenario probabilities, generating synthetic catalogues (e.g. Davies 2019). To quantify the epistemic uncertainty, multiple lists of probability of occurrence or annual rates may be provided (Selva et al. 2016; Basili et al. 2021a, 2021b). In this case, such multiple lists should be accompanied by a weight defining the subjective credibility of each list. This information may provide input models for the quantification of epistemic uncertainty, such as logic trees or ensemble modelling (Budnitz et al. 1997; Geist and Parson 2006; Bommer and Scherbaum 2008; Marzocchi and Jordan 2014; Marzocchi et al. 2015, 2021; Selva et al. 2016).

For eruptive tsunamis, eruption forecasting methods may be adopted to quantify the spatial and magnitude variability of the sources including statistical tools like event trees, belief networks, expert elicitations, Markov chains, stratified sampling, or other statistical techniques that sample probability distributions based on past data or tracking observations (Newhall and Hoblitt 2002; Aspinall et al. 2003; Marzocchi et al. 2008; Marzocchi and Bebbington 2012; Newhall and Pallister 2015; Hincks et al. 2014; Tierz et al. 2017; Sandri et al. 2017; Bebbington et al. 2019; Bebbington and Jenkins 2022; Selva et al. 2015; Neri et al. 2015; Grezio et al. 2020). Spatial distributions of vent opening are often modelled with spatial kernels or other methods based on the hierarchization of tracking features (e.g. Scandone et al. 1993; Selva et al. 2012b; Connor et al. 2019; Cañón-Tapia 2022). Due to the cascading process from volcanic initialisation to tsunami impact, linking tsunami hazard to the standard eruption forecasting methods requires the definition of a direct connection between typical size metrics for volcanic eruptions and potential sizes for the tsunami sources (Paris et al. 2019).

Several models for eruption forecasting and volcanic hazard are available as open software or in open-source codes (e.g. https://theghub.org/groups/v/), without specific finalizations for tsunamis.

For eruptive phenomena and for consequent tsunamis it is common to only quantify the conditional hazard, that is the probabilistic hazard conditional to the occurrence of an eruption (Biass et al. 2016, 2012; Jenkins et al. 2012; Sandri et al. 2017; Selva et al. 2018; Paris et al. 2019). On the one hand, this simplifies the computation, as it removes the eruption forecasting from the hazard quantification, concentrating on the natural variability of the sources when an eruption occurs. On the other hand, this conditional hazard may provide fundamental information to decision makers for planning interventions in case of eruption, like for example for the definition of evacuation areas and evacuation routes (e.g., DPC 2016). However, this choice complicates the comparison between the quantified hazard and the other hazards in a multi-hazard risk perspective as return period computation is omitted.

For non-eruptive sources during volcanic quiescence periods, the procedures discussed in Sects. 3.2, 3.3 and 3.5 may be applicable and the sources can be treated as standard non-volcanic sources. For flank collapses and landslides many of the methods developed for standard landslides may be applicable (see Sect. 3.3), even if the initial conditions may suddenly change due to the eruptive processes and the largest landslides are often related to the increase of volcanic activity or major eruptions (e.g., debris avalanches). Non-eruptive tsunami sources during the unrest episodes, that are periods with an increased volcanic activity with respect to the background, may be connected to both eruption forecasts and related hazards (e.g. Hincks et al. 2014; Tonini et al. 2016).

Forecasting of non-eruptive tsunamis generated at volcanoes is linked to the ability to forecast the occurrence of the source during the life cycles of volcanoes (which include quiet periods as well as unrest and eruptions). To this end, the statistics of the source and the triggering conditions are usually analysed through specific statistical models (e.g., Tierz et al. 2017). Spatial distributions are often studied through susceptibility studies (e.g. Masson et al. 2006). For non-eruptive events there are potential complexities due to the interaction between phenomena within the multi-hazard context of volcanoes. For example, landslides on the flanks of a volcano may occur independently of an eruption, because old eruptive material can be mobilized even years after the last eruption due to external triggers like heavy rainfall or tectonic earthquakes (Tinti et al. 1999; Lenti et al. 2015a, b). Such movement may be favoured during both eruptive and non-eruptive unrest periods because of deformation or alteration of materials in connection with degassing, hydrothermal changes, and local seismicity (Tommasi et al. 2018; Lenti et al. 2015a, b; Paparo et al. 2017; Ripepe et al. 2009, 2015, 2021; Selva et al. 2020). Due to the large variety of such phenomena, a link must be established between the volcanic phenomena and the occurrence (and the size) of the volcanically induced tsunami sources.

In Fig. 3.11, we report a possible event tree that may be adopted for the quantification of the time-varying probability of tsunami sources of volcanic origin, adapting the Bayesian Event Tree evaluation scheme (Marzocchi et al. 2008, 2010; Selva et al. 2014). At node 1, the event tree describes the three main states at which a volcano may be: quiescence, unrest or eruption. These states are connected to significantly different probabilities of occurrence of the sources, as well as significantly different approaches to quantify them (Marzocchi et al. 2008, 2010; Selva et al. 2014). At nodes 2, the main potentially tsunamigenic volcanic phenomena are listed, and each of them is generically parameterized through area of potential occurrence (node 3) and size (node 4). Node 5 deals with the forecast of the occurrence of the tsunami generation, while nodes 6 and 7 are targeted to the quantification of the related hazard curve (Marzocchi et al. 2010; Selva et al. 2014).

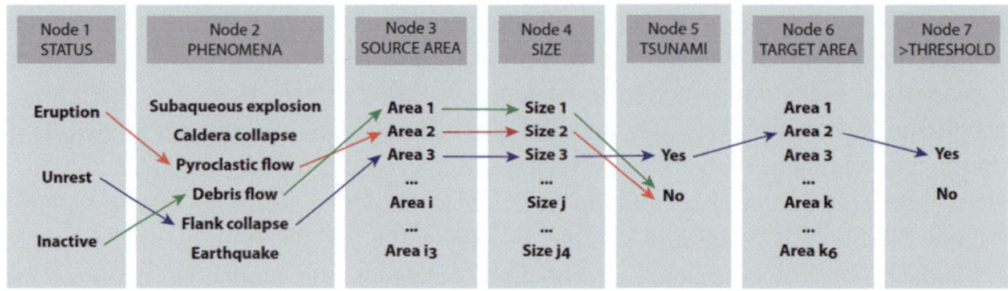

Fig. 3.11 Bayesian Event Tree for Volcanic Tsunami Hazard. The combination of all nodes is possible: arrows indicate potential paths which involve different volcanic states (node 1), phenomena (node 2), source position and size (nodes 3 and 4), and tsunami generation (node 5). In these examples, only the blue arrows indicate the path of a possible scenario generating a tsunami (node 5) and leading to the evaluation of the hazard curve at all targets (nodes 6 and 7)

The complex relationship between the different volcanic phenomena also highlights the potential for significant time-dependencies, single sources being potentially connected to the volcano state (e.g. quiescence, unrest, eruption) and to the occurrence of other volcanic or non-volcanic phenomena (e.g. deformation, degassing, rainfall, etc.). For the eruptive sources, specific time-dependency may be accounted for by switching from long-term assessment based on past data to short-term assessment based on monitoring and/or weather forecast (Marzocchi et al. 2008; Selva et al. 2014), accounting for potential changes in regime when the probability of eruption is evaluated (De la Cruz-Reyna 1993; Bebbington 2010, 2013, 2014; Selva et al. 2022). Other dependencies may be accounted for through networks or Markov chains that account for the concatenation of events (Tierz et al. 2017; Bebbington and Jenkins 2019, 2022). In general, these methods treat temporal variability by decoupling the selection of the list of scenarios from the probabilistic assessment and introducing time-dependence in the probability of individual sources.

The main inputs to probability assessments are local records of past eruptions or tsunamis and observed past unrest episodes. Geological data and maps may also play an important role, especially for determining the spatial probability, due to the relative paucity of data or lack of knowledge that affect most, if not all, individual volcanoes. An important input to probabilistic quantifications is provided by other volcanoes, especially by the so-called analogous volcanoes, that is, volcanoes that show similar dynamics to the target one (Marzocchi et al. 2008; Tierz et al. 2019). Another important contribution in this sense is provided by expert opinion, often collected through formalized expert elicitations (Aspinall 2006, 2010; Aspinall et al. 2003; Selva et al. 2012a; Aspinall and Woo 2014; Tadini et al. 2022). These approaches are of course subject to large epistemic uncertainty. In volcanology, such uncertainty is often treated by adopting Bayesian models (Aspinall 2006; Marzocchi et al. 2008; Aspinall and Woo 2014), but also methods collecting alternative implementations like logic trees or ensemble modelling have been proposed (IAEA 2012, 2016; Marzocchi et al. 2021).

The final list of scenario probabilities may be produced by discretizing the parameter space or sampling from the probability of each scenario (e.g. Basili et al. 2021a; Davies et al. 2022). In the former, the definition of the list is independent from the probability definition, while in the latter they are connected and must be performed simultaneously (Selva et al. 2016; Davies et al. 2022). When a predefined discretization is adopted, multiple statistical models may be defined independently and equally applied to all defined sources, simplifying the treatment of epistemic uncertainty (Selva et al. 2016). However, this typically leads to a large number of scenarios to be modelled individually, thus necessitating strategies to reduce the computational effort (Molinari et al. 2016; Basili et al. 2021a). Differently from seismic sources, the nonlinear hydrodynamics often characterizing volcanic tsunamis near the source do not allow for strategies based on linear superposition of unit sources. Hence, ad hoc strategies may be adopted to define representative individual scenarios. The sampling approaches may help significantly to reduce the required number of scenarios. However, if the statistical model changes, the list of scenarios must be resampled, implying new simulations to be performed (Davies et al. 2022).

3.4.5 Summary Table

See Table 3.4.

3.5 Atmospheric Sources

3.5.1 Short Description of the Method

Atmospheric sources usually generate tsunamis (hereafter referred to as meteotsunamis) that are typically an order of magnitude smaller than those triggered by other sources. They can be separated in two categories: weather and acoustic sources (Fig. 3.12). As illustrated in the schematic given in Fig. 3.12, weather-related events are directly connected to mesoscale atmospheric processes (Monserrat et al. 2006) while acoustically driven events result from the generation of planetary acoustic-gravity waves after an explosion (e.g., volcanic eruption, asteroid impact/bolide explosion). Both categories of events have the potential to generate strong surface pressure (and wind, e.g., Linares et al. 2016) disturbances inducing meteotsunamis.

The uncertainty linked to weather-related events is possibly the lowest of the entire suite of tsunami sources as it highly depends on both the relatively well constrained weather conditions (Denamiel et al. 2019a) and the geomorphology of the coastal areas (Pattiaratchi and Wijeratne 2015; Rabinovich 2020). Impact of weather-related events is mostly local as it depends on the ability of harbours and bays to amplify the offshore waves, and so-called hotspot locations—where the intensity and the

Table 3.4 Summary table: volcanic sources

Short description of method	
Theoretical/mechanical background for the methods	• All offshore and near shore volcanoes may in principle generate tsunamis • Multiple volcano-related phenomena may be tsunamigenic, during eruptions or independently from eruptions • There is a large variability of potential scenarios • Hazard is time-dependent due to the complex connections between the different volcanic phenomena during quiescence, unrest, and eruption phases • Except for large explosions or collapses, volcano tsunamis are generally dispersive with relatively short wavelengths; consequently, they are often local
Short outline of how the methods work	• In step 1, a single volcano is often pre-selected; only recently, multi-volcano analyses start being discussed • The typology of sources selected at each volcano is typically pre-selected, based on experience or recent history. Existing methods are mainly qualitative • In step 2, single scenarios are usually defined through expert opinion. In probabilistic methods, ranges are set mainly from structural considerations and past activity • In step 3, source probability is often computed conditional to eruption, while unconditional probabilities or annual rates are rarely evaluated • For eruptive tsunamis, methods for eruption forecasting and volcanic hazards are used, including event trees, belief networks, expert elicitations, Markov chains, stratified sampling, based on past data or on monitoring • For non-eruptive tsunamis, there are no consolidated methods; methods strongly vary on source type • Epistemic uncertainty is quantified with Bayesian methods, logic trees, ensemble modelling, and different flavours of expert elicitations

(continued)

frequency of such events is known to have devastating effects—have already been identified (Vilibić et al. 2021). Consequently, two main uncertainties are related to

Table 3.4 (continued)

How are the output(s) of the methods represented?	• In step 1, a list of volcanoes and source typology(-ies) is provided • In step 2, the range of variability of source parameters is provided • In step 3, a list of individual scenarios and respective probability of occurrence or annual rate in the exposure time is provided
In what way do the methods take into account time-dependent/dynamic aspects?	• The selection of sources may depend on the present state of the volcano (e.g. quiescence/unrest/eruption) • Structural changes (e..g. due to large eruptions) may induce a significant change in the parameters ranges • Other temporal modulations are due to changes in the state of the volcano or to other phenomena (flank deformation, rapidly accumulated material on flanks, heavy rains, tectonic earthquakes, climate change, land-use changes, etc.) • Probability of occurrence is strongly time-dependent at all scales, from regime changes (tens of years), to changes in the volcanoes' states (months), to quicker dynamics like heavy rain (days)
Do the methods use parallelisation, machine learning, surrogate models, advanced simulation schemes, other computational reduction strategies (e.g. cluster analysis, etc.)?	Not in present applications
What are the pros and cons and peculiarities of these methods? Are there known shortcomings / benefits of the methods? In what cases are these methods good/poor choices?	• Step 1 is qualitative. The development of more quantitative tools is required in the future • In step 2, the analysis of single scenarios may be important as a preliminary step toward the selection of the type of source, or to provide an exemplificative description of the phenomena to decision makers • For hazard/risk purposes, the selection of single scenarios is more subjective than the definition of a range of possible scenarios covering the expected natural variability. The definition of proper ranges allows quantifying the effective natural variability • Different methods exist for Step 3, with specific pros and cons, as discussed in the main text
Data and technical requirements for method	

(continued)

Table 3.4 (continued)

What input data does the method(s) use?	• Local records of past eruptions and tsunamis • Records from analogous volcanoes • Databases of active volcanoes or volcanic areas (mainly step 1) • Theoretical knowledge about potential sources of tsunamis (mainly step 1) • Geological and geophysical maps, to identify ranges for vent location, the extension and the volume of potential sources for collapse (mainly step 2) • Monitored past unrest episodes • Expert opinion
What are the technical requirements?	None
Are HPC computing resources necessary, would absence of these compromise the use?	• The lack of sufficient computing resources may be one of the reasons to select a scenario-based deterministic approach • The technique adopted at step 3 to define the list of scenarios to be propagated should be selected based on the available computational resources
Is the method(s)/workflow(s) available as a git/software/jupyter notebook?	Several tools for eruption forecasting and volcanic hazard are available as free software or open-source codes
Treatment of uncertainties for the method	
Are the method(s) itself deterministic or uncertainty based?	• Step 1: methods are usually deterministic • Step 2: methods based on one or few scenarios are deterministic, while the definition of ranges for all source parameters enables the uncertainty quantification • Step 3: methods are uncertainty based
Which are potential sources of uncertainties? Which parameters are used to characterize them?	• In step 1, the existence of unknown or never-observed types of sources • In step 2, the existence of unknown or never-observed sources, limited knowledge of the structural conditions of the volcano, or other uncertainty of this type, may lead to underestimation of the ranges, especially of their upper limits • In step 3, the paucity of data or lack of knowledge may lead to non-univocal definition of probability distributions, leading to large epistemic uncertainty

(continued)

Table 3.4 (continued)

How are uncertainties quantified for the identified parameters? Are there significant correlations between different uncertain parameters?	• Uncertainties in parameter ranges (step 2) or probabilistic formulations (step 3) are treated through an ensemble of alternative formulations, managed through logic Trees, ensemble models or Bayesian models • Significant correlations among parameters and their uncertainty may exist; they are usually treated as conditional/stratified sampling and through the use of multivariate statistics
Are the methods able to digest and propagate uncertainty in input (previous ingredient) and propagate it to the output (next ingredients)? Does the assessment of this ingredient require uncertainty propagation? If yes, how are uncertainties propagated?	• The expert selection in step 1 does not propagate uncertainty; to this end, new techniques should be developed in the future • Ranges defined through single or few scenarios do not propagate uncertainty, while the definition of full ranges allows for it • The uncertainties in source definitions and statistical properties can be treated by quantifying and propagating epistemic uncertainty
What are the consequences of lack of data in input or for boundary conditions, or for characterizing certain parameters?	• Paucity of data or scarce knowledge of the volcano may lead to ignoring important sources and underestimating variability ranges • Paucity of data increases the epistemic uncertainty in probabilistic quantifications
What are the consequences of ignoring or poorly treating uncertainties?	• Missing sources or underestimation of ranges may lead to significant underestimation of the hazard • The subjective definition of one or few scenarios introduces uncontrolled biases in hazard quantification • The scarce exploration of epistemic uncertainty does not allow the communication of existing constraints in hazard quantification

such events: (1) the detection and proper modelling of the weather sources (e.g., location, speed, period, intensity) and (2) the impact of climate change.

Acoustically driven events can have worldwide impacts but are extremely rare. They are thus associated with larger uncertainties with respect to their intensity and rate. Despite being potentially more destructive than the weather-related sources (Denamiel et al. 2023), they have not yet been widely studied. Consequently, the uncertainty linked to acoustically driven events is not well known but is yet directly tied to the explosive volcanic eruptions and asteroid impacts/bolide explosions uncertainty.

At present, Atmospheric Probabilistic Tsunami Hazard Assessment (APTHA) remains sparsely developed as only one study has produced an assessment for weather

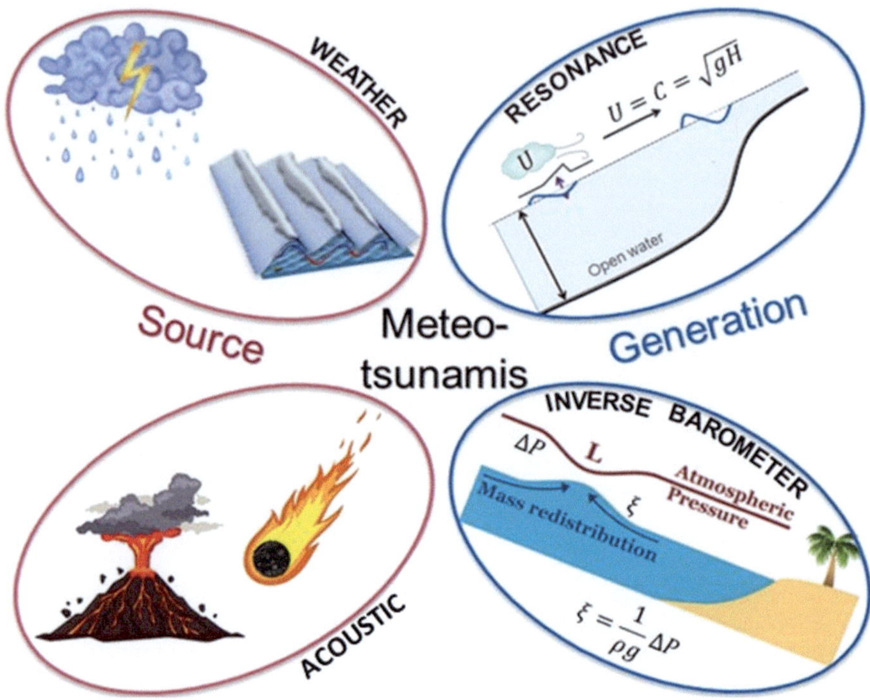

Fig. 3.12 Summary of the atmospheric sources and the generation mechanisms leading to atmospherically-driven tsunamis or meteotsunamis

related events (Geist et al. 2014c). On the contrary, most of the efforts in this field have evolved around early warning systems. However, several avenues have already been explored for the identification and modelling of the atmospheric sources. The following subsections present the different atmospheric sources as well as a family of methods specifically implemented for the quantification of atmospherically-driven tsunamis or meteotsunamis.

3.5.2 Step 1: Basis for Source Selection

Weather sources

Historically, weather-related meteotsunamis are defined as long waves in the tsunami frequency range driven by atmospheric disturbances (Monserrat et al. 2006). More generally, Non-Seismic Sea-Level Oscillations at Tsunami Timescales (NSLOTT) are generated worldwide through the transfer of energy from atmospheric waves to background ocean long waves (Vilibić and Šepić 2017). However, extreme NSLOTTs, or meteotsunamis that are, associated with flooding, structural damages, and human losses, only occur when two conditions are fulfilled. First, the transfer of energy must be increased by offshore resonance (Vilibić 2008) obtained either when the speed of the atmospheric waves is equal to the tsunami long wave speed (i.e., Proudman resonance; Proudman 1929) or when the speed of the atmospheric waves is equal to the speed of the edge waves along a coastline (Greenspan resonance; Greenspan 1956). Second, the offshore ocean long waves must be amplified nearshore, for example through shoaling and harbour resonance (Pattiaratchi and Wijeratne 2015; Rabinovich 2020). In practice, this means that weather related meteotsunamis mostly occur (1) in known hotspots (e.g., Balearic Islands, Adriatic Sea in the Mediterranean region or Great Lakes in the USA; see Fig. 3.13) favorable to the nearshore amplification of the long waves (Vilibić et al. 2016, 2021) and (2) during weather events generating atmospheric waves at a speed triggering the Proudman or Greenspan resonances.

Acoustic sources

Recently, the concept of atmospherically driven tsunamis has been extended to long waves generated by acoustic-gravity waves (Matoza et al. 2022; Amores et al. 2022; Harrison 2022; Adam 2022; Wright et al. 2022; Omira et al. 2022). These planetary meteotsunami waves spread with much greater speeds than tsunamis (Satake et al. 2020; Adam 2022) as they mostly occur during very large volcano explosions or asteroid impacts/bolide explosions capable of generating atmospheric acoustic-gravity waves with speeds far greater than the local shallow water propagation speed of tsunamis given by $\sqrt{gH}$ where g is the gravity and H is the local water depth. Their existence has been first hypothesized after the catastrophic eruption of the

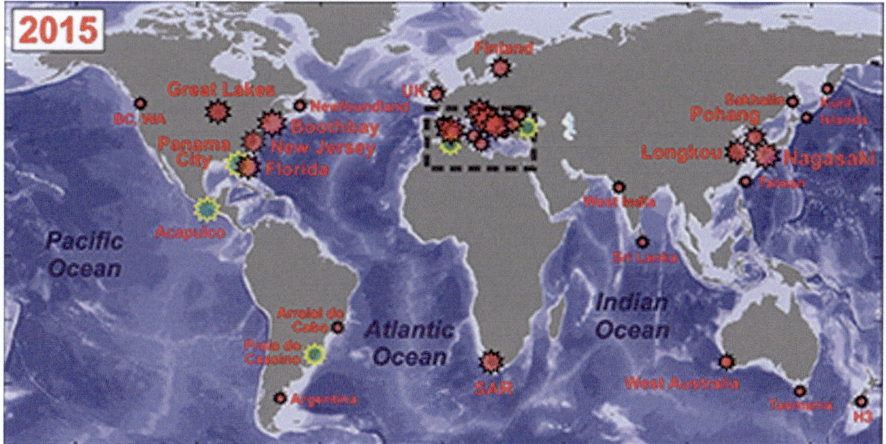

Fig. 3.13 Locations where weather related meteotsunami events are either suspected (in green) or known (in red) to have occurred until 2015. The sizes of the markers are related to the intensities of the events. The Mediterranean region, where most of these events occurred, is highlighted with a dotted box. Adapted from Vilibić et al. (2016), licensed under CC-BY 4.0

Krakatau volcano in 1883 (Press and Harkrider 1966; Self and Rampino1981) and fully confirmed with the explosive eruption of the Hunga Tonga–Hunga Ha'apai (HTHH) volcano in 2022 (e.g., Wright et al. 2022; Omira et al. 2022). This event was exceptionally well observed by satellites and recorded worldwide by thousands of microbarographs, seismometers, and tide-gauges. Based on the known meteotsunami generation mechanisms, the two processes that can amplify the acoustically driven planetary meteotsunami waves are the deep-ocean Proudman resonance and the inverse barometer effect which implies that larger meteotsunami waves should be generated by larger pressure disturbances driven by acoustic-gravity waves:

$$\xi = \frac{1}{\rho g} \Delta P$$

where ξ is the sea surface elevation, g is the gravity, ρ is the density and ΔP is the pressure disturbance. Consequently, simple numerical experiments dividing the speed of the HTHH Lamb waves by a factor 1.5 to reach the deep-ocean Proudman resonance at a speed of about 212 m/s (Experiment P; Fig. 3.14) or multiplying their amplitude by 10 to increase the inverse barometer effect (Experiment A; Fig. 3.14), lead to meteotsunami surges above 1 m (and up to 10 m) along more than 7% of the worldwide coastlines while they did not even reach 1 m during the HTHH event (Denamiel et al. 2023; see Fig. 3.14).

3.5.3 Step 2: Source Characterization

Weather sources

In contrast to the other tsunami sources, meteotsunamis are driven by atmospheric patterns and mesoscale conditions that can potentially be forecasted with numerical models. The following section is thus describing the different atmospheric sources driving meteotsunamis, at different locations of the world, and for different synoptic conditions, instead of their parameterization.

At the synoptic scale, strong mid-troposphere jets are the main atmospheric pattern causing meteotsunami events worldwide (Zemunik et al. 2022). Such events were recorded in the USA (e.g., Thomson et al. 2009), in the Mediterranean Sea (e.g., Jansà et al. 2007; Zemunik et al. 2021), in the northeast Atlantic basin (e.g., Tappin et al. 2013; Sibley et al. 2016) and in Japan (e.g., Tanaka 2010). Other recurrent synoptic patterns include (1) an inflow of warm and dry air in the lower troposphere, (2) instability of the mid-troposphere, (3) weak surface cyclones, and (4) winds that increase from the surface towards the upper levels. However, in some regions of the world, meteotsunamis may be accompanied by other atmospheric patterns depending on the mesoscale processes driving the events.

The most common mesoscale processes driving meteotsunamis are (1) atmospheric Mesoscale Gravity Waves (MGWs), (2) convective storms, (3) frontal

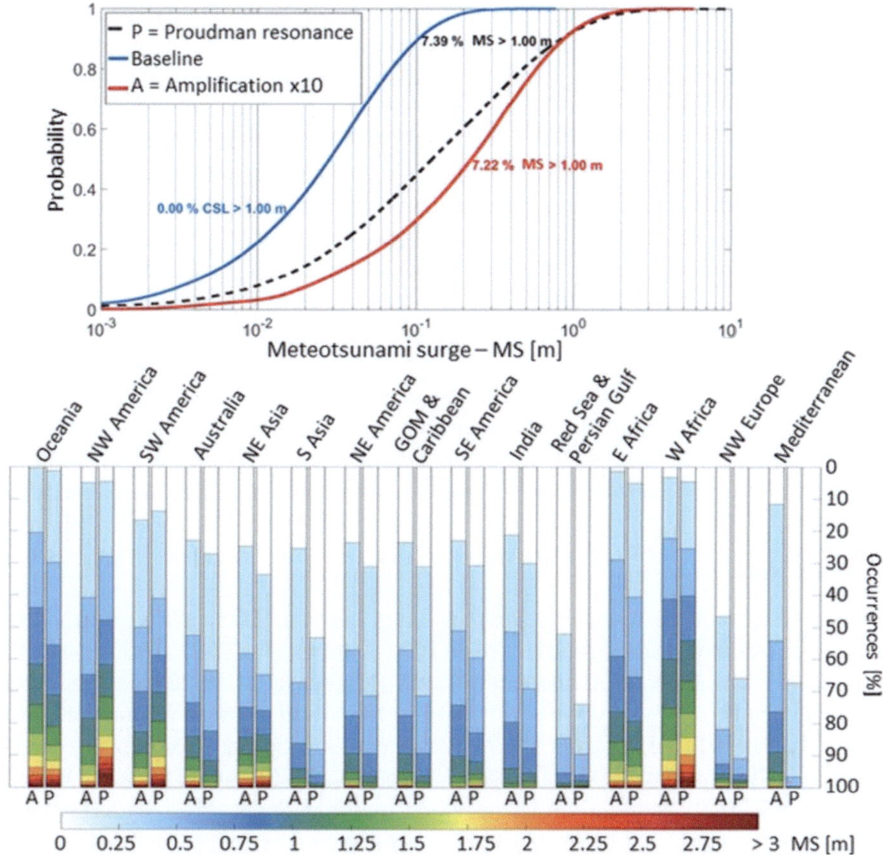

Fig. 3.14 Modelled meteotsunami surges, as cumulative density functions (top panel) and distributions by coastal sub-domains (bottom panel), obtained under deep-ocean Proudman resonance conditions (Experiment P) and tenfold amplification of the Lamb wave amplitudes (Experiment A) in the atmosphere. Adapted from Denamiel et al. (2023)

passages/squall lines and (4) tropical cyclones. Atmospheric MGWs may be generated by unbalanced jet streaks through the so-called spontaneous balance adjustment (Ruppert et al. 2022) and can interact with fronts and organized precipitation. They can cause rapid surface pressure oscillations > 5 hPa in less than an hour and can drive meteotsunamis (Bechle et al. 2016; Dusek et al. 2019). For example, Sheremet et al. (2016) measured a meteotsunami driven by atmospheric MGWs generated by a strong cold front that passed over the northeastern Gulf of Mexico on 7 March 2008. During this event, the MGWs induced rapid surface pressure falls of $\geq$ 10 hPa (including a fall of 6.7 hPa in 10 min), rapid wind vector changes (e.g., 17 m/s in 25 min), and high wind gusts (e.g., 20 m/s). Individual convective storms can be generated locally in unstable atmospheric situations and are typically associated with surface pressure perturbations that, when propagating at appropriate speeds, can cause meteotsunamis (e.g., Belušić et al. 2009). Concerning the squall lines, or lines of thunderstorms, they often form along or ahead of cold fronts and can extend laterally for hundreds of kilometres. Long-lived strong squall lines, also called "derechos" (Spanish for 'straight'), can travel many hundreds of kilometres and are associated with air pressure disturbances that can generate meteotsunamis through Proudman resonance. An example of such an event occurred on 13 June 2013 in the northeastern United States where waves impacted the New Jersey and southern Massachusetts coasts, despite clear skies and calm weather. Wertman et al. (2014) demonstrated that the waves had been generated by a derecho that had passed through the area hours earlier. Finally, during tropical cyclones, the generation of meteotsunamis can be driven by either tropical cyclone rainbands—i.e., the regions of heaviest precipitation outside of the tropical cyclone eyeball—that can develop squall-line-like characteristics or by atmospheric internal gravity waves produced in the stratosphere and the lower troposphere. Shi et al. (2020) studied tropical cyclones driving meteotsunami events in the Gulf of Mexico between 1998 and 2018. They found that Hurricane Harvey (2017) produced the largest observed meteotsunami: a single-peak wave with a maximum elevation of 0.78 m.

Acoustic sources

Catastrophic events such as volcano explosions (Choi et al. 2003) or asteroid impacts/bolide explosions (Chapman and Morrison 1994; Morgan et al. 2022; deGroot-Hedlin et al. 2011) can generate atmospheric waves (Matoza et al. 2022; see Fig. 3.12) including acoustic-gravity waves (Yeh and Liu 1974) and the most prominent Lamb waves (Lamb 1911; Bretherton 1969). Lamb waves propagate horizontally in the atmosphere with a speed close to the mean sound speed (about 318 m/s; Dragoni and Santoro 2020) and can circle the globe multiple times (Press and Harkrider 1966; Amores et al. 2022). They are associated with surface pressure oscillations of several hectopascals (hPa) per minute (Press and Harkrider 1966; Harrison 2022) and their energy is dissipated towards the thermosphere (Forbes et al. 1999) via the ionosphere. Additionally, during the first hours after the explosive eruption of the Hunga Tonga–Hunga Ha'apai volcano in 2022, several Internal Gravity Waves (IGWs)—likely generated by the complex eruption sequence and traveling with speeds between 240 and 270 m/s—were recorded by seismo-acoustic and satellite observations (Matoza et al. 2022; Wright et al. 2022). The speed of these IGWs is thought to be largely controlled by the vertical profile of the temperature perturbation introduced by the eruption (e.g., Lashkin and Cheremnykh 2023). This suggests that under favourable conditions (e.g., a steeper initial temperature anomaly profile above the tropopause), large amplitude waves with speeds favourable to Proudman resonance can be generated (Denamiel et al. 2023).

3.5.4 Step 3: Probabilistic Treatment

Weather sources

No accurate APTHA has been achieved for meteotsunami events. Consequently, their probability of occurrence or their annual rates—including also the impact of climate change on the atmospheric tsunami sources—has yet to be defined. However, several methods accounting for the uncertainty, linked to the difficulty to accurately model the atmospheric conditions driving meteotsunami events, are currently developed, mostly for operational purposes (i.e., within early warning systems). To this date, the probabilistic treatment of the atmospheric sources is thus only applied to a few methods: (1) "Synoptic Index", (2) "Deterministic Atmospheric Model", (3) "Synthetic Catalogue" and (4) "Climate Uncertainty". The derivation of synoptic indices follows the findings of Lorenz (1956, 1969) who argues that two relatively similar atmospheric synoptic situations are likely to produce similar local effects. Synoptic indices have only been successfully automatized and implemented for operational meteotsunami forecasting in the Balearic Islands, Spain (Ramis and Jansà 1983). However, they have recently been derived worldwide and strong connections between meteotsunami events and synoptic patterns have been found along most of the world's coastlines (Zemunik et al. 2022). The advantage of the approach is that no additional numerical cost is required, as synoptic indices—identifying meteotsunami events up to a week in advance—can be derived and used with relatively coarse weather models (30–80 km in resolution) provided by many weather services. The major drawback is the impossibility to quantify the intensity of the events mostly related to mesoscale processes. Consequently, the implementation of (sub)-kilometre-scale deterministic models is an unavoidable necessity. To this date, this method have only been operationally implemented in the Balearic Islands (Renault et al. 2011) and the Adriatic Sea (Denamiel et al. 2019a, 2019b). In both systems the meteotsunamigenic disturbances triggering the event mode of the early warning systems are always detected but not necessarily their exact location, speed, period, etc. However, even for a 3-day simulation, running sub-kilometre scale models has been proven to have a prohibitive numerical cost and, for the Adriatic Sea, the operational forecasting service was discontinued.

Both "Synoptic Index" and "Deterministic Model" methods are by nature deterministic but probabilistic treatment can be obtained by implementing ensemble modelling using several numerical models in order to partially account for the known uncertainty of global and regional models (i.e., the difficulty to simulate the precise location and intensity of extreme events). Additionally, weather related meteotsunami events are driven by atmospheric conditions highly influenced by ongoing global warming. Deriving APTHA for weather related events thus requires accounting for the impact of climate change on the recurrence and intensity of the meteotsunamigenic disturbances. However, little is still known about this issue as presently only two studies have been published on the subject. One used the synoptic index-based approach in the Balearic Islands (Vilibić et al. 2018) and found that meteotsunami occurrences could increase by one third under extreme warming conditions (Representative Concentration Pathway RCP 8.5 scenario). The other used the pseudo-global warming (PGW) downscaling approach (Schär et al. 1996; Denamiel et al. 2020a) for half a dozen meteotsunami events in the Adriatic Sea (Denamiel et al. 2022) and found a strong impact of climate warming on the spatial variability of the meteotsunami intensity under RCP 8.5 scenario. However, both studies were carried out with a single Regional Climate Model (RCM) and for a single climate scenario, and thus do not account for the climate uncertainty.

In the "Climate Uncertainty" method, to limit the numerical cost of APTHA for weather related events, the "Synoptic Index" method can be applied to extract the potential meteotsunami events from ensembles of global climate models and (sub)-kilometer-scale atmosphere–ocean models. Then, the "Deterministic Atmospheric Model" method can be used to simulate the mesoscale behaviour of these potential events in order to only select those leading to meteotsunamis. Finally, as meteotsunamis can be categorized as rare events, synthetic catalogues—based on the simplification and parametrization of the meteotsunamigenic disturbances and the generation of thousands of conditions—can be directly used for uncertainty quantification. Currently, such a catalogue approach has only been used in the northeast U.S. (Geist et al. 2014c) to derive APTHA, but without accounting for the impact of climate change, and in the Adriatic Sea (Denamiel et al. 2019b, 2020b, 2021) to account for the uncertainty in meteotsunami simulations. The main advantage of the approach is that it is inherently probabilistic, but it uses simplistic representations of the meteotsunamigenic disturbances that might not fully reflect the complexity of the atmospheric signal.

Acoustic sources

The detection and modelling of the atmospheric acoustic-gravity waves are currently at their exploratory stage as most studies have only focused on the recent Hunga Tonga–Hunga Ha'apai volcano eruption. But as these events are extremely rare, only the systemic reproduction of past events and the forecast of potential events could allow better understanding of the acoustic-gravity wave-driven atmosphere–ocean dynamics and the quantification of the uncertainties linked to the transfer of energy from the explosive process (volcanic eruption or asteroid impact/bolide explosion) to the acoustic-gravity wave initial conditions used in the atmospheric model. This would involve the use of both (1) "Deterministic Atmospheric Model" method that can produce the worldwide atmospheric response after a volcanic explosion or an asteroid impact/bolide explosion and (2) "Synthetic Catalogues" allowing uncertainty quantification depending on the location and intensity of the acoustically driven events. As the speed of the acoustic-gravity waves depends on the Earth's surface temperature, climate change might impact acoustically driven events. Consequently, proper APTHA should in principle also account for climate uncertainty. Finally, it is important to note that, in contrast to other tsunami sources, inland volcanic explosions or asteroid impacts/bolide explosions (e.g., Tunguska explosion; Chyba et al. 1993) can also produce worldwide meteotsunamis.

3.5.5 Summary Table

See Table 3.5.

Table 3.5 Summary table: atmospheric sources

Short description of methods	
Theoretical/mechanical background for the methods	Two categories of atmospheric tsunamis (or meteotsunamis): (1) weather related events driven by mesoscale processes and (2) acoustically driven events generating planetary acoustic-gravity waves after a volcanic explosion or an asteroid impact/bolide explosion
Short outline of how the methods work	Types of weather related events: (1) mesoscale gravity waves, (2) squall lines/derechos and (3) hurricanes Types of acoustic-gravity waves during acoustically driven events: (1) Lamb waves that can circle the globe several times and (2) internal gravity waves near the epicentre
How are the output(s) of the methods represented?	Type of studies (methods) using probabilistic treatment: (1) for weather related events: forecast from synoptic to local scales (synoptic index and operational forecast), climate warming impact from synoptic to local scales (climate model) and stochastic hazard assessments at local scale (synthetic catalogue) and (2) for acoustically driven events: remains to be performed
In what way do the methods take into account time-dependent/dynamic aspects?	Fully integrated for all atmospheric sources which highly vary in time (at both weather and climatic scales) and space (local, synoptic, and global scales)
Data and technical requirements for method	
What input data does the method(s) use?	Weather related events: (1) hotspot location database and (2) mesoscale atmospheric process identification (i.e., observation and/or modelling) Acoustically driven events: (1) explosive volcano/asteroid impact/bolide explosion location database and (2) acoustic-gravity wave identification (i.e., observation and/or modelling)
What are the technical requirements?	Step 1—mostly qualitative, assessing what is the source category and identifying hotspot locations for hazard analysis Step 2—Characterizing the different atmospheric sources Step 3—Probabilistic treatment based on the mesoscale processes/worldwide acoustic-gravity waves depending on the type of study to perform (forecast, climate, hazard assessment)
Is the method(s)/workflow(s) available as a git/software/jupyter notebook?	No

(continued)

Table 3.5 (continued)

Treatment of uncertainties for the method	
Is the method(s) itself deterministic or based on uncertainty quantification (UQ)?	The identification of the category of atmospheric source (weather or acoustic) and the types of sources (e.g., derecho, Lamb wave, etc.) as well as the choice of the type of studies are all deterministic. Uncertainty quantification can be obtained with ensemble modelling and synthetic catalogues
Which are potential sources of uncertainties? Which parameters are used to characterize them?	The main sources of uncertainties are: (1) poor/limited/sparse observational dataset and (2) limited knowledge of the mesoscale atmospheric processes/acoustic-gravity waves The parameters describing the uncertainties are: (1) for weather related events: direction, start location, amplitude, speed, period, etc. and (2) for acoustically driven events: initial condition at the epicentre and the Earth's surface temperatures
What are the consequences of lack of data in input or for boundary conditions, or for characterizing certain parameters?	Incapacity to properly identify the atmospheric process driving the meteotsunami events and improper modelling strategies
What are the consequences of ignoring or poorly treating uncertainties?	Lack of accuracy of the forecasts (false alarms/missed events) and the climate projections

3.6 Choosing What Method to Use

3.6.1 Earthquake Sources

The main drivers in the choice of the method are the instrumental coverage, historical investigations, and availability of earthquake catalogues inherited from previous efforts. Time and budget constraints control the level and granularity of the specific investigations (e.g. merging and harmonizing seismic catalogues for large regions is needed for PTHA). Regarding faults, the availability of offshore survey data to characterize the faults in detail is a first order factor in the choice of the method. PTHA projects often rely on existing fault compilations derived from literature data. Again, the choices of the project managers are controlled by time and budget constraints for detailed analysis at the scale of the single structure/system, versus the availability of fault catalogues inherited from previous projects.

The use of earthquake catalogues tends to locate future earthquakes near the locations of past earthquakes and to overlook potential sources with long recurrence intervals (in the so-called silent regions) and, in general, the larger earthquakes. The use of faults tends to locate earthquakes near known faults, but not all faults can be known with all the necessary parameters. The partition of the total moment rate into earthquakes of different sizes also depends on seismicity rate parameters such as the b-value.

Regarding the source parameterization, it is important to find the optimal trade-off between precision and accuracy. If the variability regarding, for example, the possible earthquake locations is too large, the precision can be reduced; however, if the parameter range is too narrow, there is a risk of low accuracy (bias). A similar reasoning can be applied for the MFD and time-dependence, in the sense that the more sophisticated the approach, the more data is needed to constrain it.

Dynamically consistent models of the seismic cycle are rarely used but there are promising examples. Time-dependent models are used infrequently given that simpler methods are often considered more reliable, at least on the tsunami time scales that are longer than those of the earthquakes.

3.6.2 Landslides

Available data dictates the method selection and whether a rigorous probabilistic analysis is at all possible. In areas with a relatively high amount of past large landslides, a source inventory based on past data can be established. Dating or geological time information must be available to constrain the time envelope of the events, otherwise uncertainties can be excessive. Slope-stability information gives more information about the present day situation and may hence be preferred when available. However, this approach implies that a relatively broad regional analysis of the slope movements is available, otherwise certain source areas may be overlooked. A difficulty with the slope-stability based models is that the probability of occurrence can not always be tied to the statistics of past events, unless statistical information of past landslides or probability models linking the trigger to potential slope failure and trigger is available (e.g. earthquake triggers and PSHA). Without any data constraining the volume and frequency of the landslide, either past landslides (or landslide tsunamis), or slope stability data, it is not possible to perform a probabilistic analysis or a source sensitivity analysis bounded by the data. A simpler conditional probabilistic analysis can be carried out instead. Carrying out a deterministic scenario analysis without uncertainty treatment is not recommended as the hazard is likely to be strongly biased on the chosen event.

3.6.3 Volcanoes

The driving issues that control the selection of one specific approach are (i) the computational need of the simulations (Sect. 4.4) and/or the availability of HPC resources; and (ii) the difficulty of defining the potential ranges of parameters and their probability of occurrence.

Probabilistic methods are often avoided also due to the difficulty in constraining ranges and spatio-temporal probabilities for many of the potential volcanic sources (e.g. Esposti-Ongaro et al. 2021). To this end, a few past events are often selected, typically choosing a few past scenarios in hindcasting mode. As discussed in Chap. 2, all the approaches selecting few scenarios typically imply a few tsunami simulations and allow skipping the quantification of the probability of each individual scenario (see also Step 3 in Sect. 3.4), simultaneously solving both the computational and lack of knowledge issues, and strongly simplifying the entire workflow. However, single scenarios can hardly represent the natural variability, preventing the possibility to fully quantify the hazard.

Sometimes, a few scenarios roughly covering the spatial or size variability are proposed (e.g. Tinti et al. 2011; Zaniboni et al. 2013; Necmioglu et al. 2023; Selva et al. 2019). This approach again does not allow for a full exploration of uncertainty, but at least may be considered as a preliminary step toward the probabilistic hazard, to quantify the potential of a given type of source. For example, this may provide quantitative information to feed a tool for the selection of a volcano and/or a typology of source (see Step 1 in Sect. 3.4). This information may thus feed probabilistic methods (like for example event trees), allowing an advanced exploration of source variability, as required for full hazard and risk quantifications.

The selection of a full probabilistic method allows for a full quantification of the natural variability, allowing to fully quantify hazard and risk. However, it implies the need of linking a forecast model for the source to an efficient tsunami modelling. The event tree technique has been successfully applied to this end, at least for tsunamis generated by eruptive events. Also, the treatment of time-dependency and the potential for multiple hazards acting at the same time are fundamental for all volcanic sources (e.g. IAEA 2008). This may be treated by considering, at least, the state of the volcano and potential changes of it (Marzocchi and Bebbington 2012; Selva et al. 2022). Tools like event trees have been explicitly designed to this end (Newhall and Hoblitt 2002; Newhall and Pallister 2015; Marzocchi et al. 2008, 2010). Other methods are instead devoted to treating the forecasts within a specific period, e.g. an unrest, like for example Belief Networks (e.g. Aspinal and Woo 2014, Hincks et al. 2014). In all cases, the use of expert elicitation is often recommended to treat difficult-to-constrain problems, at least to reach a consensus among experts over the results (Aspinall et al. 2002; Neri et al. 2008; Selva et al. 2012a, b).

In any case, considering the difficulty of selecting the source and to define the effective range of variability and rates, whatever is the probabilistic approach selected to explore aleatory uncertainty, methods explicitly treating epistemic uncertainty (like logic trees, ensemble, or Bayesian methods) are highly recommended.

3.6.4 Atmospheric Sources

For meteotsunamis, the selection, characterization, detection and modelling of the atmospheric sources is mostly straightforward and only depends on (1) the category of events (i.e., weather or acoustic sources), (2) the location and, hence, the type of synoptic conditions (e.g., strong mid-troposphere jets) and/or of mesoscale atmospheric processes (e.g., mesoscale gravity waves, derechos, hurricanes, etc.) driving the weather related events or the type of acoustic-gravity waves during acoustically driven events and (3) the kind of study (e.g., forecast, climate change, hazard assessments, etc.). It should be noted that, to this date, APTHA is at its very early stage and that more advanced methods are expected to be developed and implemented in the next decade with the integration of meteotsunami sources within the wider tsunami community. Consequently, the focus in meteotsunami research is presently mostly oriented towards forecasting and hindcasting (understanding past events), and long-term hazard analysis still in its infancy. Yet, several of the methods discussed above comprise possible building blocks that likely can be combined to perform future probabilistic analysis.

References

Adam D (2022) Tonga volcano eruption created puzzling ripples in earth's atmosphere. Nature 601:497. https://doi.org/10.1038/d41586-022-00127-1

Albarello D, Camassi R, Rebez A (2001) Detection of space and time heterogeneity in the completeness of a seismic catalog by a statistical approach: an application to the Italian area. Bull Seism Soc Am 91:1694–1703. https://doi.org/10.1785/0120000058

Albini P (2020) Documenting earthquakes in the United States of the Ionian Islands, 1815–1864. Seism Res Lett 91:2554–2562. https://doi.org/10.1785/0220200056

Albini P, Rovida A (2018) Earthquakes in southern Dalmatia and coastal Montenegro before the large 6 April 1667 event. J Seism 22:721–754. https://doi.org/10.1007/s10950-018-9730-4

Albini P, Musson RMW, Gómez Capera AA, Locati M, Rovida A, Stucchi M, Viganò D (2013) Global historical earthquake catalogue (GHEC) [Data set]. Istituto Nazionale di Geofisica e Vulcanologia (INGV). https://doi.org/10.13127/ghea/ghec.1.0

Albini P, Musson RMW, Rovida A, Locati M, Capera AAG, Viganò D (2014) The global earthquake history. Earthq Spec 30:607–624. https://doi.org/10.1193/122013eqs297

Allen TI, Hayes GP (2017) Alternative rupture-scaling relationships for subduction interface and other offshore environments. Bull Seism Soc Am 107:1240–1253. https://doi.org/10.1785/0120160255

Ambraseys NN, Melville C (1982) A history of Persian earthquakes. Cambridge University Press, Cambridge, ISBN-13 978-0-521-24112-0

Amores A, Monserrat S, Marcos M, Argüeso D, Villalonga J, Jordà G, Gomis D (2022) Numerical simulation of atmospheric Lamb waves generated by the 2022 Hunga-Tonga volcanic eruption. Geophys Res Lett 49:e2022GL098240. https://doi.org/10.1029/2022GL098240

Anderson JG, Luco JE (1983) Consequences of slip rate constraints on earthquake occurrence relations. Bull Seism Soc Am 73:471–496. https://doi.org/10.1785/BSSA0730020471

Ando M (1979) The Hawaii earthquake of November 29, 1975: Low dip angle faulting due to forceful injection of magma. J Geophys Res Solid Earth 84(B13):7616–7626

Annaka T, Satake K, Sakakiyama T, Yanagisawa K, Shuto N (2007) Logic-tree approach for probabilistic tsunami hazard analysis and its applications to the Japanese coasts. Pure Appl Geophys 164:577–592. https://doi.org/10.1007/s00024-006-0174-3

Aspinall WP (2006) Structured elicitation of expert judgment for probabilistic hazard and risk assessment in volcanic eruptions. Statistics Volc 1:15–30

Aspinall WP (2010) A route to more tractable expert advice. Nature 463:294–295

Aspinall WP, Woo G (2014) Santorini unrest 2011–2012: an immediate Bayesian belief network analysis of eruption scenario probabilities for urgent decision support under uncertainty. J Appl Volc 3:12. https://doi.org/10.1186/s13617-014-0012-8

Aspinall WP, Loughlin SC, Michael FV, Miller AD, Norton GE, Rowley KC et al (2002) The Montserrat volcano observatory: its evolution, organization, role and activities. Geol Soc London Memoirs 21(1):71–91

Aspinall WP, Woo G, Voight B, Baxter PJ (2003) Evidence-based volcanology: application to eruption crises. J Volc Geotherm Res 128(1–3):273–285. https://doi.org/10.1016/S0377-0273(03)00260-9

Astafyeva E, Maletckii B, Mikesell TD, Munaibari E, Ravanelli M, Coisson P et al (2022) The 15 January 2022 Hunga Tonga eruption history as inferred from ionospheric observations. Geophys Res Lett 49:e2022GL098827. https://doi.org/10.1029/2022GL098827

Bakun WH, Wentworth CM (1997) Estimating earthquake location and magnitude from seismic intensity data. Bull Seism Soc Am 87:1502–1521. https://doi.org/10.1785/BSSA0870061502

Basili R, Tiberti MM, Kastelic V, Romano F, Piatanesi A, Selva J et al (2013) Integrating geologic fault data into tsunami hazard studies. Nat Hazards Earth Syst Sci 13:1025–1050. https://doi.org/10.5194/nhess-13-1025-2013

Basili R, Brizuela B, Herrero A, Iqbal S, Lorito S, Maesano FE et al (2021a) The making of the NEAM tsunami hazard model 2018 (NEAMTHM18). Front Earth Sci 8. https://doi.org/10.3389/feart.2020.616594. Last accessed 13 Nov 2022

Basili R, Brizuela B, Herrero A, Iqbal S, Lorito S, Maesano FE et al (2021b) The making of the NEAM tsunami hazard model 2018 (NEAMTHM18). Front Earth Sci 8. https://doi.org/10.3389/feart.2020.616594

Beauval C, Yepes H, Palacios P, Segovia M, Alvarado A, Font Y, Aguilar J, Troncoso L, Vaca S (2013) An Earthquake catalog for seismic hazard assessment in Ecuador. Bull Seism Soc Am 103:773–786. https://doi.org/10.1785/0120120270

Bebbington MS (2013) Assessing spatio-temporal eruption forecasts in a monogenetic volcanic field. J Volc Geotherm Res 252:14–28

Bebbington MS, Jenkins SF (2019) Intra-eruption forecasting. Bull Volc 81:34. https://doi.org/10.1007/s00445-019-1294-9

Bebbington MS, Jenkins SF (2022) Intra-eruption forecasting using analogue volcano and eruption sets. J Geophys Res Solid Earth 127:e2022JB024343. https://doi.org/10.1029/2022JB024343

Bebbington MS (2010) Trends and clustering in the onsets of volcanic eruptions. J Geophys Res Solid Earth 115(B1)

Bebbington MS (2014) Volcanic eruptions: Stochastic models of occurrence patterns. In: Meyers RA (ed) Encyclopedia of complexity and systems science, Springer, New York, 2014, pp 1–58

Bechle AJ, Wu CH, Kristovich DAR, Anderson EJ, Schwab DJ, Rabinovich AB (2016) Meteotsunamis in the Laurentian Great Lakes. Sci Rep 6:37832. https://doi.org/10.1038/srep37832

Begét JE (2000) Volcanic tsunami. In Sigurdsson H (ed) Encyclopedia of volcanoes, Elsevier, New York, pp 1005–1013

Behrens J, Løvholt F, Jalayer F, Lorito S, Salgado-Gálvez MA, Sørensen M et al (2021) Probabilistic tsunami hazard and risk analysis: a review of research gaps. Front Earth Sci 9:628772. https://doi.org/10.3389/feart.2021.628772

Belušić D, Strelec Mahović N (2009) Detecting and following atmospheric disturbances with a potential to generate meteotsunamis in the Adriatic. Phys Chem Earth 34:918–927. https://doi.org/10.1016/j.pce.2009.08.009

Berryman K, Wallace L, Hayes GP, Bird P, Wang K, Basili R et al (2015) The GEM faulted earth subduction interface characterisation project, version 2.0, April 2015. http://itic.ioc-unesco.org/index.php?option=com_oe&task=viewDocumentRecord&docID=15688

Biass S, Frischknecht C, Bonadonna C (2012) A fast GIS-based risk assessment for tephra fallout: the example of Cotopaxi volcano, Ecuador-Part II: vulnerability and risk assessment. Nat Hazards 64:615–639

Biass S, Bonadonna C, Di Traglia F, Pistolesi M, Rosi M, Lestuzzi P (2016) Probabilistic evaluation of the physical impact of future tephra fallout events for the Island of Vulcano, Italy. Bull Volc 78:1–22

Bird P (2003) An updated digital model of plate boundaries. Geochem Geophys Geosys 4:1027. https://doi.org/10.1029/2001GC000252

Bird P, Kagan YY (2004) Plate-tectonic analysis of shallow seismicity: apparent boundary width, beta, corner magnitude, coupled lithosphere thickness, and coupling in seven tectonic settings. Bull Seism Soc Am 94:2380–2399. https://doi.org/10.1785/0120030107

Blaser L, Kruger F, Ohrnberger M, Scherbaum F (2010) Scaling relations of earthquake source parameter estimates with special focus on subduction environment. Bull Seism Soc Am 100:2914–2926. https://doi.org/10.1785/0120100111

Bommer JJ, Stafford PJ (2009) Seismic hazard and earthquake actions. In: Seismic design of buildings to Eurocode 8, CRC Press, Taylor and Francis, London, pp 6–46. https://www.taylorfrancis.com/chapters/edit/https://doi.org/10.1201/9781482266177-9/seismic-hazard-earthquake-actions-bommer-stafford?context=ubx&refId=60334b9b-5f51-4e7c-949b-25ee9baba2a4

Bommer JJ, Scherbaum F (2008) The use and misuse of logic trees in probabilistic seismic hazard analysis. Earthq Spectra 24:997–1009

Bonaccorso A, Calvari S, Garfì G, Lodato L, Patanè D (2003) Dynamics of the December 2002 flank failure and tsunami at Stromboli volcano inferred by volcanological and geophysical observations. Geophys Res Lett 30:1941. https://doi.org/10.1029/2003GL017702

Bougouin A, Paris R, Roche O (2020) Impact of fluidized granular flows into water: Implications fortsunamis generated by pyroclastic flows. J Geophys Res Solid Earth 125:e2019JB018954. https://doi.org/10.1029/2019JB018954

Brax M, Albini P, Beauval C, Jomaa R, Sursock A (2019) An earthquake catalog for the Lebanese region. Seism Res Lett 90:2236–2249. https://doi.org/10.1785/0220180292

Bretherton F (1969) Lamb waves in a nearly isothermal atmosphere. Quart J Roy Meteor Soc 95:754–757. https://doi.org/10.1002/qj.49709540608

Budnitz RJ, Apostolakis G, Boore DM (1997) Recommendations for probabilistic seismic hazard analysis: guidance on uncertainty and use of experts. United States. https://doi.org/10.2172/479072

Cañón-Tapia (2022) Kernel analyses of volcanic vent distribution: how accurate and complete are the objective bandwidth selectors? Front Earth Sci 10. https://doi.org/10.3389/feart.2022.779095

Carazzo G, Kaminski E, Tait S (2015) The timing and intensity of column collapse during explosive volcanic eruptions. Earth Plan Sci Lett 411:208–217. https://doi.org/10.1016/j.epsl.2014.12.006

Carlton B, Vanneste M, Forsberg CF, Knudsen S, Løvholt F, Kvalstad T, Holm S, Kjennbakken H, Mazhar MA, Degago S, Haflidason H (2019) Geohazard assessment related to submarine instabilities in Bjørnafjorden, Norway. Geol Soc London Special Publ 477(1):549–566. https://doi.org/10.1144/SP477.39

Chapman CR, Morrison D (1994) Impacts on the Earth by asteroids and comets: assessing the hazard. Nature 367:33–40. https://doi.org/10.1038/367033a0

Chapple WM, Forsyth DW (1979) Earthquakes and bending of plates at trenches. J Geophys Res Solid Earth 84:6729–6749. https://doi.org/10.1029/JB084iB12p06729

Choi BH, Pelinovsky E, Kim KO, Lee JS (2003) Simulation of the trans-oceanic tsunami propagation due to the 1883 Krakatau volcanic eruption. Nat Hazards Earth Syst Sci 3:321–332. https://doi.org/10.5194/nhess-3-321-2003

Chyba C, Thomas P, Zahnle K (1993) The 1908 Tunguska explosion: atmospheric disruption of a stony asteroid. Nature 361:40–44. https://doi.org/10.1038/361040a0

Clift P, Vannucchi P (2004) Controls on tectonic accretion versus erosion in subduction zones: implications for the origin and recycling of the continental crust: subduction techtonics. Rev Geophys 42. https://doi.org/10.1029/2003RG000127

Connor CB, Connor L, Germa A, Richardson J, Bebbington M, Gallant E, Saballos A (2019) How to use kernel density estimation as a diagnostic and forecasting tool for distributed volcanic vents. SIV 4:1–25. https://doi.org/10.5038/2163-338X.4.3

Cornell CA (1968) Engineering seismic risk analysis. Bull Seism Soc Am 58:1583–1606. https://doi.org/10.1785/BSSA0580051583

Davies G, Griffin J (2018) The 2018 Australian probabilistic tsunami hazard assessment: hazard from earthquake generated tsunamis. Geosci Aust. https://doi.org/10.11636/Record.2018.041

Davies G, Griffin J (2020) Sensitivity of probabilistic tsunami hazard assessment to far-field earthquake slip complexity and rigidity depth-dependence: case study of Australia. Pure Appl Geophys 177:1521–1548. https://doi.org/10.1007/s00024-019-02299-w

Davies G, Griffin J, Løvholt F, Glimsdal S, Harbitz C, Thio HK et al (2018) A global probabilistic tsunami hazard assessment from earthquake sources. Geol Soc Lond Spec Publ 456:219–244. https://doi.org/10.1144/sp456.5

Davies, G (2019) Tsunami variability from uncalibrated stochastic earthquake models: tests against deep ocean observations 2006–2016. Geophys J Int 218(3):1939–1960

Davies G, Romano F, Lorito S (2020) Global dissipation models for simulating tsunamis at far-field coasts up to 60 hours post-earthquake: multi-site tests in Australia. Front Earth Sci. https://doi.org/10.3389/feart.2020.598235

Davies G, Weber R, Wilson K, Cummins P (2022) From offshore to onshore probabilistic tsunami hazard assessment via efficient Monte Carlo sampling. Geophys J Int 230(3):1630–1651

Day SJ, Watts P, Grilli ST, Kirby JT (2005) Mechanical models of the 1975 Kalapana, Hawaii earthquake and tsunami. Mar Geol 215(1–2):59–92

Day SJ (2015) Volcanic tsunamis. In: Sigurdsson H (ed) Encyclopedia of volcanoes, Second edn, Academic Press, pp 993–1009, ISBN 9780123859389. https://doi.org/10.1016/B978-0-12-385938-9.00058-4

de Groot-Hedlin C, Hedlin MAH, Walker K (2011) Finite difference synthesis of infrasound propagation through a windy, viscous atmosphere: application to a bolide explosion detected by seismic networks. Geophys J Int 185:305–320. https://doi.org/10.1111/j.1365-246X.2010.04925.x

De la Cruz-Reyna S (1993) Random patterns of occurrence of explosive eruptions at Colima volcano. Mexico J Volc Geotherm Res 55(1–2):51–68

Denamiel C, Šepić J, Ivanković D, Vilibić I (2019a) The Adriatic sea and coast modelling suite: evaluation of the meteotsunami forecast component. Ocean Model 135:71–93. https://doi.org/10.1016/j.ocemod.2019.02.003

Denamiel C, Šepić J, Huan X, Bolzer C, Vilibić I (2019b) Stochastic surrogate model for meteotsunami early warning system in the eastern Adriatic Sea. J Geophys Res Oceans 124:8485–8499. https://doi.org/10.1029/2019JC015574

Denamiel C, Pranić P, Quentin F, Mihanović H, Vilibić I (2020a) Pseudo-global warming projections of extreme wave storms in complex coastal regions: the case of the Adriatic Sea. Clim Dyn 55:2483–2509. https://doi.org/10.1007/s00382-020-05397-x

Denamiel C, Huan X, Šepić J, Vilibić I (2020b) Uncertainty propagation using polynomial chaos expansions for extreme sea level hazard assessment: the case of the eastern Adriatic meteotsunamis. J Phys Oceanogr 50(4):1005–1021. https://doi.org/10.1175/JPO-D-19-0147.1D

Denamiel C, Huan X, Vilibić I (2021) Conceptual design of extreme sea-level early warning systems based on uncertainty quantification and engineering optimization methods. Front Mar Sci 8:650279. https://doi.org/10.3389/fmars.2021.650279

Denamiel C, Vasylkevych S, Žagar N, Zemunik P, Vilibić I (2023) Destructive potential of planetary Meteotsunami waves beyond the Hunga Tonga-Hunga Ha'apai volcano eruption. Bull Am Meteorol Soc 104(1):E178–E191. https://doi.org/10.1175/BAMS-D-22-0164.1

Denamiel C, Tojčić I, Vilibić I (2022) Meteotsunamis in orography-free, flat bathymetry and warming climate conditions. J Geophys Res Oceans 127:e2021JC017386. https://doi.org/10.1029/2021JC017386

DPC-INGV 2016–2017. Allegato B2

Dragoni M, Santoro D (2020) A model for the atmospheric shock wave produced by a strong volcanic explosion. Geophys J Int 222:735–742. https://doi.org/10.1093/gji/ggaa205

Duffy DG (1992) On the generation of oceanic surface waves by underwater volcanic explosions. J Volc Geotherm Res 50(3):323–344. https://doi.org/10.1016/0377-0273(92)90100-R

Dusek G, Di Veglio C, Licate L, Heilman L, Kirk K, Paternostro C, Miller A (2019) A meteotsunami climatology along the U.S. East coast. Bull Am Meteorol Soc 100:1329–1345. https://doi.org/10.1175/BAMS-D-18-0206.1

Dutta R, Jónsson S, Vasyura-Bathke H (2021) Simultaneous Bayesian estimation of non-planar fault geometry and spatially-variable slip. J Geophys Res Solid Earth 126:e2020JB020441. https://doi.org/10.1029/2020JB020441

Egorov Y (2007) Tsunami wave generation by the eruption of underwater volcano. Nat Hazards Earth Syst Sci 7:65–69. https://doi.org/10.5194/nhess-7-65-2007

Ekström G, Nettles M, Dziewoński AM (2012) The global CMT project 2004–2010: centroid-moment tensors for 13,017 earthquakes. Phys Earth Planet Inter 200–201:1–9. https://doi.org/10.1016/j.pepi.2012.04.002

Ellsworth WL, Matthews MV, Nadeau RM, Nishenko SP, Reasenberg PA, Simpson RW (1999) A physically-based earthquake recurrence model for estimation of long-term earthquake probabilities. US Geol Surv Open-File Rep 99(522):22

Esposti Ongaro T, de'Michieli Vitturi M, Cerminara M, Fornaciai A, Nannipieri L, Favalli M et al (2021) Modeling tsunamis generated by submarine landslides at Stromboli volcano (Aeolian Islands, Italy): a numerical benchmark study. Front Earth Sci 9:628652

Fäh D, Giardini D, Kästli P, Deichmann N, Gisler M, Schwarz-Zanetti G et al (2011) ECOS-09 Earthquake catalogue of Switzerland release 2011 report and database

Felzer KR, Cao T (2008) WGCEP historical California earthquake catalog, Appendix H in The Uniform California earthquake rupture forecast, version 2 (UCERF 2), U.S. geological survey open-file report 2007–1437H and California geological survey special report 203H, p 127. https://pubs.usgs.gov/of/2007/1437/h/

Forbes JM, Palo SE, Zhang X, Portnyagin YI, Makarov NA, Merzlyakov EG (1999) Lamb waves in the lower thermosphere: Observational evidence and global consequences. J Geophys Res 104:17107–17115. https://doi.org/10.1029/1999JA900044

Freundt A (2003) Entrance of hot pyroclastic flows into the sea: experimental observations. Bull Volc 65:144–164

Fritz HM, Hager WH, Minor HE (2004) Near field characteristics of landslide generated impulse waves. J Waterw Port Coastal Ocean Eng 130(6):287–302

Gardner JK, Knopoff L (1974) Is the sequence of earthquakes in southern California, with aftershocks removed, Poissonian? Bull Seism Soc Am 64:1363–1367. https://doi.org/10.1785/BSSA0640051363

Gasperini P, Bernardini F, Valensise G, Boschi E (1999) Defining seismogenic sources from historical earthquake felt reports. Bull Seism Soc Am 89:94–110. https://doi.org/10.1785/BSSA0890010094

Gasperini P, Vannucci G, Tripone D, Boschi E (2010) The Location and sizing of historical earthquakes using the attenuation of Macroseismic intensity with distance. Bull Seism Soc Am 100:2035–2066. https://doi.org/10.1785/0120090330

Geist EL, Parsons T (2006) Probabilistic analysis of tsunami hazards. Nat Hazards 37:277–314. https://doi.org/10.1007/s11069-005-4646-z

Geist EL, Parsons T (2014) Undersampling power-law size distributions: effect on the assessment of extreme natural hazards. Nat Hazards 72:565–595. https://doi.org/10.1007/s11069-013-1024-0

Geist EL, ten Brink US (2019) Offshore landslide hazard curves from mapped landslide size distributions. J Geophys Res Solid Earth 124(4):3320–3334. https://doi.org/10.1029/2018JB017236

Geist EL, Chaytor JD, Parsons T, ten Brink U (2013) Estimation of submarine mass failure probability from a sequence of deposits with age dates. Geosphere 9(2):287–298. https://doi.org/10.1130/GES00829.1

Geist EL, ten Brink US, Gove M (2014) Framework for the probabilistic analysis of meteotsunamis. Nat Hazards 74:123–142. https://doi.org/10.1007/s11069-014-1294-1

Geist EL, Lynett PJ (2014a) Source processes for the probabilistic assessment of tsunami hazards. Oceanography 27(2):86–93. http://www.jstor.org/stable/24862158

Gerstenberger MC, Marzocchi W, Allen T, Pagani M, Adams J, Danciu L et al (2020) Probabilistic seismic hazard analysis at regional and national scales: state of the art and future challenges. Rev Geophys 58:e2019RG000653. https://doi.org/10.1029/2019RG000653

Goda K, Yasuda T, Mori N, Maruyama T (2016) New scaling relationships of earthquake source parameters for stochastic tsunami simulation. Coast Eng J 58:1650010–1650011. https://doi.org/10.1142/S0578563416500108

Goda K (2020) Multi-hazard portfolio loss estimation for time-dependent shaking and tsunami hazards. Front Earth Science 8. https://doi.org/10.3389/feart.2020.592444

Gómez de la Peña L, Gràcia E, Maesano FE, Basili R, Kopp H, Sánchez-Serra C et al (2022) A first appraisal of the seismogenic and tsunamigenic potential of the largest fault systems in the westernmost Mediterranean. Mar Geol 445:106749. https://doi.org/10.1016/j.margeo.2022.106749

González FI, Geist EL, Jaffe B, Kânoğlu U, Mofjeld H, Synolakis CE et al (2009) Probabilistic tsunami hazard assessment at seaside, Oregon, for near- and far-field seismic sources. J Geophys Res 114:C11023. https://doi.org/10.1029/2008JC005132

Gray JP, Monaghan JJ (2003) Caldera collapse and the generation of waves. Geochem Geophys Geosyst 4(2):1015. https://doi.org/10.1029/2002GC000411

Greenspan HP (1956) The generation of edge waves by moving pressure disturbances. J Fluid Mech 1:574–592

Grezio A, Marzocchi W, Sandri L, Gasparini P (2010) A Bayesian procedure for probabilistic tsunami hazard assessment. Nat Hazards 53:159–174. https://doi.org/10.1007/s11069-009-9418-8

Grezio A, Babeyko A, Baptista MA, Behrens J, Costa A, Davies G et al (2017) Probabilistic tsunami hazard analysis: multiple sources and global applications. Revi Geophys 55:1158–1198. https://doi.org/10.1002/2017RG000579

Grezio A, Cinti FR, Costa A, Faenza L, Perfetti P, Pierdominici S et al (2020) Multisource Bayesian probabilistic tsunami hazard analysis for the Gulf of Naples (Italy). J Geophys Res Oceans 125:e2019JC015373. https://doi.org/10.1029/2019JC015373

Grilli ST, Shelby M, Kimmoun O, Dupont G, Nicolsky D, Ma G et al (2017) Modeling coastal tsunami hazard from submarine mass failures: effect of slide rheology, experimental validation, and case studies off the US East coast. Nat Hazards 86:353–391

Grilli ST, Watts P (2005) Tsunami generation by submarine mass failure. I: modeling, experimental validation, and sensitivity analyses. J Waterway Port Coast Ocean Eng 131(6):283–297. https://doi.org/10.1061/(ASCE)0733-950X(2005)131:6(283)

Grilli ST, Zhang C, Kirby JT, Grilli AR, Tappin DR, Watt SFL et al (2021) Modeling of the Dec. 22nd 2018 Anak Krakatau volcano lateral collapse and tsunami based on recent field surveys: comparison with observed tsunami impact. Mar Geol 440:106566. https://doi.org/10.1016/j.margeo.2021.106566

Grünthal G, Wahlström R (2012) The European-mediterranean earthquake catalogue (EMEC) for the last millennium. J Seismol 16:535–570. https://doi.org/10.1007/s10950-012-9302-y

Grünthal G, Mayer-Rosa D, Lenhardt W (1998) Abschätzung der Erdbebengefährdung für die D-A-CH-Staaten—Deutschland, Österreich, Schweiz. Bautechnik 75(10):753–767

Gutenberg B, Richter CF (1944) Frequency of earthquakes in California. Bull Seism Soc Am 34:185–188. https://doi.org/10.1785/BSSA0340040185

Gutenberg B, Richter CF (1949) Seismicity of the earth and associated phenomena. Princeton University Press, 273 pp

Hakimhashemi AH, Grunthal G (2012) A statistical method for estimating catalog completeness applicable to long-term nonstationary seismicity data. Bull Seism Soc Am 102:2530–2546. https://doi.org/10.1785/0120110309

Harbitz CB, Løvholt F, Bungum H (2014a) Submarine landslide tsunamis: how extreme and how likely? Nat Hazards 72:1341–1374. https://doi.org/10.1007/s11069-013-0681-3

Harbitz CB, Glimsdal S, Løvholt F, Kveldsvik V, Pedersen GK, Jensen A (2014b) Rockslide tsunamis in complex fjords: From an unstable rock slope at Åkerneset to tsunami risk in western Norway. Coastal Eng 88:101–122

Harrison G (2022) Pressure anomalies from the January 2022 Hunga Tonga-Hunga Ha'apai eruption. Weather 77:87–90. https://doi.org/10.1002/wea.4170

Hasterok D, Halpin JA, Collins AS, Hand M, Kreemer C, Gard MG et al (2022) New maps of global geological provinces and tectonic plates. Earth-Sci Rev 231:104069. https://doi.org/10.1016/j.earscirev.2022.104069

Hayes GP, Moore GL, Portner DE, Hearne M, Flamme H, Furtney M et al (2018) Slab2, a comprehensive subduction zone geometry model. Science 362:58–61. https://doi.org/10.1126/science.aat4723

Herd RA, Edmonds M, Bass VA (2005) Catastrophic lava dome failure at Soufriere hills volcano, Montserrat, 12–13 July 2003. J Volcanol Geoth Res 148:234–252

Hermanns RL, Oppikofer T, Anda E, Blikra LH, Böhme M, Bunkholt H et al (2012) Recommended hazard and risk classification system for large unstable rock slopes in Norway. NGU report, 0800–3416

Hincks TK, Komorowski J-C, Sparks RSJ, Aspinall WP (2014) Retrospective analysis of uncertain eruption precursors at La Soufrière volcano, Guadeloupe, 1975–77: volcanic hazard assessment using a Bayesian belief network approach. J Appl Volcanol 3(3) https://doi.org/10.1186/2191-5040-3-3

Holschneider M, Zöller G, Clements R, Schorlemmer D (2014) Can we test for the maximum possible earthquake magnitude? J Geophys Res Solid Earth 119:2019–2028. https://doi.org/10.1002/2013JB010319

Horspool N, Pranantyo I, Griffin J, Latief H, Natawidjaja DH, Kongko W et al (2014) A probabilistic tsunami hazard assessment for Indonesia. Nat Hazards Earth Syst Sci 14:3105–3122. https://doi.org/10.5194/nhess-14-3105-2014

Hough SE (2013) Missing great earthquakes. J Geophys Res Solid Earth 118:1098–1108. https://doi.org/10.1002/jgrb.50083

IAEA (2012) Volcanic hazards in site evaluation for nuclear installations IAEA safety standards series SSG-21. IAEA, Vienna

IAEA (2016) IAEA volcanic hazard assessments for nuclear installations. https://public.ebookcentral.proquest.com/choice/publicfullrecord.aspx?p=4853361

International Seismological Centre (2023) On-line bulletin. https://doi.org/10.31905/D808B830

Jansà A, Monserrat S, Gomis D (2007) The rissaga of 15 June 2006 in Ciutadella (Menorca), a meteorological tsunami. Adv Geosci 12:1–4. https://doi.org/10.5194/adgeo-12-1-2007

Jenkins S, Magill C, McAneney J, Blong R (2012) Regional ash fall hazard I: a probabilistic assessment methodology. Bull Volc 74:1699–1712

Jenkins SF, Biass S, Williams GT, Hayes JL, Tennant E et al (2022) Evaluating and ranking southeast Asia's exposure to explosive volcanic hazards. Nat Hazards Earth Syst Sci 22(4):1233–1265. https://doi.org/10.5194/nhess-22-1233-2022

Johnson RW (1987) Large-scale volcanic cone collapse: the 1888 slope failure of Ritter volcano, and other examples from Papua New Guinea. Bull Volc 49:669–679

Kagan YY (1999) Universality of the Seismic moment-frequency relation. Pure Appl Geophys 155:537–573. https://doi.org/10.1007/s000240050277

Kagan YY (2002a) Seismic moment distribution revisited: I. Statistical results: Seismic moment distribution: I. Geophys J Int 148:520–541. https://doi.org/10.1046/j.1365-246x.2002.01594.x

Kagan YY (2002b) Seismic moment distribution revisited: II. Moment conservation principle: Seismic moment distribution revisited: II. Geophys J Int 149:731–754. https://doi.org/10.1046/j.1365-246X.2002.01671.x

Kagan YY (2003) Accuracy of modern global earthquake catalogs. Phys Earth Planet Int 135:173–209. https://doi.org/10.1016/S0031-9201(02)00214-5

Kagan YY, Bird P, Jackson DD (2010) Earthquake patterns in diverse tectonic zones of the globe. Pure Appl Geophys 167:721–741. https://doi.org/10.1007/s00024-010-0075-3
Kanamori H, Brodsky EE (2004) The physics of earthquakes. Rep Prog Phys 67:1429–1496. https://doi.org/10.1088/0034-4885/67/8/R03
Karstens J, Berndt C, Urlaub M, Watt SF, Micallef A, Ray M et al (2019) From gradual spreading to catastrophic collapse–Reconstruction of the 1888 Ritter Island volcanic sector collapse from high-resolution 3D seismic data. Earth Planet Sci Lett 517:1–13
Kienle J, Kowalik Z, Murty TS (1987) Tsunamis generated by eruptions from Mount St. Augustine volcano, Alaska. Science 236:1442–1447
Kubota T, Saito T, Nishida K (2022) Global fast-traveling tsunamis driven by atmospheric Lamb waves on the 2022 Tonga eruption. Science 377:660. https://doi.org/10.1126/science.abo4364
Lamb H (1911) On atmospheric oscillations. Proc Roy Soc London A84:551–572. https://doi.org/10.1098/rspa.1911.0008
Lane EM, Mountjoy JJ, Power WL, Mueller C (2016) Probabilistic hazard of tsunamis generated by submarine landslides in the cook strait Canyon (New Zealand). Pure Appl Geophys 173:3757–3774. https://doi.org/10.1007/978-3-319-55480-8_6
Lashkin VM, Cheremnykh OK (2023) Acoustic-gravity waves in quasi-isothermal atmospheres with a random vertical temperature profile. Wave Motion 119:103140. https://doi.org/10.1016/j.wavemoti.2023.103140
Latter JH (1981) Tsunamis of volcanic origin: summary of causes, with particular reference to Krakatoa, 1883. Bull Volc 44:467–490. https://doi.org/10.1007/BF02600578
Le Méhauté B, Wang S (1996) Water waves generated by underwater explosion. World Scientific, vol 10
Le Méhauté B (1971) Theory of explosion-generated water waves. In: Te Chow V (eds) Advances in hydroscience, Elsevier 7:1–79, ISSN 0065–2768, https://doi.org/10.1016/B978-0-12-021807-3.50006-0.
Lee CH, Lo PHY, Shi H, Huang Z (2022) Numerical modeling of generation of landslide tsunamis: a review. J Earthq Tsunami 16(06):2241001
Lenti L, Martino S, Paciello A, Prestininzi A, Rivellino S (2015b) Recorded displacements in a landslide slope due to regional and teleseismic earthquakes. Geophys J Int 201(3):1335–1345
Lenti L, Martino S, Paciello A, Prestininzi A, Rivellino S (2015) Recorded displacements in a landslide slope due to regional and teleseismic earthquakes. Geophys J Int 201:1335–45, ISSN: 0956–540X. https://doi.org/10.1093/gji/ggv063
Leonard M (2014) Self-consistent earthquake fault-scaling relations: update and extension to stable continental strike-slip faults. Bull Seism Soc Am 104:2953–2965. https://doi.org/10.1785/0120140087
LeVeque RJ, Waagan K, González FI et al (2016) Generating random earthquake events for probabilistic tsunami hazard assessment. Pure Appl Geophys 173:3671–3692. https://doi.org/10.1007/s00024-016-1357-1
Linares Á, Bechle AJ, Wu CH (2016) Characterization and assessment of the meteotsunami hazard in northern lake Michigan. J Geophys Res Oceans 121:7141–7158. https://doi.org/10.1002/2016JC011979
Lorenz E (1969) Atmospheric predictability as revealed by naturally occurring analogues. J Atmos Sci 26:636–646. https://doi.org/10.1175/1520-0469(1969)26%3c636:APARBN%3e2.0.CO;2
Lorenz E (1956) Empirical orthogonal functions and statistical weather prediction, Tech. rep., Massachusetts Institute of Technology, Department of Meteorology, Massachusetts Institute of Technology, Department of Meteorology
Lorito S, Selva J, Basili R, Romano F, Tiberti MM, Piatanesi A (2015) Probabilistic hazard for seismically induced tsunamis: accuracy and feasibility of inundation maps. Geophys J Int 200(1):574–588. https://doi.org/10.1093/gji/ggu408
Loughlin S, Sparks S, Brown S, Jenkins S, Vye-Brown C (eds) (2015) Global volcanic hazards and risk. Cambridge University Press, Cambridge. https://doi.org/10.1017/CBO9781316276273

Løvholt F, Pedersen G, Harbitz CB, Glimsdal S, Kim J (2015) On the characteristics of landslide tsunamis. Phil Trans Roy Soc A 373:20140376. https://doi.org/10.1098/rsta.2014.0376

Løvholt F, Bondevik S, Laberg JS, Kim J, Boylan N (2017) Some giant submarine landslides do not produce large tsunamis. Geophysi Res Lett 44(16):8463–8472. https://doi.org/10.1002/2017GL074062

Løvholt F, Glimsdal S, Harbitz CB (2020) On the landslide tsunami uncertainty and hazard. Landslides 17(10):2301–2315. https://doi.org/10.1007/s10346-020-01429-z

Løvholt F, Urgeles R, Harbitz CB, Vanneste M, Carlton B (2022) Submarine landslides. In: Shroder JF (ed) Treatise on geomorphology, 2nd edn, pp 919–959. ISBN 9780128182352. https://doi.org/10.1016/B978-0-12-818234-5.00139-5

Ma KF, Kanamori H, Satake K (1999) Mechanism of the 1975 Kalapana, Hawaii, earthquake inferred from tsunami data. J Geophys Res Solid Earth 104(B6):13153–13167

Maeno F, Imamura F, Taniguchi H (2006) Numerical simulation of tsunamis generated by caldera collapse during the 7.3 ka Kikai eruption, Kyushu, Japan. Earth Planets Space 58(8):1013–1024

Maeno F, Imamura F (2011) Tsunami generation by a rapid entrance of pyroclastic flow into the sea during the1883 Krakatau eruption, Indonesia. J Geophys Res Solid Earth 116:B09205. https://doi.org/10.1029/2011JB008253

Manchuel K, Traversa P, Baumont D, Cara M, Nayman E, Durouchoux C (2018) The French seismic CATalogue (FCAT-17). Bull Earthq Eng 16:2227–2251. https://doi.org/10.1007/s10518-017-0236-1

Marzocchi W, Bebbington M (2012) Probabilistic eruption forecasting at short and long time scales. Bull Volc 74:1777–1805

Marzocchi W, Jordan TH (2014) Testing for ontological errors in probabilistic forecasting models of natural systems. Proc Natl Acad Sci USA 85:955–959

Marzocchi W, Sandri L, Selva J (2008) BET EF: a probabilistic tool for long- and short-term eruption forecasting. Bull Volc 70:623–632

Marzocchi W, Sandri L, Selva J (2010) BET_VH: a probabilistic tool for long-term volcanic hazard assessment. Bull Volc 72(6):705–716. https://doi.org/10.1007/s00445-010-0357-8

Marzocchi W, Taroni M, Selva J (2015) Accounting for epistemic uncertainty in PSHA: logic tree and ensemble modelling. Bull Seism Soc Am 105(4):2151–2159

Marzocchi W, Selva J, Jordan TH (2021) A unified probabilistic framework for volcanic hazard and eruption forecasting. Nat Hazards Earth Syst Sci 21(11):3509–3517. https://doi.org/10.5194/nhess-21-3509-2021

Masson DG, Harbitz CB, Wynn RB, Pedersen G, Løvholt F (2006) Submarine landslides: processes, triggers and hazard prediction. Phil Trans Royal Soc A 364(1845):2009–2039

Matoza RS et al (2022) Atmospheric waves and global seismoacoustic observations of the January 2022 Hunga eruption, Tonga. Science 377:95–100. https://doi.org/10.1126/science.abo7063

Mikhailova NN, Mukambayev AS, Aristova IL, Kulikova G, Ullah S, Pilz M, Bindi D (2015) Central Asia earthquake catalogue from ancient time to 2009. Annals Geophys 58:2. https://doi.org/10.4401/ag-6681

Molinari I, Tonini R, Lorito S, Piatanesi A, Romano F, Melini D et al (2016) Fast evaluation of tsunami scenarios: uncertainty assessment for a Mediterranean sea database. Nat Hazards Earth Syst Sci 16(12):2593–2602

Monserrat S, Vilibić I, Rabinovich AB (2006) Meteotsunamis: atmospherically induced destructive ocean waves in the tsunami frequency band. Nat Hazards Earth Syst Sci 6:1035–1051. https://doi.org/10.5194/nhess-6-1035-2006

Morgan JV, Bralower TJ, Brugger J, Wünnemann K (2022) The Chicxulub impact and its environmental consequences. Nat Rev Earth Env 3:338–354. https://doi.org/10.1038/s43017-022-00283-y

Muhari A, Heidarzadeh M, Susmoro H, Nugroho HD, Kriswati E, Wijanarto AB et al (2019) The December 2018 Anak Krakatau volcano tsunami as inferred from post-tsunami field surveys and spectral analysis. Pure Appl Geophys 176:5219–5233

Mulargia F, Tinti S (1985) Seismic sample areas defined from incomplete catalogues: an application to the Italian territory. Phys Earth Planet Int 40:273–300. https://doi.org/10.1016/0031-9201(85)90038-X
Murotani S, Miyake H, Koketsu K (2008) Scaling of characterized slip models for plate-boundary earthquakes. Earth Planets Space 60:987–991. https://doi.org/10.1186/BF03352855
Murotani S, Satake K, Fujii Y (2013) Scaling relations of seismic moment, rupture area, average slip, and asperity size for M~9 subduction-zone earthquakes. Geophys Res Lett 40:5070–5074. https://doi.org/10.1002/grl.50976
Murphy S, Scala A, Herrero A, Lorito S, Festa G, Trasatti E, Tonini R, Romano F, Molinari I, Nielsen S (2016) Shallow slip amplification and enhanced tsunami hazard unravelled by dynamic simulations of mega-thrust earthquakes. Sci Rep 6(1):35007. https://doi.org/10.1038/srep35007
Murru M, Akinci A, Falcone G, Pucci S, Console R, Parsons T (2016) $M \geq 7$ earthquake rupture forecast and time-dependent probability for the sea of Marmara region, Turkey. J Geophys Res Solid Earth 121:2679–2707. https://doi.org/10.1002/2015JB012595
Musson RMW, Grünthal G, Stucchi M (2010) The comparison of macroseismic intensity scales. J Seis 14:413–428. https://doi.org/10.1007/s10950-009-9172-0
Musson RMW, Cecić I (2012) Intensity and intensity scales. In: New manual of seismological observatory practice 2 (NMSOP2) 999 kb, p 41. https://doi.org/10.2312/GFZ.NMSOP-2_CH12
NCEI Global Historical Tsunami Database (2023) National Geophysical Data Center/World Data Service: NCEI/WDS. NOAA National Centers for Environmental Information. https://doi.org/10.7289/V5PN93H7
Necmioglu O, Heidarzadeh M, Vougioukalakis, Selva J (2023) Landslide induced tsunami hazard at volcanoes—the case of Santorini Island, Pure Appl Geophys in press
Necmioglu O, Özel NM (2015) Earthquake scenario-based tsunami wave heights in the eastern Mediterranean and connected seas. Pure Appl Geophys 172:3617–3638. https://doi.org/10.1007/s00024-015-1069-y
Neri A, Aspinall WP, Cioni R, Bertagnini A, Baxter PJ, Zuccaro G et al (2008) Developing an event Tree for probabilistic hazard and risk assessment at Vesuvius. J Volc Geotherm Res 178:397–415. https://doi.org/10.1016/j.jvolgeores.2008.05.014
Neri A et al (2015) Quantifying volcanic hazard at Campi Flegrei caldera (Italy) with uncertainty assessment: II. Pyroclastic density current invasion maps. J Geophys Res Solid Earth 120:2330–2349. https://doi.org/10.1002/2014JB011776
Newhall CG, Pallister JS (2015) Using multiple data sets to populate probabilistic volcanic event trees. In Shroder JF, Papale P (eds) Hazards and disasters series, volcanic hazards, risks and disasters, Elsevier. Pp 203–232. https://doi.org/10.1016/B978-0-12-396453-3.00008-3
Newhall C, Hoblitt R (2002) Constructing event trees for volcanic crises. Bull Volc 64:3–20
NGU (2012) Recommended hazard and risk classification system for large unstable rock slopes in Norway. NGU report 2012.029, ISSN 0800–3416. https://www.ngu.no/upload/Publikasjoner/Rapporter/2012/2012_029.pdf
Ogata Y (1998) Space-time point-process models for earthquake occurrences. Annals Institute Statist Math 50(2):379–402
Omira R, Ramalho RS, Kim J, González PJ, Kadri U, Miranda JM et al (2022) Global Tonga tsunami explained by a fast-moving atmospheric source. Nature 609(7928):734–740
Orsi G, Civetta L, D'Antonio M, Di Girolamo P, Piochi M (1995) Step-filling and development of a three-layer magma chamber: the Neapolitan yellow tuff case history. J Volc Geotherm Res 67(4):291–312, ISSN 0377–0273. https://doi.org/10.1016/0377-0273(94)00119-2
Paparo MA, Armigliato A, Pagnoni G, Zaniboni F, Tinti S (2017) Earthquake-triggered landslides along the Hyblean-Malta Escarpment (off Augusta, eastern Sicily, Italy)—assessment of the related tsunamigenic potential. Adv Geosci 44:1–8. https://doi.org/10.5194/adgeo-44-1-2017
Paris R (2015) Source mechanisms of volcanic tsunamis. Phil Trans Royal Soc A 373:20140380. https://doi.org/10.1098/rsta.2014.0380

Paris R, Switzer AD, Belousova M et al (2014) Volcanic tsunami: a review of source mechanisms, past events and hazards in Southeast Asia (Indonesia, Philippines, Papua New Guinea). Nat Hazards 70:447–470. https://doi.org/10.1007/s11069-013-0822-8

Paris R, Ulvrova M, Selva J, Brizuela B, Costa A, Grezio A, Lorito S, Tonini R (2019) Probabilistic hazard analysis for tsunamis generated by subaqueous volcanic explosions in the Campi Flegrei caldera, Italy. J Volc Geotherm Res 379:106–116. https://doi.org/10.1016/j.jvolgeores.2019.05.010

Pattiaratchi CB, Wijeratne EMS (2015) Are meteotsunamis an underrated hazard? Philos Trans Royal Soc A 373:20140377. https://doi.org/10.1098/rsta.2014.0377

Pelinovsky E, Zahibo N, Dunkley P, Edmonds M, Herd R, Talipova T (2004) Tsunami generated by the volcano eruption on July 12–13, 2003 at Montserrat, lesser Antilles. Sci Tsunami Hazards 22(1):44–57

Pelinovsky E, Choi BH, Stromkov A, Didenkulova I, Kim HS (2005) Analysis of tide-gauge records of the 1883 Krakatau tsunami. In: Satake K (ed) Tsunamis. Advances in natural and technological hazards research 23. Springer, Dordrecht. https://doi.org/10.1007/1-4020-3331-1_4

Press F, Harkrider D (1966) Air-sea waves from the explosion of Krakatoa. Science 154:1325–1327. https://doi.org/10.1126/science.154.3754.1325

Proudman J (1929) The effects on the sea of changes in atmospheric pressure. Geophys Suppl 2:197–209. https://doi.org/10.1111/j.1365-246X.1929.tb05408.x

Provost L, Scotti O (2020) QUake-MD: open-source code to quantify uncertainties in magnitude-depth estimates of earthquakes from Macroseismic intensities. Seism Res Lett 91:2520–2530. https://doi.org/10.1785/0220200064

Rabinovich AB (2020) Twenty-seven years of progress in the science of meteorological tsunamis following the 1992 Daytona Beach event. Pure Appl Geophys 177:1193–1230. https://doi.org/10.1007/s00024-019-02349-3

Ramis C, Jansà A (1983) Condiciones meteorológicas simultáneas a la aparición de oscilaciones del nivel del mar de amplitud extraordinaria en el Mediterráneo occidental. Rev Geofisica 39:35–42

Reasenberg P (1985) Second-order moment of central California seismicity, 1969–1982. J Geophys Res 90:5479–5495. https://doi.org/10.1029/JB090iB07p05479

Renault L, Vizoso G, Jansà A, Wilkin J, Tintoré J (2011) Toward the predictability of meteotsunamis in the Balearic Sea using regional nested atmosphere and ocean models. Geophys Res Lett 38:L10601. https://doi.org/10.1029/2011gl047361

Ripepe M, Delle Donne D, Lacanna G, Marchetti E, Ulivieri G (2009) The onset of the 2007 Stromboli effusive eruption recorded by an integrated geophysical network. J Volc Geotherm Res 182(3–4):131–136

Ripepe M, Delle Donne D, Genco R, Maggio G, Pistolesi M, Marchetti E (2015) Volcano seismicity and ground deformation unveil the gravity-driven magma discharge dynamics of a volcanic eruption. Nature Comm 6(1):6998

Ripepe M, Delle Donne D, Legrand D et al (2021) Magma pressure discharge induces very long period seismicity. Sci Rep 11:20065. https://doi.org/10.1038/s41598-021-99513-4

Roche O, Druitt TH, Merle O (2000) Experimental study of caldera formation. J Geophys Res 105:395–416

Rotondi R, Garavaglia E (2002) Statistical analysis of the completeness of a seismic catalogue. Nat Hazards 25:245–258. https://doi.org/10.1023/A:1014855822358

Rotondi R, Meroni F, Zonno G (1994) A different intensity recording for reducing the uncertainty in its assessment: an application to the completeness analysis of earthquake catalogues. Nat Hazards 10:47–58. https://doi.org/10.1007/BF00643440

Rovida A, Locati M, Camassi R, Lolli B, Gasperini P (2020) The Italian earthquake catalogue CPTI15. Bull Earthq Eng 18:2953–2984. https://doi.org/10.1007/s10518-020-00818-y

Ruppert JH, Koch SE, Chen X, Du Y, Seimon A, Sun YQ, Wei J, Bosart LF (2022) Mesoscale gravity waves and midlatitude weather: a tribute to Fuqing Zhang. Bull Am Meteor Soc 103:E129–E156. https://doi.org/10.1175/BAMS-D-20-0005.1

Salazar Monroy EF, Melgar D, Jaimes MA, Ramirez-Guzman L (2021) Regional probabilistic tsunami hazard analysis for the Mexican subduction zone from stochastic slip models. J Geophys Res Solid Earth 126. https://doi.org/10.1029/2020JB020781

Sandanbata O, Watada S, Satake K, Kanamori H, Rivera L, Zhan Z (2022) Sub-Decadal volcanic tsunamis due to submarine trapdoor faulting at Sumisu Caldera in the Izu–Bonin Arc. J Geophys Res Solid Earth 127(9):e2022JB024213

Sandri L, Acocella V, Newhall C (2017) Search Patterns Caldera Unrest Geochem Geophys Geosyst 18:2748–2768. https://doi.org/10.1002/2017GC006870

Satake K, Tanioka Y (1999) Sources of Tsunami and Tsunamigenic earthquakes in subduction zones. Pure Appl Geophys 154:467–483. https://doi.org/10.1007/s000240050240

Satake K, Heidarzadeh M, Quiroz M, Cienfuegos R (2020) History and features of trans-oceanic tsunamis and implications for paleo-tsunami studies. Earth-Sci Rev 202:103112. https://doi.org/10.1016/j.earscirev.2020.103112

Sawires R, Santoyo MA, Peláez JA, Corona Fernández RD (2019) An updated and unified earthquake catalog from 1787 to 2018 for seismic hazard assessment studies in Mexico. Sci Data 6:241. https://doi.org/10.1038/s41597-019-0234-z

Scala A, Lorito S, Romano F, Murphy S, Selva J, Basili R, Babeyko A, Herrero A, Hoechner A, Løvholt F, Maesano FE (2020) Effect of shallow slip amplification uncertainty on probabilistic tsunami hazard analysis in subduction zones: use of long-term balanced stochastic slip models. Pure Applied Geophys 177(3):1497–1520. https://doi.org/10.1007/s00024-019-02260-x

Scandone R, Giacomelli L, Gasparini P (1993) Mount Vesuvius: 2000 years of volcanological observations. J Volc Geotherm Res 58(1–4):5–25. https://doi.org/10.1016/0377-0273(93)90099-D

Scandone R, Bellucci F, Lirer L, Rolandi G (1991) The structure of the Campanian plain and the activity of the Neapolitan volcanoes (Italy). J Volc Geotherm Res 48(1–2):1–31, ISSN 0377–0273. https://doi.org/10.1016/0377-0273(91)90030-4

Schär C, Frei C, Luthi D, Davies HC (1996) Surrogate climate-change scenarios for regional climate models. Geophys Res Lett 23:669–672. https://doi.org/10.1029/96GL00265

Self S, Rampino MR (1981) The 1883 eruption of Krakatau. Nature 294:699–704. https://doi.org/10.1038/294699a0

Selva J, Costa A, Marzocchi W, Sandri L (2010) BET_VH: exploring the infuence of natural uncertainties on long-term hazard from tephra fallout at Campi Flegrei (Italy). Bull Volcanol 72 (6):717–733. https://doi.org/10.1007/s00445-010-0358-7

Selva J, Marzocchi W, Papale P, Sandri L (2012a) Operational eruption forecasting at high-risk volcanoes: the case of Campi Flegrei, Naples. J Appl Volcanol 1:1–14. https://doi.org/10.1186/2191-5040-1-5

Selva J, Orsi G, Di Vito M, Marzocchi W, Sandri L (2012b) Probability hazard map for future vent opening at the Campi Flegrei caldera, Italy. Bull Volc 74:497–510. https://doi.org/10.1007/s00445-011-0528-2

Selva J, Costa A, Sandri L, Macedonio G, Marzocchi W (2014) Probabilistic short-term volcanic hazard in phases of unrest: a case study for tephra fallout. J Geophys Res Solid Earth 119:8805–8826. https://doi.org/10.1002/2014JB011252

Selva J, Tonini R, Molinari I, Tiberti MM, Romano F, Grezio A, Melini D, Piatanesi A, Basili R, Lorito S (2016) Quantification of source uncertainties in seismic probabilistic tsunami hazard analysis (SPTHA). Geophys J Int 205(3):1780–1803

Selva J, Costa A, De Natale G, Di Vito MA, Isaia R, Macedonio G (2018) Sensitivity test and ensemble hazard assessment for tephra fallout at Campi Flegrei, Italy. J Volc Geotherm Res 351:1–28. https://doi.org/10.1016/j.jvolgeores.2017.11.024

Selva J, Acocella V, Bisson M, Costa A, Caliro S, De Martino P, Della Seta M, de Vita S, Federico C, Giordano G, Martino S, Cardaci C (2019) Multiple natural hazards at volcanic islands: a review for the Ischia volcano (Italy). J Applied Volcanol. https://doi.org/10.1186/s13617-019-0086-4

Selva J, Bonadonna C, Branca S et al (2020) Multiple hazards and paths to eruptions: a review of the volcanic system of Vulcano (Aeolian Islands, Italy). Earth-Sci Rev 207:103186. https://doi.org/10.1016/j.earscirev.2020.103186

Selva J, Marzocchi W, Sandri L, Costa A (2015) Chapter 9—Operational short-term volcanic hazard analysis: methods and perspectives. In Shroder JF, Papale P (eds) Hazards and disasters series, volcanic hazards, risks and disasters. Elsevier, pp 233–259. https://doi.org/10.1016/B978-0-12-396453-3.00009-5

Selva J, Sandri L, Taroni M, Sulpizio R, Tierz P, Costa A (2022) A simple two-state model interprets temporal modulations in eruptive activity and enhances multivolcano hazard quantification. Sci Adv 8:eabq4415. https://doi.org/10.1126/sciadv.abq4415

Serra CS, Martínez-Loriente S, Gràcia E, Urgeles R, Gómez de la Peña L, Maesano FE et al (2021) Sensitivity of tsunami scenarios to complex fault geometry and heterogeneous slip distribution: case-studies for SW Iberia and NW Morocco. J Geophys Res Solid Earth 126. https://doi.org/10.1029/2021JB022127

Sheremet A, Gravois U, Shrira V (2016) Observations of meteotsunami on the Louisiana shelf: a lone soliton with a soliton pack. Nat Hazards 84(2):471–492. https://doi.org/10.1007/s11069-016-2446-2

Shi L, Olabarrieta M, Nolan DS et al (2020) Tropical cyclone rainbands can trigger meteotsunamis. Nature Comm 11:678. https://doi.org/10.1038/s41467-020-14423-9

Sibley A, Cox D, Long D, Tappin D, Horsburgh K (2016) Meteorologically generated tsunami-like waves in the North Sea on 1/2 July 2015 and 28 May 2008. Weather 71(3):68–74. https://doi.org/10.1002/wea.2696

Sibson RH (1985) A note on fault reactivation. J Struct Geol 7:751–754. https://doi.org/10.1016/0191-8141(85)90150-6

Skarlatoudis AA, Somerville PG, Thio HK (2016) Source-scaling relations of interface subduction earthquakes for strong ground motion and tsunami simulation. Bull Seism Soc Am 106:1652–1662. https://doi.org/10.1785/0120150320

Sørensen MB, Spada M, Babeyko A, Wiemer S, Grünthal G (2012) Probabilistic tsunami hazard in the Mediterranean sea. J Geophys Res Solid Earth 117:B01305. https://doi.org/10.1029/2010JB008169

Stepp JC (1972) Analysis of completeness of the earthquake sample in the Puget Sound area and its effect on statistical estimates of earthquake hazard. In: Proceedings 1st International Confernati Microzonazion, Seattle v2, pp 897–910

Stix J, Kobayashi T (2008) Magma dynamics and collapse mechanisms during four historic caldera forming events. J Geophys Res 113:B09205. https://doi.org/10.1029/2007JB005073

Storchak DA, Harris J, Brown L et al (2017) Rebuild of the bulletin of the international seismological centre (ISC), part 1: 1964–1979. Geosci Lett 4:32. https://doi.org/10.1186/s40562-017-0098-z

Storchak DA, Harris J, Brown L, Lieser K, Shumba B, Di Giacomo D (2020) Rebuild of the bulletin of the international seismological centre (ISC)—part 2: 1980–2010. Geosci Lett 7:18. https://doi.org/10.1186/s40562-020-00164-6

Strasser FO, Arango MC, Bommer JJ (2010) Scaling of the source dimensions of interface and intraslab subduction-zone earthquakes with moment magnitude. Seism Res Lett 81:941–950. https://doi.org/10.1785/gssrl.81.6.941

Stucchi M, Albini P, Mirto C, Rebez A (2004) Assessing the completeness of Italian historical earthquake data. Annals Geophys 47:28. https://doi.org/10.4401/ag-3330

Tadini A, Harris A, Morin J, Bevilacqua A, Peltier A, Aspinall et al (2022) Structured elicitation of expert judgement in real-time eruption scenarios: an exercise for Piton de la Fournaise volcano, La Réunion island. Volcanica. https://doi.org/10.30909/vol.05.01.105131

Tanaka K (2010) Atmospheric pressure-wave bands around a cold front resulted in a meteotsunami in the east China sea in February 2009. Nat Hazards Earth Syst Sci 10(12):2599–2610. https://doi.org/10.5194/nhess-10-2599-2010

Tappin DR, Sibley A, Horsburgh K, Daubord C, Cox D, Long D (2013) The english channel "tsunami" of 27 June 2011—a probable meteorological source. Weather 68(6):144–152. https://doi.org/10.1002/wea.2061

Taroni M, Selva J (2021) A testable worldwide earthquake faulting mechanism model. Seism Res Lett 92:3577–3585. https://doi.org/10.1785/0220200445

Ten Brink US, Andrews BD, Miller NC (2016) Seismicity and sedimentation rate effects on submarine slope stability. Geol 44(7):563–566

Thingbaijam KKS, Mai PM, Goda K (2017) New empirical earthquake source scaling laws. Bull Seism Soc Am 107:2225–2246. https://doi.org/10.1785/0120170017

Thomson RE, Rabinovich AB, Fine IV et al (2009) Meteorological tsunamis on the coasts of British Columbia and Washington. Phys Chem Earth 34:971–988. https://doi.org/10.1016/j.pce.2009.10.003

Tierz P, Loughlin SC, Calder ES (2019) VOLCANS: an objective, structured and reproducible method for identifying sets of analogue volcanoes. Bull Volc 81:1–22. https://doi.org/10.1007/s00445-019-1336-3

Tierz P, Woodhouse MJ, Phillips JC, Sandri L, Selva J, Marzocchi W, Odbert HM (2017) A framework for probabilistic multi-hazard assessment of rain-triggered lahars using Bayesian belief networks. Front Earth Sci 5. https://doi.org/10.3389/feart.2017.00073

Tinti S, Bortolucci E, Armigliato A (1999) Numerical simulation of the landslide-induced tsunami of 1988 on Vulcano Island, Italy. Bull Volc 61:121–137

Tinti S, Zaniboni F, Pagnoni G, Manucci A (2008) Stromboli Island (Italy): scenarios of tsunamis generated by submarine landslides. Pure Appl Geophys 165:2143–2167

Tinti S, Tonini R, Bressan L et al (2011) Handbook of tsunami hazard and damage scenarios. JRC Sci Tech Rep 1–41. https://hdl.handle.net/11585/112670

Tommasi P, Rotonda T, Verrucci L, Graziani A, Boldini D (2018) Geotechnical analysis of instability phenomena at active volcanoes: two case histories in Italy. In: Landslides and engineered slopes. Experience, theory and practice. CRC Press, pp 53–78

Tonini R, Sandri L, Selva J, Marzocchi W (2016) Bayesian event tree for volcanic hazard. https://theghub.org/resources/betvh

Tonini R, Basili R, Maesano FE et al (2020) Importance of earthquake rupture geometry on tsunami modelling: the Calabrian Arc subduction interface (Italy) case study. Geophys J Int 223:1805–1819. https://doi.org/10.1093/gji/ggaa409

Tonini R, Sandri L, Rouwet D, Caudron C, Marzocchi W, Suparjan (2016) A new Bayesian EVENT-TREE tool to track and quantify volcanicunrest and its application to KawahIjen volcano. Geochem Geophys Geosyst 17:2539–2555. https://doi.org/10.1002/2016GC006327

Torsvik T, Paris R, Didenkulova I, Pelinovsky E, Belousov A, Belousova M (2010) Numerical simulation of a tsunami event during the 1996 volcanic eruption in Karymskoye lake, Kamchatka, Russia. Nat Hazards Earth Syst Sci 10:2359–2369. https://doi.org/10.5194/nhess-10-2359-2010

Ulvrova M, Paris R, Nomikou P, Kelfoun K, Leibrandt S, Tappin DR, McCoy FW (2016) Source of the tsunami generated by the 1650AD eruption of Kolumbo submarine volcano (Aegean Sea, Greece). J Volc Geotherm Res 321:125–139. https://doi.org/10.1016/j.jvolgeores.2016.04.034

Urgeles R, Camerlenghi A (2013) Submarine landslides of the Mediterranean sea: trigger mechanisms, dynamics, and frequency magnitude distribution. J Geophys Res Earth Surf 118(4):2600–2618. https://doi.org/10.1002/2013JF002720

Utsu T (1999) Representation and analysis of the earthquake size distribution: a historical review and some new approaches. Pure Appl Geophys 155:509–535. https://doi.org/10.1007/s000240050276

van Zelst I, Rannabauer L, Gabriel AA, van Dinther Y (2022) Earthquake Rupture on multiple splay faults and its effect on tsunamis. J Geophys Res Solid Earth 127. https://doi.org/10.1029/2022JB024300

Vilibić I (2008) Numerical simulations of the Proudman resonance. Cont Shelf Res 28:574–581. https://doi.org/10.1016/j.csr.2007.11.005

Vilibić I, Šepić J (2017) Global mapping of nonseismic sea level oscillations at tsunami imescales. Sci Rep 7:40818. https://doi.org/10.1038/srep40818

Vilibić I, Šepić J, Dunić N, Sevault F, Monserrat S, Jordà G (2018) Proxy-based assessment of strength and frequency of meteotsunamis in future climate. Geophys Res Lett 45:10501–10508. https://doi.org/10.1029/2018GL079566

Vilibić I, Denamiel C, Zemunik P, Montserrat S (2021) The Mediterranean and black sea meteotsunamis: an overview. Nat Hazards 106:1223–1267. https://doi.org/10.1007/s11069-020-04306-z

Vilibić I, Šepić J, Rabinovich AB, Monserrat S (2016) Modern approaches in meteotsunami research and early warning. Front Mar Sci 3:57. https://doi.org/10.3389/fmars.2016.00057

Viroulet S, Cébron D, Kimmoun O, Kharif C (2013) Shallow water waves generated by subaerial solid landslides. Geophys J Int 193(2):747–762

Wang K, Hu Y, He J (2012) Deformation cycles of subduction earthquakes in a viscoelastic earth. Nature 484:327–332. https://doi.org/10.1038/nature11032

Ward SN (2001) Landslide tsunami. J Geophys Res Solid Earth 106(B6):11201–11215. https://doi.org/10.1029/2000JB900450

Ward SN, Day S (2003) Ritter Island volcano—lateral collapse and the tsunami of 1888. Geophys J Int 154(3):891–902

Watts P, Waythomas CF (2003) Theoritical analysis of tsunami generation by pyroclastic flows. J Geophys Res 108(B112):2563

Wells DL, Coppersmith KJ (1994) New empirical relationships among magnitude, rupture length, rupture width, rupture area, and surface displacement. Bull Seism Soc Am 84(4):974–1002

Wertman C, Yablonsky R, Shen Y et al (2014) Mesoscale convective system surface pressure anomalies responsible for meteotsunamis along the U.S. east coast on June 13th, 2013. Sci Rep 4:7143. https://doi.org/10.1038/srep07143

Woessner J, Wiemer S (2005) Assessing the quality of earthquake catalogues: estimating the magnitude of completeness and its uncertainty. Bull Seism Soc Am 95:684–698. https://doi.org/10.1785/0120040007

Woo G (1994) Seismic area zones: myth or reality? Terra Nova 6:535–536. https://doi.org/10.1111/j.1365-3121.1994.tb00897.x

Wright CJ et al (2022) Surface-to-space atmospheric waves from Hunga Tonga-Hunga Ha'apai eruption. Nature 609:741–746. https://doi.org/10.1038/s41586-022-05012-5

Yavari-Ramshe S, Ataie-Ashtiani B (2016) Numerical modeling of subaerial and submarine landslide-generated tsunami waves—recent advances and future challenges. Landslides 13:1325–1368. https://doi.org/10.1007/s10346-016-0734-2

Yeh KC, Liu CH (1974) Acoustic-gravity waves in the upper atmosphere. Rev Geophys 12:193–216. https://doi.org/10.1029/RG012i002

Yeh H, Titov V, Gusiakov V et al (1995) The 1994 Shikotan earthquake tsunamis. Pure Appl Geophys 144:855–874. https://doi.org/10.1007/BF00874398

Youngs RR, Coppersmith KJ (1985) Implications of fault slip rates and earthquake recurrence models to probabilistic seismic hazard estimates. Bull Seism Soc Am 75:939–964. https://doi.org/10.1785/BSSA0750040939

Zaniboni F, Pagnoni G, Tinti S, Della Seta M, Fredi P, Marotta E, Orsi G (2013) The potential failure of Monte Nuovo at Ischia Island (Southern Italy): numerical assessment of a likely induced tsunami and its effects on a densely inhabited area. Bull Volc 75:1–13

Zare M, Amini H, Yazdi P, Sesetyan K, Demircioglu MB, Kalafat D, Erdik M, Giardini D, Khan MA, Tsereteli N (2014) Recent developments of the Middle East catalog. J Seismol 18:749–772. https://doi.org/10.1007/s10950-014-9444-1

Zemunik P, Bonanno A, Mazzola S et al (2021) Observing meteotsunamis ("Marrobbio") on the southwestern coast of Sicily. Nat Hazards 196:1337–1363. https://doi.org/10.1007/s11069-020-04303-2

Zemunik P, Denamiel C, Williams J, Vilibić I (2022) High-frequency sea-level analysis: global correlations to synoptic atmospheric patterns. Weather Clim Extrem 38(7):100516. https://doi.org/10.1016/j.wace.2022.100516

Zengaffinen-Morris T, Urgeles R, Løvholt F (2022) On the inference of tsunami uncertainties from landslide run-out observations. J Geophys Res Oceans 127:e2021JC018033. https://doi.org/10.1029/2021JC018033

Chapter 4
Source Modelling

Finn Løvholt, Shane Murphy, Stefano Lorito, Fabrizio Romano, Carl B. Harbitz, Jacopo Selva, Raphaël Paris, Anita Grezio, and Cléa Denamiel

4.1 Introduction

This chapter presents modeling procedures for simulating all possible source processes that are used in scenario based PTHA and PTRA (see Chap. 2 for a description of different frameworks). The methods for selecting the sources, their parameters, and probabilities, are described in Chap. 3. The methods described herein capture the mechanistic nature of each source process to different degree of complexity. For most of the sources, the uncertainty can be treated by varying the source parameters through Monte Carlo type analysis, a rigorous description of methods for analysing the model sensitivity is not included here. We describe for each model the parameters expected to influence uncertainty in tsunami generation the most. The source process simulation is used as input to modeling tsunami generation (Chap. 5). The chapter is

F. Løvholt (✉)
Offshore Energy, Norwegian Geotechnical Institute, Oslo, Norway
e-mail: finn.lovholt@ngi.no

S. Murphy
Geo-Ocean, Ifremer, Plouzané, France
e-mail: shane.murphy@ifremer.fr

S. Lorito · F. Romano
Istituto Nazionale di Geofisica e Vulcanologia, Roma, Italy
e-mail: stefano.lorito@ingv.it

F. Romano
e-mail: fabrizio.romano@ingv.it

C. B. Harbitz
Natural Hazards, Norwegian Geotechnical Institute, Oslo, Norway
e-mail: carl.bonnevie.harbitz@ngi.no

J. Selva
Dipartimento di Scienze della Terra, dell' Ambiente e delle Risorse (DiSTAR), Università di Napoli Federico II, Naples, Italy
e-mail: jacopo.selva@unina.it

M. B. Sørensen et al. (eds.), *Probabilistic Tsunami Hazard and Risk Analysis*, Mathematics of Planet Earth 14, https://doi.org/10.1007/978-3-031-98115-9_4

Table 4.1 Source modeling

What does the ingredient describe?	Modeling procedures for earthquake, landslide, volcanic, and atmospheric tsunami sources
Definitions of the ingredient	Source models that can be used to model tsunami generation
What is/are the metric(s) used for measuring/evaluating the ingredient?	Interface information used by the tsunami models, such as changes in the bathymetric depth as a function of time
Possible uses/applications of the ingredient	Modeling of tsunami generation. The models can also be relevant for modeling the source processes on their own right (independently of the tsunami generation)
Possible multi-hazard dimensions/applications? (if applicable)	Modeling of cascading / coupled geophysical phenomena
Possible time-dependent/ dynamic dimensions (if applicable)	Time dependency effects on tsunami due to finite generation time
List of the available methods to be described below	Various methods for modeling tsunami generation from earthquakes (kinematic and dynamic modeling), landslides (initial conditions, block sources, landslide dynamics models), volcanoes (pyroclastic flows, caldera collapse, underwater explosions), and atmospheric pressure fluctuations (synoptic index, synthetic catalogs, operational forecasts, climate modeling)

subdivided into four sections comprising the most frequent tsunami sources, namely earthquake sources, landslides sources, volcanic sources, and atmospheric sources. Volcanic sources may also be linked to landslides and atmospheric pressure, and in these cases the source description is given only in the landslide or atmospheric related subsections. Table 4.1 gives a brief description of the content of the ingredient.

This chapter describes the different sources in the separate Sections: Earthquakes in Sect. 4.2, including kinematic (slip distribution methods, and seafloor deformation models), and dynamic models; Landslides in Sect. 4.3, including descriptions of initial condition models, block models, and dynamic landslide models; Volcanoes in Sect. 4.4, which comprises pyroclastic flow models, caldera collapses, and underwater explosions; Atmospheric sources in Sect. 4.5, which includes methods based on

R. Paris
CNRS, IRD, OPGC, Laboratoire Magmas et Volcans, Université Clermont Auvergne, Aubière, France
e-mail: raphael.paris@uca.fr

A. Grezio
Istituto Nazionale di Geofisica e Vulcanologia, Bologna, Italy
e-mail: anita.grezio@ingv.it

C. Denamiel
Division of Marine and Environmental Research, Ruđer Bošković Institute, Zagreb, Croatia
e-mail: cdenami@irb.hr

synoptic indices, deterministic models, synthetic catalogs, and climate uncertainty. For each source type, a table that summarizes the method's description, requirements, and uncertainty treatment is provided. Finally, Sect. 4.6 briefly outlines the pros and cons of the various methods in order to help users select the appropriate models.

4.2 Earthquakes

A description of the scientific state of the art and gaps associated with modelling earthquake sources in Seismic-Probabilistic Tsunami Hazard Analysis (SPTHA) has been provided by several authors (Geist and Oglesby 2014; Davies et al. 2015; Grezio et al. 2017; Behrens et al. 2021).

Section 4.2 deals with the two main approaches to the mechanical earthquake modelling: kinematic and dynamic. The kinematic source modelling starts with a prescribed slip distribution on the fault interface, from which the seafloor displacement can be calculated. The dynamic approach models the full mechanical process starting from pre-stress and frictional state on the fault interface, and the mechanical properties of the surrounding materials, to the seismic rupture evolution and the seafloor deformation.

The kinematic model is frequently used in PTHA or PTRA given its relative simplicity and smaller computational weight in comparison with dynamic modelling, which, conversely, provides a more physics-based representation of earthquakes.

4.2.1 Method: Kinematic Rupture Modeling

Short Description of the Method

Generally, in SPTHA, the earthquakes are modeled as individual ruptures by considering the kinematic parameters of the source in terms of fault dislocation (slip and rake). Often, the additional assumption that the rupture occurs quasi-instantaneously is introduced. This means that the speed of the rupture front and the dislocation rise-time on a given portion of the fault are much faster than the tsunami propagation. This assumption may not hold in all circumstances, for example, in the case of large earthquakes were rupture occurs over large areas (e.g., 2004 Sumatra-Andaman), or slow ruptures like *tsunami earthquakes* (e.g., 2006 Java, 2010 Mentawai).

Here, we describe the two main components of the kinematic rupture process, that is (i) the slip distribution modeling (Sect. 4.2.1.1) and (ii) the associated seafloor deformation which causes the tsunami initiation by moving the water column above it (Sect. 4.2.1.2).

4.2.1.1 Method–Slip Distribution Modeling

The slip distribution modeling can be broken down into three steps: (i) defining the fault plane, (ii) linking it to the global earthquake characteristics, (i.e. total area, average slip) (iii) assigning the slip distribution (local slip value, i.e. intensity, and direction, and possibly the time-history). The distribution of the slip vector in space (and time) then constitutes the input for the second component of the method, that is the seafloor deformation.

Fault Plane Representation

Within the variability ranges prescribed by the earthquake source model characterization as discussed in Chap. 3, modelling individual ruptures is the first step to simulate tsunamis for each considered scenario. The simplest way to model a seismic scenario, with relatively little (e.g., seismic and/or geological) information, and limited computational resources, is a rectangle with uniform slip (see Fig. 4.1a). However, with increasing computational resources and higher resolution imaging of fault interfaces available (e.g., Slab2.0, that maps the subduction zone at the global scale, Hayes 2018), fault geometry can be better approximated using structured or unstructured meshes (e.g., discretized with quadrangular or triangular cells as in Fig. 4.1b).

Earthquake Scaling Relations

As discussed in Chap. 3, there are several empirical scaling relations (see Fig. 4.2) expressing the earthquake area and average slip as a function of the magnitude (e.g., Wells and Coppersmith 1994; Strasser et al. 2010; Allen and Hayes 2017; Murotani et al. 2013; Blaser et al. 2010; Skarlatoudis et al. 2016; Goda et al. 2016).

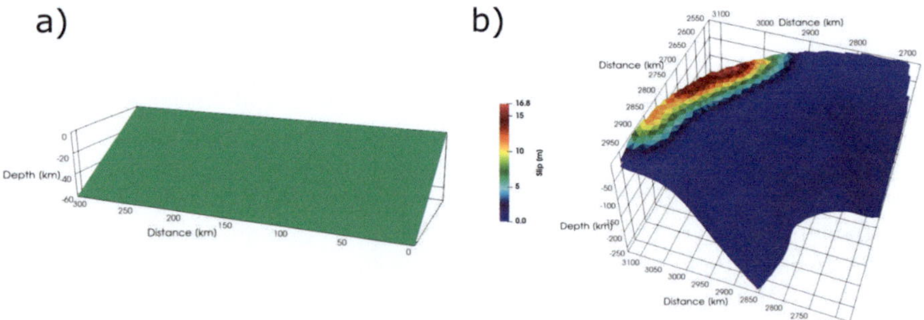

Fig. 4.1 Examples of different fault representations for a magnitude 8.5 event. Panel a) shows the simplest case: the earthquake rupture is approximated by a single planar fault with a uniform slip of 6 m. Panel b) shows a more complex modelling: the Scotia subduction plate interface as taken from Slab 2.0, and with a heterogeneous slip distribution (discussed towards the end of Sect. 4.2.1) featuring a maximum slip of 16.8 m. The same colorbar has been used for both representations

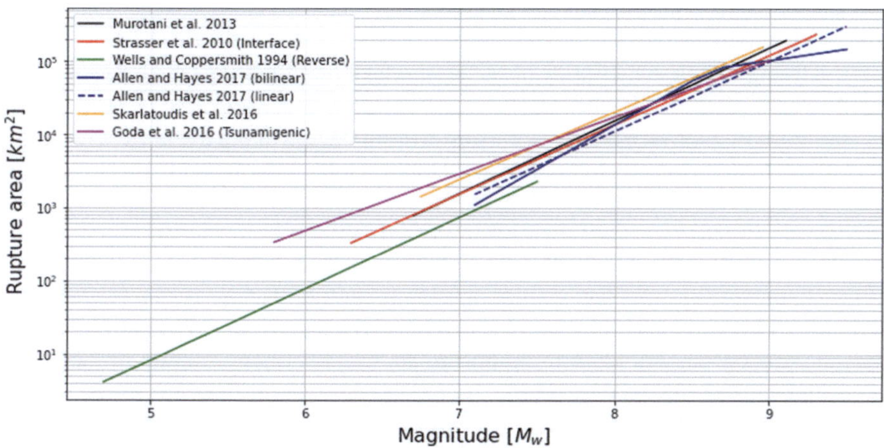

Fig. 4.2 Example of different scaling relations between magnitude and rupture area. The magnitude range of the different scaling relations is based on the magnitude range used by the authors when deriving their best fit line to the data

A comprehensive discussion on and a comparison between different scaling relations are provided by Stirling et al. (2013). An estimation on the uncertainty on the scaling is generally also provided as standard error and/or deviation which can be incorporated into PTHA.

Regardless of the scaling relation, the energy release from the simulated earthquake must match the prescribed magnitude that obeys the moment magnitude equation (Hanks and Kanamori 1979):

$$M_w = 2/3 log M_o - 6.03$$

where M_o is the seismic moment (in Nm) defined as:

$$M_o = \mu A \delta$$

With A being the slipping area (in m^2), μ the average rigidity (in GPa) of the rock, and δ the average slip (in m). A typical rock rigidity value is 30 GPa (Kanamori and Brodsky 2004), yet in subduction zones the rigidity variation with depth is significant and at shallow depth smaller than 30 GPa (e.g., Geist and Bilek 2001). In order to preserve this relation, the slip distribution is multiplied by a constant to correct the mean slip so that the desired magnitude is produced.

Slip Distribution

After defining the characteristics of the earthquake, the final step is to assign a slip distribution onto the fault surface.

The simplest approach is to assume uniform slip values along the fault (Fig. 4.1a). A method better representig reality, although more complicated, is the use of heterogeneous slip distributions (Fig. 4.1b).

Andrews (1980) proposed the first static self-similar (e.g., fractal) slip distributions based on the observation that the self-similar nature of far-field displacement is consistent with a power law scaling of seismic rupture (Hanks 1979). Herrero and Bernard (1994) introduced the *k-squared model* to generate stochastic slip distributions in the frequency domain by imposing an inverse square scaling on the radial wavenumber. The description, in the frequency domain, is then converted to slip distribution using an inverse Fourier transform. As the phase is randomly assigned, each slip distribution is different.

With time, different methods for generating stochastic models have been proposed (e.g., Lavelee and Archuleta 2003; Song and Somerville (2010); Davies et al. 2015). For example, Mai and Beroza (2002), developed a slip model based on the fitting of a von Karman autocorrelation function with a database of finite fault slip inversions; the slip distribution is self-affine rather than self-similar (as in the k-squared model previously mentioned).

A limitation of many of these slip models is the use of discrete Fourier transform which requires a regularly spaced cartesian grid limiting their use to planar faults. However new slip models are emerging for use on non planar surfaces. For example Sepulveda (2017) used a combination of the Karhunen-Lòeve expansion method with the von Karman function to produce stochastic slip on a curved north Chilean subduction zone (Leveque et al. 2016). Another method is the combination of the composite source model (Zeng et al. 1994; Ruiz et al. 2007) which involves the placement of hierarchical sets of circular slip functions on a surface, with an accurate method of calculating the geodesic distance across the non-planar surfaces (Herrero and Murphy 2018) - an example of which is shown in Fig. 4.1b.

The advantage of this method is that it is fast. The disadvantage is that the physics of the earthquake rupture on a specific fault is not fully accounted for, as the stochastic slip retains just global statistical rupture properties. Moreover, the displacement is assumed instantaneous. However, while admitting the limitations in constraining the kinematic rupture parameters like the rupture nucleation point, the rupture propagation velocity, and the rise time in this kind of modeling, a simple kinematic stochastic rupture model including these parameters can be generated by following the approach proposed by Goda et al. (2018).

Regarding the slip direction in subduction zones (where most large tsunamis originate), it is generally assumed that the fault movement occurs dominantly with a thrust mechanism whereby the hanging wall (i.e., the medium above the fault) moves upwards relative to the footwall (rake = 90 degrees), or antiparallel to the direction of plate convergence. However, other faulting mechanisms, while much less frequent, have generated tsunamis also, for example the 2018 Palu strike-slip earthquake (Ulrich et al. 2019b), splay faulting (Wendt et al. 2009) or outer rise earthquakes (Baba et al. 2020). The forecasting of the faulting mechanisms was briefly addressed in Chap. 3.

4.2.1.2 Method–Static Coseismic Seafloor Deformation Modeling

The seafloor deformation is computed based on the faulting mechanism (i.e., strike-slip, reverse, normal) and the amount of slip on the fault. This is known as the *dislocation model,* and in the simplest case it assumes that the Earth consists of one layer in which the fault is buried, with a planar surface. The advantage of this technique is that analytical solutions can be solved rapidly. Analytical equations have been developed for point and rectangular dislocations in half space media (Chinnery 1963; Mansinha and Smylie 1971; Iwasaki and Sato 1979), which Okada (1985) provided in a compact form, widely used nowadays with the code available in the public domain. Recently, similar analytical solutions with triangular surfaces have been developed (Meade 2007; Nikkhoo and Walter, 2015, the latter reference fixes some issues present in the former). These two have become popular as they can provide Green's functions from the triangular elements suited for describing non-planar fault surfaces, but also because the numerical code is available in the public domain. Care is required in all cases when cells touch the seafloor surface due to the creation of singularities along cell edges.

To model heterogeneous slip distributions, the fault plane is subdivided into smaller rectangles or triangles (the subfaults) with different slip placed in each cell based on a stochastic slip distribution (see Sect. 4.2.1.1). The total deformation is then computed by summing up the contribution from each of the individual subfaults (the Green's functions) using the principle of linear superposition which applies in the framework of linear elasticity. In the case of dislocation point sources, a Gaussian quadrature is used to numerically integrate the different sources (Segall 2010).

Alternatives to the dislocation model also exist. The *crack model* where slip is tapered towards the edge of the slipping surface (Geist and Oglesby 2014b). This method produces a constant stress drop across the cell, with no stress singularities produced at the edge of the cells. Like the dislocation model, analytical solutions for the crack model exist for sources in a half space media. The advantage of using the analytical models is the speed of computation, and only a description of the fault plane is required; however, variations in the mechanical properties of the subsurface media are not represented. To account for more complex media with multiple layers and accretionary wedges, Finite Element Models (FEM) are used (Zhao et al. 2004; Romano et al. 2014). Such models require detailed information on the media surrounding the fault, generally acquired from seismic imaging techniques as well as the meshing of the model (fault and surrounding medium). Therefore, more time is taken to build the model and calculate the deformation, but the deformation will include more information.

The output is the static seafloor displacement, assuming that the displacement occurs within a timescale much shorter than the tsunami wave propagation. In some cases, the slip on the sub-faults is assumed to occur at different instants to mimic a time-dependent rupture evolution (e.g., Satake et al. 2013). In most cases, only the vertical seafloor displacement is considered, as the horizontal one is considered to have a negligible tsunamigenic capability, except when it occurs on steep oceanic slopes and it is larger or comparable to the vertical one as for low-angle subduction interfaces. This parameter is treated as an input in Chap. 5. To consider this contribution, a largely used technique is to project the horizontal displacement onto the bathymetric gradient (Tanioka and Satake 1996).

Data and Technical Requirements for Method

The input for the slip distribution modeling is the parametrization of the fault plane. This can vary from a basic description of the width, length, position, and orientation in space of a planar fault, to a point cloud description of a non-planar surface. This parameterization does not require High-Performance Computing (HPC) resources.

The input for computing the coseismic seafloor deformation is the slip distribution defined on the fault surface. The general assumption is that the surrounding medium is elastic and homogeneous, but the level of complexity of the modeling can increase depending on the amount of information available and the type of numerical solver used; for example, using Finite Element Modeling (FEM) allows for using 3D crustal velocity models, topography, etc.).

For individual slip distributions, the seafloor deformation can be computed on standard computers without the requirement of additional computational resources. However, in PTHA, where large (thousands to millions) numbers of seafloor deformation calculations are sometimes used to model the natural (aleatory) variability of the seismic parameters and to quantify epistemic uncertainties with alternative model formulations, HPC usage may become almost mandatory. To save computational effort, Green's functions can be pre-computed by calculating the surface deformation caused by unitary slip (e.g. 1 m) on each individual cell, to be linearly combined to obtain virtually any desired scenario. This process is particularly useful when computationally demanding software (e.g. FEM) is adopted and/or the number of cells describing the fault is much less than the number of scenarios being simulated.

Treatment of Uncertainties for the Method

The uncertainty in the location and description of the fault plane is highly dependent on the amount of data available. High-resolution seismic profiles and/or well located seismic catalogs can be used when available to reduce the uncertainty. Indeed, simplification of the fault into a planar surface has been shown (e.g., Tonini et al. 2020) to affect the generated tsunami height. In the absence of a good characterization of individual faults, the geometrical fault parameters should be treated as aleatory variables like the other earthquake parameters.

For empirical scaling relations, estimates of the uncertainty related to the fit of the equations to the data are provided in the literature. This uncertainty can be incorporated into PTHA by treating the corresponding parameters as aleatory. Additionally, logic trees or other approaches can be used to account for the existence of different (epistemic, alternative) scaling relations, as discussed in Chapter 3.

Given all the other earthquake parameters, a certain number of stochastic slip distributions can be computed to account for a range of potential future scenarios, treating them as aleatoric uncertainty. Accounting for heterogeneous slip is particularly important for capturing the large variability in tsunami wave height in cases where the coast is in the near (Geist 2002) and far field (Davies and Griffin 2020). Davies (2019) demonstrated statistically that uniform slip models have a tendency to produce small tsunami's too often, on the contrary, variable rupture areas or heterogeneous models do not give rise to statistical underestimation.

An additional complexity that is only sometimes considered is the near-surface slip amplification as observed in Tohoku 2011 or tsunami earthquakes, which can lead to higher tsunami wave heights and hence hazard (Murphy et al. 2016). A simplified way of handling this phenomenon is scaling the slip with the inverse of shear modulus, which increases with depth (Bilek and Lay 1999; Geist and Bilek 2001). Scala et al. (2020) incorporated this depth dependent scaling as well as variable fault coupling into a PTHA for the Mediterranean and found that the mean hazard curves exhibited a higher probability for larger inundation when depth dependent scaling is accounted for.

The main advantage of the kinematic method is that it is fast, allowing for the uncertainty in the spatial distribution of slip to be propagated in PTHA. Analytical solutions which can be rapidly applied on standard computers assume simplified media. The displacement can be computed with different degrees of complexity as far as the elastic media parameterization is concerned (layers, 3D variations; e.g., Zhao et al. 2004). However, highly compliant material with inelastic behaviour, such as an accretionary wedge, has been shown to alter the orientation of surface deformation (Ma 2012).

4.2.2 Method: Dynamic Rupture Simulations

Governing Equations

Dynamic rupture simulations can be used to provide a tsunami source that takes into account the properties of the medium surrounding the fault, the lithospheric plate, as well as stress conditions, while kinematic modelling at least in principle can produce unphysical scenarios. The elastic wave equations are solved numerically. The relationship between stress and displacement of an isotropic elastic material is described by Hooke's law. On the fault surface, an empirical relationship between the shear stress acting on the fault surface and the normal stress (Byerlee 1978) is applied through a coefficient of friction which controls the fault strength. The evolution of the friction coefficient during an earthquake is described by friction laws based on experiments in rock physics laboratories and is generally expressed as a function of slip, slip-rate, or state variables (e.g., Ida 1972; Ohnaka 1973; Di Toro et al. 2011;

Ruina 1983; Dieterich 1979). Defining an all-encompassing friction law remains an active field of research in laboratory rock-physics.

There are two approaches to couple tsunami simulations and dynamic rupture models. The first uses the ground motion resulting from the dynamic simulation as input into a tsunami simulator (e.g., Ryan et al. 2015; Madden et al. 2021). In the second approach tsunami and dynamic rupture simulation are fully coupled using equations of motion in both the elastic and fluid domains (e.g., Lotto and Dunham 2015; Abrahams et al. 2023). Modelling the "interface" between the source and the ensuing tsunami is dealt with in Chapter 5.

There are several numerical methods that have been used to solve the elastodynamic equations incorporating the earthquake rupture (a detailed overview in Igel 2017). However, it is beyond the scope of this section to provide an in-depth description of all the numerical techniques. Rather, a discussion on the use of the different numerical methods in relation to dynamic rupture, suitable to model non-planar faults and heterogeneous media as for example in subduction zones, is provided here with references to the relevant literature.

Numerical techniques used for dynamic rupture simulations in subduction zone environments are based on the spatial discretization of the media by geometrical elements that can describe complex fault geometries and velocity structures. Rupture codes using Finite Element (FEM; Barall 2009; Ma 2009); and Finite Volume Methods (FVM; Benjemaa 2009) have been developed and successfully applied to dynamic simulations in subduction zones (Ryan et al. 2015; Ma 2012). Spectral Element Method (SEM; Festaand Villote 2005; Galvez et al. 2014) and the Discontinuous Galerkin (DG) Method (de la Puente et al. 2009; Pelties et al. 2012) are higher order versions of the FEM and FVM techniques, provide higher accuracy while still leveraging the advantages of using adaptable meshing. The application of Finite Difference (FD) methods on non-planar or non-vertical faults is more complicated and requires the use of particular techniques such as (a) support operators (Ely et al. 2009) or (b) high-order summation by parts combined with simultaneous approximation term method (Lotto and Dunham 2015). Additionally, for FD, accuracy is a function of the spatial extent of the stencil size, and consequently care is required at boundaries, particularly for high order schemes.

Initial conditions for dynamic simulations include: the geological structure/media, fault geometry, frictional parameters, shear and normal stress on the fault (which can be heterogeneous). Outputs from the modeling are temporal and spatial evolution of (a) the seismic rupture on the fault plane, (b) the seismic waves including the resulting ground motion and (c) tsunami surface elevation.

The advantage of this method is that it can provide physical insight into the observed phenomena. Dynamic modelling can of course be used also for tsunamigenic earthquakes occurring on crustal faults. Examples of this include the role of regional stress in 2016 M 7.8 Kaikoura earthquake where a number of individual faults ruptured in cascade (Ulrich et al. 2019a); the effect of rupture dynamic and fault geometry in near surface amplification (Murphy et al. 2016); the role of the frictional variation with depth in earthquake type (Murphy et al. 2018); and the role of supershear rupture in tsunami generation during the 2018 Palu earthquake

(Ulrich et al. 2019b). One of the main disadvantages with dynamic modelling is that it is computationally demanding, making it challenging to investigate uncertainty by performing numerous simulations. This obstacle will likely decrease in the near future as computational resources continue to increase allowing the possibility to incorporate a variety of different geological conditions. Additionally, a full description of the physical environment around the fault must be constructed and meshed which can be labor intensive and is subject to data availability. Care is required in construction of the numerical model in terms of nodal spacing / element size so that the simulation accurately models wave propagation and rupture process (Day et al. 2005). Finally, care is also required in coupling the solid earth and water domains, which is described in Chapter 5. In non depth-averaged fully coupled models this is accounted for with seismic waves being converted into acoustic waves at the seafloor and propagating within the water column. In order to accommodate the acoustic tsunami generation, inclusion of compressible effects are customary. In traditional depth averaged tsunami models (e.g. shallow water or Boussinesq), this phenomenon is rarely considered.

Data and Technical Requirements for the Method

Dynamic simulations require a complete description of the geophysical setting that the simulation occurs in. This includes the density as well as the P- and S-wave velocities throughout the model. The fault surface must be defined within the model, as well as the friction law on the fault plane and the stress distribution on it, which is often constrained by the regional stress field. Once the layers and fault plane are defined, meshing software (e.g., Gmsh, CUBIT) is needed to discretise the model domain into discrete elements (i.e., triangles, quadrangles, tetrahedrons, polyhedrons). Community based codes such as Seissol (https://seissol.org/) are optimized for running on HPC with inbuilt parallelism (MPI, and/or OpenMP) and/or written for GPU processors. 2D dynamic simulations are less computationally onerous than 3D simulations and can be run on standard computers but they can still take a large amount of time to run.

Treatment of Uncertainties for the Method

This method is deterministic, particularly in 3D simulations due to large computational resources required to run one simulation. Consequently, at present, extensive sensitivity analysis is not possible; however, limited analysis running a small number of scenarios is feasible (Madden et al. 2021).

It is often assumed that the rock around the fault acts elastically. Some, more computationally expensive, numerical models account for plastic deformation of the material close to the fault (Andrews 2005; Ma 2012, Ulrich et al. 2019ab), which is a phenomenon that can be important for shallow earthquakes, as deformation in a compliant material, such as the accretionary wedge, slows down the rupture on the fault.

2D simulations are less computationally intensive and may be used on HPCs to run a large number of simulations to assess the uncertainty of certain phenomena, for example the dependence of slip with depth, which can be then used to inform stochastic models (Murphy et al. 2016). However, due to the highly non-linear nature of rupture, small changes in the stress distribution or location of the nucleation can have a large effect on the subsequent slip distribution and ground motion. This nonlinearity makes it very challenging to account for uncertainty in dynamic simulations, especially in 3D environments.

4.2.3 Summary Table

See (Table 4.2).

4.3 Landslides

A description of the scientific state of the art and gaps associated with modeling landslide sources can be found in Løvholt et al. (2015), Yavari-Ramshe and Ataie-Ashtiani (2016), while adaptation to PTHA and challenges related to this are found in Grezio et al. (2017) and Behrens et al. (2021).

The landslide volume or frontal area (in the case of subaerial landslides) are the main factors influencing tsunami-genesis. In addition, the landslide dynamics and the water depth of generation also influence the tsunami generation strongly. Landslide parameters such as the local slope, external friction forces, rheological parameters, and fluid resistance parameters (e.g. viscous drag, skin friction, and added mass) influence the sensitivity to the tsunami-genesis (e.g. Grilli and Watts 2005; Watts et al. 2005; Heller and Hager 2010; Kim et al. 2019; Snelling et al. 2020).

Their sensitivity can be modeled by Monte Carlo type models systematically varying the parameters. The application of methods as described in Løvholt et al. (2020) provides examples embedding source parameter uncertainty into PTHA (see also the R8 Lyngen recipe on LPTHA). The influence of the water depth implies that landslide kinematic parameters such as the initial acceleration and Froude number correlate strongly with the tsunami-genesis (e.g. Ward 2001; Grilli and Watts 2005; Løvholt et al. 2015; Yavari-Ramshe and Ataie-Ashtiani 2016; Zengaffinen et al. 2020).

This section outlines the main classes of methods presently used for modeling landslide tsunamis: We describe first the simplest one that is the static initial condition, where an analytical shape is provided as input to the tsunami model. The second method we describe is the block model, using a fixed shape moving block as a tsunami generator. Finally, we describe landslide dynamics methods, which may inherit different degrees of sophistication (depth averaged or vertically resolved) or different rheologies (e.g. granular or viscoplastic).

Table 4.2 Summary table: earthquake sources

	Kinematic models	Dynamic models
Short description of method(s)		
Theoretical / mechanical background for the method(s)	Phenomenologically-based computationally lightweight way of simulating the earthquake kinematics by prescribing a stochastic slip distribution on a particular fault plane as input for the calculation of Green's functions of the static seafloor deformation	Various models which account for the time-dependent dynamics of earthquake rupture based on elastic/plastic theory, using friction laws and pre-stress on the fault surface
Short outline of how the method(s) works	Spatial description of an earthquake as one or a series of cells which move instantaneously as input to the elastic static seafloor deformation. The slip distribution is stochastic with a spatial probability on the fault plane which is usually uniform, yet different scaling relations for the rupture dimension and average slip, and different stochastic or physics-based slip models can be used (e.g. self-similar, self-affine, shallow slip amplification)	Numerical methods are used to solve elastodynamic equations with rupture physics represented by boundary conditions. Models differ widely in complexity with some modeling solely the solid earth while others incorporate hydrodynamics
How is/are the output(s) of the method(s) represented?	Residual static or time-history of the vertical seafloor displacement. A more accurate output includes contributions from the horizontal displacement times the gradient of the bathymetry	Dependent on the model complexity. All involve time dependent variation of slip evolution on fault, seismic wave field, ground motion, and in coupled models also the tsunami generation. Sometimes only the residual static displacement (as described in the kinematic case) is used as a tsunami initial condition
In what way do(es) the method(s) take into account time-dependent/dynamic aspects?	Generally the earthquake and the ensuing seafloor displacement are assumed to occur instantaneously, but a kinematic representation is possible	The evolution of the stress, slip-rate and slip are calculated across the fault plane as well as ground motion at the sea floor and in some cases tsunami wave generation

(continued)

Table 4.2 (continued)

	Kinematic models	Dynamic models
Do(es) the method(s) use parallelisation, machine learning, surrogate models, advanced simulation schemes, other computational reduction strategies (e.g. cluster analysis, etc.)?	Parallelism is not generally required. A common strategy used to reduce computational resources is to precalculate the displacement Green's functions caused by unit slip on subfaults (i.e. smaller spatial elements that describe the fault surface) which can then be rapidly used in subsequent simulations of different heterogeneous slip distributions	Generally, parallelisation is used due to large computational resources required for each simulation
What are the pros and cons and peculiarities of this specific method(s)? Are there known shortcomings / benefits of the method(s)? In what cases is this method(s) a good/poor choice?	Pros: Stochastic slip generation techniques and analytical solutions are fast; pre-computed Green's functions are a computationally efficient way to simulate a large number of seafloor displacements using the same fault after upfront cost Cons: Physical consistency of random nature of models is not verified; analytical solutions greatly simplify mechanical properties of the subsurface. Time-dependent ruptures are often simplified to be instantaneous	Pros: Models the full physics of an earthquake Cons: Construction of model can be labor intensive; simulations are often computationally demanding; parametric uncertainties regarding friction, pre-stress and structure can also be problematic due to lack of specific observational constraints
Data and technical requirements for method		
What input data does the method(s) use?	Description of fault surface and faulting mechanism, mechanical properties of subsurface, and bathymetry	Complete physical description of the fault geometry, frictional parameters and surrounding material: density, Lame's parameters, bathymetry
What are the technical requirements?	Software/scripts for fault discretization, stochastic slip generation, seafloor displacement calculation	Software/scripts for fault discretization, softwares for dynamic modelling
Are HPC computing resources necessary, would absence of these compromise the use?	Can be used in batch mode for uncertainty quantification. Also, complex FEM models may require more computational resources	In most cases large HPC facilities required, although simple models (e.g. 2D) can be run on less powerful computers

(continued)

Table 4.2 (continued)

	Kinematic models	Dynamic models
Is the method(s)/workflow(s) available as a git/software/jupyter notebook?	Various softwares are freely available online	Some softwares are freely available online
Treatment of uncertainties for the method		
Is the method(s) itself deterministic or uncertainty based?	Deterministic	Deterministic
Which are potential sources of uncertainties? Which parameters are used to characterize them?	Lack of data for fault and surrounding medium characterization, slip randomness and depth-independence, uncertainty on earthquake mechanism and scaling relations, simplified treatment of the elastostatic problem (e.g. homogeneous medium)	Initial conditions on fault plane (initial stress, nucleation), fault geometry, material parameters, friction law and its spatial variation
How are uncertainties quantified for the identified parameters? Are there significant correlations between different uncertain parameters?	Aleatory uncertainty in slip location and earthquake scaling is accounted for by generated large ensembles; the procedure can be repeated using different assumptions (e.g. depth dependent rigidity, coupling conditions), or with different material properties and bathymetries, to quantify the epistemic uncertainty. The earthquake scaling relations also define the correlation among the different earthquake parameters	Uncertainties are rarely accounted for because of practical and computational limitations. Sensitivity tests are sometimes performed especially with 2D models

(continued)

Table 4.2 (continued)

	Kinematic models	Dynamic models
Is the method(s) able to digest and propagate uncertainty in input (previous ingredient) and propagate it to the output (next ingredients)? Does the assessment of this ingredient require uncertainty propagation? If yes, how are uncertainties propagated?	Yes, through parameter variation and multiple simulations slip distributions and of the ensuing seafloor displacement. Uncertainty is naturally propagated to the next step which will require simulating a tsunami for each initial condition in each ensemble	Complex models rarely account for uncertainty due to computational cost for running each simulation, simpler 2D models can be used to test some initial conditions in the model by running large batches of simulations with varying conditions, in a similar fashion to the kinematic models
What are the consequences of lack of data in input or for boundary conditions, or for characterizing certain parameters?	Method highly dependent on definition of fault geometry and material properties. For example, using a wrong fault dip or neglecting features such as compliant accretionary wedges may cause large differences	Large epistemic uncertainty in describing conditions on the fault plane which produces large model bias and uncertainty in results
What are the consequences of ignoring or poorly treating uncertainties?	First order biases or underestimation of the uncertainty may occur, which are comparable to those regarding the earthquake occurrence characterization (Chap. 3)	The nonlinear nature of earthquake rupture makes it challenging to account for uncertainty in dynamic simulations

4.3.1 Method: Static Initial Conditions

Short Description of the Method

Static initial conditions represent the simplest method for modeling landslide tsunamis, and are often used in the absence of landslide dynamics models or possibilities of including general time-dependent sources in tsunami models. An example of an static initial condition model was first presented by Watts (1998), combining theoretical (constant slope) block motion with dimensional scale analysis. Later model derivations have followed a more rigorous mathematical development to yield analytical solutions for dispersive waves (Lo and Liu 2017, 2021).

Static initial conditions cannot take into account the local geometry where the slide occurs, and may therefore represent local generation effects and effect of the complex bathymetry poorly, which is a shortcoming compared to the more sophisticated generation models described below. In other words, the dynamic evolution of the landslide over time and complex topography is only represented by a single instance, which is a simplified representation. The advantage of using static initial models is that they are simple, on the other hand they may have significant biases related to the tsunami generation, both in terms of tsunami heights and the wave characteristics as the model equations are generally not adapted to the local bathymetry, nor take time-dependent landslide dynamics with deformation into account.

Data and Technical Requirements for the Method

The model of Watts (1998) yields as output both the surface elevation and an estimate of the wavelength along a transect. Necessary inputs include the dimensions of the landslide block (length, width and thickness), landslide density, slope angle, water depth, as well as values for the skin friction, pressure drag, and added mass coefficients. It is necessary to make further assumptions about the azimuthal distribution of the wave field using ad hoc assumptions, such as analytical shape functions (e.g. Synolakis et al. 2002). This provides a 2D output of the surface elevation. The 2D wave field was included in the later development of Lo and Liu (2017) who took into account the spreading of the wave. These models involve analytical expressions that only require small computational power, thus HPC is not needed.

Treatment of Uncertainties for the Method

The main parameters influencing the uncertainty of the tsunami generation are described in the intro paragraph of Sect. 4.3, and they can be modeled using Monte Carlo type methods varying the parameters. The lack of direct inclusion of physics and time-dependency in the generation (see above) results in the possibility of significant biases in the simulations. They are also less tested than the two other model classes when it comes to forecasting ability. Hence, there are also more considerable uncertainties and biases associated with these methods than with the more sophisticated methods described below.

4.3.2 *Method: Block Sources*

Short Description of the Method

The method uses moving blocks as sources in the tsunami model. Block sources are fixed shaped 3D objects that move as flexible blankets over the seafloor, and the time-dependent motion is used for displacing the water mass in tsunami models (see Fig. 4.3). Harbitz (1992) described the first applications of such sources, and later analysis is given in e.g. Løvholt et al. (2005), Løvholt et al. (2015).

The block has a given shape determined by the user and can either have prescribed motion or be modeled according to simple ordinary differential equations of a sliding block on an inclined plane (e.g. Gylfadottir et al. 2017). The moving block is then given as a time dependent thickness field as input to the wave model, either as direct input in the volume displacement terms in the tsunami model (e.g. Løvholt et al. 2005) or subject to time-dependent source filtering of the seabed displacement before being used as input to the tsunami model (see Løvholt et al. 2015, and further discussion in Chap. 5). As described in Chap. 5, low-pass filtering is necessary to remove unphysical wave components during generation. This can also be partly mitigated by rounding the shape of the block. The frequency of feeding time-dependent mass into the wave model must be tested for convergence in a similar fashion as for grid refinement tests for the time step (e.g. Zengaffinen-Morris et al. 2022), by updating the seafloor deformation sufficiently frequent in the tsunami model. The block sources have the advantage over static initial conditions to more rigorously capture the bathymetric effect and the time dependent tsunami generation, while still being relatively simple. It is further more broadly tested with respect to modelling real events (e.g. Løvholt et al. 2005; Harbitz et al. 2014). But they do require the tsunami model to take into account time dependent forcing, which represents increased complexity for the user compared to the simpler static initial conditions. It can also generate large datasets since the model input may need to be fed to the tsunami model with short time steps. When combined with source filtering, the computational time can increase significantly, but otherwise this method is computationally relatively cheap. In most cases, block models do not take into account the landslide deformation (an

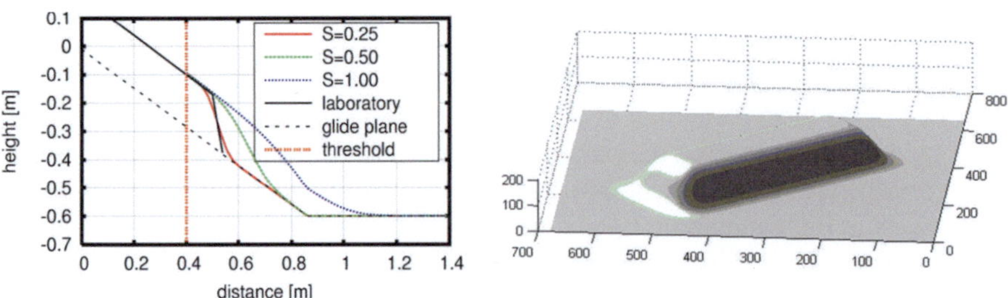

Fig. 4.3 Examples of block sources: Left panel, transect of the front of a smoothed block representing a subaerial block sliding down a slope. Right panel, shape of block used in early estimates of the Storegga slide tsunami

exception is the retrogressive block model by Løvholt et al. 2016, 2017), which can be important, in particular for submarine landslides.

Data and Technical Requirements for the Method

This method uses a grid file with topo-bathymetry as input, and gives time-dependent fields of the water depth deformation as a function of time in the same reference system as the tsunami model as output. As the governing equations are simple, they do not require much computational power or HPC use themselves (although the source filtering may require significant computational resources).

Treatment of Uncertainties for the Method

The main parameters influencing the uncertainty of the tsunami generation are described in the intro paragraph of Sect. 4.3, and they can be modeled using Monte Carlo type methods varying the parameters. The landslide motion (e.g. landslide kinematic or dynamic parameters) cannot so far be measured in the field, and there is a limited number past events to calibrate the model against. Hence, for testing models to establish plausible parameter ranges, it may be necessary to rely on the experience from modeling representative events where the field evidences are comprehensive (such examples could for instance comprise Lake Askja 2014: Gylfadottir et al. 2017, Rauter et al. 2022; Tyndall Glacier 2015: George and Iverson 2014; Anak Krakatoa 2018: Grilli et al. 2021), to generate plausible parameter ranges as input to the uncertainty analysis. It should be stressed that such evidence represents global data that do not necessarily have local source information. Tailoring inputs to benchmark tests (e.g. Kirby et al. 2022) can further help the calibration process. On the other hand, most of the community benchmark tests proposed by Kirby et al. (2022) are composed only of laboratory tests. Forecasting full-scale hazard events then necessitates an assessment of the upscaling from laboratory to field conditions, and landslide dynamics is not scale invariant (Lucas et al. 2014), hence testing against field events is imperative (e.g. Rauter et al. 2022).

4.3.3 Method: Landslide Dynamics Models

Short Description of the Method

Numerical landslide models represent a large family of tools that can be used to model the time-dependent landslide dynamics as source input in tsunami modeling. Due to the large number of such numerical models that are available, we will not describe them in detail, but rather go through the main principles of how they work (some more details are given in the reviews by Yavari-Ramshe and Ataie-Ashtiani 2016; Løvholt et al. 2022). The main difference compared to the abovementioned block models is that these landslide models may take into account the landslide material behavior (considering the rheology) and deformation. They comprise many different categories, but here we separate only between depth averaged and vertically resolved (generalised) CFD type models:

1. **Depth averaged landslide models (model type 1)**: These are based on the shallow water approximation (as also most tsunami models are) typically including non-Newtonian fluid rheologies. The models will in most cases assume that the landslide is completely fluidised after release, although some models also consider the failure process by heuristically taking material remolding into account (e.g. BingClaw, Løvholt et al. 2017; Kim et al. 2019). Fluid resistance can be included by one-way forcing via viscous drag terms and added mass in the model (e.g. Kim et al. 2019). Examples of output and induced tsunami due to the landslide volume displacement is shown in Fig. 4.4. Depth-averaged models can further include hydrostatic (e.g. George and Iverson 2014; Macias et al. 2015) or non-hydrostatic (e.g. Zhang et al. 2021) layers of both the landslide and the ambient fluid including tsunami modeling in coupled layers. It is stressed that due to the depth-averaging and hydrostatic assumptions not all the physics in the wave propagation is necessarily included, this will depend on the model formulation. Some of the listed models may capture dispersion and wave-breaking as well as depth-averaged landslide rheology to different degrees of sophistication depending on the model complexity, but they cannot capture more complex 3D features involving vorticity, such as landslide-water mixing, wave plunging, or crater formations during wave generation.
2. **Vertically resolved Computational Fluid Dynamics (CFD) type models (model type 2)**: A family of numerical models that resolve the vertical structure of the flow, includes the ambient fluid in addition to the landslide itself, and to different extent allows for interaction and mixing between the phases. These models capture more of the physics of the tsunami generation for subaerial landslides, and in particular to account for the 3D effects that require coupling with the water column such as impact cratering, vertical jet formations etc. They inherit few simplifications (and are hence often coined primitive models, based on primitive equation sets). On the other hand they will most often need extensive HPC resources and parallelisation in order to solve real world problems, are generally too computationally intensive to be used in PTHA and hence not discussed in further detail here. Examples of primitive models from the literature include e.g. Abadie et al. (2012), Ma et al. (2013), Horillo et al. (2013), Chen et al. (2020), Si et al. (2018), Rauter et al. (2022), and they include complexities related to the representation of the landslide rheology to various degrees. Various numerical implementations exists (e.g. continuum methods, Rauter et al. (2022), or particle methods such as SPH, Panizzo et al. (2007)) Just a few of these models incorporate complex time dependent evolution of the rheology such as of $\mu(I)$ type (see below), e.g. Si et al. (2018) and Rauter et al. (2022).

Numerical Treatment of the Depth Averaged Models

The depth-averaged types of landslide dynamics models (model type 1) share some similarities with the block models as they both provide time-dependent displacement

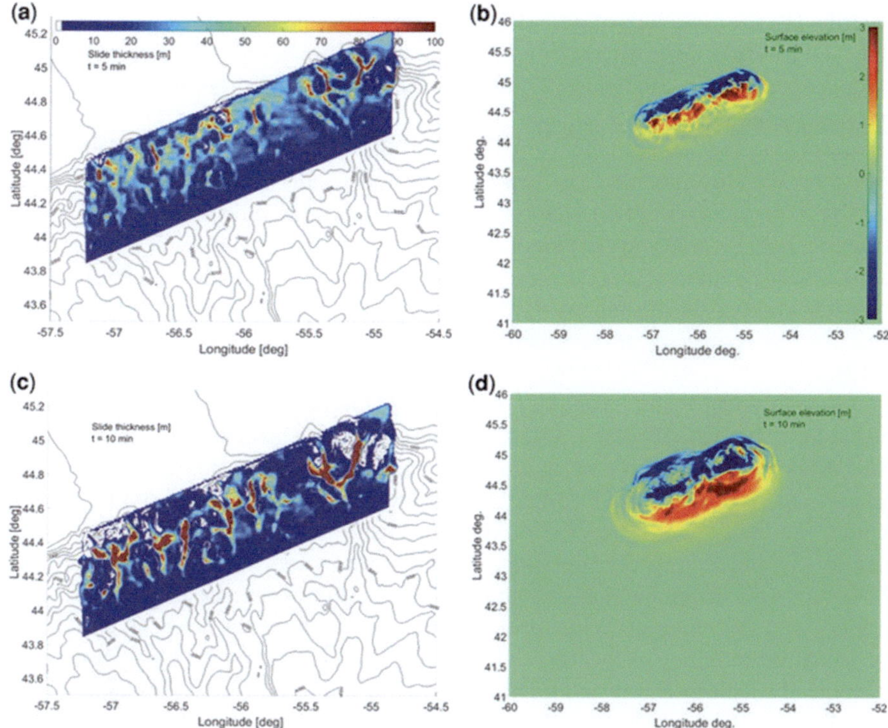

Fig. 4.4 Simulated landslide dynamics for the translational part of the 1929 Grand Banks landslide (thickness as a function of time) using the depth-averaged code bingclaw (left panels), and the induced wave field (right panels). Snapshots (from top to bottom) are shown after 5 and 10 min, respectively. Figure modified from Løvholt et al. (2019)

fields as input to the tsunami models, but they are computationally more intensive. However, HPC resources are usually not needed for single model runs. The equations are nonlinear, and Finite Volume schemes or similar are needed for robust treatment of shock-front formation and improved stability. The time resolution for feeding the depth-dependent information frequently enough into the tsunami models needs convergence testing, similar to the application of the block models (above), and discussed by Zengaffinen-Morris et al. (2022).

Model type 1 can accurately model the displacement pattern and the landslide dynamics needed for the tsunami analysis, as they consider both the material behavior and the complex terrain. They are suitable to be combined with PTHA, but when combined with source filtering (described in Chapter 5), the computational time and data need can potentially increase significantly.

Rheology of the Depth Averaged Models

There are two main classes of landslide rheologies. The largest group consists of the frictional-collisional (granular type) models, where the internal stresses are mainly described by the collisions of particles, sometimes also incorporating pore-fluid pressure and internal drag forces in addition to the granular-induced stresses. These models can describe landslides with relatively high concentrations and large particles, e.g. sands or rock avalanches. The models differ with respect to complexity, depending on the mathematical model description, e.g. relatively simple Savage-Hutter or Voellmy models (e.g. *VolcFlow* Kelfoun and Druitt 2005; *VoellmyClaw* Kim2014; *Landslide-HySEA* Macias et al. 2015), to more advanced Coulomb-type materials (e.g. Iverson and Denlinger 2001; *DClaw*, George and Iverson 2014; Pitman and Le 2005; Pudaisini 2012). Another class is the cohesive (visco-plastic) type of landslides that primarily represent more clayey material, such as e.g. the two-layer *BingClaw* described by a Herschel-Bulkley fluid (Løvholt et al. 2017; Kim et al. 2019).

A development within granular rheology, $\mu(I)$ (Jop 2006), show that the granular stresses can be described as a viscosity parameterized through the inertia number (labeled I) in dry granular media. This and related rheologies unify the granular and cohesive material description. More recent developments also included the effects of pore-fluid into the $\mu(I)$ description (Guazzelli and Poliquen 2018). $\mu(I)$ rheologies may be combined consider time-dependent material effects due to contraction and dilatancy in the granular material and may thus include time-dependent frictional properties of the material. While $\mu(I)$ of shallow-water type has not been used much for tsunami modeling so far, there are still depth-averaged models available using this type of rheology (Gray and Edwards 2014), and we expect that they will be used more frequently in future implementations for tsunami hazard analysis.

Data and Technical Requirements for the Method

The depth-averaged landslide models need the material parameters for the landslide, and topo-bathymetric grids, as input. They also need several model-dependent settings, such as the selection of the numerical scheme, this will depend on the model (some libraries such as GeoClaw, Leveque 2002, allow for adapting the numerical scheme as input to the model). The main output to the tsunami model is the time dependent seafloor displacement. The landslide simulations typically require similar or less computational resources compared to typical tsunami propagation codes. They often operate on a much smaller domain than the computational domain for the tsunami model and need shorter simulation times. On the other hand, they often require finer grids to resolve the undulations in the slide material and the slide front. When used as input to tsunami models, the necessary data output may be considerable.

The input to primitive models is similar to the depth-averaged models, but setting up the numerical models and gridding the input are generally much more time-consuming. For feasibility, the codes need to be efficient and parallelized and the codes must run on large clusters (typically, at least hundreds of CPUs are required).

Treatment of Uncertainties for the Method

The treatment of uncertainties shares the same challenges described for the block models, and would necessitate the calibration process in the same way. Hence, using past tsunami events to calibrate the uncertainty range is also necessary here. For more details, see Sect. 4.3.2 on block sources.

4.3.4 Summary Table

See (Table 4.3).

4.4 Volcanoes

Numerical modeling of volcanic tsunamis is still challenging because these sources are often complex and partly not well understood, for reasons that we discussed in Chap. 3.4. There is currently no up-to-date review on the modeling of volcanic tsunamis. In the following section we provide some guidelines for three types of volcanic tsunami sources: pyroclastic flows, caldera collapse, and underwater explosion. The proposed modeling approach will be different from one type of source to another, because of the different physical processes implied, and because the mechanisms of wave initialization are intrinsically different.

4.4.1 Methods for Tsunamis from Pyroclastic Flows

Short Description

The phenomenon of tsunami generation by a pyroclastic flow is complex, and different from landslide tsunamis for several reasons: (1) Pyroclastic flows are highly-mobile, (2) they are hot (typically 200–800 °C) and might produce steam explosion when entering the water, (3) they are internally structured in a lower dense component and an upper dilute component, (4) and they are often heterometric (from block to ash-size clasts). Watts and Waythomas (2003) demonstrated that the most energetic and coherent water waves are produced by the dense, basal debris flow component of the pyroclastic flow. Preliminary reviews on tsunami generation by pyroclastic flows were presented by deLange (2001), Watts and Waythomas (2003), and Maeno and Imamura (2007), Maeno and Imamura (2011). Direct observations of tsunamis generated by pyroclastic flows are rare, and the dense part of the flow (i.e., the one that is likely to generate a tsunami) is masked by the dilute cloud. Hence, the tsunami-genesis from pyroclastic flows is less well understood than for (dense flow)

Table 4.3 Summary table: Summary of basis for the different landslide models

	Static initial conditions	Block sources	Landslide dynamics models
Short description of methods			
Theoretical / mechanical background for the method	Analytical models for landslide shape	Time dependent motion / kinematics of a box with a fixed shape	Depth averaged or fully resolved numerical models describing the equations of motion, with or without water layers embedded
Short outline of how the methods work	Analytical shape of the wave as an initial condition based on landslide parameter inputs	Analytical or numerical description of a moving flexible blanket that displaces water as a function of time	Numerical models solving Partical Differential Equations for the time-dependent dynamics of landslides, with or without taking into account the hydrodynamic resistance or coupling to the fluid layer (air and / or water)
How is/are the output(s) of the method represented?	Shape function of initial tsunami surface elevation	Time-dependent perturbations of the slide thickness as a function of time and space	Time-dependent perturbations of the slide thickness as a function of time and space. For some models, the coupling to wave dynamics is also incorporated
In what way do the methods take into account time-dependent/ dynamic aspects?	Static output, no time dependency in the generation model	Time-dependent effects of the landslide motion and the wave generation is included. Landslide dynamics models also take into account time dependent slide deformation	Time-dependent effects of the landslide motion and the wave generation is included. Landslide dynamics models also take into account time dependent slide deformation

(continued)

Table 4.3 (continued)

	Static initial conditions	Block sources	Landslide dynamics models
Do the methods use parallelisation, machine learning, surrogate models, advanced simulation schemes, other computational reduction strategies (e.g. cluster analysis, etc.)?	No	No	Parallelisation is most often crucial for the vertically resolved models
What are the pros and cons and peculiarities of these methods? Are there known shortcomings / benefits of the methods? In what cases are these methods good/poor choices?	Pros: Simple to use. Robust with respect to instabilities Cons: Severely simplify the generation process, the time-dependent landslide motion and local bathymetry is not taken into account	Pros: Robust method with respect to noise or instabilities Cons: Does not take into account landslide deformation. Spanning the uncertainty space for multiple parameters can be difficult to convey into PTHA for the time dependent models. Can lead to large data outputs when coupled to tsunami generation models	Pros: More accurate, they take into account complex landslide deformation, rheology, and bathymetry effects Cons: Stability can be an issue. Need for sophisticated filtering. Spanning the uncertainty space for multiple parameters can be difficult to convey into PTHA for the time-dependent models. Can lead to large data outputs when coupled to tsunami generation models. Vertically resolved models are too resource intensive to be used in PTHA
Data and technical requirements for method			
What input data does the method(s) use?	Water depth, slope, density, friction, hydrodynamic resistance parameters (pressure drag, skin friction, added mass). Initial volume parameters (length, width, thickness)	Topo-bathymetric data (water depth, slope), rheological information (e.g. density, friction angle, yield strengths, etc.), and hydrodynamic resistance parameters (pressure drag, skin friction, added mass). Initial volume parameters (length, width, thickness). Sometimes also prescribed motion	Topo-bathymetric data (water depth, slope), rheological information (e.g. density, friction angle, yield strengths, etc.), and hydrodynamic resistance parameters (pressure drag, skin friction, added mass). Initial volume parameters (length, width, thickness).

(continued)

Table 4.3 (continued)

	Static initial conditions	Block sources	Landslide dynamics models
What are the technical requirements?	None	Topo-bathymetric data	Topo-bathymetric data
Are HPC computing resources necessary, would absence of these compromise the use?	No	No	HPC is obligatory only for the vertically resolved models, but on the other hand, they are too computationally intensive for use in PTHA / PTRA
Is the method(s)/workflow(s) available as a git/software/jupyter notebook?	No		
Treatment of uncertainties for the method			
Are the method(s) itself deterministic or uncertainty based?	Deterministic	Deterministic	Deterministic
Which are potential sources of uncertainties? Which parameters are used to characterize them?	Potential bias due to model simplifications. Scale-effects if empirical data from laboratory experiments are used as the basis for the model	Lack of material parameters, or knowledge of the landslide material / frictional behavior and scaling. Model parameters are often set based on the experience of retrofitting past events as several parameters from real landslides in the field cannot be measured. In many settings a relatively large epistemic uncertainty related to this is expected	Lack of material parameters, or knowledge of the landslide material / frictional behavior and scaling. Model parameters are often set based on the experience of retrofitting past events as several parameters from real landslides in the field cannot be measured. In many settings a relatively large epistemic uncertainty related to this is expected
How are uncertainties quantified for the identified parameters? Are there significant correlations between different uncertain parameters?	For certain parameters, such as the hydrodynamic resistance factors, there might be correlations between them (e.g. for coefficients for pressure drag, skin friction, added mass)		

(continued)

Table 4.3 (continued)

	Static initial conditions	Block sources	Landslide dynamics models
Are the method(s) able to digest and propagate uncertainty in input (previous ingredient) and propagate it to the output (next ingredients)? Does the assessment of this ingredient require uncertainty propagation? If yes, how are uncertainties propagated?	No, only through parameter variation in the probabilistic framework, i.e. representing different model realizations (and parameters) with different probabilities		
What are the consequences of lack of data in input or for boundary conditions, or for characterizing certain parameters?	There is rarely local data available describing the parameters. The lack of data implies that epistemic uncertainties should be modeled through parameter variation. With lack of data, use of benchmark tests and calibration towards other events may be necessary to get a grip of the uncertainty, even if they do not represent local conditions		
What are the consequences of ignoring or poorly treating uncertainties?	Model bias. Vice versa, such models would often necessitate broad parameter variations to cover epistemic uncertainties		

landslides, and an understanding of the respective roles of the different parameters implied in the wave generation is yet to be firmly established.

The important parameters controlling the interactions between pyroclastic flows and water bodies are the bulk density of the flow and its preservation offshore, the mass flux, the momentum angle, and the angle of incidence (Cas and Wright 1991; Carey et al. 2000; Paris et al. 2014). Experimental granular flows with a bulk density similar to water (e.g., pumice-rich flows) generate waves, regardless of their temperature (Freundt 2003). Experiments on fluidised granular flows by Bougouin et al. (2020) demonstrated that the wave features are mostly controlled by mass flux and the volume of the flow, unlike the water depth (Fig. 4.5). This is confirmed by the simulations presented in Fig. 4.6, where a 10^6 m^3/s flow generates waves considerably larger than a 10^5 m^3/s flow. The grain size may also affect the wave amplitude, as it controls the ability of water to penetrate the granular material (Bougouin et al. 2020). Interestingly, experimental fine-grained granular flows (e.g., ash-rich flows) behave as single-phase water flows in terms of wave generation.

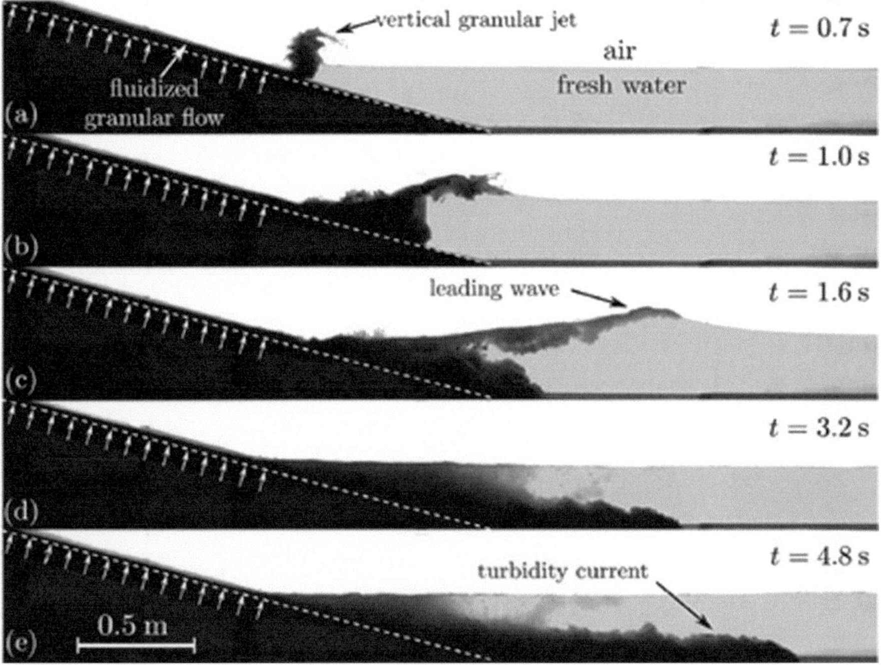

Fig. 4.5 Snapshots of an experimental fluidized granular flow (i.e. pyroclastic flow) entering water at different times after gate opening (t = 0; modified from Bougouin et al. 2020). ©2020. American geophysical union

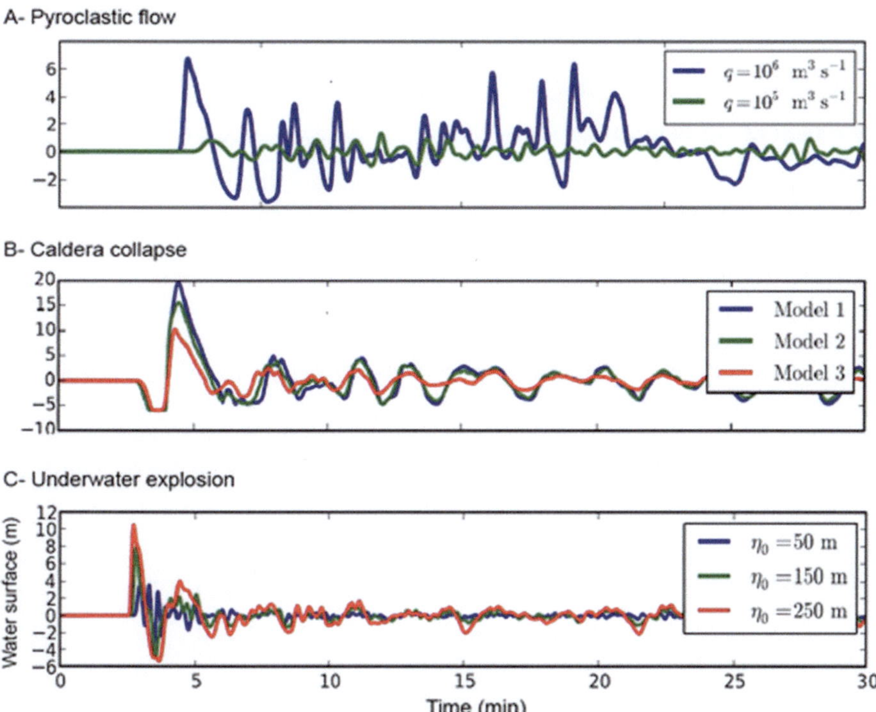

Fig. 4.6 Simulated water surface elevations in Santorini for different types of tsunamis generated by an eruption at Kolumbo volcano, Aegean Sea. **a** pyroclastic flow (plume collapse) with a volume $V = 0.1$ km^3 and a discharge rate $q = [10^5: 10^6]$ m^3/s entering the sea; **b** caldera collapse with different geometries (model 1 is a full-volume single-block collapse, model 2 is a half-volume single-block collapse, and model 3 corresponds to the deepening of a pre-existing caldera); **c** underwater explosions with different energies and initial water surface elevations $\eta_0 = [50: 250]$ m (modified from Ulvrova et al. 2016)

Data and Technical Requirement

For such a complex tsunami source, well-controlled laboratory experiments are necessary to validate the numerical simulations. Facing the diversity and complexity of the physical processes acting during the impact of a pyroclastic flow with a water body, there is no single type of model that could be recommended. Different numerical models can be developed depending on the scale of interest and objectives of the simulations. Coupled CFD-DEM models (Computational Fluid Dynamics - Discrete Element Method) can simulate the discrete grain motion and its interaction with water, but their high computational cost restricts their use to very small scale (and not to natural cases). Multi-fluid CFD models based on the Navier–Stokes equations are probably the best option to simulate the generation of a tsunami by a pyroclastic flow. Laboratory experiments can also provide empirical correlations or scaling laws. As an example, Bougouin et al. (2020) proposed a dimensionless number that predicts the initial wave amplitude from experimental data. As stressed in Sect. 4.6, CFD type analysis tools are well suited for characterizing the phenomenon, but cannot presently be used in PTHA due their computational intensity.

Treatment of Uncertainty

The level of epistemic uncertainty remains high until we better understand the process of tsunami generation by pyroclastic flows. The main uncertainties are both related to the physics of the phenomenon itself, and to the parameterization of the flow in the numerical models. In terms of understanding the physics, open questions are: (1) What are the consequences of the impact with water on the characteristics of the flow and most particularly on flow segregation (dense vs. dilute components)? (2) How to address the role of temperature (200–800 °C before the impact with water)? Ongoing experimental studies will help better understand these complex flow-water interactions, and define a reasonable level of approximation in the numerical models, which in any case will struggle to reproduce the complexity of such a phenomenon.

4.4.2 Methods for Tsunamis from Caldera Collapse

Short Description

The diameter of a caldera typically ranges from hundreds of meters to several kilometers. The duration of a caldera collapse is poorly constrained (from minutes to hours, keeping in mind that large collapses typically last hours) and the geometry varies from a single block to multi-stage subsidence of sub-blocks (Fig. 4.6). The collapse first generates a subsidence of the water surface, which is followed by the formation of a dome of water (e.g., Nomanbhoy and Satake 1995; Maeno et al. 2006; Ulvrova et al. 2016). Consequently, the wave train is characterized by a leading trough, followed by much larger positive crests. The initial subsidence of the water surface depends on the geometry (diameter, depth, single collapse or multi-blocks collapse) and duration of the collapse. Using a piston-like plunger model, Maeno and Imamura (2011) found that the water elevation above the caldera was the largest for a dimensionless collapse speed VC* of 0.01 ($VC^* = VC/\sqrt{gh}$). With a similar approach, Ulvrova et al. (2016) tested different collapse geometries and durations (Fig. 4.7 - right), and they found that only fast (and probably unrealistic) caldera collapses (<5 min) are efficient in terms of tsunami generation. Real-case caldera collapses usually last from ~ 30 min (e.g., the Pinatubo 1991 event, Philippines) to 12 days (e.g., the Fernandina 1968 Galapagos event).

Data and Technical Requirement

As explained above, a caldera collapse is not an instantaneous process. Consequently, it is required to adopt a finite time initialization for the tsunami generation. The motion of the sea floor (i.e. the caldera collapse) is used as a forcing term. There is no specific technical requirement to treat this kind of initialization, except that the grid should

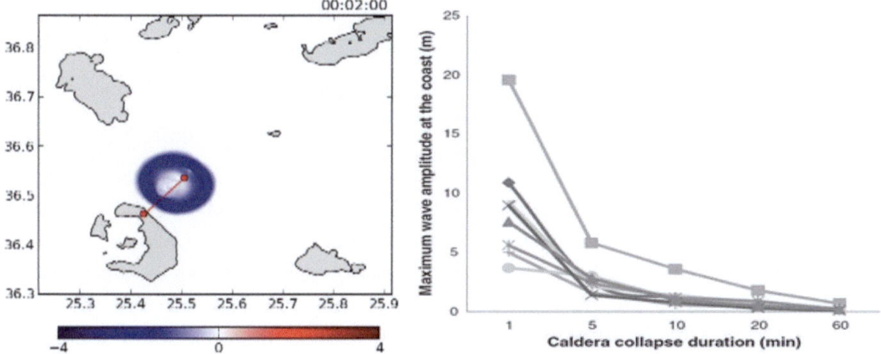

Fig. 4.7 Example of a simulated tsunami generated by a caldera collapse at kolumbo volcano, Aegean Sea. (Left) map of the initial water subsidence following the caldera collapse, the colorbar depict the simulated tsunami water levels. The red line shows the distance from the volcano to Santorini Island where the maximum amplitudes are recorded; (Right) The graph represents the maximum simulated tsunami wave amplitudes extracted in different sites of the eastern and southern coasts of Santorini versus duration of the caldera collapse (modified from Ulvrova et al. 2016). © 2016 Elsevier B.V

have a resolution high enough to accurately reproduce the contours of the caldera. For the tsunami propagation, a dispersive model might be necessary, depending on the water depth and diameter of the collapse area, and the duration of the sea floor motion.

Treatment of Uncertainty

The main source of uncertainty is the duration of the caldera collapse, although it is also difficult to predict the geometry of the collapse. There are few well-documented examples of submarine caldera collapse. It is thus recommended to define a range of different collapse volumes to investigate their impact. Examples of caldera collapses (mostly subaerial), their volumes and kinematics, can be found in the literature. The duration of the collapse should exceed 5 min, and different values of duration should be tested until a threshold is reached; threshold for which the collapse does not generate waves higher than a given value (determined by the operator).

4.4.3 *Methods for Tsunami from Underwater Explosions*

Short Description

Compared to other sources of underwater explosions (e.g. industrial, military explosions), the dynamics of phreatomagmatic eruptions is complex. The physics of magma-water interactions is controlled by many parameters: water depth, geometry of the vent and magma-water interface, transfer of thermal energy, processes of intermingling and mixing between magma and water, metastability of superheated water, and quantity of gas in the ascending magma (Kokelaar 1986; Wohletz 1986). Field observations of underwater eruptions and laboratory experiments show two different types of fountains at the water surface, dome-regime and finger-regime fountains, depending on explosion intensity and water depth (Shen et al. 2021a). There is a critical water depth at which an explosion with a given intensity generates the largest waves (Shen et al. 2021b). Underwater explosions in the ocean typically generate waves of shorter period and greater dispersion compared to earthquakes, thus reducing their far-field impact (e.g., the Myojin-Sho event 1952). On the contrary, violent explosions in shallow waters, and more particularly in lakes, have the potential to produce high-runup local tsunamis (e.g., the Karymskoye Lake event 1996).

Tsunami generated by underwater explosions typically consists of two main positive waves followed by smaller undulations propagating radially from the source, as demonstrated by experiments (Le Méhauté and Wang 1996; Kedrinskii 2005) and numerical simulations (Torsvik et al. 2010; Ulvrova et al. 2016; Paris and Ulvrova 2019). The water is initially pushed upward, forming a crater with a cylindrical bore that expands radially to form the leading wave, followed by a wave trough. Initial surface displacement (i.e., maximum height of the bore) can be estimated directly as a function of explosion energy at a given water depth (Le Méhauté 1971, Fig. 4.8 - right panel). The water crater collapses as in a dam break model, generating a steep dome of water that turns to a second cylindrical bore (Le Méhauté and Wang 1996).

For simulating the tsunami source, we recommend a semi-analytical approach that neglects complex interactions between dispersed fragments of lava, gas bubbles and water (which would require extremely complex multi-phase models). The underwater explosion is approximated by imposing a specific initial water surface displacement (η_0) that depends on the explosion depth and energy released during the explosion (e.g., Torsvik et al. 2010; Ulvrova et al. 2014).

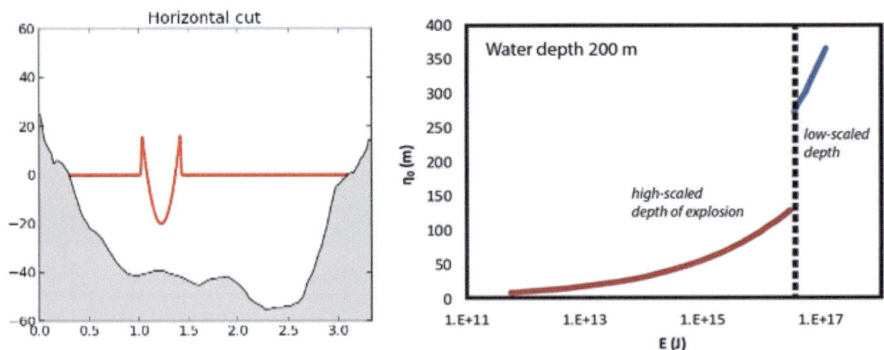

Fig. 4.8 (Left) Profile of initial water surface displacement for an underwater explosion. (Right) relationship between the explosion energy and the initial water surface displacement (calculated from the empirical equation of LeMéhauté 1971). Two cases of explosions should be distinguished: deep explosions with a high-scaled depth of explosion (compared to a water depth of 200 m in this case), and shallow explosions with a low-scaled depth (LeMéhauté and Wang 1996)

Data and Technical Requirement

The source of the tsunami is considered as instantaneous here (i.e. instantaneous formation of a crater in the water, as an initial water surface deformation). Therefore, the modeling of tsunami generation is an independent step and does not require HPC. In order to have a homogenous circular source at the apron of the explosion, it is recommended to use fine grid resolution (typically < 50 m) or to refine the available grid. A poor resolution of the source (few grid points) might generate artificial directivity of propagation (Paris and Ulvrova 2019). Underwater explosions typically generate short-period waves.

Treatment of Uncertainty

In the case of a tsunami generated by a volcanic subaqueous explosion, the source of uncertainty comes from the estimation of the size of the explosion (i.e., its energy and diameter), or the definition of a range of possible explosions with a given energy and probability (Fig. 4.6). The explosion energy can be estimated indirectly using the size of the volcanic craters formed by previous explosions (Sato and Taniguishi 1997; Goto et al. 2001), assuming that the diameter of the crater corresponds to the size of the largest explosion. In fact, the morphology of the crater often results from multiple vertically and horizontally-migrating explosions, and post-eruptive erosional processes.

4.4.4 Summary Table

See (Table 4.4).

4.5 Atmospheric Sources

To this date, modeling of atmospheric sources has mostly been implemented for either the hindcast or the forecast of weather related events. However, the methods developed within this context, and previously introduced in Sect. 3.5, form the basis needed for Atmospheric Probabilistic Tsunami Hazard Analysis (APTHA).

Section 4.5 deals with three main approaches to the atmospheric sources modeling mostly developed for weather related events: "Synoptic Index" which allows to select meteotsunamigenic conditions from global atmospheric models at resolution of the order of 100 km, "Deterministic Atmospheric Model" which is used to downscale the global atmospheric fields (about 100 km resolution) to the kilometer-scale needed to represent the mesoscale conditions driving the meteotsunami events and "Synthetic Catalog" that can be used to account for the uncertainty associated with the deterministic models. Finally, the last method, "Climate Uncertainty", presents how

Table 4.4 Summary table: Volcanic sources: pyroclastic flows, caldera collapses and underwater explosions

	Pyroclastic flows	Caldera collapses	Underwater explosions
Short description of method			
Theoretical / mechanical background for the method	CFD model (full Navier–Stokes)	Finite-time sea floor motion to simulate the collapse of a submarine caldera	Instantaneous formation of a water crater as initial water surface deformation
Short outline of how the method(s) works and references to existing case studies implementing this method	The described methods were applied to several case studies mentioned in the main text		
How is/are the output(s) of the method represented?	Full or layered velocity field and displacement of the water surface as a function of time	Velocity field and displacement of the water surface as a function of time	Velocity field and displacement of the water surface as a function of time
In what way does the method take into account time-dependent/dynamic aspects?	Time-dependent motion of the coupled flow, water body, and air	Time-dependent motion of the caldera floor	Static (instantaneous initialization)
Does the method use parallelisation, machine learning, surrogate models, advanced simulation schemes, other computational reduction strategies (e.g. cluster analysis, etc.)?	Parallelisation and access to cluster resources is recommended but not mandatory	No	No

(continued)

Table 4.4 (continued)

	Pyroclastic flows	Caldera collapses	Underwater explosions
What are the pros and cons and peculiarities of this specific method? Are there known shortcomings / benefits of the method? In what cases is this method a good/poor choice?	Pros: few approximations, reproduce interactions between the different fluids, reliable if benchmarked by laboratory experiments Cons: high computational cost, difficult to set up PTHA. It is recommended to limit the use of the Full Navier–Stokes model to the near-field (tsunami initiation/ generation)	Pros: low computational cost Cons: other processes involved in a caldera collapse are not taken into account (e.g. rise and collapse of the eruptive column during the on-going caldera collapse itself)	Pros: low computational cost Cons: other processes involved in a volcanic subaqueous explosion are not taken into account (e.g. upward flux of fragmented magma), the geometry of the initial water crater is simplified (e.g. there exist different fountain regimes)
Data and technical requirements for method			
What input data does the method use?	Bathymetry and coastal topography data, flow parameters	Bathymetry and coastal topography data, collapse geometry and timing	Bathymetry and coastal topography data, explosion parameters
What are the technical requirements?	CFD model (full Navier–Stokes) for tsunami generation, and Depth-averaged weakly dispersive model for tsunami propagation	Depth-averaged weakly dispersive model	Depth-averaged weakly dispersive model
Are HPC computing resources necessary, would absence of these compromise the use?	Recommended but not mandatory	No	No
Is the method/workflow available as a git/software/jupyter notebook?	No	No	No
Treatment of uncertainties for the method			
Is the method itself deterministic or uncertainty based?	Deterministic	Deterministic	Deterministic

(continued)

Table 4.4 (continued)

	Pyroclastic flows	Caldera collapses	Underwater explosions
Which are potential sources of uncertainties? Which parameters are used to characterize them?	Uncertainties on flow parameters (density, mass flux, structure, segregation dense-dilute, temperature, etc.) and its interactions with air and water	Uncertainties on the geometry (single block vs multiple blocks) and timing (duration, incremental vs massive) of the caldera collapse	Uncertainties on the explosion energy (it is recommended to test different values of energy, taken from the literature), and the explosion depth (keeping in mind that there is a critical water depth at which an eruption with a given source intensity will generate the largest waves)
How are uncertainties quantified for the identified parameters? Are there significant correlations between different uncertain parameters?	Uncertainties on the physical processes are not quantified		
Is the method able to digest and propagate uncertainty in input (previous ingredient) and propagate it to the output (next ingredients)? Does the assessment of this ingredient require uncertainty propagation?	No	No	No
What are the consequences of lack of data in input or for boundary conditions, or for characterizing certain parameters?	Uncertainty may result in a low reliability of the results. This can be compensated by an intermediate stage of numerical simulation of laboratory experiments, which helps estimating the influence of parameters	Uncertainty may result in a low reliability of the results. An overestimation of the level of hazard if a short (< 5 min) and unrealistic collapse duration is selected	Uncertainty may result in a low reliability of the results. The reliability of results decrease if the parameters are arbitrarily defined (e.g. overestimation of the possible range of energies)
What are the consequences of ignoring or poorly treating uncertainties?	Poor reliability of the results if the source parameters are not well-defined		

these three approaches can be combined within an atmospheric source modeling framework needed for APTHA.

4.5.1 Method: Synoptic Index

Short Description

For weather related events, the derivation of meteotsunami synoptic indices is based on the identification of atmospheric conditions favorable to meteotsunami events using both synoptic conditions (e.g., vertical profiles and spatial distributions of wind, temperature, etc.) extracted from numerical models, or echo sounding data, and long-term records of Non-Seismic Sea-Level Oscillations at Tsunami Timescales (NSLOTT) at coastal stations. In practice, the synoptic index approach is implemented at a given coastal station depending only on a combination of several wisely chosen atmospheric variables extracted from numerical models. The main advantage of the method is its low computational cost. However, its main drawbacks are that (1) it produces a large number of false positive events (detected events that are not leading to meteotsunamis) as the local and mesoscale effects are ignored and (2) it cannot be derived for acoustically driven events. Consequently, the primary use of the synoptic index method for hazard assessments is to extract potential weather related meteotsunami events from existing ensembles of global climate models (e.g., Coupled Model Intercomparison Project Phase 6 or CMIP6, Eyring et al. 2016). At this time, two different approaches have been successfully implemented: (1) Linear Combination (LC) and (2) Neural Network (NN).

Linear Combination (LC): In this framework, the synoptic index (μ_n), as defined in Zeminik et al. (2022) for stations worldwide (Fig. 4.9), is given by: $\mu_n(t) = \sum_{i=1}^{n} a_i x_i(t)$, with n the number of synoptic variables, x_i the synoptic variables, a_i the coefficient obtained by LC, as described below, and t the time. The derivation of the synoptic index is as follows:

1. *Derivation of the NSLOTT ranges*—hereafter $\boldsymbol{h(t)}$. Resampling NSLOTT to the temporal sampling of the synoptic variables;
2. *Choice of the first synoptic variable.* Solving the equation $h(t) = \mu_1(t)$ with least square method (Barrett et al. 1994) to obtain a_1 for $x_1(t)$ with the highest correlation with $h(t)$;
3. *Choice of the second synoptic variable.* Finding $x_2(t)$ with the highest cost function for the remaining $n - 1$ synoptic variables and calculating a_2;
4. *Choice of third synoptic variable.* If synoptic index not optimal, finding the highest cost function for the remaining $n - 2$ variables and calculating a_3;
5. *Next synoptic variables***.** Carrying out iteratively the procedure for the remaining $n - 3$ synoptic variables till the optimal synoptic index is derived.

The LC approach can easily be implemented without parallelization using softwares like Python, Matlab or R. However, depending on the temporal and spatial

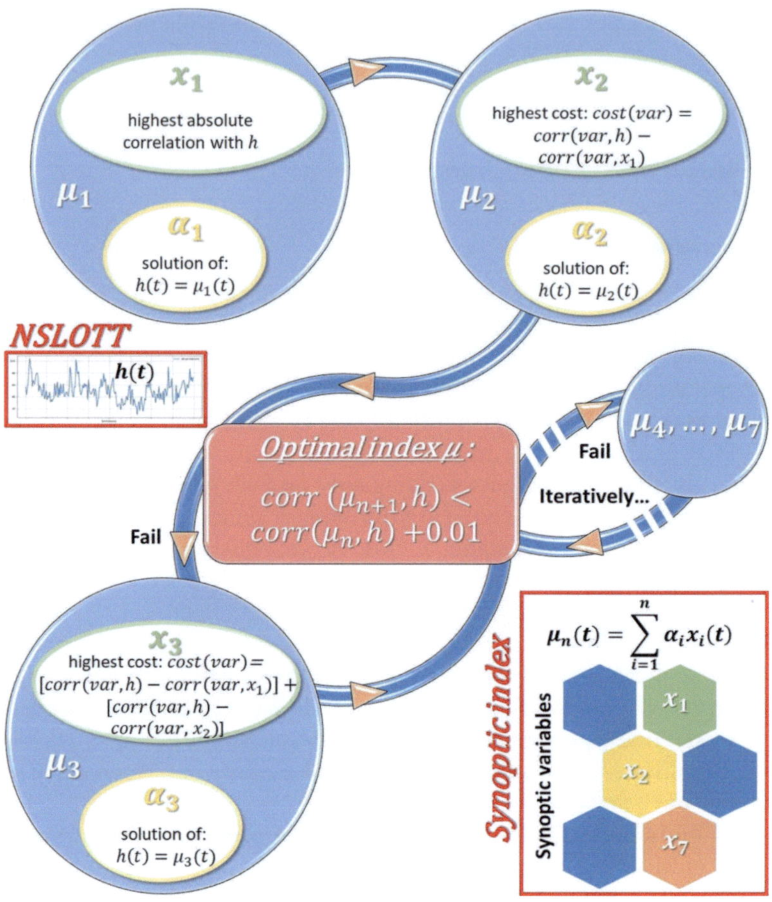

4.9 Derivation of the meteotsunami synoptic index using a linear combination (LC) approach. Adapted from Zemunik et al. (2022), licensed under CC-BY 4.0. See text for explanations of variables

resolution of the reanalysis products used to extract the synoptic variables, it might require substantial storage.

Neural Network (NN): In this framework, the synoptic index is defined following the procedure presented in Vich and Romero (2021). A Neural Network is a computational system capable of reproducing reality without a rulebook (Bishop et al. 1995; Haykin 2004). It is formed by at least three layers: (1) input, (2) output and (3) hidden. For the derivation of the meteotsunami synoptic index, the input consists of atmospheric variables on several vertical layers while the output layer is the daily maximum NSLOTT range. In order to create and test the NN for the synoptic index, two associated datasets are needed: (1) observed daily maximum NSLOTT ranges in the harbor where meteotsunamis occur and (2) observations (or reanalysis product) of the atmospheric vertical structure in the vicinity of this harbor. The maximum daily

NSLOTT range dataset (and the associated atmospheric dataset) is then randomly split into two sets: a training database (70% of the cases) and a testing database (the remaining 30%). The best NN configuration is based on three criteria: (1) limited spread (i.e., NN prediction as close as possible to the NSLOTT range observations), (2) symmetry (i.e., as few over-predictions as under-predictions) and (3) low degree of overfitting (i.e., discarding configurations that fit the training subset too closely). The NN approach can easily be implemented using softwares like Python, Matlab or R including artificial intelligence packages.

Data and Technical Requirements

NSLOTT data can be extracted from tide gauge observations provided with at least a 5-min temporal resolution. The tide gauge observations are filtered with a 2-h Kaiser-Bessel filter to produce high-frequency NSLOTT signal which has then to be quality-checked. To this date, the MInute SEa-Level Analysis (MISELA) dataset (Zemunik et al. 2021a, 2021b) is the only publicly available quality-checked NSLOTT worldwide dataset. It includes 331 tide gauges mostly extracted from the Intergovernmental Oceanographic Commission Sea Level Station Monitoring Facility (IOC SLSMF, http://www.ioc-sealevelmonitoring.org).

Synoptic variables can be extracted from global or regional reanalysis products or observational data. However, the forecast of weather related events with synoptic indices requires extracting the synoptic variables from global or regional weather forecasts (e.g., produced by the European Centre for Medium-Range Weather Forecasts - ECMWF or by the National Centers for Environmental Prediction–NCEP). The extraction of the synoptic variables does not require the use of High Performance Computing (HPC) facilities in operational mode.

The LC approach described in Zemunik et al. (2022) derives synoptic indices for 307 worldwide stations of the MISELA dataset. The seven synoptic variables are extracted from the hourly ECMWF/ERA5 reanalysis dataset (Hersbach et al. 2020) and consist in: (1) vertical profiles of wind speed, temperature and relative humidity from the surface up to a pressure level of 400 hPa at the ERA5 grid point closest to the corresponding MISELA station, and (2) horizontal fields of mean sea-level pressure, temperature at 850 hPa and geopotential at 550 hPa in a rectangular area of [-10°, 10°]x[-10°, 10°] in longitude and latitude centered over the point closest to the MISELA station.

The NN approach described in Vich and Romero (2021) derives a synoptic index for meteotsunamis occurring in Ciutadella (Balearic Islands, Spain). The NN is trained with two datasets obtained between July 1981 and July 2018: (1) a maximum daily NSLOTT range dataset available in Ciutadella and (2) an associated echo sounding dataset (i.e., temperature, zonal and meridional wind components as well as dew point interpolated on 21 fixed vertical levels) measured at 00 and 12 UTC in Palma station (100 km west of Ciutadella). The NN is created and tested with 126 meteotsunami days (i.e., NSLOTT ranges above 70 cm) and 549 days without meteotsunamis extracted from these datasets. In its current setting, the forecast of the meteotsunami events in Ciutadella is envisioned to be derived, at nearly no numerical cost, with the operational outputs of the NCEP/Global Forecasting System (GFS; Wang et al. 2019).

Treatment of Uncertainties

The synoptic index method, whether obtained with a LC or a NN approach, is by nature deterministic. However, for hazard assessments, it can be used with an ensemble of global or regional climate models and can thus produce a time-varying distribution of potential meteotsunami events at a given coastal location. Switching from a deterministic to a probabilistic hazard assessment at the climate scale is crucial as, even at the synoptic scale, the climate models are different due to their resolutions, their numerical schemes, etc. Consequently, using a single climate model increases the probability to miss potential meteotsunami events and, hence, decreases the reliability of the hazard assessments.

4.5.2 Method: Deterministic Atmospheric Models

Short Description

Weather related events: Deterministic atmospheric models allow the downscaling of global models to resolve the mesoscale conditions during meteotsunami events at hotspot locations. The equations of the atmospheric models used in state-of-the-art weather/climate models are usually based on the fully compressible Euler nonhydrostatic equations, the scalar-conserving flux form for prognostic variables and the complete Coriolis and curvature terms. They often include the possibility to use multiple nests and to set up mass-based terrain following coordinates with a vertical grid-spacing varying with height. The set of equations used in atmospheric models are: (1) the mass continuity equation depending on the wind velocity and the density; (2) the momentum equations depending on the wind velocity, the potential temperature, the water vapor, the atmospheric pressure and the forcing terms arising from model physics, turbulent mixing and the Earth's rotation; (3) the ideal gas law and (4) the moisture equation depending on the transport of the water vapor. The atmospheric model physics can be represented by many numerical schemes and parametrizations and include (1) microphysics, (2) cumulus parameterizations (for coarser grids), (3) surface physics, (4) planetary boundary layer physics and (5) atmospheric radiation physics.

In the framework of meteotsunami hazard assessments, atmospheric models should be able to capture meteotsunamigenic processes such as mesoscale gravity waves (MGWs), convective clouds, squall lines, etc. This requires the use of (1) telescopic nested grids downscaling the global climate models to high horizontal spatial resolution – at least 2 km (Horvath and Vilibić 2014), as MGWs typically range between 10 and 1000 km (Plougonven and Zhang 2014), (2) terrain following coordinates with an increased resolution in the first 2 km above the Earth surface, and (3) substantially reduced model time-steps (below 1-min), in order to properly represent the generation and the propagation of the tsunamigenic processes (characterized by atmospheric variations of the order of a few minutes) and to preserve the numerical stability of the model. In terms of numerical resources, kilometer-scale atmospheric forecasts require fully parallelized models that can run in a timely manner to produce 3 to 5 day-long simulations.

Acoustically driven events: In the barotropic framework (Kasahara 1977), a horizontally propagating surface gravity wave has a phase speed $c = \sqrt{gD}$, where g is the gravity and D is the mean fluid depth. This approach extends into the baroclinic atmosphere, where D becomes the equivalent depth (Cohn and Dee 1989). Consequently Lamb waves can be simulated following the rotating shallow-water equations on the sphere (Vasylkevych and Žagar 2021). To properly represent the deflection of Lamb waves by major mountain ranges, such models should include a realistic orography on a global computational grid. For example, the T170 grid (about 70 km horizontal resolution) which corresponds to the regular Gaussian grid of 680 and 320 grid points in the longitudinal and latitudinal directions has proven to have an adequate resolution (Denamiel et al. 2023a). Further, acoustic internal gravity waves can also be generated by volcanic explosions or asteroid impacts but would require a different modeling framework (e.g., baroclinic high-resolution models). This framework has not been described in detail, to this date. It remains to be properly developed and implemented for practical applications to acoustically driven events. Due to their relatively low resolution and their barotropic nature, global Lamb wave models are far less demanding in terms of numerical resources and storage and much faster to run than the local kilometer-scale models used for weather related events.

Data and Technical Requirements

Weather related events: In the meteotsunami community, the Weather Research and Forecasting model (WRF; Skamarock et al. 2005) is the atmospheric model of choice for the forecast of weather related events. For example, it is used in the Mediterranean within both Balearic Islands (Mourre et al. 2021) and Adriatic Sea (Denamiel et al. 2019b) early warning systems (see Fig. 4.10) while it is also tested in Korea (Kim et al. 2022). Additionally, the atmospheric model needs (1) pre-defined nested grids adapted to the region of interest, (2) initial conditions and lateral boundary conditions provided within a relaxation zone, (3) top boundary conditions at constant pressure level, and (4) bottom boundary conditions like sea-surface temperature, soil moisture and temperature, etc. In climate studies, these boundary conditions are extracted from the three dimensional fields of global or regional climate models. Finally, the use of kilometer-scale atmospheric models requires access to High Performance Computing (HPC) facilities. For example, in the Adriatic Sea, the WRF model at 3-km resolution runs on 64 CPUs to produce 5 day-long simulations on the ECMWF ATOS HPC (https://www.ecmwf.int/en/about/media-centre/news/2022/taking-ecmwfs-new-high-performance-computing-facility-operation; see Adriatic APTH Recipe, R10).

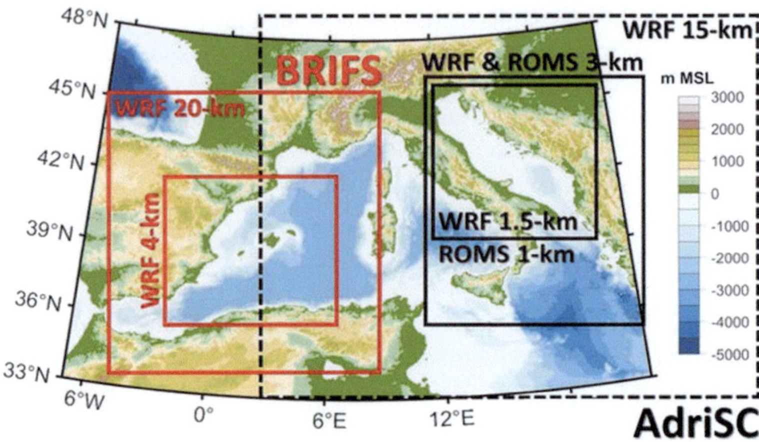

Fig. 4.10 Existing meteotsunami modeling systems in the Mediterranean Sea: BRIFS (in red) in the Balearic Islands and AdriSC (in black) in the Adriatic Sea. Adapted from Vilibić et al. (2021)

Acoustically driven events: Modeling of acoustic-gravity waves following for instance volcanic explosions is still at its very beginning. However, after the Hunga Tonga–Hunga Ha'apai volcano eruption on 15 January 2022, several models were used to reproduce the generated Lamb wave (e.g., Amores et al. 2022; Themens et al. 2022). For example, a few days after the event, the Lamb wave propagation was reproduced with the Transient Inertia–Gravity And Rossby wave dynamics model (TIGAR; Vasylkevych and Žagar 2021) solving the presented primitive equations with realistic orography using terrain-following coordinates and the Hough harmonics as the basis functions. A large amplitude Gaussian geopotential height perturbation centered on the volcano location with an amplitude of $\Delta p = 35.1$ hPa and a scale parameter of $\sigma = 2.7°$ was specified as the initial condition of the simulation. The parameters of the initial condition were optimized to match the observed wavelength of the wave response, while the dissipation time scale $\tau = 3.4$ days was obtained by a best fit of the wave amplitude to the observed pressure fluctuations at evaluation sites (Themens et al. 2022). It should be noted that, to this date, none of the Lamb wave models have been used operationally which would require some significant developments in terms of coupling with the hydrodynamic models and quantification of the uncertainty related to the initial condition parameters.

Treatment of Uncertainties

Deterministic modeling of meteotsunami events often fails due to the uncertainty linked to the modeling of the atmospheric conditions. Consequently, to derive meteotsunami hazards, using "Ensemble Modeling" (for weather related events only) or "Uncertainty Quantification (UQ) techniques" (presented under the "Synthetic Catalog" method) to account for the atmospheric condition uncertainties is necessary.

In the Balearic Islands, following Mourre et al. (2021), the "Ensemble Modeling" approach, using more than one atmospheric model configuration, improves the forecast of meteotsunami events. Mourre et al. (2021) thus implemented an ensemble of nine simulations varying the schemes and parametrizations within the WRF model. The ensemble was used to reproduce the nine largest past meteotsunamis in the Balearic Islands between 2014 and 2019. All simulations were found to realistically represent the meteotsunamigenic disturbances but their mesoscale characteristics (e.g., location, speed) were significantly modified with the change of parameterization. It should also be noted that no set of parameterizations overall led to a better forecast of the weather related events. In fact, by using this ensemble of nine simulations, a large range of meteotsunami surges – including the observed surges in 70% of the cases – was obtained. Consequently, in the probabilistic sense, ensemble modeling was found to improve the accuracy of meteotsunami predictions compared to single deterministic forecasts. However, ensembles of kilometer-scale atmospheric models have enormous computational cost. For meteotsunami hazard assessments, the "Ensemble Modelling" approach can be used to downscale meteotsunami events from an ensemble of global or regional climate models in order to account for the climate uncertainty.

4.5.3 Method: Synthetic Catalog

Short Description

Meteotsunamis are rare events and reliable observations of such events are consequently scarce. This particularly applies to air pressure (or wind) high-resolution measurements as 1-min records are not the standard in meteorological observations. Consequently, synthetic catalogs should be used to produce any kind of statistics, synthetic scenarios as input to hazard analysis, or sensitivity studies. Hereafter, the following steps of the synthetic catalog methodology are presented for meteotsunamis driven by pressure disturbances:

1. **Definition of the pressure field as a stochastic process (P_s).** The pressure field should first be defined as a stochastic process $P_s(x, y, t, \omega) = P_0 + P_A(x, y, t, \omega)$ with the horizontal coordinates following longitude and latitude (x, y), the time coordinate t, the stochastic variable ω, the mean atmospheric pressure P_0 and the stochastic process representing the analytical formulation of the atmospheric disturbance $P_A(x, y, t, \omega)$.
2. **Definition of the idealized atmospheric disturbance as a function of stochastic variables.** Several analytical formulations can be used for the different types of atmospheric disturbances driving meteotsunami events worldwide. Generally the analytical formulation of the pressure disturbance includes several stochastic variables such as the direction of propagation $\theta(\omega)$, the amplitude $P_A(\omega)$, the period $T(\omega)$, and the propagation speed $c(\omega)$. The number of

stochastic variables is highly variable depending on the choice of the analytical formulation and can be further reduced when a parameter is proven to have negligible impact on the final results (Denamiel et al. 2018, 2020).

3. **Definition of the stochastic variable distributions**. The stochastic variables used to define the meteotsunamigenic disturbance should be mutually independent and associated with a specific distribution. These distributions can be based on pressure measurements recorded during meteotsunami events but, if these measurements are too scarce, uniform distributions can be used for all variables. Both uniform distribution and mutual independence assumptions follow the principle of maximum entropy. This principle states that when little is known about a distribution, then the distribution with the largest entropy should be chosen as the least-informative default since it involves the fewest assumptions (Jaynes 1957).
4. **Sampling of the stochastic variable distributions**. Different strategies can be adopted concerning the sampling of the distributions (e.g., Monte Carlo, Markov Chain Monte Carlo, sparse grids, adaptive sparse grids, etc.). The choice of the sampling methodology highly depends on the final usage of the catalog and the numerical resources available. Indeed, every single Mesoscale Gravity Wave (MGW) defined with the combination of parameters sampled from the stochastic variable distribution has to be used as a forcing of a numerical hydrodynamic model to calculate the meteotsunami surges.

Overall, the generation of synthetic catalogs is numerically expensive, particularly when the number of stochastic variables is high (e.g., like in Denamiel et al. 2020). On the other hand, such catalogs can bring valuable insights concerning the potential hazards caused by rare events like meteotsunamis. Further, this methodology can also be applied for acoustically driven events whose intensity and location (in the case of asteroid impacts) can be highly uncertain.

Data and Technical Requirements

Synthetic catalogs are generally implemented when observations are scarce and are designed to work even with little knowledge of the prior distributions of the stochastic variables. However, the use of 1-min atmospheric records can help better constrain these prior distributions and hence, build more targeted synthetic catalogs – i.e., catalogs mostly formed of atmospheric disturbances that will lead to meteotsunami events at hotspot locations.

In terms of numerical resources, the generation of the atmospheric disturbances does not require the use of HPC facilities. However, to produce the final meteotsunami surges, sub-kilometer-scale hydrodynamic models are run for thousands of simulations forced by the sampled atmospheric disturbances. Consequently, when creating synthetic catalogs, HPC facilities are needed for both parallelization and storage.

Treatment of Uncertainties

Synthetic catalogs are essential to the implementation of Uncertainty Quantification - (UQ) - methods. For APTHA, UQ methods are useful as simulations of weather- and acoustically- driven meteotsunami waves at hotspot locations are highly influenced by the exact reproduction of the atmospheric disturbance parameters (e.g., start location, speed, period, etc.) which might not be captured deterministically. To this date two different UQ methods have been developed for weather related events: (1) Probabilistic Analysis and (2) Surrogate Model. However, these methods are based on a single application for a single geographical location.

The "Probabilistic Analysis" approach was implemented in the northeast U.S. by Geist et al. (2014) for isolated atmospheric pressure disturbances from squall lines and derechos. It is based on multiple synthetic catalogs and a Monte Carlo (MC) scheme. First, the atmospheric disturbances related to the squall lines in the northeast U.S. are parameterized and the observations from Automated Surface Observing System stations between 2000 and 2013 are used to establish a continuous analytic distribution of each parameter as well as the overall Poisson rate of occurrence. Then, the probabilistic equations are implemented using a MC scheme and a synthetic catalog of squall lines is compiled by sampling the parameter distributions. Finally, multiple synthetic catalogs are generated by resampling from the parent parameter distributions.

The "Surrogate Model" approach was implemented in the Adriatic Sea by Denamiel et al. (2019a, 2020, 2021). As the "Probabilistic Analysis", it is based on the generation of synthetic catalogs but for atmospheric disturbances assumed to be MGWs. The main difference between the two methods is that the sampling of the uniform distributions follows a sparse grid rule in order to build the surrogate model following the generalized Polynomial Chaos Expansion (gPCE) method that can be found at: https://github.com/cdenamiel/Adriatic_Sea_meteotsunami_surrogate_model.

Further, ANalysis Of VAriance (ANOVA), based on the gPCE method and providing the sensitivity to the atmospheric stochastic variables, was also performed at hotspot locations (e.g., Vela Luka and Stari grad) along the eastern Adriatic coast (e.g., Total and Main Sensitivities presented in Fig. 4.11). It reveals that the most important parameters for the generation of meteotsunamis are the speed and the period of the atmospheric disturbances.

To this date, UQ has not yet been developed for, nor applied to, acoustically driven events but the above presented approaches could also be used to derive the APTHA of planetary meteotsunami waves driven by volcanic eruptions or asteroid impacts.

4.5.4 Method: Climate Uncertainty

Short Description

Contrarily to the other tsunami sources, the long-term frequency and intensity of the atmospherically-driven tsunami sources are impacted by climate change (Vilibić et al. 2018; Denamiel et al. 2022). However, accounting for the climate uncertainties in meteotsunami modeling present two main challenges. On one hand, weather

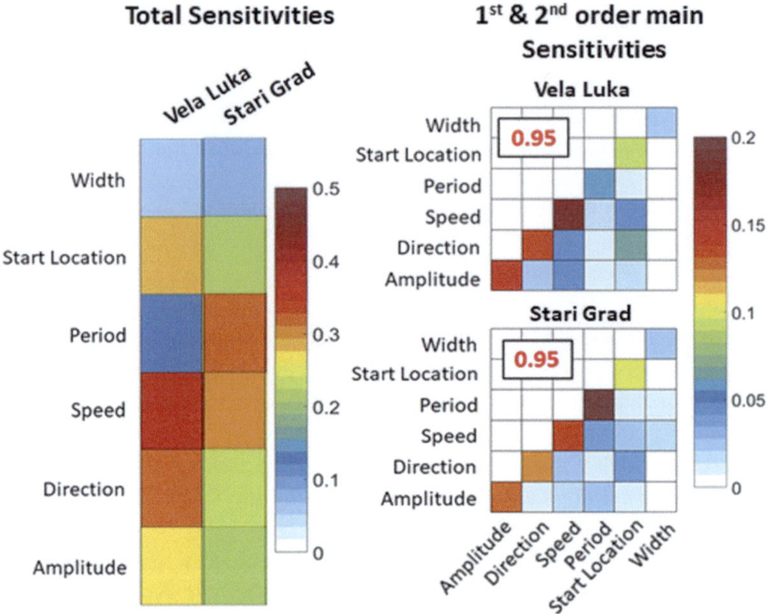

Fig. 4.11 Sensitivity analysis at Vela Luka and Stari Grad, two meteotsunami hotspot locations of the Adriatic Sea. Total (left panel) and 1st and 2nd order main sensitivities which represent 95% (0.95) of the meteotsunami impacts (right panels) calculated for the six parameters of the atmospheric disturbance used in the "Surrogate Model" approach. Adapted from Denamiel et al. (2020)

related meteotsunami events can not be dynamically reproduced by the available atmospheric Global or Regional Climate Models (GCMs or RCMs) with resolutions of the order of 100 km and 10 km respectively but can be extracted from these models at the synoptic scale. On the other hand, atmospheric kilometer-scale models, which downscale the results of the GCMs or RCMs, can represent the meteotsunamigenic disturbances but their numerical cost and slowness is an obstacle for long-term climate studies. For example, the coupled atmosphere–ocean AdriSC model running in the Adriatic Sea (Denamiel et al. 2019a) would take 7 years continuously using 256 CPUs to downscale one single climate scenario from one single GCM or RCM for the 1950–2100 period. Consequently, the methodology proposed by Denamiel et al. (2023b) to derive the frequency and intensity of the meteotsunamigenic disturbances in the future, as well as their associated uncertainties, at an affordable numerical cost is based on (1) synoptic indices that preselect a substantial number of short-term meteotsunami episodes from the GCMs and (2) a suite of atmospheric models that downscale these episodes from an ensemble of GCMs to the sub-kilometre-scale.

To this date, the methodology has not yet been fully tested but could practically be applied anywhere in the world by following 6 steps (Fig. 4.12):

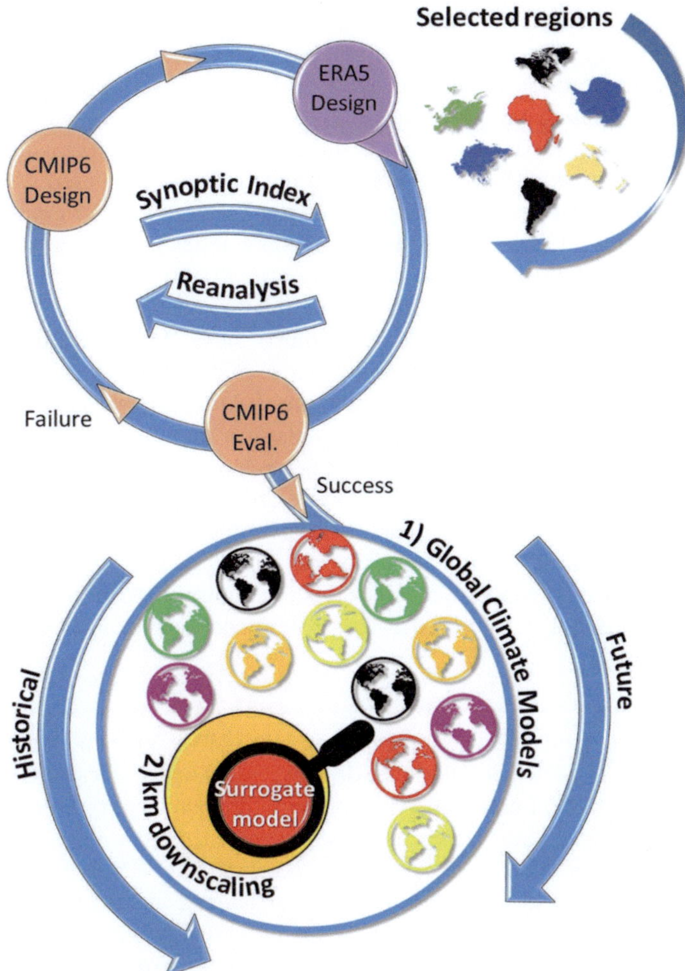

Fig. 4.12 Workflow of the methodology for projecting meteotsunami hazards using synoptic indices designed with global reanalysis products (ERA5 Design) and GCM historical simulations (CMIP6 Design) to extract meteotsunamigenic atmospheric disturbances including the climate uncertainty by downscaling (km downscaling) historical (CMIP6 Eval.) and future GCM simulations and using surrogate models to account for the atmospheric model uncertainties. Adapted from Denamiel et al. (2023b), licensed under CC-BY 4.0

1. **Synoptic Index Design**: Implementation of a synoptic index based on atmospheric reanalysis fields (e.g., ERA5) and available multi-year observations including high-frequency sea level measurements
2. **Synoptic Index Evaluation**: First, the synoptic index derived in step 1 is used with the historical simulations of the selected GCMs (e.g., CMIP6). Second, the sea-level distributions derived with the synoptic index for the reanalysis product

and the GCM historical simulations are compared. Then, if the comparison is successful – e.g., sea-level distributions significantly similar following the Z-test – the extracted meteotsunami events from the reanalysis product and the GCM historical simulations are downscaled to the kilometer-scale. Finally, the atmospheric and sea-level distributions obtained at hotspot locations by downscaling reanalysis products and GCM historical simulations are compared.

3. **Synoptic Index Redesign**: if the previous step fails – e.g., distributions significantly different following the Z-test, the synoptic index is redesigned by considering both GCM historical simulations (e.g., CMIP6) and reanalysis product (e.g., ERA5) until success of the Synoptic Index Evaluation step.
4. **Kilometer-scale Downscaling**: Downscaling of the GCM future climate scenario ensembles for the meteotsunami events extracted with the synoptic index.
5. **Selection of meteotsunamigenic conditions**: False positive events are excluded from the ensemble of kilometer-scale simulations for both historical and future simulations. The selection of the events is based on the automatic detection of meteotsunamigenic disturbances from the 1 km resolution atmospheric results – e.g., as presented in Denamiel et al. (2019b) for the Adriatic Sea.
6. **Uncertainty Quantification**: It is extremely likely that meteotsunamigenic disturbances do not generate meteotsunamis at hotspot locations due to a shift in location, speed, etc. In this case, surrogate models can be used to account for the uncertainty.

In this methodology, the derivation of the synoptic indices for meteotsunami events has been well tested and validated (Ramis and Jansà 1983; Vich and Romero 2021; Zemunik et al. 2022). Further, kilometer-downscaling – where telescopically nested domains at a decreasing horizontal grid spacing with boundary conditions provided by GCMs or reanalysis until kilometer-scales are reached – can also be considered state-of-the-art (Prein et al. 2015). However, the combination between climate and modeling uncertainties has never really been envisioned in the climate modeling community. Despite these untested aspects, the presented methodology remains, to this date, the best avenue to derive and project meteotsunami hazards (i.e., proper APTHA) under the on-going climate warming at an affordable numerical cost.

Data and Technical Requirements

Three main sources of data are needed to implement the method: (1) an ensemble of GCMs, (2) a reanalysis product and (3) sea-level observations. The best available ensemble of GCMs is, to this date, provided by the Coupled Model Intercomparison Project Phase 6 (CMIP6, Eyring et al. 2016) which includes between 14 and 16 members providing 6-hourly pressure level results at 100 km resolution for both Shared Socioeconomic Pathways (SSP) 2–4.5 and 5–8.5 scenarios. For the synoptic index derivation, the 6-hourly ERA-Interim atmospheric reanalysis (Dee et al. 2011) or the hourly ERA5 reanalysis at 25 km resolution (Hersbach et al. 2020) can be used. Indeed, ERA-Interim has a horizontal resolution, and thus a representation of atmospheric physics, closer to the above mentioned CMIP6 GCMs while ERA5 is the latest and probably the best reanalysis product from ECMWF. Finally, for the long-term quality-checked high-frequency sea-level observations, the MISELA dataset (Zemunik et al. 2021a, 2021b) can be used.

In terms of the numerical resources needed to implement the methodology, kilometer-scale models should be fully parallelized in order to efficiently downscale the GCMs. In addition, surrogate models used to account for the atmospheric model uncertainties require the generation of synthetic catalogs which are numerically expensive to implement. Finally, the use of GCM simulations as well as the generation of ensembles of kilometer-scale simulations require access (even temporarily) to nearly unlimited storage resources (e.g., tape system). Consequently, for the "Climate Uncertainty" method, even within the presented framework, access to HPC facilities is mandatory. For example, in the Adriatic Sea, the kilometer-scale modeling suite runs on 256 CPUs on the new ECMWF ATOS HPC (https://www.ecmwf.int/en/about/media-centre/news/2022/taking-ecmwfs-new-high-performance-computing-facility-operation).

Treatment of Uncertainties

The presented method deals with two sources of uncertainties: (1) the climate uncertainty linked to the downscaling of an ensemble of GCMs and (2) the uncertainty linked to the atmospheric disturbance modeling.

Although GCM ensembles do not necessarily represent the full uncertainty in climate projections (Knutti 2010), they allow investigating uncertainty in future projections and particularly examining the maximum–minimum climate change scenarios. To avoid erroneous interpretations of the obtained climate impact assessments, several strategies have been derived to select subsets of GCMs (e.g., smaller biases in the historical climate simulations, coverage of the widest possible uncertainty range of future projections; Reichler and Kim 2008; Cannon 2015; Mendlik and Gobiet 2016). Within the proposed method, all selected meteotsunami events from all appropriate GCMs (i.e., all GCMs with 100 km resolution) are downscaled. Consequently, the subsampling procedure is not applied to select GCMs but to detect meteotsunami events. Provided that the synoptic index is efficient (i.e., that enough meteotsunami events are extracted), the range of the climate uncertainty derived with the proposed approach will thus be similar to the one derived with the GCM ensemble. However, a bias can be introduced due to the specific setup of the kilometer-scale modeling suite implemented to downscale the GCMs.

The treatment of the atmospheric model uncertainty is already described within the "Synthetic Catalog" method.

4.5.5 Summary Table

See (Table 4.5).

Table 4.5 Summary table: atmospheric sources

	Synoptic Index	Deterministic Model	Synthetic Catalog	Climate Uncertainty
Short description of methods				
Theoretical/mechanical background for the method	Used to identify meteotsunami events with synoptic conditions extracted from global atmospheric models	Used to downscale the fields from global atmospheric models for weather sources or to generate the acoustic-gravity waves for acoustic sources	Used to generate all potential scenarios of a given atmospheric forcing (at a given location)	Used to quantify the long-term variability and frequency of meteotsunamigenic atmospheric disturbances under climate change
Short outline of how the methods work	Based on local sea level observations and atmospheric reanalyses/observations	Based on complex atmospheric models	Based on synthetic atmospheric forcing and catalogs of events	Based on downscaling global climate models only during meteotsunami events
How is/are the output(s) of the method represented?	Provides the dates of potential meteotsunami events at hotspot locations	Provides 2D/3D fields like pressure and winds needed to force meteotsunamis but also temperature, humidity, etc	Provides a catalog of events that can be used to derive/ implement statistics, hazard curves, surrogate models and sensitivity studies	Provides long-term projections of the intensity and frequency of the meteotsunamigenic atmospheric disturbances
In what way do the methods take into account time-dependent/dynamic aspects?	Included in the atmospheric global models/observations	Included in the complex kilometer-scale atmospheric models	Included in the time and space varying synthetic forcing/ long-term observations,	Included in the complex kilometer-scale atmospheric models
Do the methods use parallelisation, machine learning, surrogate models, advanced simulation schemes, other computational reduction strategies (e.g. cluster analysis, etc.)?	Uses machine learning tools	Uses parallelisation and optimization of the numerical schemes as well as large storage (e.g., tapes)	Used to implement surrogate models	Uses parallelisation and optimization of the numerical schemes as well as large storage (e.g., tapes)

(continued)

Table 4.5 (continued)

	Synoptic Index	Deterministic Model	Synthetic Catalog	Climate Uncertainty
What are the pros and cons and peculiarities of these methods? Are there known shortcomings / benefits of the methods? In what cases are these methods good/poor choices?	Pros: cheap to build and simple to use Cons: qualitative method often missing strongest events and providing a great number of false positives	Pros: dynamical solution based on model physics Cons: numerically expensive and not highly reliable	Pros: used for hazard assessments and uncertainty quantification Cons: synthetic forcing not as complex as real atmospheric forcing	Pros: full climate uncertainty by only downscaling meteotsunamigenic disturbances Cons: cumulates all problems of the other methods
Data and technical requirements for method				
What input data does the method(s) use?	Needs 1-min tide gauge observations and atmospheric reanalyses/observations	Needs initial/boundary conditions and/or topography/surface layer characterization from global models	Needs prior distributions of the atmospheric synthetic forcing stochastic parameters	Needs all of the inputs of the other methods
What are the technical requirements?	Uses machine learning	Set-up of kilometer-scale and/or global atmospheric models	Sampling techniques and surrogate model methods	All of the technical requirements from the other methods
Are HPC computing resources necessary, would absence of these compromise the use?	No	Yes. Without HPC the method is impossible to implement	Yes. Without HPC, oversimplification of the synthetic forcing or decrease of the ocean model resolution are required	Yes. Without HPC the method is impossible to implement
Is the method(s)/workflow(s) available as a git/software/jupyter notebook?	No	No	Yes	No

(continued)

Table 4.5 (continued)

	Synoptic Index	Deterministic Model	Synthetic Catalog	Climate Uncertainty
Treatment of uncertainties for the method				
Are the method(s) itself deterministic or uncertainty based?	Mostly deterministic	Mostly deterministic	Fully uncertainty based	Fully uncertainty based
Which are potential sources of uncertainties? Which parameters are used to characterize them?	Sources: local effects not accounted for and knowledge of future climate changes Parameters: atmospheric temperature, wind, pressure at the synoptic scale	Sources: inaccurate topography and surface layer characterizations or initial and boundary conditions as well as model parameterizations Parameters: pressure and wind at the mesoscale for weather sources	Sources: sparse observations and/or poor knowledge of the atmospheric process shape Parameters: direction, start location, amplitude, speed, period, attenuation in time, Earth's surface temperatures	Sources: cumulates all of the previous method sources Parameters: cumulates all of the previous method parameters
How are uncertainties quantified for the identified parameters? Are there significant correlations between different uncertain parameters?	Based on ensemble modeling	Based on ensemble modeling	Based on prior distributions of mutually independent synthetic forcing stochastic parameters	Based on ensembles of climate models and future climate change scenarios
Are the method(s) able to digest and propagate uncertainty in input (previous ingredient) and propagate it to the output (next ingredients)? Does the assessment of this ingredient require uncertainty propagation? If yes, how are uncertainties propagated?	Uncertainty propagation uses different indices as an ensemble of conditions	Uncertainty propagation uses ensembles of kilometer-scale atmospheric models	Uncertainty propagation uses probabilistic and surrogate model approaches	Uncertainty propagation uses all the previous method approaches.

(continued)

Table 4.5 (continued)

	Synoptic Index	Deterministic Model	Synthetic Catalog	Climate Uncertainty
What are the consequences of lack of data in input or for boundary conditions, or for characterizing certain parameters?	Consequences: poorly built synoptic index	Consequences: incapacity of reproducing the atmospheric disturbances	Consequences: inaccurate prior distributions and synthetic atmospheric forcing	Consequences: all of the previous method consequences as well as wrong climate uncertainties
What are the consequences of ignoring or poorly treating uncertainties?	Inaccurate: selection of the meteotsunamigenic atmospheric disturbances	Inaccurate: modeling of the atmospheric mesoscale conditions driving meteotsunami events	Inaccurate: meteotsunami hazard quantification	Inaccurate: climate hazard assessments

4.6 Choosing What Method to Use

4.6.1 Earthquakes

In PTHA modeling the earthquake scenario parameterization plays a key role. The choice of which method to use (rectangular or 3D faults, homogeneous or heterogeneous slip) depends on the amount of available information and the size of the earthquake. As a rule of thumb, the larger the earthquake, the larger the rupture area and thus 3D faults and heterogenous slip distributions are more suitable to describe these scenarios. Rectangular geometries can be used for crustal earthquakes, whereas the modeling of the ruptures along subduction faults requires a complex fault parameterization that honors as much as possible the slab geometry. For the same reasons, particular attention should be paid on using appropriate earthquake scaling relations depending on the magnitude range (see e.g. Figure 4.2) and the type of tectonic regime/faulting type.

The selection of the method to model the seafloor deformation also depends on the available information and the computational resources. The use of analytical methods is valid for both planar (rectangular) or 3D complex fault geometries; the latter requires a tessellation of the fault surface. If information such as 3D seismic tomography is available, the medium surrounding the fault can be characterized by the elastic parameters (e.g., shear modulus, Young modulus) and the deformation computed by means of FEM. Inclusion of the bathymetry allows computation of contribution of horizontal displacement to the tsunami source which can be significant in cases of shallow rupture.

Dynamic rupture models require the use of well-known faults and additional data related to the mechanical properties of the fault embedding rock volume that are often unavailable, or available just as average estimates. Assumptions are also required to the type of friction law, the spatial variability of its parameters as well as the distribution of initial stress across the fault plane.

4.6.2 Landslides

For landslides, the volume or frontal area, as well as rheology and landslide deformation with time control the tsunami generation. The simplest model is the static initial condition, while the block models and finally the depth averaged landslide models introduce better physical source representation but also more complexity. To embed the tsunami generation most correctly, it is preferential to use the block and landslide dynamics rather than the static initial conditions if resources allow it as these models better represent the source physics, whereas static initial conditions introduce large simplifications. In all cases, the models are associated with significant uncertainties related to the tsunami generation that require calibration towards past events before using them in PTHA.

Static initial condition methods for tsunami elevation are by far the least resource intensive. They may be easily embedded into PTHA models as they are instantaneous much like Okada-type sources for earthquakes. Benefits of these models are that they are easy to use and require almost no compute power. On the other hand they cannot integrate the effect of local bathymetry and the time dependence of the landslide into the tsunami model, and this limits the accuracy compared to all other models reviewed here. They may involve significant bias due to the many simplifying assumptions when describing a more complex type of source factors.

Block models are the simplest alternative when it comes to time-dependent source models that incorporate bathymetric effects. They are cheaper, more stable, and somewhat simpler to deploy than landslide dynamics models that require solving partial differential equations. Yet, projecting them into tsunami models needs careful effort, and this may be a challenge in mass-production simulations such as PTHA.

Numerical depth averaged landslide models include the deformation pattern of the landslide and can include rheology as well, and incorporate more physics than static initial conditions and block models. They may be feasible to include in a PTHA although they require more computational resources than static initial conditions and block models. Full three dimensional models are presently computationally too expensive to be embedded into PTHA, but are important for testing the representation of accurate physics, in particular for subaerial landslide tsunamis.

4.6.3 Volcanoes

There exist different approaches to model tsunamis induced by volcanic activity, depending on the type of source. The choice of the method depends on the targeted tsunami source (explosion, pyroclastic flow, etc.), the quality of available grids used for calculation (keeping in mind that volcanic tsunami sources often require high-resolution grids to be correctly defined, unlike seismic sources), and the objectives of the simulations. The strategy might be different between, for instance, (a) a deterministic approach aiming at accurately reproducing a past event at a local scale, and (b) a probabilistic approach aiming at simulating a database of scenarios for a hazard study.

Instantaneous initialization (static source model) is suitable if tsunami generation happens rapidly compared to the shallow water wave speed. This works for underwater explosions only (note that an instantaneous caldera collapse initiation is not realistic because a caldera collapse typically lasts minutes to hours). If the forcing happens over a specified time at the start of the modeling, then a time dependent initialization is required. In the case of a caldera collapse, a common approach is to apply the time dependent sea floor motion as a forcing term. Our understanding of tsunamis generated by pyroclastic flows is still in its infancy, but in the current state of knowledge, it appears strongly recommended to use multi-fluid CFD models (full Navier–Stokes models). These models are computationally too expensive for PTHA, therefore, it is often necessary to limit their use to the modeling of the source, and then use less advanced models for the propagation and integration into PTHA. Another case of finite time generation is an ongoing forcing by a pressure anomaly, such as Lamb waves produced by large explosions in the atmosphere (see also atmospheric sources).

4.6.4 Atmospheric Sources

In the context of PTHA, contrary to other tsunami sources, weather related meteotsunami events are dependent on the climatic changes and are thought to be highly influenced by the on-going climate warming. Hence classical time dependent PTHA methods are likely to have shortcomings. Instead, a climate change approach attempting to extract the past and future events from climate models in order to derive statistics on their behavior and characteristics should be used. However, to this date, only few generic methods are implemented for operational forecast or research purposes within the meteotsunami community, but on the other hand not yet put to use for Atmospheric PTHA.

Most of the generic methods (e.g., synoptic index, deterministic model) so far only deal with short term forecast meteotsunami events with the aim to issue public warnings at hotspot locations. Synoptic indices can predict the occurrence of meteotsunami events up to five days in advance but not their intensity and they are known to generate many false alarms. In terms of providing input to APTHA, synoptic indices can be used to extract meteotsunami events from global climate models in order to derive the past and future frequency of weather related events depending on the chosen climate scenarios. Concerning the deterministic atmospheric numerical models, they have been proven to be able to successfully simulate the meteotsunamigenic disturbances. In terms of providing input to APTHA, they can thus be used to downscale the global climate models to the kilometer-scale needed to represent the intensity of the atmospheric disturbances driving the meteotsunami events. However, deterministic models have difficulties to properly reproduce the exact location, speed, period, etc., of these disturbances. Consequently, the need to account for the uncertainty in the atmospheric simulations (e.g., ensemble modeling, surrogate models) has become increasingly important in recent years. APTHA should thus also include the representation of the atmospheric model uncertainties.

Practically, the only method that can be used for APTHA of weather related events is the climate uncertainty method combining synoptic indices, deterministic models and uncertainty quantification linked to both climate projections and atmospheric modeling. However, the climate uncertainty method has not yet been successfully implemented due to both the lack of reliability of the synoptic indices and the extreme numerical costs of kilometer-scale atmosphere–ocean modeling.

References

Abadie SM, Harris, JC, Grilli ST, Fabre R (2012) Numerical modeling of tsunami waves generated by the flank collapse of the Cumbre Vieja Volcano (La Palma, Canary Islands): Tsunami source and near field effects. J Geophys Res: Oceans 117(C5). https://doi.org/10.1029/2011JC007646

Abrahams LS, Krenz L, Dunham EM, Gabriel AA, Saito T (2023) Comparison of methods for coupled earthquake and tsunami modelling. Geophys J Int 234(1):404–426. https://doi.org/10.1093/gji/ggad053

Allen TI, Hayes GP (2017) Alternative rupture-scaling relationships for subduction interface and other offshore environments. Bull Seism Soc Am 107(3):1240–1253. https://doi.org/10.1785/0120160255

Amores A, Monserrat S, Marcos M, Argüeso D, Villalonga J, Jordà G, Gomis D (2022) Numerical simulation of atmospheric Lamb waves generated by the 2022 Hunga-Tonga volcanic eruption. Geophys Res Lett 49:e2022GL098240. https://doi.org/10.1029/2022GL098240

Andrews DJ (1980) A stochastic fault model: 1. Static case. J Geophys Res: Solid Earth 85(B7):3867–3877. https://doi.org/10.1029/JB085iB07p03867

Andrews DJ (2005) Rupture dynamics with energy loss outside the slip zone. J Geophys Res: Solid Earth 110(B1). https://doi.org/10.1029/2004JB003191

Baba T, Chikasada N, Nakamura Y, Fujie G, Obana K, Miura S, Kodaira S (2020) Deep investigations of outer-rise tsunami characteristics using well-mapped normal faults along the Japan Trench. J Geophys Res: Solid Earth 125(10):e2020JB020060. https://doi.org/10.1029/2020JB020060

Barall M (2009) A grid-doubling finite-element technique for calculating dynamic three-dimensional spontaneous rupture on an earthquake fault. Geophys J Int 178(2):845–859. https://doi.org/10.1111/j.1365-246X.2009.04190.x

Barrett R, Berry M, Chan T, Demmel J, Donato J, Dongarra J, Eijkhout V, Pozo R, Romine C, van der Vorst, HA (1994) Templates for the solution of linear systems: Building blocks for iterative methods (2nd ed.). SIAM, Philadelphia, PA. https://doi.org/10.2307/2153507

Behrens J, Løvholt F, Jalayer F, Lorito S, Salgado-Gálvez MA et al (2021) Probabilistic tsunami hazard and risk analysis: a review of research gaps. Front Earth Sci 9:p628772. https://doi.org/10.3389/feart.2021.628772

Benjemaa M, Glinsky-Olivier N, Cruz-Atienza VM, Virieux J (2009) 3-D dynamic rupture simulations by a finite volume method. Geophys J Int 178(1):541–560. https://doi.org/10.1111/j.1365-246X.2009.04088.x

Bilek SL, Lay T (1999) Rigidity variations with depth along interplate megathrust faults in subduction zones. Nature 400(6743):443–446

Bishop CM et al (1995) Neural networks for pattern recognition. Oxford University Press, Oxford

Blaser L, Krüger F, Ohrnberger M, Scherbaum F (2010) Scaling relations of earthquake source parameter estimates with special focus on subduction environment. Bull Seism Soc Am. 100(6):2914–2926

Bougouin A, Paris R, Roche O (2020) Impact of fluidized granular flows into water: Implications fortsunamis generated by pyroclastic flows. J Geophys Res: Solid Earth 125:e2019JB018954. https://doi.org/10.1029/2019JB018954

Byerlee JD (1978) Friction of rocks. In: Byerlee JD and Wyss M (eds) Rock friction and earthquake prediction. Birkhäuser, Basel, p 615–626

Cannon AJ (2015) Selecting GCM scenarios that span the range of changes in a multimodel ensemble: application to CMIP5 climate extremes indices. J Clim 28:1260–1267. https://doi.org/10.1175/JCLI-D-14-00636.1

Carey S, Sigurdsson H, Mandeville C, Bronto S (2000) Volcanic hazards from pyroclastic flow discharge into the sea: examples from the 1883 eruption of Krakatau, Indonesia. Geol Soc Am Special Publ 345:1–14

Cas RAF, Wright JV (1991) Subaqueous pyroclastic flows and ignimbrites: an assessment. Bull Volc 53:357–380

Chen F, Heller V, Briganti R (2020) Numerical modelling of tsunamis generated by iceberg calving validated with large-scale laboratory experiments. Adv Water Res 142:103647. https://doi.org/10.1016/j.advwatres.2020.103647
Chinnery MA (1963) The stress changes that accompany strike-slip faulting. Bull Seism Soc Am 53(5):921–932
Cohn SE, Dee DE (1989) An analysis of the vertical structure equation for arbitrary thermal profiles. Quart J Roy Meteor Soc 115:143–171. https://doi.org/10.1002/qj.49711548508
Cornell CA (1968) Engineering seismic risk analysis. Bull Seism Soc Am 58:1583–1606. https://doi.org/10.1785/BSSA0580051583
Davies G (2019) Tsunami variability from uncalibrated stochastic earthquake models: tests against deep ocean observations 2006–2016. Geophys J Int 218(3):1939–1960. https://doi.org/10.1093/gji/ggz260
Davies G, Griffin J (2020) Sensitivity of probabilistic tsunami hazard assessment to far-field earthquake slip complexity and rigidity depth-dependence: case study of Australia. Pure Appl Geophys 177:1521–1548
Davies G, Horspool N, Miller V (2015) Tsunami inundation from heterogeneous earthquake slip distributions: Evaluation of synthetic source models. J Geophys Res: Solid Earth 120(9):6431–6451
Day SM, Dalguer LA, Lapusta N, Liu Y (2005) Comparison of finite difference and boundary integral solutions to three-dimensional spontaneous rupture. J Geophys Res: Solid Earth 110(B12). https://doi.org/10.1029/2005JB003813
Dee DP, Uppala SM, Simmons AJ, Berrisford P, Poli P, Kobayashi S et al (2011) The ERA-Interim reanalysis: configuration and performance of the data assimilation system. QJR Meteor Soc 137:553–597. https://doi.org/10.1002/qj.828
de Lange WP, Prasetya GS, Healy TR (2001) Modelling of tsunamis generated by pyroclastic flows (ignimbrites). Nat Haz 24(3):251–266
de la Puente J, Ampuero JP, Käser M (2009) Dynamic rupture modeling on unstructured meshes using a discontinuous Galerkin method. J Geophys Res: Solid Earth 114(B10). https://doi.org/10.1029/2008JB006271
Denamiel C, Šepić J, Vilibić I (2018) Impact of geomorphological changes to harbor resonance during meteotsunamis: The Vela luka bay test case. Pure Appl Geophys 175:3839–3859. https://doi.org/10.1007/s00024-018-1862-5
Denamiel C, Šepić J, Huan X, Bolzer C, Vilibić I (2019a) Stochastic surrogate model for meteotsunami early warning system in the eastern Adriatic Sea. J Geophys Res Oceans 124:8485–8499. https://doi.org/10.1029/2019JC015574
Denamiel C, Šepić J, Ivanković D, Vilibić I (2019b) The Adriatic Sea and coast modelling suite: evaluation of the meteotsunami forecast component. Ocean Model 135:71–93. https://doi.org/10.1016/j.ocemod.2019.02.003
Denamiel C, Huan X, Šepić J, Vilibić I (2020) Uncertainty propagation using polynomial chaos expansions for extreme sea level hazard assessment: the case of the eastern Adriatic meteotsunamis. J Phys Oceanogr 50(4):1005–1021. https://doi.org/10.1175/JPO-D-19-0147.1D
Denamiel C, Huan X, Vilibić I (2021) Conceptual design of extreme sea-level early warning systems based on uncertainty quantification and engineering optimization methods. Front Mar Sci 8:650279. https://doi.org/10.3389/fmars.2021.650279T
Denamiel C, Tojčić I, Vilibić I (2022) Meteotsunamis in orography-free, flat bathymetry and warming climate conditions. J Geophys Res Oceans 127:e2021JC017386. https://doi.org/10.1029/2021JC017386
Denamiel C, Vasylkevych S, Žagar N, Zemunik P, Vilibić I (2023a) Destructive potential of planetary meteotsunami waves beyond the hunga tonga-hunga ha'apai volcano eruption. Bull Am Meteo Soc 104(1):E178–E191. https://doi.org/10.1175/BAMS-D-22-0164.1
Denamiel C, Belušić D, Zemunik P, Vilibić I (2023b) Climate projections of meteotsunami hazards. Front Mar Sci 10:1167863. https://doi.org/10.3389/fmars.2023.1167863

Dieterich JH (1979) Modeling of rock friction: 1. Experimental results and constitutive equations. J Geophys Res Solid Earth 84(B5):2161–2168

Dietrich JC, Tanaka S, Westerink JJ, Dawson CN, Luettich RA Jr, Zijlema M et al (2012) Performance of the unstructured-mesh, SWAN+ADCIRC model in computing hurricane waves and surge. J Sci Comput 52:468–497. https://doi.org/10.1007/s10915-011-9555-6

Di Toro G, Han R, Hirose T, De Paola N, Nielsen S, Mizoguchi K, Ferri F, Cocco M, Shimamoto T (2011) Fault lubrication during earthquakes. Nature 471(7339):494–498

Ely GP, Day SM, Minster J-B (2009) A support-operator method for 3-D rupture dynamics. Geophys J Int 177(3):1140–1150. https://doi.org/10.1111/j.1365-246X.2009.04117.x

Esposti Ongaro T, De'Michieli Vitturi M, Fornaciai A et al. (2021) Modelling tsunamis generated by submarine landslides at Stromboli volcano (Aeolian Islands, Italy): a numerical benchmark study. Front Earth Sci 9. https://doi.org/10.3389/feart.2021.628652

Eyring V, Bony S, Meehl GA, Senior CA, Stevens B, Stouffer RJ et al (2016) Overview of the coupled model intercomparison project phase 6 (CMIP6) experimental design and organization. Geosci Model Dev 9:1937–1958. https://doi.org/10.5194/gmd-9-1937-2016

Festa G, Vilotte JP (2005) The Newmark scheme as velocity–stress time-staggering: an efficient PML implementation for spectral element simulations of elastodynamics. Geophys J Int 161(3):789–812

Freundt A (2003) Entrance of hot pyroclastic flows into the sea: experimental observations. Bull Volc 65:144–164

Galvez P, Ampuero JP, Dalguer LA, Somala SN, Nissen-Meyer T (2014) Dynamic earthquake rupture modelled with an unstructured 3-D spectral element method applied to the 2011 M 9 tohoku earthquake. Geophys J Int 198(2):1222–1240

Geist EL (2002) Complex earthquake rupture and local tsunamis. J Geophys Res Solid Earth 107(B5). https://doi.org/10.1029/2000JB000139

Geist EL, Bilek SL (2001) Effect of depth-dependent shear modulus on tsunami generation along subduction zones. Geophys Res Lett 28(7):1315–1318

Geist EL, Oglesby DD (2014) Earthquake mechanism and seafloor deformation for tsunami generation. In: Beer M, Kougioumtzoglou I, Patelli E, Au IK (eds) Encyclopedia of Earthquake Engineering. Springer, Berlin, Heidelberg. https://doi.org/10.1007/978-3-642-36197-5_296-1

Geist EL, Oglesby DD, Ryan KJ (2022) Tsunamis: stochastic models of occurrence and generation mechanisms. In: Tilling RI (ed) Complexity in Tsunamis, Volcanoes, and their Hazards. Encyclopedia of Complexity and Systems Science Series. Springer, New York, NY. https://doi.org/10.1007/978-1-0716-1705-2_595

Geist EL, ten Brink US, Gove MA (2014) A framework for the probabilistic analysis of meteotsunamis. Nat Haz 74:123–142. https://doi.org/10.1007/s11069-014-1294-1

George DL, Iverson RM (2014) A depth-averaged debris-flow model that includes the effects of evolving dilatancy. II. Numerical predictions and experimental tests. Proc Royal Society A: Math, Phys Eng Sci 470(2170):20130820. https://doi.org/10.1098/rspa.2013.0820

Goda K, Yasuda T, Mori N, Maruyama T (2016) New scaling relationships of earthquake source parameters for stochastic tsunami simulation. Coastal Eng J 58(3):1650010–1650011

Goda K, Yasuda T, Mai PM, Maruyama T, Mori N (2018) Tsunami simulations of mega-thrust earthquakes in the Nankai-Tonankai Trough (Japan) based on stochastic rupture scenarios. In: Scourse EM, Chapman NA, Tappin DR, Wallis SR (eds) Tsunamis: Geology, Hazards and Risks. Geological Society, London. Special Publications 456:55–74. https://doi.org/10.1144/SP456.1

Goto A, Taniguchi H, Yoshida M, Ohba T, Oshima H (2001) Effects of explosion energy and depth to the formation of blast wave and crater: field explosion experiment for the understanding of volcanic explosion. Geophys Res Lett 28(22):4287–4290

Gray JMNT, Edwards AN (2014) A depth-averaged-rheology for shallow granular free-surface flows. J Fluid Mech 755:503–534. https://doi.org/10.1017/jfm.2014.450

Grezio A, Babeyko A, Baptista MA, Behrens J, Costa A, Davies G et al (2017) Probabilistic tsunami hazard analysis: multiple sources and global applications. rev Geophys 55:1158–1198. https://doi.org/10.1002/2017RG000579

Grilli ST, Watts P (2005) Tsunami generation by submarine mass failure. I: Modeling, experimental validation, and sensitivity analyses. J Waterway Port Coast Ocean Eng 131(6):283–297. https://doi.org/10.1061/(ASCE)0733-950X(2005)131:6(283)

Grilli ST, Zhang C, Kirby JT, Grilli AR, Tappin DR, Watt SFL et al. (2021) Modeling of the Dec. 22nd 2018 Anak Krakatau volcano lateral collapse and tsunami based on recent field surveys: Comparison with observed tsunami impact. Mar Geol 440:106566. https://doi.org/10.1016/j.margeo.2021.106566

Guazzelli É, Pouliquen O (2018) Rheology of dense granular suspensions. J Fluid Mech 852:P1. https://doi.org/10.1017/jfm.2018.548

Gutenberg B, Richter CF (1944) Frequency of earthquakes in California. Bulletin Seism Soc Am 34:185–188. https://doi.org/10.1785/BSSA0340040185

Gutenberg B, Richter CF (1949) Seismicity of the earth associated phenomena. Princeton University Press, Princeton

Gylfadóttir SS, Kim J, Helgason JK, Brynjólfsson S, Höskuldsson Á, Jóhannesson T, Harbitz CB, Løvholt F (2017) The 2014 Lake askja rockslide-induced tsunami: optimization of numerical tsunami model using observed data. J Geophys Res Oceans 122:4110–4122. https://doi.org/10.1002/2016JC012496

Hanks TC, Kanamori H (1979) A moment magnitude scale. J Geophys Res Solid Earth 84(B5):2348–2350

Harris RA, Barall M, Andrews DJ, Duan B, Ma S et al (2011) Verifying a computational method for predicting extreme ground motion. Seism Res Lett 82(5):638–644

Harris RA, Barall M, Aagaard B, Ma S, Roten D et al (2018) A suite of exercises for verifying dynamic earthquake rupture codes. Seism Res Lett 89(3):1146–1162

Harbitz CB (1992) Model simulations of tsunamis generated by the Storegga slides. Mar Geol 105(1–4):1–21. https://doi.org/10.1016/0025-3227(92)90178-K

Harbitz CB, Glimsdal S, Løvholt F, Kveldsvik V, Pedersen GK, Jensen A (2014) Rockslide tsunamis in complex fjords: from an unstable rock slope at Åkerneset to tsunami risk in western Norway. Coastal Eng 88:101–122

Hayes G (2018) Slab2 - a comprehensive subduction Zone Geometry Model. U.S. Geological Survey data release. https://doi.org/10.5066/F7PV6JNV

Haykin S (2004) Neural Networks: A comprehensive foundation. Prentice Hall, Upper Saddle River NJ

Heller V, Hager WH (2010) Impulse product parameter in landslide generated impulse waves. J Waterway Port Coastal Ocean Eng 136(3):145–155. https://doi.org/10.1061/(ASCE)WW.1943-5460.0000037

Herrero A, Bernard P (1994) A kinematic self-similar rupture process for earthquakes. Bull Seism Soc Am 84(4):1216–1228

Herrero A, Murphy S (2018) Self-similar slip distributions on irregular shaped faults. Geophys J Int 213(3):2060–2070

Hersbach H, Bell B, Berrisford P, Hirahara S, Horányi A, Muñoz-Sabater J et al (2020) The ERA5 global reanalysis. QJ Roy Meteorol Soc 146:1999–2049. https://doi.org/10.1002/qj.3803

Horillo J, Wood A, Kim GB, Parambath A (2013) A simplified 3-D Navier-Stokes numerical model for landslide-tsunami: application to the Gulf of Mexico. J Geophys Res Oceans 118(12):6934–6950. https://doi.org/10.1002/2012JC008689

Horvath K, Vilibić I (2014) Atmospheric mesoscale conditions during the Boothbay meteotsunami: a numerical sensitivity study using a high-resolution mesoscale model. Nat Hazards 74:55–74. https://doi.org/10.1007/s11069-014-1055-1

Ida Y (1972) Cohesive force across the tip of a longitudinal-shear crack and Griffith's specific surface energy. J Geophys Res 77(20):3796–3805

Igel H (2017) Computational seimology—a practical introduction. Oxford University Press, Oxford, pp 365

Iverson RM, Denlinger RP (2001) Flow of variably fluidized granular masses across three-dimensional terrain: 1. Coulomb mixture theory. J Geophys Res Solid Earth 106(B1):537–552. https://doi.org/10.1029/2000JB900329

Iwasaki T, Sato R (1979) Strain field in a semi-infinite medium due to an inclined rectangular fault. J Phys Earth 27(4):285–314

Jaynes ET (1957) Information theory and statistical mechanics. Phys Rev 106:620–630. https://doi.org/10.1103/PhysRev.106.620

Jop P, Forterre Y, Pouliquen O (2006) A constitutive law for dense granular flows. Nature 441:727–730. https://doi.org/10.1038/nature04801

Kanamori H, Brodsky EE (2004) The physics of earthquakes. Rep Prog Phys 67:1429–1496. https://doi.org/10.1088/0034-4885/67/8/R03

Kasahara A (1977) Numerical integration of the global barotropic primitive equations with hough harmonic expansions. J Atmos Sci 34:687–701. https://doi.org/10.1175/1520-0469(1977)034%3c0687:NIOTGB%3e2.0.CO;2

Kedrinskii VK (2005) Hydrodynamics of Explosion: experiments and models. Springer, Berlin, Heidelberg. https://doi.org/10.1007/3-540-28563-6

Kokelaar P (1986) Magma-water interactions in subaqueous and emergent basaltic volcanism. Bull Volc 48:275–289

Kelfoun K, Druitt TH (2005) Numerical modeling of the emplacement of Socompa rock avalanche, Chile. J Geophys Res: Solid Earth 110(B12). https://doi.org/10.1029/2005JB003758

Kim J (2014) Finite volume methods for tsunamis generated by submarine landslides. PhD Thesis, Univ Washington. http://hdl.handle.net/1773/25374

Kim J, Løvholt F, Issler D, Forsberg CF (2019) Landslide material control on tsunami genesis—the storegga slide and tsunami (8,100 years BP). J Geophys Res: Oceans 124(6):3607–3627. https://doi.org/10.1029/2018JC014893

Kim M-S, Woo S-B, Eom H, You SH, Lee HM (2022) Towards observation- and atmospheric model-based early warning systems for meteotsunami mitigation: a case study of Korea. Weather Clim Extrem 37:100463. https://doi.org/10.1016/j.wace.2022.100463

Kirby JT, Grilli ST, Horrillo J, Liu PLF, Nicolsky D, Abadie S et al (2022) Validation and inter-comparison of models for landslide tsunami generation. Ocean Model 17:101943. https://doi.org/10.1016/j.ocemod.2021.101943

Knutti R (2010) The end of model democracy? Clim Change 102:395–404. https://doi.org/10.1007/s10584-010-9800-2

Lavallée D, Archuleta RJ (2003) Stochastic modeling of slip spatial complexities for the 1979 Imperial Valley, California, earthquake. Geophys Res Lett 30(5):1245. https://doi.org/10.1029/2002GL015839

Le Méhauté B (1971) Theory of Explosion-Generated Water Waves. In: Te Chow V (ed) Advances in Hydroscience. Elsevier/Academic Press, New York, 7:1–79. https://doi.org/10.1016/B978-0-12-021807-3.50006-0

Le Méhauté B, Wang S (1996) Water waves generated by underwater explosion (Vol. 10). World Scientific, Singapore

LeVeque RJ (2002) Finite volume methods for hyperbolic problems (Vol. 31). Cambridge university press, Cambridge

LeVeque RJ, Waagan K, González FI et al (2016) Generating random earthquake events for probabilistic tsunami hazard assessment. Pure Appl Geophys 173:3671–3692. https://doi.org/10.1007/s00024-016-1357-1

Lo H-Y, Liu P (2017) On the analytical solutions for water waves generated by a prescribed landslide. J Fluid Mech 821:85–116. https://doi.org/10.1017/jfm.2017.251

Lo PH-Y, Liu P (2021) On water waves generated by a bottom obstacle translating at a subcritical speed. J Fluid Mech 923:A26. https://doi.org/10.1017/jfm.2021.537

Lotto GC, Dunham EM (2015) High-order finite difference modeling of tsunami generation in a compressible ocean from offshore earthquakes. Comp Geosci 19(2):327–340

Lucas A, Mangeney A, Ampuero JP (2014) Frictional velocity-weakening in landslides on Earth and on other planetary bodies. Nature Comm 5(1):3417. https://doi.org/10.1038/ncomms4417

Løvholt F, Harbitz CB, Haugen KB (2005) A parametric study of tsunamis generated by submarine slides in the Ormen Lange/Storegga area off western Norway. Mar Petr Geol 22(1–2):219–231. https://doi.org/10.1016/B978-0-08-044694-3.50023-8

Løvholt F, Pedersen G, Harbitz CB, Glimsdal S, Kim J (2015) On the characteristics of landslide tsunamis. Phil Trans Roy Soc A 373:20140376. https://doi.org/10.1098/rsta.2014.0376

Løvholt F, Pedersen G, Harbitz CB (2016) Tsunami-genesis due to retrogressive landslides on an inclined seabed. In: Lamarche G et al. (eds) Submarine mass movements and their consequences. Advances in Natural and Technological Hazards Research 41. Springer, Cham. https://doi.org/10.1007/978-3-319-20979-1_57

Løvholt F, Bondevik S, Laberg JS, Kim J, Boylan N (2017) Some giant submarine landslides do not produce large tsunamis. Geophys Res Lett 44(16):8463–8472. https://doi.org/10.1002/2017GL074062

Løvholt F, Schulten I, Mosher D, Harbitz C, Krastel S (2019) Modelling the 1929 Grand Banks slump and landslide tsunami. Geo Soc Spec Publ, London, 477(1), 315–331

Løvholt F, Glimsdal S, Harbitz CB (2020) On the landslide tsunami uncertainty and hazard. Landslides 17(10):2301–2315. https://doi.org/10.1007/s10346-020-01429-z

Løvholt F, Urgeles R, Harbitz CB, Vanneste M, Carlton B (2022) Submarine Landslides. In Shroder JF (ed) Treatise on Geomorphology (Second Edition). Academic Press, London, p 919–959. https://doi.org/10.1016/B978-0-12-818234-5.00139-5

Ma G, Kirby JT, Shi F (2013) Numerical simulation of tsunami waves generated by deformable submarine landslides. Ocean Modeling 69:146–165. https://doi.org/10.1016/j.ocemod.2013.07.001

Ma S (2009) Distinct asymmetry in rupture-induced inelastic strain across dipping faults: An off-fault yielding model. Geophys Res Lett 36(20):L20317. https://doi.org/10.1029/2009GL040666

Ma S (2012) A self-consistent mechanism for slow dynamic deformation and tsunami generation for earthquakes in the shallow subduction zone. Geophys Res Lett 39:L11310. https://doi.org/10.1029/2012GL051854

Macías J, Vázquez JT, Fernández-Salas LM, González-Vida JM, Bárcenas P, Castro MJ, Díaz-del-Río V, Alonso B (2015) The Al-Borani submarine landslide and associated tsunami. A Modelling Approach. Mar Geol 361:79–95. https://doi.org/10.1016/j.margeo.2014.12.006

Madariaga R (1976) Dynamics of an expanding circular fault. Bull Seism Soc Am 66(3):639–666

Madden EH, Bader M, Behrens J, van Dinther Y, Gabriel AA, Rannabauer L, Ulrich T, Uphoff C, Vater S, van Zelst I (2021) Linked 3-D modelling of megathrust earthquake-tsunami events: from subduction to tsunami run up. Geophys J Int 224(1):487–516. https://doi.org/10.1093/gji/ggaa484

Maeda T, Furumura T (2013) FDM simulation of seismic waves, ocean acoustic waves, and tsunamis based on tsunami-coupled equations of motion. Pure Appl Geophys 170:109–127

Maeno F, Imamura F, Taniguchi H (2006) Numerical simulation of tsunamis generated by caldera collapse during the 7.3 ka Kikai eruption, Kyushu, Japan. Earth Planets Space 58(8):1013–1024

Maeno F, Imamura F (2007) Numerical investigations of tsunamis generated by pyroclastic flows from the Kikai caldera. Japan. Geophys Res Lett 34:L23303. https://doi.org/10.1029/2007GL031222

Maeno F, Imamura F (2011) Tsunami generation by a rapid entrance of pyroclastic flow into the sea during the1883 Krakatau eruption. Indonesia. J Geophys Res: Solid Earth 116:B09205. https://doi.org/10.1029/2011JB008253

Mansinha LA, Smylie DE (1971) The displacement fields of inclined faults. Bull Seism Soc Am 61(5):1433–1440

Mai PM, Beroza GC (2002) A spatial random field model to characterize complexity in earthquake slip. J Geophys Res: Solid Earth, 107(B11):ESE-10

Marzocchi W, Bebbington M (2012) Probabilistic eruption forecasting at short and long time scales. Bull Volc 74:1777–1805

Meade BJ (2007) Algorithms for the calculation of exact displacements, strains, and stresses for triangular dislocation elements in a uniform elastic half space. Comp Geosci 33(8):1064–1075

Mendlik T, Gobiet A (2016) Selecting climate simulations for impact studies based on multivariate patterns of climate change. Clim Change 135:381–393. https://doi.org/10.1007/s10584-015-1582-0

Mourre B, Santana A, Buils A, Gautreau L, Ličer M, Jansà A et al (2021) On the potential of ensemble forecasting for the prediction of meteotsunamis in the Balearic Islands: Sensitivity to atmospheric model parameterizations. Nat Hazards 106:1315–1336. https://doi.org/10.1007/s11069-020-03908-xO

Murotani S, Satake K, Fujii Y (2013) Scaling relations of seismic moment, rupture area, average slip, and asperity size for M 9 subduction-zone earthquakes. Geophys Res Lett 40(19):5070–5074

Murphy S, Scala A, Herrero A, Lorito S, Festa G, Trasatti E, Tonini R, Romano F, Molinari I, Nielsen S (2016) Shallow slip amplification and enhanced tsunami hazard unravelled by dynamic simulations of mega-thrust earthquakes. Sci Rep 6(1):35007. https://doi.org/10.1038/srep35007

Murphy S, Di Toro G, Romano F, Scala A, Lorito S, Spagnuolo E, Aretusini S, Festa G, Piatanesi A, Nielsen S (2018) Tsunamigenic earthquake simulations using experimentally derived friction laws. Earth Planet Sci Lett 486:155–165. https://doi.org/10.1016/j.epsl.2018.01.011

Nikkhoo M, Walter TR (2015) Triangular dislocation: an analytical, artefact-free solution. Geophys J Int 201(2):1119–1141

Nomanbhoy N, Satake K (1995) Generation mechanism of tsunamis from the 1883 Krakatau eruption. Geophys Res Lett 224:509–512

Ohnaka M (1973) Experimental studies of stick-slip and their application to the earthquake source mechanism. J Phys Earth 21(3):285–303

Okada Y (1985) Surface deformation due to shear and tensile faults in a half-space. Bull Seism Soc Am 75:1435–1154

Panizzo A, Cuomo G, Dalrymple RA (2007) 3D-SPH simulation of landslide generated waves. Coast Eng 2006:1503–1515. https://doi.org/10.1142/9789812709554_0128

Paris R, Switzer AD, Belousova M et al (2014) Volcanic tsunami: a review of source mechanisms, past events and hazards in Southeast Asia (Indonesia, Philippines, Papua New Guinea). Nat Hazards 70:447–470. https://doi.org/10.1007/s11069-013-0822-8

Paris R, Ulvrova M (2019) Tsunamis generated by subaqueous volcanic explosions in Taal Caldera Lake, Philippines. Bull Volc 81:1–14. https://doi.org/10.1007/s00445-019-1272-2

Pelties C, De la Puente J, Ampuero JP, Brietzke GB, Käser M (2012) Three-dimensional dynamic rupture simulation with a high-order discontinuous Galerkin method on unstructured tetrahedral meshes. J Geophys Res: Solid Earth 117:B02309. https://doi.org/10.1029/2011JB008857

Pitman EB, Le L (2005) A two-fluid model for avalanche and debris flows. Phil Trans Roy Soc A 363(1832):1573–1601. https://doi.org/10.1098/rsta.2005.1596

Plougonven R, Zhang F (2014) Internal gravity waves from atmospheric jets and fronts. Rev Geophys 52:33–76. https://doi.org/10.1002/2012RG000419

Prein AF, Langhans W, Fosser G, Ferrone A, Ban N, Goergen K et al (2015) A review on regional convection-permitting climate modeling: demonstrations, prospects and challenges. Rev Geophys 53:323–361. https://doi.org/10.1002/2014RG000475

Pudasaini SP (2012) A general two-phase debris flow model. J Geophys Res: Earth Surf 117(F3). https://doi.org/10.1029/2011JF002186

Ramis C, Jansà A (1983) Condiciones meteorológicas simultáneas a la aparición de oscilaciones del nivel del mar de amplitud extraordinaria en el Mediterráneo occidental. Rev Geofisica 39:35–42

Rauter M, Viroulet S, Gylfadóttir SS, Fellin W, Løvholt F (2022) Granular porous landslide tsunami modelling–the 2014 Lake Askja flank collapse. Nature Comm 13(1):678. https://doi.org/10.1038/s41467-022-28296-7

Reichler T, Kim J (2008) How well do coupled models simulate Today's climate? Bull Am Meteorol Soc 89:303–312. https://doi.org/10.1175/BAMS-89-3-303

Renault L, Vizoso G, Jansà A, Wilkin J, Tintoré J (2011) Toward the predictability of meteotsunamis in the Balearic Sea using regional nested atmosphere and ocean models. Geophys Res Lett 38:L10601. https://doi.org/10.1029/2011gl047361

Romano F, Trasatti E, Lorito S, Piromallo C, Piatanesi A, Ito Y, Zhao D, Hirata K, Lanucara P, Cocco M (2014) Structural control on the Tohoku earthquake rupture process investigated by 3D FEM, tsunami and geodetic data. Sci Rep 4(1):5631. https://doi.org/10.1038/srep05631

Ruina A (1983) Slip instability and state variable friction laws. J Geophys Res: Solid Earth 88(B12):10359–10370

Ruiz J, Baumont D, Bernard P, Berge-Thierry C (2007) New approach in the kinematic k− 2 source model for generating physical slip velocity functions. Geophys J Int 171(2):739–754

Ryan KJ, Geist EL, Barall M, Oglesby DD (2015) Dynamic models of an earthquake and tsunami offshore Ventura. California. Geophys Res Lett 42(16):6599–6606

Satake K, Fujii Y, Harada T, Namegaya Y (2013) Time and space distribution of coseismic slip of the 2011 Tohoku earthquake as inferred from tsunami waveform data. Bull Seism Soc Am 103(2B):1473–1492. https://doi.org/10.1785/0120120122

Sato H, Taniguchi H (1997) Relationship between crater size and ejecta volume of recent magmatic and phreato-magmatic eruptions: Implications for energy partitioning. Geophys Res Lett 24(3):205–208

Scala A, Lorito S, Romano F, Murphy S, Selva J, Basili R, Babeyko A, Herrero A, Hoechner A, Løvholt F, Maesano FE (2020) Effect of shallow slip amplification uncertainty on probabilistic tsunami hazard analysis in subduction zones: use of long-term balanced stochastic slip models. Pure Applied Geophys 177(3):1497–1520. https://doi.org/10.1007/s00024-019-02260-x

Segall P (2010) Earthquake and volcano deformation. Princeton University Press, Princeton, In Earthquake and Volcano Deformation. https://doi.org/10.1515/9781400833856

Sepúlveda I, Liu PLF, Grigoriu M, Pritchard M (2017) Tsunami hazard assessments with consideration of uncertain earthquake slip distribution and location. J Geophys Res: Solid Earth 122(9):7252–7271

Shen Y, Whittaker CN, Lane EM, White JDL, Power W, Nomikou P (2021a) Laboratory experiments on tsunamigenic discrete subaqueous volcanic eruptions. Part 1: Free surface disturbances. J Geophys Res: Oceans 126: e2020JC016588. https://doi.org/10.1029/2020JC016588

Shen Y, Whittaker CN, Lane EM, White JD, Power W, Nomikou P (2021b) Laboratory experiments on tsunamigenic discrete subaqueous volcanic eruptions. Part 2: Properties of generated waves. J Geophys Res: Oceans 126(5):e2020JC016587. https://doi.org/10.1029/2020JC016587

Skarlatoudis AA, Somerville PG, Thio HK (2016) Source-scaling relations of interface subduction earthquakes for strong ground motion and tsunami simulation. Bull Seism Soc Am 106(4):1652–1662

Snelling B, Neethling S, Horsburgh K, Collins G, Piggott M (2020) Uncertainty quantification of landslide generated waves using gaussian process emulation and variance-based sensitivity analysis. Water 12(2):416. https://doi.org/10.3390/w12020416

Song SG, Somerville P (2010) Physics-based earthquake source characterization and modeling with geostatistics. Bulletin Seism Soc Am 100(2):482–496

Si P, Shi H, Yu X (2018) Development of a mathematical model for submarine granular flows. Phys Fluids 30(8):083302. https://doi.org/10.1063/1.5030349

Skamarock WC, Klemp JP, Dudhia J, Gill DO, Barker DM (2005) A description of the Advanced Research WRF Version 2. NCAR Technical Note NCAR/TN-468+STR. https://doi.org/10.5065/D6DZ069T

Synolakis CE, Bardet JP, Borrero JC, Davies HL, Okal EA, Silver EA et al (2002) The slump origin of the 1998 Papua New Guinea tsunami. Proc Royal Soc London A 458(2020):763–789. https://doi.org/10.1098/rspa.2001.0915

Strasser FO, Arango MC, Bommer JJ (2010) Scaling of the source dimensions of interface and intraslab subduction-zone earthquakes with moment magnitude. Seism Res Lett 81(6):941–950

Stirling M, Goded T, Berryman K, Litchfield N (2013) Selection of earthquake scaling relationships for seismic-hazard analysis. Bull Seism Soc Am 103:2993–3011. https://doi.org/10.1785/0120130052

Tanioka Y, Satake K (1996) Tsunami generation by horizontal displacement of ocean bottom. Geophys Res Lett 23:861–864. https://doi.org/10.1029/96GL00736

Themens D et al. (2022) Global propagation of ionospheric disturbances associated with the 2022 Tonga volcanic eruption. Geophys Res Lett 49:e2022GL09815. https://doi.org/10.1029/2022GL098158

Tonini R, Basili R, Maesano FE, Tiberti MM, Lorito S, Romano F, Scala A, Volpe M (2020) Importance of earthquake rupture geometry on tsunami modelling: the Calabrian Arc subduction interface (Italy) case study. Geophys J Int 223(3):1805–1819. https://doi.org/10.1093/gji/ggaa409

Torsvik T, Paris R, Didenkulova I, Pelinovsky E, Belousov A, Belousova M (2010) Numerical simulation of a tsunami event during the 1996 volcanic eruption in Karymskoye lake, Kamchatka, Russia. Nat Hazards Earth Syst Sci 10:2359–2369. https://doi.org/10.5194/nhess-10-2359-2010

Ulrich T, Gabriel AA, Ampuero JP, Xu W (2019a) Dynamic viability of the 2016 Mw 7.8 Kaikōura earthquake cascade on weak crustal faults. Nature Comm 10(1):1213. https://doi.org/10.1038/s41467-019-09125-w

Ulrich T, Vater S, Madden EH, Behrens J, van Dinther Y, Van Zelst I, Fielding EJ, Liang C, Gabriel AA (2019b) Coupled, physics-based modeling reveals earthquake displacements are critical to the 2018 Palu, Sulawesi tsunami. Pure Appl Geophys 176:4069–4109. https://doi.org/10.1007/s00024-019-02290-5

Ulvrova M, Paris R, Kelfoun K, Nomikou P (2014) Numerical simulations of tsunamis generated by underwater volcanic explosions at Karymskoye lake (Kamchatka, Russia) and Kolumbo volcano (Aegean Sea, Greece). Nat Hazards Earth Syst Sci 14:401–412. https://doi.org/10.5194/nhess-14-401-2014

Ulvrova M, Paris R, Nomikou P, Kelfoun K, Leibrandt S, Tappin DR, McCoy FW (2016) Source of the tsunami generated by the 1650AD eruption of Kolumbo submarine volcano (Aegean Sea, Greece). J Volc Geotherm Res 321:125–139. https://doi.org/10.1016/j.jvolgeores.2016.04.034

Vasylkevych S, Žagar N (2021) A high-accuracy global prognostic model for the simulation of rossby and gravity wave dynamics. Quart J Roy Meteor Soc 147:1989–2007. https://doi.org/10.1002/qj.4006

M-d-M V, Romero R (2021) Forecasting meteotsunamis with neural networks: the case of Ciutadella harbour (Balearic Islands). Nat Hazards 106:1299–1314. https://doi.org/10.1007/s11069-020-04041-5

Vilibić I, Šepić J, Dunić N, Sevault F, Monserrat S, Jordà G (2018) Proxy-based assessment of strength and frequency of meteotsunamis in future climate. Geophys Res Lett 45:10501–10508. https://doi.org/10.1029/2018GL079566

Vilibić I, Denamiel C, Zemunik P, Montserrat S (2021) The Mediterranean and Black Sea meteotsunamis: an overview. Nat Hazards 106:1223–1267. https://doi.org/10.1007/s11069-020-04306-z

Wang JA, Sardeshmukh PD, Compo GP, Whitaker JS, Slivinski LC, McColl CM, Pegion PJ (2019) Sensitivities of the NCEP global forecast system. Mon Wea Rev 147:1237–1256. https://doi.org/10.1175/MWR-D-18-0239.1

Ward SN (2001) Landslide tsunami. J Geophys Res: Solid Earth 106(B6):11201–11215. https://doi.org/10.1029/2000JB900450

Watts P (1998) Wavemaker curves for tsunamis generated by underwater landslides. J Waterway Port Coast Ocean Eng 124(3):127–137

Watts P, Grilli ST, Tappin DR, Fryer GJ (2005) Tsunami generation by submarine mass failure. II: Predictive equations and case studies. J Waterway Port Coast Ocean Eng 131(6):298–310

Watts P, Waythomas CF (2003) Theoritical analysis of tsunami generation by pyroclastic flows. J Geophys Res 108(B112):2563

Wells DL, Coppersmith KJ (1994) New empirical relationships among magnitude, rupture length, rupture width, rupture area, and surface displacement. Bull Seism Soc Am 84(4):974–1002

Wendt J, Oglesby DD, Geist EL (2009) Tsunamis and splay fault dynamics. Geophys Res Lett 36:L15303. https://doi.org/10.1029/2009GL038295

Wohletz KH (1986) Explosive magma-water interactions: thermodynamics, explosion mechanisms, and field studies. Bull Volc 48:245–264

Yavari-Ramshe S, Ataie-Ashtiani B (2016) Numerical modeling of subaerial and submarine landslide-generated tsunami waves—recent advances and future challenges. Landslides 13:1325–1368. https://doi.org/10.1007/s10346-016-0734-2

Zeng Y, Anderson JG, Yu G (1994) A composite source model for computing realistic synthetic strong ground motions. Geophys Res Lett 21(8):725–728

Zhang C, Kirby JT, Shi F, Ma G, Grilli ST (2021) A two-layer non-hydrostatic landslide model for tsunami generation on irregular bathymetry. 2. Numerical discretization and model validation. Ocean Modelling 160:101769. https://doi.org/10.1016/j.ocemod.2021.101769

Zhao S, Müller RD, Takahashi Y, Kaneda Y (2004) 3-D finite-element modelling of deformation and stress associated with faulting: effect of inhomogeneous crustal structures. Geophys J Int 157(2):629–644

Zemunik P, Vilibić I, Šepić J, Pellikka H, Ćatipović L (2021a) MISELA: 1-minute sea-level analysis global dataset. Earth Sys Sci Data 13:4121–4132. https://doi.org/10.5194/essd-13-4121-2021

Zemunik P, Vilibić I, Šepić J, Pellikka H, Ćatipović L (2021b) MISELA: Minute Sea-Level Analysis, Marine Data Archive. https://doi.org/10.14284/456

Zemunik P, Denamiel C, Williams J, Vilibić I (2022) High-frequency sea-level analysis: global correlations to synoptic atmospheric patterns. Weather Clim Extrem 38(7):100516. https://doi.org/10.1016/j.wace.2022.100516

Zengaffinen T, Løvholt F, Pedersen GK, Harbitz CB (2020) Effects of rotational submarine slump dynamics on tsunami genesis: new insight from idealized models and the 1929 Grand Banks event. Geol Soc London Special Publ 500(1):41–61. https://doi.org/10.1144/SP500-2019-201

Zengaffinen-Morris T, Urgeles R, Løvholt F (2022) On the inference of tsunami uncertainties from landslide run-out observations. J Geophys Res: Oceans 127:e2021JC018033. https://doi.org/10.1029/2021JC018033

Chapter 5
Hydrodynamic Modelling

Íñigo Aniel-Quiroga, Maria Ana Baptista, Jörn Behrens, Miquel Canals, Cléa Denamiel, Utku Kânoğlu, Galderic Lastras, Finn Løvholt, Jorge Macías, Shane Murphy, and Raphaël Paris

5.1 Introduction

This chapter presents several methods to simulate tsunami propagation, inundation, and impact on land. The chapter is subdivided into five sections: Preparation of Data, Interfaces, Propagation, Inundation, and Testing. The order of the sections represents a typical sequence of tsunami modelling in a probabilistic tsunami hazard assessment (PTHA) workflow.

The first step is the preparation of the data for use in numerical models. The second step involves the interface between the source modelling described in Chap. 4 and the modelling of tsunami propagation and inundation. Finally, inundation modelling is described in the fourth step providing the output for exposure modelling, discussed in Chap. 6.

Tsunami models predict wave height, surface elevation, flow depth and inundation areas. These parameters are crucial for either deterministic or probabilistic tsunami hazard assessment methods. Moreover, they can compute real-time tsunami travel time and offshore wave heights for early warning and evacuation purposes. The propagation and inundation of tsunamis is generally governed by equations

Í. Aniel-Quiroga
Instituto de Hidráulica Ambiental (IH Cantabria), Universidad de Cantabria, Santander, Spain
e-mail: anieli@unican.es

M. A. Baptista
Instituto Superior de Engenharia de Lisboa, Instituto Politécnico de Lisboa, Lisboa, Portugal
e-mail: mavbaptista@gmail.com

J. Behrens (✉)
Department of Mathematics, Universität Hamburg, Hamburg, Germany
e-mail: joern.behrens@uni-hamburg.de

M. Canals · G. Lastras
Department of Earth and Ocean Dynamics, CRG Marine Geosciences, Universitat de Barcelona, Barcelona, Spain
e-mail: miquelcanals@ub.edu

M. B. Sørensen et al. (eds.), *Probabilistic Tsunami Hazard and Risk Analysis*, Mathematics of Planet Earth 14, https://doi.org/10.1007/978-3-031-98115-9_5

of motion—the Navier-Stokes equations—and simplifications thereof. Due to the complexity of the equations and the geometry of the domain needed in practical simulations, no general analytical solutions exist. Therefore, numerical codes for tsunami simulation are used to solve the relevant equations. To do this, the general process includes simplification/elimination of terms of the Navier Stokes equations, generation of numerical grids incorporating boundary conditions and bathymetry/topography, application of numerical methods such as finite differences, finite elements or finite volumes to the equations of motion, solution of the inherent discrete equation sets—often by iterative solvers.

Tsunami inundation models predict tsunami propagation over land (onshore). These models rely heavily on topography and land use data and to date still comprise uncertainty, when considering detailed impact. For large-scale hazard assessments heuristic amplification factors or statistically based emulators, based on wave heights near to the shore and derived from tsunami propagation models, are frequently used to estimate inundation depths.

Code validation and verification is of paramount importance to produce reliable and reproducible results. As data from real time events are scarce, modellers recur to analytical solutions, laboratory experimental and observed data (Synolakis et al. 2008; NTHMP 2012, 2016; Lynett et al. 2017) to numerically benchmark the models.

G. Lastras
e-mail: glastras@ub.edu

C. Denamiel
Division of Marine and Environmental Research, Ruđer Bošković Institute, Zagreb, Croatia
e-mail: cdenami@irb.hr

U. Kânoğlu
Department of Aerospace Engineering, Dokuz Eylül University, Izmir, Turkey
e-mail: utku.kanoglu@deu.edu.tr

F. Løvholt
Offshore Energy, Norwegian Geotechnical Institute, Oslo, Norway
e-mail: finn.lovholt@ngi.no

J. Macías
Departamento de Análisis Mathemático, Estadística e Investigación Operativa y Mathemática Aplicada, Universidad de Málaga, Malaga, Spain
e-mail: jmacias@uma.es

S. Murphy
Geo-Ocean, Ifremer, Plouzané, France
e-mail: shane.murphy@ifremer.fr

R. Paris
CNRS, IRD, OPGC, Laboratoire Magmas et Volcans, Université Clermont Auvergne, Aubière, France
e-mail: raphael.paris@uca.fr

A summary of this chapter's contents and intention can be found in condensed form in Table 5.1.

5.2 Preparation of Data

Diverse types of data are important to tsunami modelling. Mathematically, tsunami modelling can be described as an initial and boundary value problem, thus results sensitively depend on the input data.

Most importantly, boundary data influencing the modelling results are topography and bathymetry data. A more in-depth description of their inclusion in tsunami modelling is given below. Initial data, more precisely an accurate representation of the tsunami source, is another crucial data type for accurate modelling results, mostly covered in the next section. Finally, topography, sometimes supplemented by land use data and corresponding tsunami modelling parameterizations are important for a realistic inundation behaviour of tsunami models. A condensed list of contents of this section is presented in Table 5.2.

5.2.1 Bathymetric and Topographic Data

Bathymetric and topographic data are key ingredients for the numerical modelling of tsunami propagation and its impact on target coastlines. A typical continental margin shallows shoreward from the trench up to the shelf. This configuration produces a shoaling effect in tsunami propagation, which is most noticeable over the continental shelf. Shoaling implies the decrease of tsunami wavelength and velocity and the increase of wave height when the wave propagates from deep to shallow water. The significance of shoaling depends on the shape and depth of each margin: the width and depth of its shelf, the gradient of its continental slope, the presence of submarine canyons dissecting the slope, the incision of these canyons into the continental shelf eventually with their heads close to the coastline, and the presence of seamounts, shallow banks or wide troughs, among many other features.

Such topographic features induce significant local changes to the wave characteristics. Tsunami arrival time to a coastline stretch located shoreward of a highly incised submarine canyon is short, and wave height is less, whereas enhanced wave setup occurs along the flanks of the canyon, with an opposite effect at both sides of the relevant coastal stretch (Iglesias et al. 2014). Extremely incised canyons favour the formation of trapped edge waves along the coast. In addition, seafloor and coastline features such as semi-enclosed bays, estuaries, and fjords can lead to resonance phenomena enhancing tsunami effects. Moreover, the complexity of bathymetric features, such as a sand spit, necessitates accurate topographic data for the correct estimation of runup. Thus, tsunami numerical modelling is sensitive to both the resolution and the accuracy of bathymetric and topographic data. The appropriate data

Table 5.1 Hydrodynamic modelling

What does the ingredient describe?	Tsunami modelling is at the core of each PTHA/PTRA workflow. This ingredient describes the different steps of setting up a modelling workflow
Definitions of the ingredient	Input data: What are the inputs available or required Sources: Different phenomena triggering tsunamis such as earthquake, volcano, landslide, meteorological sources Interfaces: Sources trigger tsunamis via interfaces such as initial, boundary or forcing data Equations: A tsunami model is represented by mathematical equations of the wave propagation and fluid dynamics Discretization: The transformation of mathematical equations into a form consisting of elementary computational formulas, digestible by a computer Coupling: Exchange of appropriate data between different models at their interfaces Propagation: The phase in a tsunami model approach, in which tsunami waves travel through the ocean basin Inundation: The phase, in which a tsunami penetrates land Verification: Test that a model implementation is free of errors and behaves as it should according to its defined specification Validation: Ensure that a model behaves in accordance with the physical world (up to its model assumptions/limitations)
What is/are the metric (s) used for measuring/ evaluating the ingredient?	On input (disturbances from a resting sea): These may be sea surface elevation, ocean bottom displacement, mass displacements at the boundaries, or forcing terms On output: Hazard metrics such as sea surface elevation offshore and at the coast, inundation height, run-up, inundation area, momentum flux, flow velocity, etc
Possible uses/applications of the ingredient	Except for event-based hazard assessment (based on field data catalogues), modelling is one of the best-available ways to assess hazard from tsunami activity
Possible multi-hazard dimensions/ applications?	The multi-hazard aspect is represented through the source, or after the tsunami has hit in the risk analysis parts

(continued)

Table 5.1 (continued)

What does the ingredient describe?	Tsunami modelling is at the core of each PTHA/PTRA workflow. This ingredient describes the different steps of setting up a modelling workflow
Possible time-dependent/dynamic aspects	Sources may be instantaneous or dynamic. The modelling outputs may be maximum values or time-series
List of the available methods to be described below	Data Preparation Interfacing Propagation Inundation Validation

representation depends on the complexity of the shoreline. On a considerably plain and very long beach such as those of Southern California, a 100-m-grid resolution may be sufficient. Small offshore and onshore features likely to affect tsunami wave behaviour should be reflected in the numerical grids. If a location is fronted by a sand spit of width 100 m, a sufficient number of grid points are needed to provide accurate resolution of the flow over the spit, thus higher resolution is necessary. On the other hand, deep ocean propagation does not require such fine grid spacings. Therefore, adaptive or nested grids are often used for representing bathymetry and topography adequately.

5.2.2 Technical Requirements for and Availability of Topo-Bathymetry Data

Since most tsunami propagation tools employ low to medium order numerical methods (i.e. lower than 3rd order consistency), as a rule of thumb one may use that the relevant wavelength or feature size needs to be sampled by at least 10–20 grid points or cells for being adequately resolved, although this should be verified by sensitivity tests. To produce a good Digital Elevation Model (DEM) to simulate tsunami inundation, we need to merge high-resolution ocean depth data (bathymetry) and height data (topography). This merging process requires a common definition for the vertical reference surface. It is important to note that bathymetric and topographic maps use different vertical references. The bathymetric maps use a local reference to the main harbour in the area, whereas topographic maps use the mean sea level referred to a specific datum.

Tsunami numerical modelling requires computational grids, and for inundation modelling often also several nested grids, to model generation, propagation, and inundation phases. While the tsunami propagation requires accurate bathymetric data,

Table 5.2 Summary table modelling data

Short description of method	
Theoretical/mechanical background for the method	Considered in this section are data that are not related to the tsunami source, such as topography and bathymetry data
Short outline of how the method works	Thorough consideration of accuracy requirements and compatibility with modelling assumptions is required. Guidelines are given in this section
How is/are the output (s) of the method represented?	The outputs of this method are input files to tsunami modelling tools, usually in standardized formats
In what way does the method take into account time-dependent/dynamic aspects?	Most of the data are time-independent relative to the time-scales of tsunami modelling
Does the method use parallelisation, machine learning, surrogate models, advanced simulation schemes, other computational reduction strategies (e.g. cluster analysis, etc.)?	No
What are the pros and cons and peculiarities of this specific method? Are there known shortcomings/benefits of the method? In what cases is this method a good/poor choice?	N/A
Data and technical requirements for method	
What input data does the method use?	Diverse sources of elevation data
What are the technical requirements?	Vertical accuracy is more important than horizontal accuracy, elevations around mean sea level are most critical
Are HPC computing resources necessary, would absence of these compromise the use?	No
Is the method/workflow available as a git/software/jupyter notebook?	Usually no unified workflows available, but sometimes individual/proprietary scripts, since the data processing depends on specific model requirements
Treatment of uncertainties for the method	
Is the method itself deterministic or uncertainty based?	Deterministic
Which are potential sources of uncertainties? Which parameters are used to characterize them?	Data accuracy is a big issue. Definition of mean sea level may not be uniform, elevation data may include canopies and may not represent ground elevation
How are uncertainties quantified for the identified parameters? Are there significant correlations between different uncertain parameters?	Difficult to quantify, sensitivity is large, but little systematic assessment available

(continued)

Table 5.2 (continued)

Short description of method	
Is the method able to digest and propagate uncertainty in input (previous ingredient) and propagate it to the output (next ingredients)? Does the assessment of this ingredient require uncertainty propagation? If yes, how are uncertainties propagated?	This represents input uncertainty, there is no propagation
What are the consequences of lack of data in input or for boundary conditions, or for characterizing certain parameters?	Relevant hazard metrics may be rendered incorrect
What are the consequences of ignoring or poorly treating uncertainties?	Results, in particular inundation maps, may be very incorrect

the inundation phase requires both bathymetric and topographic data in consistent combination. Several issues can be listed in setting up DEM for tsunami numerical modelling (Venturato 2007) that might contribute to cumulative error in the data. These can be listed as follows:

1. Most often, high-resolution bathymetric and topographic data are proprietary, or potentially restricted, specifically close to the shoreline where the highest accuracy is needed.
2. Bathymetric and topographic data can be obtained from different sources. Hence, there might be problems matching the data, especially for the shoreline.
3. Inherent error within the original data, conversion error due to several data sources (for example, converting from a map projection to geographic coordinates), interpolation error of spatially varying data sources, and error due to subjective interpretation.
4. A common source of error is the different datum (e.g. mean sea level definition) of sea maps versus topography maps, and this mismatch must be corrected for.

In terms of bathymetric data, global bathymetric grids such as ETOPO2022 (2022), GEBCO (2022), and others are derived from satellite altimetry, ship soundings, and multibeam bathymetry. The Nippon Foundation GEBCO Seabed 2030 project aims to bring together all available bathymetric data sets and produce a definitive, 100% complete seafloor map by 2030. These compilations are fed with seafloor bathymetry data, even from remote regions, and are useful to compute large-scale offshore tsunami propagation. Nevertheless, resolutions of such global maps may not be high enough to resolve tsunami propagation near the coastline, where shoaling effects take place. GEBCO (2022) has a 15-arc second grid size, which translates to approximately 450 m at the Equator or 300 m at 50° of latitude. As an example, continental shelves in the Mediterranean Sea, with a mean width of approximately 20 km, are described by half a hundred bathymetric points, and canyon heads incised in the shelf may be represented by half a dozen of soundings. This is not sufficiently fine for resolving the details needed for the near shore propagation.

EMODnet Bathymetry provides a service for viewing and downloading the best available harmonised Digital Terrain Model (DTM) for the European sea regions, together with a range of other bathymetric data, products, and services (https://emodnet.ec.europa.eu/en/bathymetry). Topographic data or Digital Elevation Models (DEMs) are available freely or commercially. DEMs are available in two forms: as Digital Terrain Models (DTM), which contain bare-earth elevation data, i.e., buildings, vegetation, and others removed; or as Digital Surface Models (DSM), which include all these features. Light Detection and Ranging (LiDAR), Airborne Interferometric Synthetic Aperture Radar (IFSAR), Advanced Spaceborne Thermal Emission and Reflection Radiometer (ASTER), Shuttle Radar Topography Mission (SRTM), and Multifunctional camera—High-Resolution Stereoscopic Camera (HRSC) are possible sources of topographic data (Griffin et al. 2015).

It is important to note that for both bathymetric as well as topographic data, the accuracy with respect to the vertical dimension is more important than the horizontal accuracy. To make this clear, it makes a big difference if the topographic height includes an uncertainty of the order of metre, because this can determine inundation or safe grounds. If a hill is relocated by one meter horizontally, this will hardly affect the result of an inundation model.

There are several data formats in use for bathymetric and topographic data sets. The use of self-describing binary data formats, such as NetCDF, HDF5 or GRIB data formats are helpful for replicability. However, ASCII formats as well as diverse geographical information system (GIS) file formats are also often in use. In many cases a rectangular grid format is used, but there are also triangulated bathymetry files or point data sets in use. In any case, it is common to interpolate from a given dataset to the computational mesh, this needs care.

5.2.3 Other Data Types

In many tsunami models the bottom roughness parameter is used to calibrate the model for realistic applications. This parameter is used to represent subgrid-scale processes, such as turbulent flows, wave breaking, interaction with structures in inundation modelling, etc. on the larger modelling scale. It is most common to use a Manning roughness between 0.25 and 0.35 but heterogeneous roughness parameterization covering a wider range of friction parameters for different types of land use are available in certain models.

Another parameter used in some tsunami models is an eddy diffusivity parameter, accounting for turbulent processes. In meteotsunami applications, a wind-stress tensor is used to parameterize the effect of wind shear as a forcing to the water wave and flow build-up. All these parameters are subject to uncertainty and usually some testing and calibration will need to be applied, based on field data. This indicates the importance of test cases, as described in the last section of this chapter.

5.3 Interface

This section is titled interface, since the simulation of tsunami propagation and inundation can be understood as a mapping of data to outcomes such as inundation height, fluxes, or wave amplitudes. So, tsunami simulation needs to be interfaced with models providing accurate input data. This interfacing needs to be consistent with the physical as well as numerical characteristics of the target model. Again, we condensed the contents of this chapter into a table for better overview and orientation, which can be found in Table 5.3.

5.3.1 Tsunami Modelling as an Initial Boundary Value Problem

In the input mapping there are several stages that need to be interfaced, these stages are described here. The core of the tsunami simulation is formed by an equation set representing the fluid motion, and these equations are further controlled by initial and boundary values as well as parameters and source terms. So, data as described in the previous section need to be interfaced appropriately with the modelling environment. It is paramount to adhere to the physical consistency of all stages and interface conditions to obtain accurate results.

In an abstract sense, the wave propagation and inundation of a tsunami is simulated by the equations of motion which in a very general way can be given in the following abstract form:

$$\frac{\partial \mathrm{q}}{\partial t} + \nabla \cdot F(\mathrm{q}) = S(\mathrm{q}) \tag{5.1}$$

where q contains the conserved quantities (density ρ, momentum ρu with u the velocity, total energy e), F(q) represents the flux terms, and S(q) represents the source terms. Adding to this set of equations, initial conditions are required:

$$\mathrm{q}(t = 0) = I(\mathrm{q})$$

and possibly boundary conditions for the domain, in which the simulation is intended to be conducted.

Euler Equations of fluid motion often serve as the master equations from which simplifications are derived and are given by the following definitions for q, F, and S (where P is pressure)

$$\mathrm{q} = \begin{bmatrix} \rho \\ \rho\mathrm{u} \\ \rho e \end{bmatrix}, F(\mathrm{q}) = \begin{bmatrix} \rho\mathrm{u} \\ \rho\mathrm{u} \times \mathrm{u} + P \\ (\rho e + P)\mathrm{u} \end{bmatrix}, S(\mathrm{q}) = \begin{bmatrix} f_C \\ -\rho\nabla\phi + f_M \\ -\rho\mathrm{u}\nabla\phi + f_E \end{bmatrix}$$

Table 5.3 Summary table interfaces

Short description of method	
Theoretical/mechanical background for the method	Source modelling is considered as an interface problem, which may either be be an input as is, or needs additional treatment for consistent representation
Short outline of how the method works	Input from heuristic or physical models for diverse sources is taken and processed such that it is compatible with propagation modelling
How is/are the output (s) of the method represented?	The outputs of this method are usually directly linked to tsunami propagation models. They are often integrated into the models themselves
In what way does the method take into account time-dependent/dynamic aspects?	Dynamical modelling is often considered and poses one of the challenges to the compatibility requirements. So processing data, such that temporal scale differences can be bridged is one of the method's main purposes
Does the method use parallelisation, machine learning, surrogate models, advanced simulation schemes, other computational reduction strategies (e.g. cluster analysis, etc.)?	Some approaches use ML for a sophisticated interpolation of data. Parallel processing may also be applied, in particular, if the method is implemented in the same software framework as the target model
What are the pros and cons and peculiarities of this specific method? Are there known shortcomings/benefits of the method? In what cases is this method a good/poor choice?	N/A
Data and technical requirements for method	
What input data does the method use?	Diverse source models
What are the technical requirements?	Compatibility with modelling assumptions, as well as with simulation software infrastructure is required
Are HPC computing resources necessary, would absence of these compromise the use?	Not necessary, but often utilized
Is the method/workflow available as a git/software/jupyter notebook?	No
Treatment of uncertainties for the method	
Is the method itself deterministic or uncertainty based?	Usually deterministic, but may consider uncertainties by varying the source parameters
Which are potential sources of uncertainties? Which parameters are used to characterize them?	Uncertainty may arise from improper treatment of sources
How are uncertainties quantified for the identified parameters? Are there significant correlations between different uncertain parameters?	Difficult to quantify, sensitivity may be large

(continued)

Table 5.3 (continued)

Short description of method	
Is the method able to digest and propagate uncertainty in input (previous ingredient) and propagate it to the output (next ingredients)? Does the assessment of this ingredient require uncertainty propagation? If yes, how are uncertainties propagated?	This method may consider and propagate uncertainty by sampling the possible input parameter space
What are the consequences of lack of data in input or for boundary conditions, or for characterizing certain parameters?	Relevant hazard metrics may be rendered incorrect
What are the consequences of ignoring or poorly treating uncertainties?	Results, in particular inundation maps, may be very incorrect

Here we do not discuss all the terms in these equations, because we do not want to focus on the mathematical representation. This only serves as an illustration. To summarise, there are several parts in the mathematical representation of a tsunami model that need to be specified:

- q represents the set of variables that need consideration, usually mass (water height/density), momentum (including horizontal and possibly vertical velocities) and optionally energy.
- F represents the inherent physical balances and is usually only determined by the choice of the model representation and needs no further specification
- S contains forcing terms that may be externally given and needs various types of data, depending on the modelling case.
- I contains the initial conditions, which can be resting states or predefined conditions. These conditions apply only instantaneously and will not carry a time dependency.
- Finally, boundary conditions will need specification of the domain and will usually also not be time-dependent, except for inundation boundary conditions or tidal boundary forcing.

Turning to the conditions, note that S(q) represents source terms that partly are fed from data and partly from physical processes. f_M and f_E represent friction terms that either relate to internal friction, to bottom friction or (in case of meteorological forcing) to wind stress. f_C is usually set to zero (and then corresponds to mass conservation), but is used as a forcing term in mass movement induced tsunamis. The gradient terms often include pressure forcing (from atmospheric disturbances) or time dependent changes in the bathymetry. However, an instantaneous uplift from e.g. an earthquake rupture can also be modelled by an initial condition and will then enter in the term I(q) above.

5.3.2 *Interfacing Sources of Tsunamis in the Modelling Process*

While most modelling approaches separate the source modelling from the tsunami propagation and inundation modelling steps, there are still many examples where source and tsunami models are integrated. Examples with 3D finite difference schemes using staggered pressure and velocity grids have been used by Saito and Furumura (2009), and Maeda and Furumura (2013). A variation on this is presented by Lotto and Dunham (2015) who implemented a different numerical scheme using summation by parts (SBP) finite difference operators with simultaneous approximation terms (SAT) for implementing boundary conditions. Panza et al. (2000) provided an alternative technique using the Haskell method to solve the equations of motion for a planar multilayer ocean and elastic layered model. In Chap. 4, other examples of coupled landslide tsunami models are described.

Interfacing sources with tsunami models often involves transferring physical processes that may not be representable with respect to the modelling approach in one of the parts. For example, fast seismic Rayleigh waves are not straightforwardly representable in a shallow water approximation to the hydrodynamic tsunami wave propagation models partly because they would result in supercritical wave speeds. Another example is the dispersion of the uplift signal through the vertical water column, which is not representable in a one-layer model without correction terms for the hydrodynamic response. The simplest method for modelling the hydrodynamic response is to copy the seabed deformation directly to the sea surface. This may work well for long wavelength seafloor deformations (e.g. wavelengths of more than 20 water depths), but for shorter seafloor disturbance, it is necessary to model the hydrodynamic response.

A commonly used technique to handle this later problem is a parameterization of this process. By considering the velocity potential, φ, which is related to the fluid velocity by $v = \nabla\varphi$. For non-breaking waves, it is possible to model this process using for instance boundary integral methods (e.g. Grilli et al. 1989; Grilli and Watts 2005; Pedersen 2008) taking into account full potential theory (for fully non-linear, fully dispersive non-breaking waves), but this is often not feasible due to the computational cost.

A frequently used method is to use close form expressions assuming linearity and instantaneous seabed deformation. A celebrated method for this purpose is the analytical solution of Kajiura (1963), which assumes constant depth and instantaneous deformation for a point source using linear potential theory. A similar solution assuming a rectangular deformation and slowly varying topography was further proposed by Nosov and Kolesov (2011). By summing different unit sources for an earthquake source, it is possible to take into account the hydrodynamic response for compound sources. Glimsdal et al. (2013) updated this to also take into account time dependent sources such as landslides, but this method could in principle also take into account time dependent earthquakes or volcano processes as well. To reduce computational efforts further, it is possible to simply assume smoothing length on the depth deformations and utilise any kind of analytical smoothing function (in a similar way as done for the box shaped slides, Harbitz et al. 1991; Harbitz 1992), e.g. through Gaussian unit sources for earthquake tsunamis (e.g. Molinari et al. 2016). This process is faster than Kajiura filtering or Nosov and Kolesov's method, but will not take into account the correct hydrodynamic response as the smoothing length and shape is fixed.

5.3.3 Approximations and Simplifying Assumptions

The most common simplification to the equations of motion in tsunami modelling is the shallow water assumption, i.e. that the relevant length scales in the horizontal dimensions are large compared to those in the vertical, that the wave height is small compared to the water depth and that the wavelength is large compared to the water depth. Under these conditions, the vertical motion within the water column can be neglected and the two-dimensional shallow water equations can be derived by averaging over the depth. The shallow water equations omit the third energy-related equation in (5.1) and have simplified terms in the first and second equations.

$$\mathrm{q} = \begin{bmatrix} h \\ h\mathrm{u} \end{bmatrix}, F(\mathrm{q}) = \begin{bmatrix} h\mathrm{u} \\ h\mathrm{u} \times \mathrm{u} + \frac{1}{2}h^2\mathbb{I}_2 \end{bmatrix}, S(\mathrm{q}) = \begin{bmatrix} 0 \\ -gh\nabla b + f_M \end{bmatrix}$$

Under this assumption frequency dispersion is not possible. Also, non-hydrostatic velocity amplification over bathymetry is not well represented. It is also important to note that in this regime, only one homogeneous water layer exists, and thus a horizontal displacement of the bottom uplift shifts the whole water column (which may be an unrealistically large transfer of momentum). Some of the above-mentioned deficiencies may be parameterized. For example, the frequency dispersion effect can be parameterized to some extent (assuming constant water depth) by an artificial diffusive approach through the numerical resolution. The missing smoothing of the sea surface response may be parameterized by the above-mentioned Kajiura filter.

Alternatively, it can be assumed that only vertical deformation is considered in the generation of the tsunami. However, in the presence of a bathymetric slope this is not the case. The horizontal movement of the slope will cause a vertical displacement of the water column which contributes to the tsunami source. Tanioka and Satake (1996) calculated the total vertical displacement of water, D, using:

$$D = u_x \frac{\partial H}{\partial x} + u_y \frac{\partial H}{\partial y} + u_z$$

where u_x and u_y are the horizontal deformation of the seafloor, u_z is the vertical displacement and H is water depth (i.e. positive downwards). The effect of the additional uplift caused by accounting for the bathymetric gradient can be significant for tsunami generation, particularly in the case of shallow earthquakes near the trench (Tanioka and Seno 2001).

A common further simplification is to assume the fluid is incompressible ($\nabla \cdot v = 0$), (e.g. Saito and Furumura 2009; Maeda and Furumura 2013; Kajiura 1963); however this is not universal (e.g. Lotto and Dunham 2015; Nosov and Kolesov 2007).

A more complex, but still simplifying approximation to the full equations of motion are Boussinesq-type models (see e.g. Fuhrman and Madsen 2009). These modelling approaches provide more realistic wave frequency dispersion relations and a better representation of wave-induced flow patterns, compared to (hydrostatic) shallow water theory, while maintaining a certain amount of efficiency, compared to the full 3-dimensional Navier-Stokes or Euler Equations.

Note that for the presented simplifications of equations there are no significant differences in the input requirements. However, the representable physical processes differ substantially. While a fully 3-dimensional Navier-Stokes equation model is capable of representing (and thus are accepted as input) fast Rayleigh waves, as well as horizontal dislocations, this is not the case for shallow water equation based models, as described above.

5.3.4 Input

Modelling is generally performed in a two-step approach: in the first step the seafloor deformation or—in case of meteotsunamis—the sea-surface perturbation is simulated; in the second step the output from this simulation is used as an input for the hydrodynamic modelling. In this case there is no interaction between the two models. The advantage of this approach is that the complexity of the simulation is reduced, often decreasing the possibility of instabilities (e.g. at the fluid-solid interface). The disadvantage to this technique is the removal of the interaction between fluid and solid domains, consequently the full physics of the phenomena are not represented. See Chap. 4 for a description of source modelling. Coupled physics-based models which incorporate both source and hydrodynamic modelling do exist with a discussion on coupled source simulations provided in Chap. 4.

With this "separation of concerns" type of approach in mind, the interface may simply represent a layer between the models to interpolate between different mesh layouts or even just a driver to couple correct file formats. However, if different physical processes are involved, then the interfaces layer may also add some sort of filtering to provide the receiving model with appropriately modified information. This layer is also responsible for modifying outputs from one model to be suitable for the other. An example is the computation of wind stress values from wind velocities in meteotsunami modelling. A second example is the conversion of rupture velocity in complex earthquake modelling into bottom uplift. A third example can be a time dependent full potential filter conveying seabed deformation to sea surface perturbations due to various types of sources (i.e. earthquakes, landslides, volcanoes).

Note, that this filtering computation may be embedded directly into one of the major models. An analytical half space model based on a dislocation on a fault surface is usually included in tsunami hydrodynamic wave models (see Sect. 5.4.2), so that these can be initialised with simple text files containing the (possibly time-dependent) parameter sets.

5.3.5 *Outputs*

The output from the interface modelling is dependent on the source phenomena in question; it can be in the form of:

- *source terms* in the case of atmospheric tsunamis
- *initial conditions* such as sudden sea surface displacement for instantaneous earthquake or volcanic explosions
- *boundary conditions* like a time-dependent sea surface displacement for landslide, volcano (pyroclastic flow, eruptive column collapse, caldera collapse), and earthquake sources.

5.3.6 *Data Requirements for the Interfaces*

The input data varies depending on the source and the type of model. For the following discussion, it is assumed that source simulations are separate from the hydrodynamic models (i.e. uncoupled models).

- Earthquake: in most cases the static seafloor deformation is used (see Sect. 5.4.2). This assumes that the earthquake process is instantaneous and the rupture velocity (i.e. the velocity at which the slipping area spreads across the fault) is much faster than the tsunami wave propagation. Alternatively, the time series of the dynamic seafloor deformation can be used, which would account for the temporal evolution of the earthquake but will increase computational resources. It is worth mentioning that time-dependent source interfacing may be important in particular in large source area situations. Sensitivity of the exact source location and timing is large for near-field tsunami simulations. In case of an instantaneous seafloor deformation initial conditions may be defined. Otherwise, the seafloor deformation is given as a forcing term to the equations.

  In cases where ground motions from dynamic simulations are being used, care needs to be taken when interfacing to simple (e.g. shallow water) approximations to the wave propagation simulation. In particular Rayleigh waves may create numerical artefacts, since their propagation speed is much faster than the shallow water gravity wave speed and can therefore be removed in such simulations. A viable approach may be to filter such waves (Madden et al. 2021). Bathymetric data is required to accurately account for the contribution that horizontal deformation will make to the tsunami source.

- Landslide (submarine, subaerial): A variety of methods are used (most are time dependent, while only some assume an initial condition). For time dependent models it is necessary to employ an interface model at a given time step in the propagation. Then it is necessary to make sure that time stepping related to updating the (volumetric) source contribution to the wave model is sufficiently frequent (Zengaffinen et al. 2020; Zengaffinen-Morris et al. 2022).

  For each individual update of the source model feeding into the tsunami model, it is clear that the choice of the interface model will have a strong flavour of the selected source model (see Chap. 4). For initial conditions, the shape should consider integral contribution of both the source mechanics and the water response, and no filtering or interface model should in principle be needed. For block models and depth averaged landslide models, smoothing filters are needed, either of Boundary Integral Methods (BIM) type or Kajiura type taking into account the physics, or simpler models that have as the main purpose to remove noise. For block models, smoothing the box shape may render interface models unnecessary if the sole purpose is to remove noise. For coupled models the tsunami modelling is already embedded, and interface models are in principle not necessary. On the other hand, it is stressed that the physics in some coupled models may involve simplified coupling terms (e.g. copying the landslide response to the wave model without filtering), and that the user must assess if the coupling mechanism in each model is sufficiently accurate.

  For the models where the landslide modelling is carried out independently of the wave model, the source displacement is conveyed to the wave model at several output times. Ideally, this displacement pattern should be fed into the wave model at every time step. However, interface modelling may require significant computational resources. Therefore, it might be necessary to set a larger time interval for conveying the source information into the wave model. Convergence tests (e.g. Zengaffinen et al. 2020) should be undertaken to optimise this time interval, as the requirement may vary from case to case.

- Volcano: Tsunami sources in a volcanic context are diverse: underwater explosions, caldera collapses, pyroclastic flows, eruptive column collapse, landslides, etc. There is therefore no single solution for modelling the volcanic sources of tsunamis. A distinction will be made between: (1) static or instantaneous sources, such as underwater explosions, where a displacement of the surface of the water is imposed as the initial condition; (2) time-dependent sources, such as pyroclastic flows, eruptive column collapse, and caldera collapse (see Sects. 4.3.3 and 4.4.3 for more details on these source mechanisms and their modelling). Caldera collapses can be treated with a simple approach, using sea floor motion as a finite-time forcing term in a depth-averaged model, whereas pyroclastic flows and eruptive column collapses require the use of complex CFD full Navier-Stokes type models. Indeed, the physics of tsunami generation by pyroclastic flows and eruptive column collapse is so complex that it justifies the use of advanced models considering several fluids and their interactions. However, this type of advanced modelling is computationally very demanding and currently not applicable for PTHA.

- Atmosphere: There are several terms that control the tsunami behaviour of a meteorological tsunami. Most importantly, the air pressure forcing at the sea surface, provided as a time-dependent perturbation and causing an uplift or downdraft of the surface elevation, and the wind stress pushing the sea surface by a friction term. To have a significant tsunamigenic effect, these terms need to be represented with sufficient spatial and temporal resolution. The air pressure as well as the wind stress are tsunamigenic only if they correspond to resonance frequencies related to the topo-bathymetric situation. Therefore, exact location, timing, extent and magnitude of the air pressure perturbation and wind speed anomaly need to be used to force the tsunami model.

  Therefore, the continuous time-dependent mean sea level atmospheric pressure and wind stress forcing must be introduced within the model equations. Consequently, adding two supplementary terms in the equations, reading the atmospheric forcing at a given time-step (e.g., every 1-min to 5-min) and interpolating this forcing at every tsunami model time-step is required. Further, for weather related events (i.e., events driven by mesoscale atmospheric processes), meteotsunamis are generally an order of magnitude smaller than tsunamis driven by other sources. Tides and storm surges can thus play a critical role in the final extreme sea level hazard. It is also worth mentioning that up to 70% of the meteotsunami wave transformation and amplification occur nearshore where the resolution of the tsunami models should have an order of just tens of metres. Finally, for sonic boom related events (i.e., events driven by planetary acoustic-gravity waves, such as the Hunga Tonga-Hunga Ha'apai event, see Lynett et al. 2022), the tsunami model should be able to run globally (see propagation section below).

5.3.7 Technical Requirements

Technical requirements are dependent on the type of coupling between source and interface simulations. In the case of direct coupling, the output from the source simulation will be automatically included. Separate simulations for the source and hydrodynamics will involve file transfer in the form of the output from the source model used as an input of the hydrodynamic model. In this case either self-defining binary data formats such as NetCDF or HDF5 or standard ASCII formats such as Esri™ ASCII are employed (the binary formats will be preferable because of their storage requirements and standardised inclusion of metadata).

An alternative approach is to include a kinematic model of the source embedded directly in the tsunami propagation model. The last technique is commonly employed when modelling earthquake-generated tsunamis with the earthquake represented as a slipping surface (see Sect. 4.1 for a discussion on the topic). HPC computing resources are generally not used for this method.

Interfacing to landslide or meteorological sources is usually more involved. The inputs used in the interface to a landslide forcing are typically binary or ASCII formatted files in the local source area. As the source domain is typically smaller and more finely resolved, interpolation in time and space is needed. The time steps of the inputs from the source model typically range from a few seconds (e.g. Zengaffinen et al. 2020) to about one minute (e.g. Kim et al. 2019). Typically, one needs to employ smoothing (see Sect. 4.3) either using box type sources or physically based full potential type filters to remove short wavelength components. The inputs used in the interface to meteorological forcing are 1-min to 5-min mean sea level pressure (and/or wind stress) time and space varying forcing that is interpolated on the tsunami model grid/mesh. This forcing is either generated by atmospheric models (realistic forcing: see Deterministic Atmospheric Models in Sect. 4.5.2) or consisting of parametric equations (synthetic forcing; see Synthetic Catalogue in Sect. 4.5.3). Due to the amount of data, binary formats (NetCDF or HDF5) should be used to store the multidimensional inputs when realistic forcings are used. Additionally, tidal information and appropriate forcing needs to be considered.

5.3.8 Treatment of Uncertainties for the Method

In this section first we discuss the propagation of uncertainty in the representation of the source, afterwards uncertainties within the method are analysed. Single realisations of the interface methods are deterministic. However, the method is able to propagate uncertainty in the input due to the small computational requirements in the case where the interface between seafloor and sea surface is represented by a Kajiura filter. This allows uncertainty representation from the source (e.g. heterogenous slip for earthquake) by including multiple source representations in probabilistic analysis, by assigning conditional probabilities to each model realisation.

A potential source of uncertainty in this method is the effect of simplifying the physics of hydrodynamics. As mentioned above, a common assumption used in many studies is that sea water is an incompressible fluid. An additional uncertainty may be triggered in the way different physics interact (e.g. between the elastic and acoustic domains). Several full potential based filter models assume constant depth and deep-water tsunami generation (linearity). Uncertainty in parameters such as the density of the water or the kinematic viscosity are generally not accounted for, but methodologies based on multiple simulations using Monte-Carlo type variations or a probability density function for expected ranges of parameters could conceivably be applied to account for such uncertainty.

The consequences of a lack or poor resolution of input data depend on the sensitivity of each case. For example, deep bathymetric data may be safely interpolated while the resolution of disturbance data in atmospheric tsunami sources are critical.

In the case of seismic sources, the sensitivity of the tsunami source to the bathymetric gradient is dependent on geometry and the depth of the slip, i.e. large slip at shallow depths and fault dip can be highly sensitive to gradient. The representation of the physics and the treatment of uncertainty between the source and subsequent models need to be consistent with particular importance on the interface between models where sensitivity may be high (e.g. bathymetric gradient). Poor treatment of source uncertainty will propagate through PTHA, this has been demonstrated for seismic sources where simplification of earthquakes to fixed size rectangles with uniform slip produces lower hazard appraisals compared to more complicated models (Davies 2019; Davies and Griffin 2020).

5.4 Propagation

5.4.1 Introduction

As discussed above, the modelling and simulation of tsunami propagation is based on mostly non-linear partial differential equations. Three equation sets are commonly used to model the tsunami propagation and inundation, namely the (linear or non-linear) shallow water models, Boussinesq type long wave models that consider non-linearity and frequency dispersion, and complete fluid dynamic (i.e. Navier-Stokes) type models.

Numerical tsunami propagation models describe the propagation of tsunamis from the source area towards coastal waters. Forecasts of tsunami sea surface elevation and first wave arrival along a coastline and approximations for inundation maps use the outputs of those simulations to serve warning purposes and coastal management and planning.

The equations are solved using numerical methods, such as finite difference, finite volume, or Galerkin-type methods, to obtain a solution to the tsunami wave propagation problem. Hydrodynamic models solve empirical and governing equations at different complexities. The limitations of tsunami modelling result from the limitations of the input data, which may contain errors, interpolations, or estimates. An overview of the contents of this section can be found in Table 5.4.

5.4.2 Physical Approximations

Linear and non-linear Shallow Water Models

The shallow water equations (SWE) describe the evolution of the water surface elevation with time and the depth-averaged water particle velocity, or alternatively, the momentum evolution. They are derived from the Euler equations by depth averaging and neglecting vertical components of acceleration and velocity. To be valid, such

Table 5.4 Summary table propagation

Short description of method	
Theoretical/mechanical background for the method	This method solves hydrodynamic equations to propagate the tsunami wave, initiated by a source to the coast
Short outline of how the method works	The method works as a partial differential equation solver and may utilize diverse discretization schemes, such as finite differences, finite volumes, finite elements, etc.
How is/are the output (s) of the method represented?	The outputs of this method may be direct lists of hazard metrics (such as wave height at the coast) or maps
In what way does the method take into account time-dependent/dynamic aspects?	This method usually is fully dynamical and processes large numbers of timesteps
Does the method use parallelisation, machine learning, surrogate models, advanced simulation schemes, other computational reduction strategies (e.g. cluster analysis, etc.)?	Most propagation models use parallel processing and HPC infrastructure. Some models use GPU acceleration. ML is rarely used for this method, however surrogate models have been proposed
What are the pros and cons and peculiarities of this specific method? Are there known shortcomings/benefits of the method? In what cases is this method a good/poor choice?	There is no alternative method
Data and technical requirements for method	
What input data does the method use?	Input from interface method and data method
What are the technical requirements?	Well-tested modelling software needs to be used
Are HPC computing resources necessary, would absence of these compromise the use?	Not strictly necessary, but almost always utilized. In particular if probabilistic approaches are intended then large numbers of simulations require the use of HPC infrastructure
Is the method/workflow available as a git/software/jupyter notebook?	There are many different models available, and listed in the chapter
Treatment of uncertainties for the method	
Is the method itself deterministic or uncertainty based?	Fully deterministic, but used for probabilistic assessment by Monte-Carlo approaches
Which are potential sources of uncertainties? Which parameters are used to characterize them?	Uncertainty may arise from improper choice of model for sources
How are uncertainties quantified for the identified parameters? Are there significant correlations between different uncertain parameters?	Sensitivity analysis should be applied to quantify uncertainty

(continued)

Table 5.4 (continued)

Short description of method	
Is the method able to digest and propagate uncertainty in input (previous ingredient) and propagate it to the output (next ingredients)? Does the assessment of this ingredient require uncertainty propagation? If yes, how are uncertainties propagated?	This method may consider and propagate uncertainty only by ensemble modelling
What are the consequences of lack of data in input or for boundary conditions, or for characterizing certain parameters?	Relevant hazard metrics may be rendered incorrect
What are the consequences of ignoring or poorly treating uncertainties?	Results may be very incorrect

simplifications require the surface elevation to be small and the wavelength to be large compared to the total water depth, which is true for tsunami offshore propagation.

These equations constitute a good approximation for tsunamis generated by subduction zone earthquakes since the seafloor deformation over a large region generally leads to tsunamis with a wavelength of several tens of kilometres or more, compared to the average ocean depth of only few kilometres. A gravity wave behaviour can represent tsunami propagation in the deep ocean, with negligible particle velocities (Behrens and Dias 2015). The shallow water equations describe the conservation of mass and momentum and are solved numerically.

The negligible particle velocity allows for a further simplification: by omitting the (velocity driven) advection term linear shallow water models (LSW) are obtained. Linearity leads to better numerical stability properties enabling faster computing times and simpler discretization schemes. In any case the two prognostic variables of such models are the wave height and the momentum or (particle) velocity.

As these models are limited to a single layer of water, they can only account for depth-averaged motions (which might be too simplistic for complex tsunami problems). They are only valid for relatively small amplitude waves.

LSW models can adequately predict tsunami waves in deep water, up to a water depth corresponding to near-shore regions (say 50 m), while non-linear models including moving boundary treatment may be used to simulate inundation as well (Behrens and Dias 2015). When the simulation involves a large region and propagation time (transoceanic propagation), the Coriolis' effect should not be neglected (see e.g. Omira et al. 2022). Also, earth's curvature needs to be taken into account in this modelling regime.

The computational stability depends on the numerical method used, and in case of explicit methods on the relationship between the time step and grid cell size. Temporal and spatial accuracy should be maintained to ensure the accuracy and stability of the computations which in turn can minimise unrealistic effects on the solution. The numerical stability can be assessed with the CFL (Courant-Friedrichs-Lewy) stability criterion. This criterion states that wave velocity, in other words the information propagation velocity, does not exceed the discretely representable velocity. Other stability criteria may also be applicable, but the discussion of numerical stability goes beyond the scope of this chapter.

It is worth noting that explicit time stepping methods are mandatory for tsunami modelling, since implicit methods are damping the gravity wave propagation too severely (an early analysis of the effect of explicit vs. implicit numerical methods on shallow water gravity wave propagation can be found in Harding 1982). While in other disciplines, where the flow regime is the important feature, this smoothing effect is desirable for improved stability, in tsunami wave propagation, it would render the simulation results unrealistically smoothed.

The inputs to the model include the source model and bathymetric grid. The results of the simulations depend heavily on the quality of bathymetric data (see Sect. 5.2) and on the numerical scheme used to solve the equations. Non-linear shallow water models (NLSW) including near-shore and on-shore (inundation) computations often necessitate additional input parameters related to treating the variable shoreline and bed friction, in particular for the inundation modelling. Usually, the complex small-scale behaviour of a tsunami inundating land is represented by a roughness parameterization, given by (a generalised) Manning roughness. Computational results are sensitive to this parameter (see Sect. 5.5).

Most SWE codes are computationally fast and inexpensive and run on a simple PC. High-speed processors (e.g. GPUs, de la Asunción et al. 2013) or compute clusters may be used to shorten the computation time. Non-linear and linear shallow water models are commonly used for early warning purposes. They provide—within the range of uncertainties given by the limited forecast time (see Greenslade et al. 2014)—accurate forecasts for expected wave heights, arrival times and possibly inundation areas.

For commonly used and validated codes refer to Lynett et al. (2017). Another set of models specific for landslide induced tsunamis can be found in Kirby et al. (2022).

Bussiness Models

There are situations in which the shallow-water equations are inadequate, in particular for short wavelength tsunamis and when non-linear wave-to-current interactions are of importance (e.g. in ports, see Glimsdal et al. 2013; Ma et al. 2012; Løvholt et al. 2015; Lynett et al. 2012).

Boussinesq type models are depth-averaged models accounting for nonlinear and frequency dispersion. These models are based on the Boussinesq equations, and can to different extent consider friction, Coriolis force, and consider wave breaking and energy dissipation. Tsunamis travelling long distances can be significantly modified by effects of frequency dispersion (Glimsdal et al. 2013). Dispersion is also relevant for certain non-seismic sources such as landslides and volcano flank collapses (Løvholt et al. 2015). The inputs of the code are the source model and bathymetric grid, similar to shallow water models. In general, these models provide accurate predictions of wave height, arrival time, and momentum flux.

Like shallow water models, Boussinesq type models are affected by the method used to discretize the model equations. Due to the existence of higher order and mixed derivatives in the equation set, stability is much more difficult to achieve in Boussinesq-type models (e.g. Løvholt et al. 2009; Løvholt et al. 2013) compared to shallow water models. While higher order Boussinesq equations comprise mixed higher derivatives and are often difficult to discretize in a stable way, an alternative (and for some equation sets equivalent, see Jeschke et al. 2017) approach is given by a non-hydrostatic correction method (Casuli and Stelling 1998; Yamazaki et al. 2009). The correction then empowers shallow water models to represent wave frequency dispersion and non-linear velocity amplification. To ensure the stability of the model it is important to use suitable numerical techniques and appropriate high resolution meshes and time stepping schemes. Combining finite volume techniques, use of limiters, and turn off dispersion terms in shallow water have also improved stability in the near shore propagation (e.g. Shi et al. 2012; Kim et al. 2009, 2017).

A related class of models are the layered models, that assume depth averaging in each layer (e.g. Ma et al. 2012; Macias et al. 2021). They may extend the validity range compared to Boussinesq models, but will generally require more computational resources.

These models are adequate to simulate tsunami propagation in coastal regions, to properly describe transoceanic tsunami propagation, and to some extent even to represent tsunami structure interaction.

Full Navier Stokes models

The three-dimensional or Full Navier-Stokes equations are the most complete description of fluid movement. This description is needed to describe complex wave interactions and general treatment of vorticity. Full Navier-Stokes simulation is a complex sophisticated process. It involves solving the incompressible Navier-Stokes equations in a two- or three-dimensional domain to model the flow of water. The use of these models requires sophisticated numerical methods on parallel computers to calculate the models accurately. In most cases this is computationally much too costly, and for most tsunami applications the vertical structure of the flow does not need to be modelled directly (Grezio et al. 2017).

The physical parameters like the water density, gravity, water surface elevation, and boundary conditions must be properly specified in the model. After that, the model can be used to simulate the evolution of a tsunami from the source over a certain duration of time. The simulated results can be used to study the effects of tsunamis on coastal regions and assess the potential local hazards posed by tsunamis.

It should be noted that Navier-Stokes equations are usually not adequate for simulating large scale geophysical phenomena due to the geometry and relatively thin layer characteristic of such processes. Therefore, certain approximations (e.g. hydrostatic layering, spherical geometry, etc.) can be made to adjust the equation set to such applications. Currently, the authors are not aware of any successful application of a three-dimensional Navier-Stokes-based model to global or large-scale tsunami modelling.

However, local phenomena, such as fluid structure interaction, debris transport, inundation processes, or fluvial landslide tsunamis, often require a sophisticated equation set given by Navier-Stokes equations (see Sect. 5.5).

5.4.3 Discretization Methods

Numerical methods used in tsunami simulations must consider the physical approximations used to model wave behaviour. The most used numerical methods for solving the shallow water wave equations are the finite-difference (FD) and finite-volume (FV) methods. Besides these, also Galerkin methods—finite element (FE) or discontinuous Galerkin (DG) methods—and smoothed particle hydrodynamics (SPH) methods are in use. Except for SPH, most of these methods rely on discretizing the domain into a grid of small cells. In the finite-difference method, the equations are approximated using discrete differences between cell centres or cell vertices, while the finite-volume method involves integrating the differential equations over each cell volume.

Finite Differences

The finite difference method uses approximations to the derivatives by finite differences to approximate the analytical equations of motion and the associated exact solutions. This means that the domain is discretized into grid points (nodes) and the differential terms of the equation are converted into an algebraic equation, the so-called finite-difference equation. The dependent variable values are stored at the nodes only. Early tsunami simulation codes used the FD method (see Imamura 1996). It should be noted that FD discretization methods are suitable for modelling large domains, but they require fine meshes to capture the wave behaviour accurately.

Several tsunami simulation codes use FD discretization schemes, including for instance: TUNAMI-N2 (Imamura et al. 2006) MOST - Method of Splitting Tsunamis (Titov and Gonzalez 1997), COMCOT (Liu et al. 1998), NSWING Nonlinear Shallow Water (Miranda et al. 2014).

Finite Volumes

The finite volume method is based on the Eulerian perspective of fluid flows. The domain of interest is partitioned into (small) discrete volumes (cells) and the conserved variables in cells are evolved in time using the fluxes with neighbouring cells. By this approach FV methods achieve exact discrete conservation of the main conservation properties of the governing equations, such as mass and momentum conservation for the shallow water equations.

Since FV methods assume averaged quantities within cells, they are usually of low convergence order. On the other hand, because no rigorous assumptions on the shape and size of the cells are necessary, FV methods are well suited to represent complicated geometries and are also suitable for locally refined meshes. Nevertheless, rectangular meshes are mostly used and nested grids implemented for high-resolution inundation.

Examples of numerical tsunami codes using Finite Volumes: Tsunami-HySEA (Macías et al. 2017), Geoclaw (LeVeque et al. 2010), or ANUGA from *Australian* National University and Australia Geosciences (Nielsen et al. 2005).

Galerkin Methods

Galerkin methods, such as FE and DG, are based on the idea of representing the solution functions by polynomial approximations. So, instead of looking for the true continuous solution function, a simpler approximation function—a polynomial, which has nice properties with respect to derivatives and integrals—is derived. To be efficient, the domain is again partitioned into small cells (the elements) in order to provide local support for the approximating polynomials and eventually leading to an algebraic equation with a sparse matrix. In FE methods a piecewise continuous polynomial representation of the solution is achieved. DG methods on the other hand use the same polynomial approximation within cells, allow for discontinuities between elements and represent their coupling by numerical flux functions. Therefore, DG methods can be viewed as high-order FV type methods.

Finite element methods are conservative in an integral sense, so mass and momentum are conserved, when these quantities are integrated over the whole domain. DG methods—as well as FV—are locally conservative, meaning that balances between cells are exactly represented discretely. Galerkin type methods are well suited for local mesh refinement, even dynamic adaptive mesh refinement implementations have been presented.

Examples of Galerkin type codes are TsunAWI (Behrens 2008; Rakowsky et al. 2013), TsunaFlash (Pranowo et al. 2008), StormFlash (Vater et al. 2019).

Smoothed Particle Hydrodynamics

Smoothed-Particle Hydrodynamics (SPH) is a computational method for simulating fluid flow, based on the Lagrangian representation of fluids by particles. The equations of motion and properties of these particles are determined from the continuum equations of fluid dynamics taking interactions of particles into account by interpolation from neighbouring particles (Monagan 1992, 2005, 2012; Rogers and Dalrymple 2008). SPH is meshless, it does not require the use of a grid to calculate spatial derivatives obtained by analytical differentiation of interpolation formula.

The SPH algorithm has several advantages over classical computational methods. Firstly, it can efficiently simulate complex flows involving moving and deforming objects. Secondly, the SPH method naturally conserves mass, momentum, and energy by satisfying the continuity and momentum equations. It has the standard features of a robust numerical solver and is highly accurate with no numerical dissipation along Lagrangian paths (see Sarfaraz and Pak 2017; Pringgana et al. 2021; Reis et al. 2022). SPH can handle complex geometries and simulate the fluid flow around obstacles being particularly useful in problems of fluid structure interaction or in our case in tsunami structure interaction. The method can be computationally demanding but offers high accuracy and versatility in modelling complex fluid flows, such as those involving turbulence, shocks, or free surfaces. SPH methods are mostly used for near field simulations, since in far field propagation the particle density tends to decrease due to spreading, leading to more and more diffusive results.

Well-established SPH software for tsunami simulation are not yet available to the knowledge of the authors.

5.5 Inundation

5.5.1 Introduction

The aim of this method is to present methods for near shore propagation and inundation modelling, i.e. interaction between the tsunami and the coast includes near shore propagation and inundation.

The numerical simulation of inundation aims at determining the areas that are inundated, as well as describing the variables that characterize such inundation and the process in which the inundation occurs.

The inundated area is determined by the interaction between waves and elements near-shore and onshore. This interaction can be assessed following several approaches, mainly depending on the way in which the numerical model solves the frontier from offshore to onshore, from bathymetry to topography. An overview of this section's contents is included in Table 5.5.

5.5.2 Calculation Approaches

A first approach is the application of models that solve the inundation as a continuation of, or coupled to, the propagation, using the same equations or codes applied for the propagation. This is the case of the models mentioned in the previous propagation section: in particular the Non-Linear-Shallow-Water-Equations (NSWE) models, such as Tsunami-HySEA (Macías et al. 2017), GeoClaw (Berger et al. 2011) COMCOT (Wang and Liu 2011), MOST (Titov et al. 2016), StormFlash (Vater et al. 2019), and many others.

Other approaches separate the processes that occur before and after the shoreline, by coupling two different models. The reasons for this separation can be several, such as using the models that are best suited for each process, or avoiding the high computational cost that is usually linked to the simulations of this type of process. There are two ways in which this coupling is addressed. First, by sharing files between models, meaning that the output of the propagation model is processed to be used as input for an inundation model. And second, an actual coupling in which the whole process occurs in a continuous calculation.

In addition, there are other approaches to inundation calculation. Analytical solutions remain as an alternative, especially in cases where not all the necessary data for numerical modelling is available (Park 2015; Madsen 2010; Aniel-Quiroga et al. 2018). The method of amplification factors is another simplified method that uses look up tables for extracting inundation heights globally or regionally (Løvholt et al. 2012; Glimsdal et al. 2019). These methods incorporate the uncertainty treatment of the maximum inundation height. A related but more elaborate method is the Energy Grade line approach (Kriebel et al. 2017). Machine Learning and emulators trained

Table 5.5 Summary Table Inundation

Short description of method	
Theoretical/mechanical background for the method	This method computes the inundation by tsunami waves on land. It is either based on hydrodynamic modelling or on heuristically relating wave height at the coast, terrain characteristics to inundation height
Short outline of how the method works	The method works locally or regionally and computes some of the most important and relevant hazard metrics, such as flow depth, flux, or inundation height
How is/are the output(s) of the method represented?	Either as metric values at points of interest or as maps
In what way does the method take into account time-dependent/dynamic aspects?	This method may output time dependent values, but is mostly used to compute maximum or statistical (fixed) values
Does the method use parallelisation, machine learning, surrogate models, advanced simulation schemes, other computational reduction strategies (e.g. cluster analysis, etc.)?	Most hydrodynamic inundation models use parallel processing and HPC infrastructure. Heuristic inundation schemes are often implemented as postprocessing schemes and may use parallelism in probabilistic (Monte-Carlo) settings
What are the pros and cons and peculiarities of this specific method? Are there known shortcomings / benefits of the method? In what cases is this method a good/poor choice?	This method has to be used with care, since inundation results are very sensitive with respect to accuracy of topography data as well as land use structure
Data and technical requirements for method	
What input data does the method use?	Input from hydrodynamic propagation models, and topographic data
What are the technical requirements?	Very accurate and validated topographic data sets are required, as well as accurate and validated parameterizations (usually roughness parameterization) to account for land use characteristics
Are HPC computing resources necessary, would absence of these compromise the use?	For hydrodynamic modelling HPC resources are required
Is the method/workflow available as a git/software/jupyter notebook?	There are many different models available and listed in the chapter
Treatment of uncertainties for the method	
Is the method itself deterministic or uncertainty based?	Hydrodynamic modelling is fully deterministic, but heuristic methods are often based on probabilistic approaches
Which are potential sources of uncertainties? Which parameters are used to characterize them?	Uncertainty may arise from incorrect data or improper calibration of roughness parameterization

(continued)

Table 5.5 (continued)

Short description of method	
How are uncertainties quantified for the identified parameters? Are there significant correlations between different uncertain parameters?	Sensitivity analysis should be applied to quantify uncertainty. Land use and topography data may contain interdependent errors, since data may be spoiled by canopy effects
Is the method able to digest and propagate uncertainty in input (previous ingredient) and propagate it to the output (next ingredients)? Does the assessment of this ingredient require uncertainty propagation? If yes, how are uncertainties propagated?	This method may consider and propagate uncertainty, e.g. by ensemble modelling
What are the consequences of lack of data in input or for boundary conditions, or for characterizing certain parameters?	Relevant hazard metrics may be rendered incorrect
What are the consequences of ignoring or poorly treating uncertainties?	Results may be very incorrect

using simulations are becoming increasingly popular for rapid inundation height estimation (e.g. Mulia et al. 2022). A typical simplification that has been traditionally applied consists of the so-called *Bathtub* approach, in which the maximum run-up (maximum elevation gained in the flooding process) of the inundation is directly assumed as the tsunami height on the coast. This simplification provides a conservative approach. The bathtub approach is often combined with simplified inundation models for extrapolating the inundation height over topography.

5.5.3 Calculation Scales

The tsunami inundation analysis can be conducted at regional or local scale, which is determined by the aim of the assessment, the resolution of the analysis and the quality of the available data. Due to the scale differences between the wider ocean basin and local structures at the coast, variable resolution techniques play a crucial role for inundation simulations. Many tools employ a nested grid technique, whereas others use non-uniform unstructured meshes or adaptive mesh refinement (Harig et al. 2008; Pranowo et al. 2008; LeVeque et al. 2010). Thus, the numerical simulation of the tsunami inundation starts with the preparation of topo-bathymetric data into appropriate high-resolution grids required by the numerical model. The grid-spacing must be high enough to resolve coastal features and to guarantee accurate results. Some previous sensitivity analysis and calibration with test cases (see the following Sect. 5.6) may be necessary. To build these grids, the topographic and bathymetric data is interpolated. Special care needs to be taken in the definition of the coastline since it is common to observe a gap between the last topographic value and the first bathymetric one (see Sect. 5.2 on data preparation, above).

5.5.4 Interaction Between the Tsunami and the Coastal Elements

Inundation models, or those parts of the model that evaluate the inundation, involve solving the interaction between the flooding wave and the land. In this interaction the roughness of the type of land onshore is included by means of a friction coefficient (usually de Chezy or Manning roughness is considered). These friction coefficients incorporate the way in which the interaction retards the wave in the inundation process (e.g. Gibbons et al. 2022). Different kinds of land surfaces can be represented by different values of the roughness coefficient to mimic differences in surface resistance (see e.g. Kaiser et al. 2011 for a discussion).

In addition, if the topographic data, given by a Digital Terrain Model (DTM) or Digital Elevation model (DEM), are accurate enough, buildings or other small-scale obstacles existing in the potentially flooded area can be included in the grid, working as a barrier for the inundation.

Similarly, coastal protection structures can be represented in the small-scale grids. These structures are the first barrier that the coast presents to the tsunami, and their performance is crucial for diminishing tsunami hazard (Rosetto 2011; Aniel-Quiroga 2019). As they are part of the coastline, coastal structures are included directly into the topographic grid. In addition, some models calculate this interaction in a vertically two-dimensional approach, such as volume-of-fluid models. Finally, both coastal buildings and structures can be modelled by means of full three-dimensional Navier-Stokes based models (e.g. OpenFOAM, Guler et al. 2018), that may represent the interaction of structure and fluid flow, including debris, more accurately, although the computational cost is huge, and normally not feasible for implementation in PTHA.

5.5.5 Outputs

The final result of the inundation is the characterization of the inundated area. This is commonly represented by a tsunami inundation map containing the calculated variables.

These variables are normally related to the following definitions (see also IOC Tsunami Glossary 2019):

- Inundation depth (I_{Depth}), more often referred to as *flow depth*: Depth, or height of the tsunami above the ground, at a specific grid point of the output map.
- Inundation length or inundation distance (I_{Length}): The horizontal distance inland that a tsunami penetrates, generally measured perpendicularly to the shoreline.
- Maximum run-up (*Ru*): Difference between the elevation of maximum tsunami penetration (inundation line) and the sea level at the time of the tsunami.
- Maximum inundation height (MIH): The maximum elevation reach above sea level along a transect between the coastline and the trim line.

- Tsunami travel time: Time required for the first tsunami wave to propagate from its source to a given point on a coastline.
- Trim line: the iso-line of maximum inundation lengths.
- Flow velocity (u): the particle velocity of the fluid, mostly relevant only in inundation simulations.
- Momentum flux (d_f): velocity squared multiplied with flow depth, relevant for loads on structures in inundation simulation.
- Inundation Area: Area flooded with water by the tsunami.
- Forces on structures: Buildings, coastal structures

Commonly, these variables are directly obtained from the numerical simulations and represented in tsunami inundation maps that have several applications. One of the most important ones is its use as input for evacuation models towards the proposal and design of evacuation routes and plans. While the evaluation of the tsunami travel time in deep water can be computed accurately, tsunami behaviour onshore is a quite more complex problem, involving friction coefficients, land use, presence of coastal structures and buildings, etc. For determining it accurately, an accurate near shore bathymetry is necessary. Tsunami traversal, offshore and onshore, provides important information used to design potential evacuation routes in a tsunami-prone area.

Finally, the calculation of the inundated area by a tsunami opens the door to setting the limits of the exposed assets and helps to properly assess their vulnerability. Their combination allows addressing the analysis towards tsunami risk assessment.

5.6 Testing, Validation, and Verification

5.6.1 Short Description of Method

The aim of the present method is to characterise the models: their behaviour to uncertain input data and parameterizations, the accuracy/precision of models compared to mathematically or numerically derived benchmark solutions, in reproducing laboratory data and, finally, compared to observed data. This is necessary to ensure the suitability of the model for physically robust and sound simulations in parameter regimes that lack previous measured data. Or in other words, a validated model is reliable for forecasting purposes.

Corresponding methods for testing and validation are:

1. Sensitivity to input data (in particular to uncertainties in the generating source, to bathymetry/topography uncertainties and errors, to the numerical resolution, or bottom friction parameterizations, friction coefficient, etc.).
2. Verification, i.e. testing the model specifications. In other words, does the program work without errors and does it produce the expected results for specified inputs.

3. Validation, or testing the model against the real examples, making sure that it represents the modelled phenomenon in a useful way, which can be tested using experimental data or field data.
4. Testing of a complete workflow in particular applications as, for example, early warning systems.

Sensitivity analysis to model parameters

Sensitivity analysis consists in testing the behaviour and variation in model results when the different model parameters are varied within a range of reasonable values. When model parameters are too many this leads to a large number of combinations to be simulated in order to cover the full set of parameter combinations. This also includes sensitivity to bathymetry as it is a source of uncertainty (possible errors and gaps, coarse resolution, inhomogeneities, etc.). The role of numerical resolution is also relevant here and must be addressed as simulations in general need to have a compromise between precision and computational cost. We also mention parameterisation of friction parameters, in particular when inundation is addressed. Nevertheless, when only propagation is computed this is not a key issue. The use of constant friction vs variable friction coefficients is also an important issue for high-resolution inundation studies.

Verification

Verification is the first step in model benchmarking, this is defined as producing model results that conform expected behaviour in a defined and known context (within model specifications). This is achieved by comparison of model results against analytical solutions, quasi-analytical approximate solutions and model properties, such as mass and momentum conservation or other metrics. This means verification can to some extent be automatized. It is good practice to use automatic code checking and analysis as well as reproducibility standards.

Validation

Validation is a second step in benchmarking where experimental and observed field data are used to check model results. The formal definition of validation is to ensure that model results compare reasonably with the physical world the model is designed for. This part is crucial as we want to reproduce real world events and at the same time represents a great challenge as high-quality observed data are scarce and they will always include larger measurement errors. Expert knowledge is necessary for model validation, since often uncertainty in representing the real world needs to be considered in setting up quality metrics.

Testing

This part consists in checking the correct functioning of a workflow including simulations in the process. In a general framework, a complete workflow could include acquisition and preprocessing of input data, filtering, model simulations, post-processing of output data and visualisation. While each component of the workflow needs to be verified and validated, this alone does not guarantee the functional efficiency and operational reliability of the whole system, because the interplay of the components needs to be tested as well.

For all the above steps it is necessary to formulate useful testing protocols with appropriately defined metrics for quality control. This involves how much deviation from predefined expected output values are allowed in the verification process, and involves decision of the importance of testing metrics, e.g. whether the inundation area of the model should correspond to the inundation area of the real event at specified points or in an integral sense. It should also be ensured that the model works for different tests without changing the parameter set for individual test cases.

5.6.2 Examples and Case Studies

Sensitivity analysis studies in seismic tsunami propagation and inundation modelling may usually try to investigate the effects of different fault plane parameters on different hazard measures (for example, maximum wave amplitude or inundated area).

We first make an example for earthquake tsunamis: As the Okada model for the bottom deformation (see Sect. 4.2) is mostly used, the parameters studied are usually the location of the epicentre, focal depth, rake, dip and strike angles, fault plane dimensions, and slip displacement (Gica et al. 2007). It is also a way to deal with uncertainties in tsunami modelling. In González et al. (2022) this is done to estimate the economic cost associated with tsunamis in Spain for the insurance sector.

For landslide generated tsunamis, the level of variation due to different values for the parameters involved is larger and associated uncertainties in model results, sensitivity studies for this kind of tsunamis are many (Grilli and Watts 2005; González-Vida et al. 2019; Zengaffinen et al. 2020; Zengaffinen-Morris et al. 2022, to name few).

Synolakis et al. (2008) propose a suite of increasingly complex benchmarks for verifying and validating tsunami propagation and inundation models. All such models involved in hazard assessment and used as tools for tsunami mitigation plans need to be tested and benchmarked. This process must include both verification and validation. This must be a continuous process as new knowledge and data are acquired or are made available.

A pioneering effort on model benchmarking was proposed in the early 1990s and manifested itself in a first joint National Science Foundation (NSF) and National Oceanographic and Atmospheric Administration (NOAA) National Tsunami Hazard Mitigation Program (NTHMP) workshop, documented in Yeh et al. (1997). A number of workshops followed, including one in 2011 where participating models were approved for tsunami inundation modelling and results were documented in the "Proceedings and results of the 2011 NTHMP Model Benchmarking Workshop"

(NTHMP 2012). Besides NTHMP (2012) and references therein, for NTHMP-benchmarked tsunami models, other authors have performed similar benchmarking efforts, as is the case of Nicolsky et al. (2011), Apotsos et al. (2011), Tolkova (2014), Macías et al. (2017) or Vater et al. (2019). Similar benchmarking initiatives exist for hazard modelling of tsunami induced currents in harbors (Lynett et al. 2017) or landslide induced tsunamis (Kirby et al. 2022).

5.6.3 Further Testing Considerations

The usual output in sensitivity analysis studies is the range of model values for the variable or variables for which the sensitivity study is performed (for example, maximum wave height and/or run-up at several locations). They are usually represented as bar plots.

In validation/verification studies comparison of model results with analytical, measured or observed values is the final aim. Model goodness is assessed using maximal differences or NRMSD (Normalised Root Mean Square Deviation).

Sensitivity analysis studies and model verification mainly use maximum values of some variables (usually run-up or wave height) at several locations. For validation and in some cases for verification, when data exists, time series are also used. In any case, sensitivity studies can use any variable being maximum values or time series as they only rely on computed data.

Any method requiring numerical simulations is susceptible to use parallel computational resources. Many scenarios used in a Monte-Carlo sensitivity approach as well as operationally in scenario-based probabilistic hazard assessment were computed on parallel clusters in (Rakowsky et al. 2013; Gibbons et al. 2020).

We see these methods as mandatory for reliable, reproducible and robust numerical modelling. However, verification and validation test cases can face shortcomings regarding the quality or quantity of observed data, often not avoidable. Care must be taken to collect suitable data and make best use of available data to compare and verify numerical model results.

5.6.4 Treatment of Uncertainties for the Method

Tsunami simulations are often subject to various sources of uncertainty, including uncertainties in the initial conditions, boundary conditions, topography, bathymetry, and model parameters.

Aleatory variability refers to the unpredictability (or randomness) of a physical process. This randomness is expressed through the objective probability of the PTHA while epistemic uncertainty expresses limitations in our understanding of natural processes and different opinions regarding these processes, resulting in uncertainty on quantifying these objective probabilities (Grezio et al. 2017). Sources of uncertainties include:

- Uncertainties in the initial condition. The computation of the initial condition includes uncertainties in the bottom displacement, and unknowns on the earthquake mechanism in case of earthquake generated tsunamis, and uncertainties in the bottom movement in case of landslide generated tsunamis. For all sources of tsunamis (earthquake, volcanic, meteorological, and atmospheric tsunamis) uncertainties on source should be accounted for (Fukutani et al. 2015; Cienfuegos et al. 2018; Davies et al. 2018; Basili et al. 2021; Gibbons et al. 2020).
- Uncertainties in bathymetric data constitute another source of perturbations in tsunami propagation models (Behrens et al. 2009; Griffin et al. 2015). The quality of the data is crucial to compute the bathymetric domain to run the simulation. GEBCO and EMODNET data sets are examples of free sources of bathymetry data suited to run deep water simulations but not suitable to assess the inundation.
- Uncertainties in model physics. The shallow water equations are derived with the assumptions that the wavelength is larger than the water depth and that the vertical velocity component is constant in the water column. These assumptions are no longer valid for example for water flowing through the structures of a harbour or through streets (see Sect. 5.5), for impact tsunamis, or dispersive waves.
- Uncertainties in parameters such as the bottom friction. This issue is of particular importance when dealing with high-resolution inundation simulations.
- Uncertainties introduced by limitations in mesh resolution.

To account for these uncertainties, various methods and techniques can be employed in tsunami simulations, some of which are:

1. Ensemble simulations: This involves running multiple simulations with different sets of initial conditions, model parameters, or boundary conditions. The results of these simulations can be used to estimate the uncertainties and provide probabilistic forecasts of the tsunami wave height, inundation, and arrival time.
2. Sensitivity analysis: This involves varying one or more input parameters of the numerical model and observing its impact on the simulation results. This can help identifying the most critical parameters affecting the simulation results and hence uncertainties and model bias.
3. Data assimilation: This involves integrating observational data, such as tsunami wave height, sea level, or GPS data, into the numerical model to reduce the uncertainties in the input data and improve the accuracy of the simulation results.
4. Uncertainty quantification: This involves estimating the uncertainty in the simulation output using statistical methods such as Monte Carlo simulations and Bayesian inference.

5.7 Choosing What Method to Use

Modelling the tsunami obviously is one of the key components of a PTHA/PTRA workflow. It is therefore paramount to select an appropriate modelling approach. The following considerations may help to decide on the appropriate model.

1. What is the aim of the analysis? Is the analysis directed towards a global, regional, or local area of coverage? For a global analysis it may suffice to neglect inundation computation, while for a local analysis it may be necessary to compile detailed inundation data.
2. What data is available? Are accurate topography and bathymetry data available? If so, then a detailed inundation model may be viable, otherwise the uncertainty in the data will be much larger than the suggested model accuracy.
3. What computational resources are available? Some models require high-performance computing devices or graphics processing units, while others may run on laptops. Detailed non-linear modelling usually requires much larger computational effort than linear shallow water theory.
4. How large is the uncertainty in the sources? The larger the source uncertainty (i.e. the epistemic uncertainty in the possible triggering events) the less severe is a model error. While model error should in general be omitted, some inaccuracy in discretizing the equations, or solving the discrete model can usually be accepted. More accurate models are usually also more demanding with respect to computational resources.
5. How detailed is the model approach? Are the physical processes accurately represented? More precisely, are the approximations to the real world in each of the modelling steps balanced? If one step in the analysis workflow represents physical processes that the others cannot capture, it may not be adequate to include such processes.
6. Which software is available? Some tools may be very accurate and efficient; however, they may not be easily available. Others are easily downloadable from open repositories but may not be fit for the purpose. It is by any means necessary to use well validated tsunami modelling tools. Community standards for validation are described in Synolakis et al. (2008).

References

Aniel-Quiroga Í, Quetzalcóatl O, González M, Guillou L (2018) Tsunami run-up estimation based on a hybrid numerical flume and a parameterization of real topobathymetric profiles. Nat Hazards Earth Syst Sci 185194:1469–1491. https://doi.org/10.5194/nhess-18-1469-2018

Aniel-Quiroga Í, Vidal C, Lara JL, González M (2019) Pressures on a rubble-mound breakwater crown-wall for tsunami impact. Coastal Eng 152. https://doi.org/10.1016/j.coastaleng.2019.103522

Apotsos A, Buckley M, Gelfenbaum G, Jaffe B, Vatvani D (2011) Pure Appl Geophys 168(11):2097–2119. https://doi.org/10.1007/s00024-011-0291-5

Babeyko A, Hoechner A, Sobolev SV (2010) Source modeling and inversion with near real-time GPS: a GITEWS perspective for Indonesia. Nat Hazards Earth Syst Sci 10:1617–1627. https://doi.org/10.5194/nhess-10-1617-2010

Basili R, Brizuela B, Herrero A, Iqbal S, Lorito S, Maesano FE et al (2021) The making of the NEAM tsunami hazard model 2018 (NEAMTHM18). Front Earth Sci 753. https://doi.org/10.3389/feart.2020.616594

Behrens J (2008) Unstrukturierte finite Elemente Simulationen für die Tsunami Frühwarnung in Indonesien. NAFEMS Magazin 10:41–51. http://hdl.handle.net/10013/epic.32423.d001

Behrens J, Pranowo WS, Wekerle C (2009) Credible worst case tsunami scenario simulation for Padang. http://hdl.handle.net/10013/epic.51625

Behrens J, Dias F (2015) New computational methods in tsunami science. Phil Trans R Soc A 373:20140382. https://doi.org/10.1098/rsta.2014.0382

Berger MJ, George DL, LeVeque RJ, Mandli KT (2011) The GeoClaw software for depth-averaged flows with adaptive refinement. Adv Water Res 34(9):1195–1206

Casulli V, Stelling G (1998) Numerical simulation of 3D quasi-hydrostatic, free-surface flows. J Hydraul Eng 124(7):678–686

Cienfuegos R, Catalán PA, Urrutia A, Benavente R, Aránguiz R, González G (2018) What can we do to forecast tsunami hazards in the near field given large epistemic uncertainty in rapid seismic source inversions? Geophys Res Lett 45(10):4944–4955

Davies G, Griffin J, Løvholt F, Glimsdal S, Harbitz C, Thio HK et al (2018) A global probabilistic tsunami hazard assessment from earthquake sources. Geol Socty, London 456(1):219–244

Davies G (2019) Tsunami variability from uncalibrated stochastic earthquake models: tests against deep ocean observations 2006–2016. Geophys J Int 218(3):1939–1960

Davies G, Griffin J (2020) Sensitivity of probabilistic tsunami hazard assessment to far-field earthquake slip complexity and rigidity depth-dependence: case study of Australia. Pure Appl Geophys 177(3):1521–1548

Dalrymple RA, Rogers BD (2006) Numerical modeling of water waves with the SPH method. Coastal Eng 53(2–3):141–147

de la Asunción M, Castro MJ, Fernández-Nieto ED, Mantas JM, Acosta SO, González-Vida JM (2013) Efficient GPU implementation of a two waves TVD-WAF method for the two-dimensional one layer shallow water system on structured meshes. Com Fluids 80:441–452

ETOPO2022 (2022) NOAA National Centers for Environmental Information. 2022: ETOPO 2022 15 Arc-Second Global Relief Model. NOAA National Centers for Environmental Information. https://doi.org/10.25921/fd45-gt74. Accessed 23 Mar 2023

Fuhrman DR, Madsen PA (2009) Tsunami generation, propagation, and run-up with a high-order Boussinesq model. Coastal Eng 56(7):747–758. https://doi.org/10.1016/j.coastaleng.2009.02.004

Fukutani Y, Suppasri A, Imamura F (2015) Stochastic analysis and uncertainty assessment of tsunami wave height using a random source parameter model that targets a Tohoku-type earthquake fault. Stoch Env Res Risk Assessment 29:1763–1779

Furumura T (2009) Integrated ground motion and tsunami simulation for the 1944 Tonankai earthquake using high-performance super computers. J Disaster Res 4:118–126

GEBCO 2022: URL: https://www.gebco.net. Accessed 23 Mar 2023

Gibbons SJ, Lorito S, Macías J, Løvholt F, Selva J, Volpe M (2020) Probabilistic tsunami hazard analysis: high performance computing for massive scale inundation simulations. Front Earth Sci 8:591549

Gibbons SJ et al (2022) The sensitivity of tsunami impact to earthquake source parameters and manning friction in high-resolution inundation simulations. Front Earth Sci 9:757618. https://doi.org/10.3389/feart.2021.757618

Gica E et al (2007) Sensitivity analysis of source parameters for earthquake-generated distant tsunamis. J Waterway Port Coast Ocean Eng 133(6):429–441

Glimsdal S, Pedersen GK, Harbitz CB, Løvholt F (2013) Dispersion of tsunamis: does it really matter? Nat Hazards Earth Syst Sci 13(6):1507–1526

Glimsdal S, Løvholt F, Harbitz CB, Romano F, Lorito S, Orefice S et al (2019) A new approximate method for quantifying tsunami maximum inundation height probability. Pure Appl Geophys 176:3227–3246

Goda K, Mai PM, Yasuda T et al (2014) Sensitivity of tsunami wave profiles and inundation simulations to earthquake slip and fault geometry for the 2011 Tohoku earthquake. Earth Planet Space 66:105. https://doi.org/10.1186/1880-5981-66-105

González A et al (2022) Pseudo-probabilistic design for high-resolution tsunami simulations in the southwestern Spanish coast. GeoHazards 3:294–322. https://doi.org/10.3390/geohazards3020016

González-Vida JM et al (2019) The Lituya Bay landslide-generated mega-tsunami. Numerical simulation and sensitivity analysis. Nat Hazards Earth Syst Sci 19:369–388. https://doi.org/10.5194/nhess-19-369-2019

Greenslade DJM, Annunziato A, Babeyko A, Burbidge DR, Ellguth E et al (2014) An assessment of the diversity in scenario-based tsunami forecasts for the Indian ocean. Cont Shelf Res 79:36–45. https://doi.org/10.1016/j.csr.2013.06.001

Grezio A, Babeyko A, Baptista MA, Behrens J, Costa A et al (2017) Probabilistic tsunami hazard analysis: multiple sources and global applications. Rev Geophys 55(4):1158–1198

Griffin J, Latief H, Kongko W, Harig S, Horspool N et al (2015) An evaluation of onshore digital elevation models for modeling tsunami inundation zones. Front Earth Sci 3:1–32

Grilli ST, Skourup J, Svendsen IA (1989) An efficient boundary element method for nonlinear water waves. Eng Anal Boundary Elem 6(2):97–107

Grilli S, Watts P (2005) Tsunami generation by submarine mass failure. I: Modeling, experimental validation, and sensitivity analyses. J Waterway Port Coast Ocean Eng 131(6):283–297. https://doi.org/10.1061/(ASCE)0733-950X(2005)131:6(283)

Guler HG, Baykal C, Arikawa T, Yalciner AC (2018) Numerical assessment of tsunami attack on a rubble mound breakwater using OpenFOAM®. Appl Ocean Res 72:76–91. https://doi.org/10.1016/j.apor.2018.01.005

Harbitz CB, Pedersen G, Gjevik B (1991) Numerical simulations of slide generated water waves. Sci Tsunami Hazards 9(1):15–23

Harbitz CB (1992) Model simulations of tsunamis generated by the Storegga slides. Mar Geol 105(1–4):1–21

Harding JM (1982) Relative numerical dispersion of shallow water gravity waves caused by three differencing techniques. Technical Note ADA115296, Defense Technical Information Center. https://apps.dtic.mil/sti/citations/ADA115296

Harig S, Chaeroni PWS, Behrens J (2008) Tsunami simulations of several scales-comparison of approaches with unstructured meshes and nested grids. Ocean Dyn 58:429–440. https://doi.org/10.1007/s10236-008-0162-5

Iglesias O, Lastras G, Souto C, Costa S, Canals M (2014) Effects of coastal submarine canyons on tsunami propagation and impact. Mar Geol 350:39–51. https://doi.org/10.1016/j.margeo.2014.01.013

Imamura F (1996) Review of Tsunami Simulation with a Finite Difference Method. In: Yeh H, Liu P, Synolakis C (eds), Long-Wave Run-Up Models. World Scientific 25–42

Imamura F, Yalciner AC, Ozyurt G (2006) Tsunami modelling manual. UNESCO IOC international training course on Tsunami numerical modelling

Intergovernmental Oceanographic Commission (2019) Tsunami Glossary. Paris, UNESCO. IOC Technical Series, 85. (English, French, Spanish, Arabic, Chinese) (IOC/2008/TS/85 rev. 4. http://itic.ioc-unesco.org/images/stories/about_tsunamis/tsunami_glossary/tsunami_glossary_en_v19.pdf

Jeschke A, Pedersen GK, Vater S, Behrens J (2017) Depth-averaged non-hydrostatic extension for shallow water equations with quadratic vertical pressure profile: equivalence to Boussinesq-type equations. Int J Num Meth Fluids 84(10):569–583. https://doi.org/10.1002/fld.4361

Jung T-H, Son S, Lynett PJ (2016) A comprehensive sensitivity analysis of tsunami model system to the parametric and input uncertainties. J Coast Res 75(1):1117–1121. https://doi.org/10.2112/SI75-224.1

Kaiser G, Scheele L, Kortenhaus A, Løvholt F, Römer H, Leschka S (2011) The influence of land cover roughness on the results of high resolution tsunami inundation modeling. Nat Hazards Earth Syst Sci 11(9):2521–2540

Kajiura K (1963). The leading wave of a tsunami. Bull Earthq Res Inst University of Tokyo 41(3):535–571

Kim DH, Lynett PJ, Socolofsky SA (2009) A depth-integrated model for weakly dispersive, turbulent, and rotational fluid flows. Ocean Model 27(3–4):198–214

Kim J, Pedersen GK, Løvholt F, LeVeque RJ (2017) A Boussinesq type extension of the GeoClaw model-a study of wave breaking phenomena applying dispersive long wave models. Coast Eng 122:75–86

Kim J, Løvholt F, Issler D, Forsberg CF (2019) Landslide material control on tsunami genesis—the Storegga Slide and tsunami (8,100 years BP). J Geophys Res: Oceans 124(6):3607–3627. https://doi.org/10.1029/2018JC014893

Kirby JT, Grilli ST, Liu PL-F, Nicolsky D, Abadie S et al (2022). Validation and inter-comparison of models for landslide tsunami generation. Ocean Model 170, https://doi.org/10.1016/j.ocemod.2021.101943

Kirby JT, Wei G, Chen Q, Kennedy AB, Dalrymple RA (1998) FUNWAVE 1.0: fully nonlinear Boussinesq wave model-Documentation and user's manual. research report NO. CACR-98-06

Kriebel DL, Lynett PJ, Cox DT, Petroff CM, Robertson IN, Chock GY (2017) Energy method for approximating overland tsunami flows. J Waterway Port Coast Ocean Eng 143(5):04017014

Leveque RJ, Berger MJ, Mandli KT (2010) The GeoClaw software for geophysical flows. In AGU Fall Meeting Abstracts, vol 2010, pp. DI51A-1855

Liu PLF, Woo SB, Cho YS (1998) COMCOT computer program for tsunami propagation and inundation. Cornell University New York, In Tech Rep

Lynett PJ, Borrero JC, Weiss R, Son S, Greer D, Renteria W (2012) Observations and modeling of tsunami-induced currents in ports and Harbors. Earth Planet Sci Lett 327–328:68–74. https://doi.org/10.1016/j.epsl.2012.02.002

Lynett PJ, Gately K, Wilson R, Montoya L, Arcas D et al (2017) Inter-model analysis of tsunami-induced coastal currents. Ocean Model 114:14–32

Lynett P, McCann M, Zhou Z et al (2022) Diverse tsunamigenesis triggered by the Hunga Tonga-Hunga Ha'apai eruption. Nature 609:728–733. https://doi.org/10.1038/s41586-022-05170-6

Lotto GC, Dunham EM (2015) High-order finite difference modeling of tsunami generation in a compressible ocean from offshore earthquakes. Comput Geosci 19(2):327–340

Løvholt F, Pedersen G (2009) Instabilities of Boussinesq models in non-uniform depth. Int J Num Meth Fluids 61(6):606–637

Løvholt F, Glimsdal S, Harbitz CB, Zamora N, Nadim F, Peduzzi P, Dao H, Smebye H (2012) Tsunami hazard and exposure on the global scale. Earth-Sci Rev 110(1–4):58–73

Løvholt F, Lynett P, Pedersen G (2013) Simulating run-up on steep slopes with operational Boussinesq models; capabilities, spurious effects and instabilities. Nonlin Process Geophys 20(3):379–395

Løvholt F, Pedersen G, Harbitz CB, Glimsdal S, Kim J (2015) On the characteristics of landslide tsunamis. Phil Trans Royal Soc A 373:20140376

Ma G, Shi F, Kirby JT (2012) Shock-capturing non-hydrostatic model for fully dispersive surface wave processes. Ocean Modell 43:22–35

Maeda T, Furumura T (2013) FDM simulation of seismic waves, ocean acoustic waves, and tsunamis based on tsunami-coupled equations of motion. Pure Appl Geophys 170:109–127

Macías J, Castro MJ, Ortega S, Escalante C, González-Vida JM (2017) Performance benchmarking of Tsunami-HySEA model for NTHMP's inundation mapping activities. Pure Appl Geophys 174. https://doi.org/10.1007/s00024-017-1583-1

Macías J, Escalante C, Castro MJ (2021) Multilayer-HySEA model validation for landslide-generated tsunamis–Part 1: rigid slides. Nat Hazards Earth Syst Sci 21(2):775–789

Madden EH, Bader M, Behrens J, van Dinther Y, Gabriel AA et al (2021) Linked 3-D modelling of megathrust earthquake-tsunami events: from subduction to tsunami run up. Geophys J Int 224(1):487–516

Madsen PA, Schäffer HA (2010) Analytical solutions for tsunami runup on a plane beach: single waves, N-waves and transient waves. J Fluid Mech 645:27–57. https://doi.org/10.1017/S0022112009992485

Miranda JM et al (2014) Validation of NSWING, a multi-core finite difference code for tsunami propagation and run-up. American Geophysical Union (AGU) fall meeting, San Francisco. Paper. No. S21A-4390

Molinari I, Tonini R, Lorito S, Piatanesi A, Romano F et al (2016) Fast evaluation of tsunami scenarios: uncertainty assessment for a Mediterranean Sea database. Nat Hazards Earth Syst Sci 16(12):2593–2602

Monaghan JJ (1992) Smoothed particle hydrodynamics. An Rev Astronomy Astrophys 30(1):543–574

Monaghan JJ (2005) Smoothed particle hydrodynamics. Rep Prog Physi 68(8):1703

Monaghan JJ (2012) Smoothed particle hydrodynamics and its diverse applications. An Rev Fluid Mech 44:323–346

Mulia IE, Ueda N, Miyoshi T, Gusman AR, Satake K (2022) Machine learning-based tsunami inundation prediction derived from offshore observations. Nature Commun 13(1):5489

National tsunami hazard mitigation program (NTHMP) (2012) Proceedings and results of the 2011 NTHMP model benchmarking workshop. Boulder: U.S. Department of Commerce/NOAA/NTHMP; NOAA Special Report. p 436

National tsunami hazard mitigation program (NTHMP) (2016) Report on the 2015 NTHMP Current Modeling Workshop. Portland, Oregon. p 202

Nicolsky DJ, Suleimani EN, Hansen RA (2011) Validation and verification of a numerical model for Tsunami propagation and runup. Pure Appl Geophys 168(6):1199–1222

Nielsen O, Sexton J, Schneider J (2005) The ANUGA tsunami inundation model. In: proceedings of ModSIM. Australia: Australian National University, pp 518–523

Nosov MA, Kolesov SV (2007) Elastic oscillations of water column in the 2003 Tokachi-oki tsunami source: in-situ measurements and 3-D numerical modelling. Nat Hazards Earth Syst Sci 7(2):243–249

Nosov MA, Kolesov SV (2011) Optimal initial conditions for simulation of seismotectonic tsunamis. Pure Appl Geophys 168:1223–1237

Omira R, Ramalho RS, Kim J et al (2022) Global tonga tsunami explained by a fast-moving atmospheric source. Nature 609:734–740. https://doi.org/10.1038/s41586-022-04926-4

Panza GF, Romanelli F, Yanovskaya TB (2000) Synthetic tsunami mareograms for realistic oceanic models. Geophys J Int 141(2):498–508

Park H, Cox DT, Petroff CM (2015) An empirical solution for tsunami run-up on compound slopes. Nat Hazards 76:1727–1743. https://doi.org/10.1007/s11069-014-1568-7

Pedersen G (2008) Modeling runup with depth integrated equation models. In PLF Liu et al (eds) Advances in Coastal and Ocean Engineering, pp 3–41. https://doi.org/10.1142/9789812790910_0001.

Pranowo WS, Behrens J, Schlicht J, Ziemer C (2008) Adaptive mesh refinement applied to tsunami modeling: TsunaFlash. In: Proceedings of the international conference on tsunami warning (ICTW), State Ministry of Research and Technology Republic of Indonesia (RISTEK), Jakarta, Indonesia. http://hdl.handle.net/10013/epic.32425

Pringgana G, Cunningham LS, Rogers BD (2021) Influence of orientation and arrangement of structures on Tsunami impact forces: numerical investigation with smoothed particle hydrodynamics. J Waterway Port Coast Ocean Eng 147(3):04021006

Rakowsky N, Androsov A, Fuchs A, Harig S, Immerz A et al (2013) Operational tsunami modelling with TsunAWI—recent developments and applications. Nat Hazards Earth Syst Sci 13:1629–1642. https://doi.org/10.5194/nhess-13-1629-2013

Reis C, Barbosa AR, Figueiredo J, Clain S, Lopes M, Baptista MA (2022) Smoothed particle hydrodynamics modeling of elevated structures impacted by tsunami-like waves. Eng Struct 270:114851

Rogers BD, Dalrymple RA (2008) SPH modeling of tsunami waves. In Advanced numerical models for simulating tsunami waves and runup, pp 75–100

Rossetto T, Allsop W, Charvet I, Robinson DI (2011) Physical modelling of tsunami using a new pneumatic wave generator. Coastal Eng 58(6):517–527. https://doi.org/10.1016/j.coastaleng.2011.01.012

Saito T, Furumura T (2009) Three-dimensional tsunami generation simulation due to sea-bottom deformation and its interpretation based on the linear theory. Geophys J Int 178(2):877–888

Sarfaraz M, Pak A (2017) SPH numerical simulation of tsunami wave forces impinged on bridge superstructures. Coastal Eng 121:145–157

Shi F, Kirby JT, Harris JC, Geiman JD, Grilli ST (2012) A high-order adaptive time-stepping TVD solver for Boussinesq modeling of breaking waves and coastal inundation. Ocean Model 43:36–51

Synolakis CE, Bernard EN, Titov VV, Kanoglu U, Gonzalez FI (2008) Validation and verification of Tsunami numerical models. Pure Appl Geophys 165:2197–2228. https://doi.org/10.1007/s00024-004-0427-y

Tanioka Y, Satake K (1996) Tsunami generation by horizontal displacement of ocean bottom. Geophys Res Lett 23(8):861–864

Tanioka Y, Seno T (2001) Sediment effect on tsunami generation of the 1896 Sanriku tsunami earthquake. Geophys Res Lett 28(17):3389–3392

Titov V, González FI (1997) Implementation and testing of the method of splitting Tsunami (MOST) model. In: NOAA Tech Memo ERL PMEL-112 (PB98-122773), NOAA/Pacific Marine Environmental Laboratory, Seattle, WA, p 11

Titov V, Kânoğlu U, Synolakis C (2016) Development of most for real-time Tsunami forecasting. J Waterway Port Coast Ocean Eng 142(6):03116004. https://doi.org/10.1061/(ASCE)WW.1943-5460.0000357

Tolkova E (2014) Land-water boundary treatment for a tsunami model with dimensional splitting. Pure Appl Geophys 171(9):2289–2314

Vater S, Beisiegel N, Behrens J (2019) A limiter-based well-balanced discontinuous Galerkin method for shallow-water flows with wetting and drying: triangular grids. Int J Numer Meth Fluids 91(8):395–418. https://doi.org/10.1002/fld.4762

Venturato AJ (2007) Tacoma, Washington, tsunami hazard mapping project modeling tsunami inundation from Tacoma and Seattle fault earthquakes. DIANE Publishing

Wang X, Power W (2011) COMCOT: a Tsunami generation propagation and run-up model. GNS Science Report 2011/43, p 129

Yamazaki Y, Kowalik Z, Cheung KF (2009) Depth-integrated, non-hydrostatic model for wave breaking and run-up. Int J Numer Meth Fluids 61:473–497. https://doi.org/10.1002/fld.1952

Yeh H, Liu P, Synolakis C (1997) Long-wave runup models. In: Proceedings international workshop, World Scientific, p 420. https://doi.org/10.1142/9789814530330

Zaitzev A, Kurkin A, Pelinovsky E, Yalciner AC (2019) Numerical tsunami model NAMI-DANCE. Sci Tsunami Hazards 38(4)

Zengaffinen T, Løvholt F, Pedersen GK, Muhari A (2020) Modelling 2018 Anak Krakatoa flank collapse and Tsunami: effect of landslide failure mechanism and dynamics on tsunami generation. Pure Appl Geophy 177:2493–2516

Zengaffinen-Morris T, Urgeles R, Løvholt F (2022) On the inference of tsunami uncertainties from landslide run-out observations. J Geophys Res: Oceans 127: e2021JC018033. https://doi.org/10.1029/2021JC018033

Electronic supplements (notebooks, codes, references to web services, etc.)

Codes

Tsunami-HySEA. https://zenodo.org/record/6400815#.ZB8f8OzMJQJ
Landslide-HySEA. https://zenodo.org/record/6400825#.ZB8gf-zMJQI
TsunAWI. https://tsunami.awi.de/
GeoClaw. https://depts.washington.edu/clawpack/geoclaw/

Benchmarks

EDANYA benchmarks. https://edanya.uma.es/hysea/index.php/benchmarks
NOAA PMEL benchmark suite. https://nctr.pmel.noaa.gov/benchmark/

Open access topography data sets

GEBCO. https://www.gebco.net
ETOPO. https://www.ncei.noaa.gov/products/etopo-global-relief-model
EMODNET. http://www.emodnet-hydrography.eu

Useful Resources for modelling the interface are

Gareth Davie's Code for PTHA. https://github.com/GeoscienceAustralia/ptha
RuptGen (Babeyko et al. 2010)

Chapter 6
Exposure Modelling

Mario A. Salgado-Gálvez, Naveen Ragu Ramalingam, Fatemeh Jalayer, Danijel Schorlemmer, Helen Crowley, and Rade Rajkovchevski

6.1 Introduction

The effects of tsunamis to individuals, communities, the built environment, the natural environment, natural capital, local and international economies can be devastating. Two recent examples are the 2004 Indian Ocean and the Tohoku 2011 tsunamis, which recorded more than 220,000 and 20,000[1] fatalities, respectively (Satake et al. 2006), besides injuries, internal and cross-border displacement, as well as damages and losses on buildings, lifelines, and other critical infrastructure (Mori et al. 2022).

M. A. Salgado-Gálvez (✉)
CIMNE, Barcelona, Spain
e-mail: mario.sal.gal@gmail.com

ERN International, Mexico City, Mexico

N. R. Ramalingam
University School for Advanced Studies–IUSS Pavia, Pavia, Italy
e-mail: naveen.raguramalingam@iusspavia.it

F. Jalayer
Department of Disaster Risk Reduction, University College London, London, UK
e-mail: f.jalayer@ucl.ac.uk

D. Schorlemmer
Seismic Hazard and Risk Dynamics, Helmholtz Centre Potsdam GFZ German Research Centre for Geosciences, Potsdam, Germany
e-mail: ds@gfz-potsdam.de

H. Crowley
Global Earthquake Model (GEM) Foundation, Pavia, Italy
e-mail: helen.crowley@globalquakemodel.org

R. Rajkovchevski
Faculty of Security–Skopje, St. Kliment Ohridski University, Skopje, North Macedonia
e-mail: rade.rajkovcevski@uklo.edu.mk

[1] Besides more than 2,500 reported as missing.

M. B. Sørensen et al. (eds.), *Probabilistic Tsunami Hazard and Risk Analysis*, Mathematics of Planet Earth 14, https://doi.org/10.1007/978-3-031-98115-9_6

Any type of assessment of such consequences requires a good characterization of the exposure, which aims to provide a reliable and up-to-date description of the distribution of people and assets within the study area, an estimate of the cost of repairing or replacing the assets, the identification of the key structural and non-structural attributes, and their interconnections with other systems and networks.

The United Nations Office for Disaster Risk Reduction (UNDRR) defines exposure as "*the situation of people, infrastructure, housing, production capacities and other tangible human assets located in hazard-prone areas*" (UNDRR 2017). Even if as per UNDRR's, and other similar and widely used definitions, the concept of exposure is well framed, in the literature and professional practice it is sometimes interchanged with the vulnerability concept, which is described in a separate chapter of this book. The reason for this is the intertwined relationship between exposure and vulnerability components, where the first provides all relevant data and attributes that allow the vulnerability characterization for the hazard(s) of interest. Nevertheless, what vulnerability models provide are relationships between a hazard intensity (e.g., ground shaking, flood depth, wind speed), and the expected losses (in one or multiple dimensions) for the exposed population and/or assets, something that exposure models by themselves are not able to do.

Besides population and the built environment, there are other types of assets that can be impacted by tsunamis with local and regional consequences. Cropped areas within 5 to 10 kms of the shoreline can be flooded causing loss of harvesting production as well as negative impacts on the environment and resources, such as soil and coastal aquifer salination and contamination. Another example are mobile assets, such as boats, cars and airplanes, whose classification and monitoring are usually carried out by insurance companies. However, the development of exposure models for these types of assets is beyond the scope of this chapter which is mostly focused on the identification of population, buildings, and critical infrastructure at risk.

Conceptually, exposure is a necessary, but not sufficient, determinant of risk, since it is possible to be exposed but not vulnerable to a hazard (as for instance, by living in a floodplain but having sufficient means to modify the asset's characteristics to avoid or mitigate future potential losses, or by occupying a building in an earthquake prone region that was designed and built according to earthquake-resistant provisions). Within a risk assessment, this means that an exposure model is always required for estimating the losses on population and/or exposed assets. Table 6.1 shows.

Within the (re)insurance industry and at catastrophe modelling firms, several exposure models (mostly for the built environment) have been developed for many years to assess risk to one or multiple hazards. However, these databases are not public, nor accessible, and this issue has triggered the development of openly available exposure data repositories particularly for developing countries, where these types of data are rarely available and at the same time are most needed. Realising the limitations

Table 6.1 Key definitions, metrics and possible uses of the described ingredient

What does the ingredient describe?	The methodologies and datasets to develop exposure models for the built environment (i.e., buildings and infrastructure) and population
Definitions of the ingredient	This process typically consists of the identification and characterization of the built environment and population exposed to tsunamis. It includes the population of attributes that describe the characteristics of the exposed elements and populations which allow the assignation of vulnerability models to estimate the impacts
What is/are the metric(s) used for measuring/ evaluating the ingredient?	Location of the exposed assets and population Number and geolocation of people exposed in the study area Economic appraisal of the exposed physical assets
Possible uses/applications of the ingredient	Exposure models to be used in probabilistic tsunami risk assessments
Possible multi-hazard dimensions/ applications?	Multi-hazard attributes can be used in the definition of the taxonomies
Possible time-dependent/dynamic aspects	Time-dependent attributes can be used to describe the dynamic nature of human exposure, such as daily/weekly/seasonal variatons
List of the available methods to be described below	For population exposure models: Top-down approaches Bottom-up approaches Via proxy For built-environment exposure models Bottom-up approaches Top-down approaches

imposed by ad hoc and proprietary exposure data, initiatives to use a common format have been recently developed at the (re)insurance industry and catastrophe modelling firms, such as the Open Exposure Data (OED) Standards (OASIS 2022) aimed to allow the interoperability of these datasets in different proprietary and free to use risk modelling tools.

6.1.1 Development of Exposure Models

The development of exposure models has always been a challenge within the risk assessment field. This is mainly because several data sources are needed to populate all necessary attributes, and each of them can (and typically) have different update schedules, resolution, and level of detail. Generally, exposure models are specifically developed for a single hazard, an aspect that is reflected in two aspects: the resolution (aggregation or disaggregation) level, and the types of attributes assigned to each entry in the database. Exposure datasets can be visualised using GIS (e.g., QGIS) on their different discretization levels, such as administrative units, grids, tessellations, and asset-level resolution.

Even if large differences exist between the hazards for which losses are generally quantified (e.g., tsunami vs. hurricane winds), the exposed assets of interest are mostly the same: population, buildings, and different types of infrastructure. This means that the general concepts of any exposure model are peril-agnostic and therefore, have a multi-hazard characteristic (i.e., any exposure model can be used to carry out multi-hazard risk assessments). Unfortunately, this commonality has not been reflected in the development of standardised multi-hazard exposure databases and only recently, initiatives such as the UNDRR's Global Exposure Database (De Bono 2013; De Bono and Chatenoux 2014), and GED4ALL (Silva et al. 2018) have developed multi-hazard exposure models, schemas, and taxonomies.

Figure 6.1 shows the basic components of an exposure model. The main tasks for developing an exposure model can be summarised as providing a *description* of the exposed asset, to generate a *collection* of the exposed assets and their main characteristics, to be subject to a *validation* procedure, and finally, the *visualisation* and communication of the exposure model entries.

Three main methods can be distinguished for collecting exposure data: *bottom-up* (e.g., full enumeration, statistical sampling), *top-down* (e.g., remote sensing, dasymetric based on ancillary datasets, see the definition in Sect. 6.2.1.1) and the use of *proxies* (e.g., population, building footprint, mobility patterns of population, night-time lights). Validation is an important part of exposure modelling as it can lead to the identification of the uncertainties in the exposure model, which must be communicated and can be in some cases visualised. Considering that there are no established methods to develop tsunami-specific exposure models, in professional practice these have been either developed from scratch or adapted from those developed for other perils (mostly earthquake, tropical cyclones and floods). Also, it must be noted that since the tsunami intensity can drastically vary within nearby locations (i.e., the correlation length of tsunamis is very low, in the order of metres), the accurate mapping of exposed assets and population requires high-resolution exposure models (Gomez-Zapata et al. 2021).

The description of the exposed assets is usually done using *taxonomy* strings. These are directly related to definitions of the spatio-temporal database schemas, which are used for storing the exposure data (e.g., Murnane et al. 2019) and together they provide a structured framework for collecting and managing the information for the development of any exposure model. According to Pittore et al. (2018), the

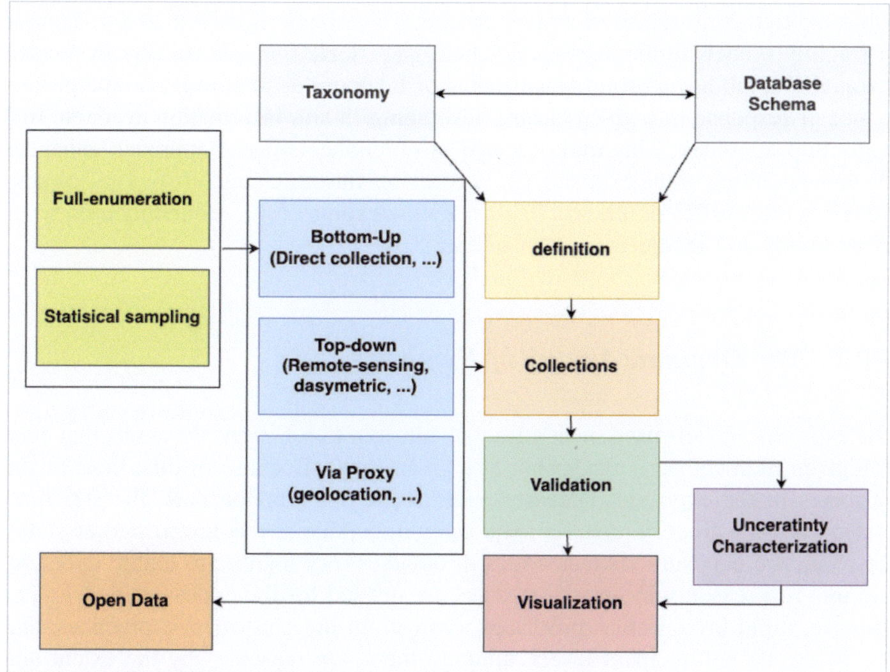

Fig. 6.1 Exposure modelling (adapted from Pittore et al. 2017)

taxonomies can be *risk-oriented (hierarchical)* or *faceted*. Risk-oriented taxonomies (e.g., PAGER-STR, HAZUS, vulnerability classes in EMS98) have been tailored to assigning fragility and/or vulnerability models. However, they tend to lose specific detailed data collected for individual buildings and are usually hazard-specific (e.g., seismic risk) and tend to have a limited geographical scope. The risk-oriented taxonomies work with a top-down approach to exposure modelling (described in more detail hereafter). Faceted taxonomies (e.g., Brzev et al. 2013; Silva et al. 2022) tend to overcome some of these limitations. A faceted taxonomy is a set of taxonomies describing alternative categories or *facets* from a different aspect. A faceted classification allows a single attribute to be linked to other attributes represented in a subject listing, which in the context of exposure databases and taxonomies provide a classification scheme to arrange knowledge in a systematic manner (Broughton 2006). Attributes can be categorical, numerical, and ordinal data, with finite, mutually exclusive and collectively exhaustive values for each considered attribute type. A faceted taxonomy lends itself well to bottom-up (e.g., based on enumeration, described in more detailed hereafter) exposure modelling.

A risk-oriented taxonomy classifies buildings, assets, or populations based on predefined typological classes that reflect their performance (or response) when subjected to the intensities of a given hazard. This system is used to assign damage or fragility models to the exposed assets in large scale risk assessments. A faceted taxonomy combines a group of attributes or taxonomies to summarise the performance or response to a given hazard, leveraging in-situ information available in a bottom-up approach. This multi-faceted or components-based approach enhances the understanding of the exposure by considering various characteristics and dimensions in a more efficient scheme. Table 6.2 shows some of the commonly used types of taxonomy and attributes for exposure models.

6.1.2 The Dynamic Nature of Exposure

The exposure models used in risk assessments aim to represent the assets that exist at a given moment (i.e., a snapshot at a given time). Because of this, most of the attributes in the exposure databases such as the construction date, the structural system, the construction material, and economic value at risk are treated as static. However, the exposure characteristics of almost every region are highly dynamic, and this is a reason why regular updates are needed for the exposure models. For instance, rapid urbanisation processes, changes in the construction practices, and building code enforcement levels, among others, are reasons why the extent and characteristics of the exposed assets can change within relatively short periods (i.e. 3 to 5 years). Therefore, collecting updated and local-scale exposure data remains as the main objective to increase the reliability of any exposure model. With this objective, some proposals regarding the development of exposure models that allow capturing spatio-temporal dynamics at a global scale have been made (Pittore et al. 2017), so that the evolution over space and time during the lifespan of the assets can be captured and monitored.

Also, in the context of different research initiatives, some proposals have been drafted around the idea of forecasting exposure for 10, 30 and up to 50 years (Cremen et al. 2023; Scaini et al. 2023a; 2023b) using different shared socioeconomic pathways (SSP), urban growth models and population growth projections. Sleeter et al. (2017), used the LUCAS model (Daniel et al. 2016) to predict future land-use changes and assigned the population based on historical and empirical land-change information. If developing a reliable model for what is exposed today is a challenge, doing the same for different time frames is even more so and uncertainties around the estimates inevitably will increase, although these projections are highly useful in long-term risk assessments (see for instance Coccia et al. 2023; Salgado-Gálvez et al. 2023).

Table 6.2 Some commonly used types of taxonomy and attributes in exposure models at various scales and use cases

Spatial scale (exposure resolution)	Type of taxonomy	Relevant attributes (Spatio-temporal Value)		Use cases	Reference dataset/ exposure models
		Population	Built-environment		
Global (1–5 km)	Risk-oriented	Population count and classification - urban and rural	Economic value, global building typology, occupancy type	Identify population and regions at high tsunami risk for global prioritisation and disaster risk policy making	GAR, GEM Global Exposure Models
	Faceted	Population count, age distribution, income levels	Aggregated building count and economic value, construction materials, occupancy type,infrastructure systems,	Identify critical infrastructure at risk - nuclear power plant and other sites of global socio-economic importance	Industry and infrastructure exposure datasets - GRIP global roads database
Regional to National (100 m-1 km)	Risk-Oriented	Population count, building occupancy, land use	Economic value, regional building typoplogy, occupancy type	To guide decision making related to disaster planning and mitigation at national level like structural efforts, early warning systems etc.	Census and National Datasets - Eurostat
	Faceted	Population count, Land use and zoning, Socio economic factors	Economic value, building age, construction type, occupancy type	To assess the resilience of regional and national infrastructure and networks - roads, power transmission etc	Public building inventory, OSM and EUBUCO

(continued)

Table 6.2 (continued)

Spatial scale (exposure resolution)	Type of taxonomy	Relevant attributes (Spatio-temporal Value)		Use cases	Reference dataset/ exposure models
		Population	Built-environment		
Community to Local (Less than 100 m or asset level)	Risk-Oriented	Population count, proximity to evacuation, routes, medical condition	Economic value or replacement cost local building typology or construction material occupancy and service type	For understanding the specific community and planning local tsunami risk communication, preparedness/ mitigation activities etc.	Local cadastral, zoning map and surverys
	Faceted	Geo-location across time, socio economic factors	Economic value or replacement cost building height, vulnerability features, mitigation measures, local elevation	To assess resilience of local urban ecosystems - evacuation site, bridges, schools factories, hospitals etc.	Seaside test dataset

6.2 Population Exposure Models

Exposure models are generally multi-layered (Aubrecht et al. 2013), where the population distribution datasets usually form the layer zero. Additional layers such as socioeconomic and demographic data are typically overlayed to enrich the exposure model. Because of this, human exposure models are particularly relevant for any tsunami risk assessment since they provide a description of the number of people exposed, together with demographic and socio-cultural characteristics that can further determine their risk (e.g., age, sex, disabled, minorities, etc.), the location of people at a given time (e.g., indoors or outdoors, at home, work, schools, parks or beaches), and daily, weekly, and/or seasonal variations (e.g., holidays) that allow estimating the impacts of disasters on different population groups.

The relevance to include as many as possible demographic indicators in the population exposure model is related to the possibility to later disaggregate and study the risk results for different groups, either divided by sex, age, mobility impairments, and income levels. These data allow the definition of comprehensive disaster risk management strategies where all the needs (e.g., evacuation routes for the mobility impaired, or financial assistance to low-income communities) are well identified.

While it is straightforward to identify and characterise the exposed physical assets using the methods later explained in Sect. 6.3, doing the same for the exposed population is a much more complex task. The flow of people is difficult to predict, since it is strongly dependent on parameters like occupancy, time, besides other socio-cultural factors. Most of the available population density datasets rely on population census as the key data source. Population census data provide a description for the number of inhabitants per administrative unit, and it has been mostly considered as a proxy for the night-time population exposure, assuming that most people are at their place of residence at that time of the day. While the census data can be considered as very accurate in different aspects, it lacks information about the exact spatio-temporal distribution of the population as a function of the time, day and/or season. Nevertheless, it must be considered that updated and reliable census data are not available for all countries. To have a reliable population exposure model, it is also necessary to understand how many people are present at a given study area and their geographical distribution within it. For the definition of population groups, there have been different proposals such as the one by Fraser et al. (2014), who defined 5 different groups (i.e. school/childcare, working-age adults, independent elderly, dependent elderly, and visitors).

Other relevant input data for the population exposure models include land use, land cover, and urban planning datasets, population census and economic indicators, cadastral and infrastructure databases, building footprints, settlement boundaries, satellite imagery, night-time lights, surveys, and interviews with locals and experts. In some cases, insurance portfolio and claims data, and other governmental and NGO datasets have been used. Also, the Visible Infrared Imaging Radiometer Suite (VIIRS) instruments aboard the joint NASA/NOAA Suomi National Polar-orbiting Partnership (Suomi NPP) and NOAA-20 satellites provide global daily measurements of nocturnal visible and near-infrared (NIR) light that are suitable for Earth system science and applications studies. VIIRS Day/Night Band (DNB) data are used for estimating population, assessing electrification of remote areas, monitoring disasters and conflict, and understanding biological impacts of increased light pollution.

6.2.1 Methods to Develop Population Exposure Models

The methods for collecting the population density data can be classified as top-down, bottom-up, or based on proxies such as direct geolocation. The latter has been used by taking advantage of almost real-time data gathered from mobile and social networks.

6.2.1.1 Top-Down

Remote sensing, and dasymetric approaches, (Bhaduri et al. 2007) are classified as top-down on which starting from the census counts at the level of census unit (top) data are disaggregated into finer resolution units (e.g., grids, tessellations) based on supporting information provided by land use, land cover, and building masks. "Dasymetric mapping is a geospatial technique that uses information such as land cover types to more accurately distribute data that have been assigned to selected boundaries like census blocks." as defined by the US Environmental Protection Agency (EPA). *Dasymetric mapping* methods use various ancillary data (e.g., satellite imagery, night-time light coverage, land-use data, urban extents, land cover, etc.) to disaggregate spatially census counts from census units (e.g., administrative units) aiming at developing higher resolution datasets. Binary dasymetric mapping uses one extra data layer, corresponding to the building's footprints, to distribute the population from census units (which are in many cases large) to areas identified as urban settlements, as for instance those determined using the Global Human Settlement Layer, (GHSL) which is a global built-up surface map product described in Fig. 6.2. Special care is required with such datasets as in the case of GHSL in which larger parts of the built area are missing depending on the vegetation and climate of the respective area.

The georeferenced statistics from the population census can be further disaggregated into finer resolutions, such as grid cells by using dasymetric mapping approaches (see Sutton 1997; Bhaduri et al. 2007; Freire et al. 2013) using ancillary datasets from remote sensing, land use maps, human settlements (built-area), building's footprints, or already available population datasets derived using top-down approaches that preserve the total value for the population from the main input dataset. Regular and orthogonal grids provide a suitable structure for modelling and representing the geographical distribution of the population, allowing a description of their location (i.e., residing, working, or visiting) and differentiating inhabited and uninhabited areas, as well as data on population in terms of counts or densities.

Multivariate dasymetric mapping methods use more than one ancillary dataset for disaggregation. The use of machine learning (ML) techniques for pattern recognition (e.g., the identification of settlements and non-inhabited areas, or the type of land cover and land use) is becoming more common. Other methods often used are kriging and adaptive kernels to perform areal interpolations. Many ML techniques have been trained on existing buildings footprint data and there are biases towards western-world building structures, which is reflected in wrongly identified buildings or the inability to distinguish single homes in densely populated areas. Further problems arise from low-resolution, or not rectified, imagery in some areas, usually at low-income countries.

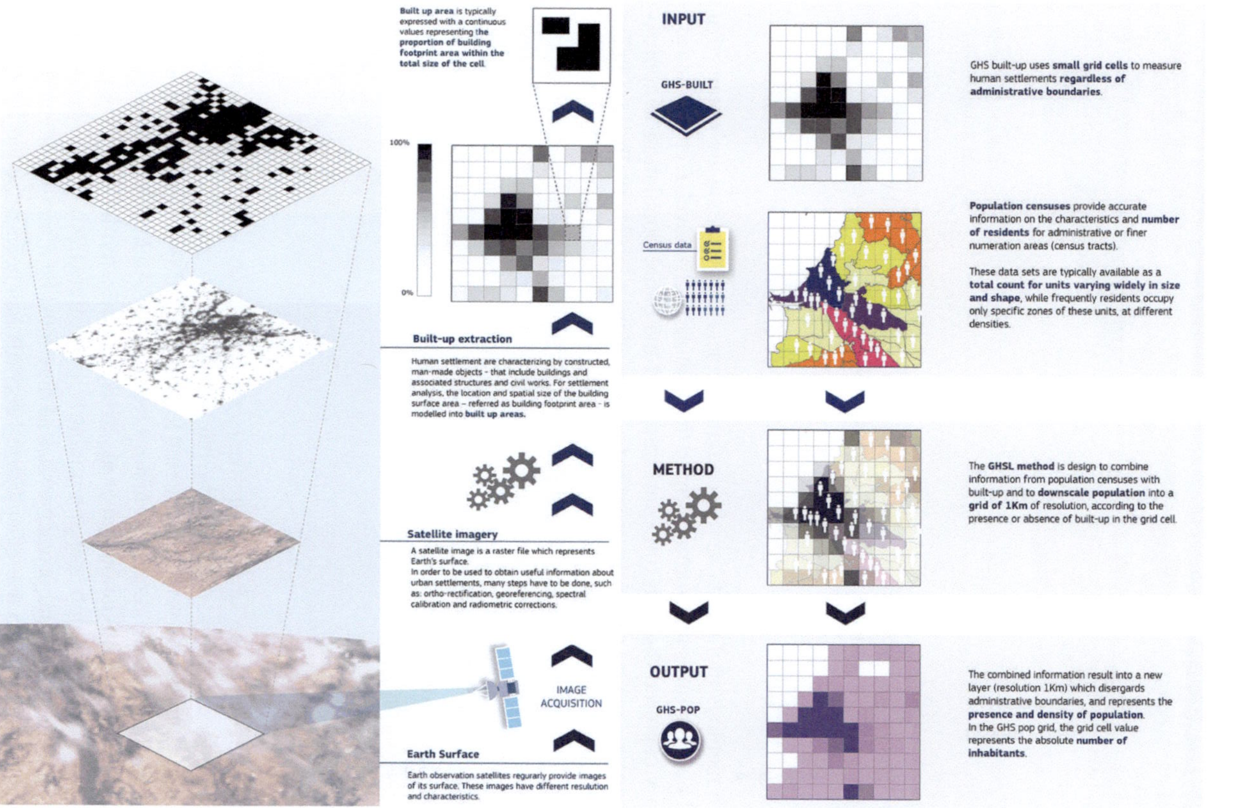

Fig. 6.2 Concept and method of deriving global human settlement layer GHSL **a** built up area and **b** population density (from https://ghsl.jrc.ec.europa.eu/data.php?sl=2#GHSLBasics, licensed under CC-BY 4.0)

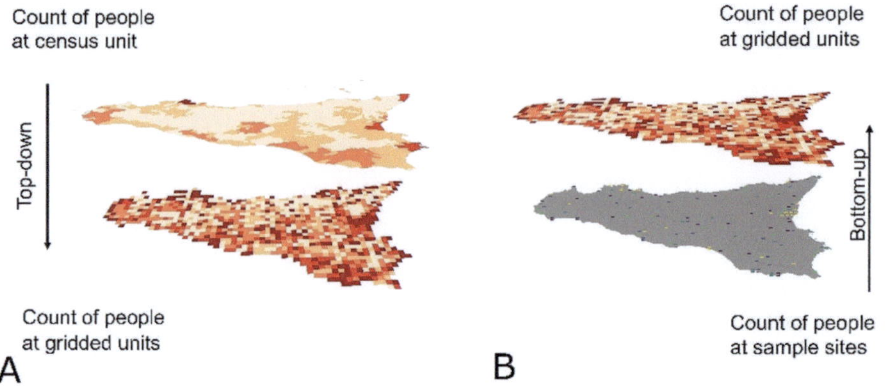

Fig. 6.3 Two common approaches in exposure data collection are illustrated with an example of population estimation for Sicily. **a** Top-down approach **b** Bottom-up approach

6.2.1.2 Bottom-Up

Bottom-up methods include those of population census, which data are obtained at individual housing units and summed up to the administrative level. Census data are usually only captured through full enumeration of the entire population in approximately regular decadal intervals, mostly because of the high costs. Besides being resource-intensive, population censuses tend to become outdated within a short time after their release, and for this, some proposals to update the information at shorter time intervals have been made, as for instance the rolling census approach by Pittore et al. (2017). The top-down and bottom-up approaches to exposure modeling are shown in Fig. 6.3, where the population at grids are estimated using primary data from large census units for the top-down approach and direct survey at limited collection sites are used for the bottom-up approach.

6.2.1.3 Via Proxy

These methods make use of measurements of population activity from mobile network operators, sensors, or social media. Geotagged posts on social networks such as Twitter and Facebook have been used to derive mobility patterns (Wesolowski et al. 2015; Lu et al. 2013). Location history recorded by Android smartphones has been used in research (Ruktanonchai et al. 2018) to assess human mobility at long temporal scales and with high spatial granularity, but suffers from limited geographical scale, sampling bias, and privacy issues as they must be directly collected from the device users. The wide use of call-detail records has been impeded by ethical (privacy) reasons when using individual users' data but aggregating the data at towers instead can help preserve their anonymity albeit at the loss of accuracy. To account for the overall population over a large area, one needs to put together data from all mobile phone service providers simultaneously (which is highly difficult), and develop methods to avoid double-counts, or to consider people who do not own a phone. However, these data are difficult to obtain and cannot be shared openly, rendering scientific reproducibility impossible.

6.2.2 Spatial Resolution

Exposure models can vary in spatial extent, resolution level and the type and number of attributes. The spatial resolution level is mostly determined by the available data and resources, as well as the purpose of use of the results (i.e., a country level study may not require building by building level of detail since hazard data are coarse-grained), usually seeking a good balance between detail and resources. Regarding the resolution level, coherence with that used for the hazard component is also needed, which in the case of tsunamis is usually high. Also, the scale of the analysis and the selection of the resolution level has also to do with the method used to develop the exposure models. Whereas for small municipalities or selected areas at urban centres the choice of high-resolution and building-by-building exposure models is usually made (Goda and De Risi 2018), to cover larger areas, as in the case of national and global assessments, regular grids with different spacings of up to 5 × 5 km have been used (Løvholt et al. 2015; De Bono and Chatenoux 2014).

Some examples of the use of different spatial resolutions are the work by Lovholt et al. (2014), who used the Land Scan exposure dataset (Bright et al. 2008) at 30 arc-seconds (1 km at the Equator) resolution to capture population data. With such resolution, capturing the local population accumulation which are often observed in coastal urban areas and places of interest is not possible. The more recent global and regional population and settlement layers datasets are available at much higher resolution referred to as "lightly modelled" as they constrain and disaggregate the population to building footprints extracted from high-resolution satellite imagery using machine learning methods–GHS-POPv2023 (100 m) and HRSL (30 m) although have not been validated (Thomson et al., 2021). As an alternative to available high-resolution gridded dataset, Jin et al., (2023) used a random forest regression kriging technique to downscale census data to multi-resolution (1 km, 500 m, 100 m) gridded population datasets for Nanjing China.

6.2.3 Modelling the Temporal Variability

The overall occupancy distribution of buildings (which can be single, multiple, or of mixed use) determines at what time of the day people are expected to be inside a given building. Depending on the time of the day, the day of the week, or the season, more people will be at home, school or at work, or commuting outdoors and enjoying public spaces. At coastal touristic areas, the population increases multi-fold during the high season, that usually lasts for some weeks, and people can be expected to concentrate in large numbers on the beaches during those periods, so that estimates of the number of tourists are needed to represent these flows.

The night-time population exposure is usually based on the building occupancies and is derived through population census data. Daytime population exposure scenarios can be defined by augmenting the night-time exposure by different factors, such as the maximum building floor capacity, employment records, land use, building's main use, season, location of institutions, schools, and working day/non-working day, among others. (Fraser et al. 2014). Indeed, in the literature, differences of an order of two were reported between the daytime and nighttime exposure in the European city centres (Batista e Silva et al. 2020). Also, on average, daytime population densities exponentially decay as a function of the distance to the urban centre (Batista e Silva et al. 2020). Some studies have used population profiles at different times of the day, in different locations, to develop specific occupancy scenarios. For instance, the U.S. Department of Transportation data on the population throughout the day has been used to estimate the changes in day spatio-temporal population distribution (Kobayashi 2011).

6.2.4 Human Behaviour

Human behaviour, as well as the interaction of humans with the environment play an important role in how people react in the event of a tsunami (and that these reactions are not static), therefore its understanding in the context of tsunami risk assessments is critical, since at many locations there is the possibility to take evacuation measures and save lives. For this, methods such as Agent-based modelling (ABM) provides a simulation based, high-resolution and dynamic human exposure prediction by relating the behaviour of individuals and groups at a micro-scale, and the constraints from their environment to find patterns and trends of the larger population at macro-scale (Mas et al. 2015). ABM ran agent-to-agent interactions that can be used to simulate the evacuation behaviour, although a high-resolution population exposure dataset is needed as input for the people's start location in the model. There are several examples of ABM used in the context of tsunami evacuation modelling (Mas et al. 2013; Wang et al. 2021; Mostafizi et al. 2019; Mohammad et al. 2021; Kim et al. 2022; Chen et al. 2022).

6.3 Built Environment Exposure Models

Any exposure model for the built environment has as the main objective providing a reliable estimation of the number of assets, their characteristics, their replacement or repairing cost, and their geographical distribution within a study area. Typically, the assets included in these exposure databases are buildings and critical infrastructure (e.g., schools, hospitals, power lines, etc.), although exposure models for other types of assets, such as heritage sites, have been developed. It is common practice when estimating risk to one or multiple hazards at national level through country risk profiles, to include in the built environment exposure model different types of assets, classified by type and sector (e.g., residential, commercial, industrial, etc.).

Population density and population counts have been widely used as the layer zero when developing exposure models for the built environment. These data are combined with other indicators, such as gross domestic product, capital stock, delimitation of urban and rural areas, and building censuses, to develop these models. Another relevant data source, where available, are the cadastral (and infrastructure) databases. As in the case of population exposure databases, the data sources for the built environment are multiple and in most cases are not homogeneous, meaning that they do not have the same degree of completeness and not all the information is publicly available (e.g., real estate and insurance/claims datasets In recent years, national and local governments have put efforts in improving the quality and accuracy of the data included in those datasets, by making use of satellite/aerial imagery to define key attributes such as the plan geometry (footprint) and height.

The geolocation process for the case of tsunami exposure databases is critical because the hazard intensity measures (e.g., run-up height, water depth, water velocity) can have large variations within relatively short distances (i.e., metres), as opposed to ground motions by earthquakes (at rock) or wind speeds (at coastal locations). Because of this, the exposure models for PTRA can highly benefit from high-resolution representations. For instance, at city level when cadastral data are available, exposure models are generally developed with a building-by-building resolution, whereas in the case of national scale assessments, the aggregation of the exposed assets is typically performed at different levels, such as regional or municipal, leaving aside relevant location details that are usually not appropriate for PTRA. An additional aspect to be considered in the development of tsunami exposure models is that besides the plan location of the building and all other relevant attributes, their relative elevation with respect to ground is also used.

6.3.1 Methods to Develop Built Environment Exposure Models

The development of an exposure database is mainly a twofold process, comprising the identification and the characterization of the built environment. On the one hand, the *identification process* involves the geolocation of each exposed asset, that can be exact or approximate; whereas on the other hand, the *characterization process* involves assigning to each asset in the database all relevant attributes that describe their performance when subjected to a hazard intensity measure (e.g., number of storeys, structural system, construction date, wall openings, etc.), besides an economic appraisal that, as will be discussed later, can be carried out in terms of different values. Table 6.3 provides a brief description of the relevance of each attribute of an exposure model within a PTRA.

Data availability is not uniform at all locations and even more, the update frequencies can vary strongly. Whereas for the case of some assets, such as water and sanitation pipelines, that are typically classified as critical infrastructure, the number of segments, location, age, and other relevant attributes of the assets is available and is needed in the day to day operation of the systems (see Salgado-Gálvez et al.,

Table 6.3 Key attributes and their relevance in exposure models for the built environment

Attribute	Relevance
Location	Provides the exact or approximate location of the exposed asset. For PTRA determines the hazard level at each site of analysis
Occupancy class	Provides a description about the main use of the building. Typical categories are residential, commercial, industrial, and governmental
Height	This attribute is generally provided in terms of the number of storyes and in the case of PTRA has high relevance when assigning the vulnerability models
Construction material and structural system	Provide information about all relevant characteristics that correlate well with the damages and losses. They allow a proper connection of the exposure data with the vulnerability models
Year of construction	Provide information about the age of the asset, which can be related to the use of a given building code, and the occurrence of previous damages
Structural value	Provides an estimate of the monetary value, usually in terms of the replacement cost, of each asset. Allows obtaining the risk results in monetary terms
Contents value	Although not mandatory, it provides the cost of the items within a building but that are not part of the main structure
Business interruption value	Provides an estimate of the loss of revenues caused by the damage or disruption of an asset caused by a direct damage

2018), or nuclear power plants for which global tsunami exposure analyses have been carried out (Løvholt et al. 2014), for other assets, such as residential buildings, the databases are rarely complete. Regarding critical infrastructure, different initiatives have been deployed for its mapping, including the Critical Infrastructure Spatial Index (Nirandjan et al. 2022) which aggregates at $0.1^\circ \times 0.1^\circ$ or $0.25^\circ \times 0.25^\circ$ cells over 30 types of critical infrastructure assets classified in 7 systems (transport, energy, telecommunications, waste, water, education, and health) collected from the high-resolution crowdsourced geospatial dataset OpenStreetMap (OSM). Depending on the characteristics of some critical infrastructure assets, their location and main attributes can be easily captured using aerial or satellite images (e.g., roads, rail lines, and airports). The most recent risk assessments for the (re)insurance industry and the development sector (e.g., World Bank, Inter-American Development Bank) have required the inclusion of transport and supply infrastructure given their relevance in terms of not only exposed value, but the services they provide, such as interconnectivity for the economies of the countries, and supply chains, among others.

6.3.1.1 Bottom-Up

The bottom-up approach has, in comparison to the top-down approach, higher accuracy. However, more resources of all types are needed since applying this method usually requires field visits and surveying activities, which are doable for areas of moderate size (e.g., such as small municipalities or specific areas within a large city), although this method has been used for small islands, as in Tonga (ADB 2023). In this case, detailed information at building/asset level is available, providing a reliable estimate of the key attributes such as the building class, replacement/repairing cost, construction date, material, and structural system. This method is rarely used in large-scale analyses because it is not practical, nor economical, to carry out building by building classifications in large-scale (e.g., country or global level) risk assessments.

Exposure databases with bottom-up approaches usually make use of cadastral databases, where available. Although cadastral databases are developed mostly for property taxing purposes, they include relevant information that can be used to estimate and infer different parameters. Cadastral databases typically have a dwelling-by-dwelling resolution level, and assets are assigned with their exact location and in several cases, with the actual plan geometry. These databases usually have restricted access because of the obvious confidentiality issues, but in some countries like Spain, the general information regarding location, plan geometry, number of storeys, main use and construction date are public, and have been used in earthquake risk assessments (Salgado-Gálvez et al. 2016; 2023). It is precisely with those data that some key parameters can be inferred. For instance, by knowing the plan geometry, the plan area can be calculated and for the same asset, the total constructed area can be approximated by multiplying the plan area by the number of storyes. The total constructed area is an attribute with the utmost importance since the value at risk, in terms of the replacement cost, can be estimated by using indicative costs per constructed square metre that are usually building class dependent. Examples of exposure databases used in a PTRA developed with the bottom-up approach can be found for Acapulco (Mexico) and Callao (Peru) in Ordaz et al. (2019).

The bottom-up approach can benefit directly from crowd-sourcing initiatives (including structured crowd-sourced datasets, such as OpenStreetMap–OSM (e.g., the crowd-sourcing Global Dynamic Exposure Model, https://git.gfz-potsdam.de/globaldynamicexposure) by relying on a community of thousands of individuals mapping for OSM with various intentions to populate the global map with a variety of features and their attributes. In the field of disaster risk management, crowd-sourced data was used mostly in post-disaster activities (i.e., to record the damages at individual household level), although its use is increasing in the development of exposure models with the possibility of collecting large amounts of data in a timely manner. Different efforts have been made with this approach in the mapping and characterization of critical assets such as schools (Scaini et al. 2022) where a dataset including over 170 schools was developed for Italy. Although some training is usually delivered before the data collection using crowd-sourcing initiatives, there will always be a large subjective component in the individual's evaluation (data collected) which also, as the number of entries in the database increases, becomes more difficult to control and/or validate. Such crowd-sourced datasets are known to suffer from varying coverage, inconsistent description or missing attributes, and they need to be combined, harmonised and validated with other available datasets (government datasets and remote sensing products). Figure 6.4 compares the OSM and EUBUCCO building polygon datasets with their occupancy attribute highlighting the incompleteness in the attribute information, for reference is the GHS-BUILT-C a built up characteristic raster from GHSL.

6.3.1.2 Top-Down

The top-down (or downscaling) approach consists of choosing and collecting different national/regional socio-economic indicators such as income, employment rates, housing statistics, health, and education levels, and combining them with the delimitation of the built environment (as for instance using GHSL as a mask) and the geographical distribution of the population within those boundaries as proxies for downscaling. The latter aspect denotes the close relationship between the built environment, population, and socio-economic/demographic indicators. In a nutshell, this method goes from the general to the particular when identifying and characterising assets. The workflow uses a multi-layered, dasymetric, spatial approach, to distribute the socio-economic data, usually available at the administrative level, to regular grids that can have different resolution levels. Mapping schemas provide a basis to link building and structures with more detailed attributes for example construction type, age, height, and area missing in regional housing statistics needed to assess the cost of the assets and to characterise their vulnerability.

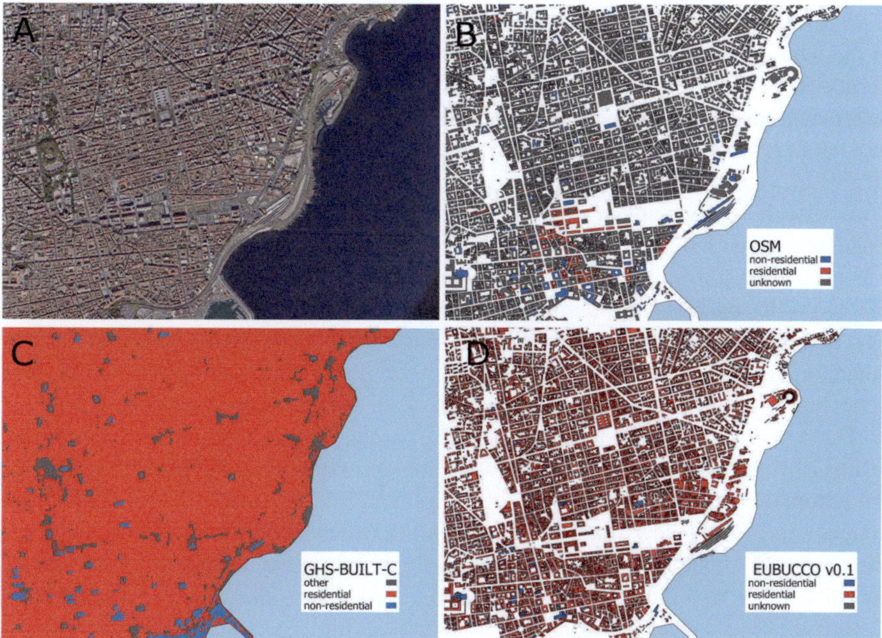

Fig. 6.4 **a** Shows an aerial picture of the coastal city of Catania, Sicily. Different dataset to characterize the built-up region **b** OSM building polygons **c** GHS built-up characteristics **d** EUCUCCO building polygons

Disaggregation weights are factors or coefficients used to distribute data from coarse to finer units, expressing the relative importance of smaller units within the larger administrative units. Population density, land-use maps, and other socio-economic proxies help defining these disaggregation weights. Local expertise can also be used to create mapping schemes and disaggregation weights to carry out the downscaling, although such assumptions should always be validated for guaranteeing the reliability of the exposure model. Some examples of these databases can be found, as for instance the GAR15 coarse-grain global exposure database by UNDRR (De Bono and Chatenoux 2014), used for a PTRA, the European and South American exposure databases by GEM (Yepes-Estrada et al. 2017), and the Central Asia exposure database by Scaini et al. (2023a, 2023b) developed for earthquake and flood risk assessments.

Depending on the purpose of the risk assessment and the available data, the entries in the exposure databases can be statistically aggregated through grids with different areal units. These grids, which can have fixed or variable resolution, are generally represented through a geospatial database like Post GIS or distributed in files with geospatial data like Shapefiles or Geo Package.

Although conceptually simple, the top-down approach presents serious challenges in the data selection since the sources are multiple, with varying update and granularity levels (i.e., the cutoff years and resolution level for the data are usually different). Population censuses are updated on average every 10 years, but other indicators regarding employment and income levels tend to be updated more regularly (e.g., yearly). Therefore, choosing what data to use at national and global scales is not trivial. The two above-mentioned exposure models are good examples for this since, for the South American earthquake exposure model only public information was used (e.g., household and population census), whereas the GAR15 model made use of public datasets and complemented it with proprietary data (e.g., LandScan which was at that time proprietary but became publicly available later). It must be noted too that depending on the selection of underlying public and private data, some licensing restrictions about the distribution of the datasets can exist.

The results of these two main methods can be enhanced and complemented by the results of remote-sensing and crowd-sourcing approaches. In the first case, through the automated processing of high-resolution satellite or aerial images (e.g. drone images) and street-view imagery, the main attributes of buildings can be captured and assigned, such as the footprint area, height, construction material, information on building facade and openings etc. (Kang et al. 2018; Hoffmann et al. 2019). In the second case, crowd-sourcing can be used to complement automated processing, they can help create a labeled training dataset necessary for machine learning based classification and lastly help directly validate the results of such indirect inferences. Biljecki et al. (2023) discuss the highly variable quality and use of OpenStreetMap for building information.

Finally, a stage that is usually overlooked in the development of exposure models for the built environment is their validation. This is critical to assess the quality of the collected, modelled, and/or integrated data and can be carried out by comparisons against other reference data values, such as those derived from housing census, or global repositories such as WAPMERR (Wyss et al. 2013) and World Housing Encyclopaedia (Greene et al. 2008), or through field visits, keeping in mind that some of these datasets were published several years ago and no updates are available. It is also worth noting that validation is one of the areas in which, for the overall PTRA, exposure data can assist, by allowing the calibration of the different components and comparing the modelled losses against the reported ones for historical events.

6.3.2 Defining the Key Attributes for the Built Environment Exposure Models

Besides the geolocation of the assets, identifying and assigning key attributes are part of the objectives of any exposure model. These allow a proper characterization of the different assets to later assign an appropriate vulnerability model in the risk assessment, and also define the monetary value at risk, which is needed to obtain the risk results in terms of the commonly used metrics such as the average annual loss, return period specific losses, or more comprehensively, the loss exceedance probability curves.

6.3.2.1 Taxonomies

Exposure databases have been widely used in the catastrophe risk field, on which the risk assessments are usually carried out for large portfolios of buildings, and for which assets that share certain characteristics are grouped into categories. These categories are denoted as building classes and are defined through a taxonomy that accounts for all relevant building parameters and attributes, and that are usually hazard dependent. A taxonomy can be defined as an exhaustive set of mutually exclusive and well described attributes (Pittore et al. 2017), that can be specific to a country or a region. Although for some hazards (e.g., earthquakes and tropical cyclones) commonly accepted taxonomies have been developed and used, such as ATC-13 (ATC 1985), EMS-98 (Grünthal et al. 1998), HAZUS (FEMA 2003) UNDRR (De Bono, 2013; De Bono and Chatenoux, 2014), PAGER (Jaiswal et al., 2010), and GEM (Brzev et al. 2013; Silva et al. 2022), for the case of tsunamis no widely accepted taxonomy exists and the processes to characterize the assets in the currently available exposure models for this hazard are mostly ad-hoc. However, some taxonomies are exhaustive and cover, at least in a partial manner, the relevant building attributes to tsunamis.

The number of building classes is usually determined by the scale of the analysis and the available data. Whereas in the case of local analysis where a building-by-building characterization can be achieved a considerably large number of building classes can be used, if needed, to capture all the differences between the assets, in the case of national, regional, and global analyses, the number of building classes tends to be limited to a manageable number and can be supported by the available data. Also, related to the grouping of assets into a large portfolio, the attribute characterization relies on the law of large numbers, meaning that the over- or underestimation of certain parameters at the asset level, tends to the expected number and ends up compensating. It is precisely the reason why the results of a risk assessment carried out for a portfolio that includes several (up to thousands) buildings, should not be disaggregated to an individual asset scale.

For the study area, or subregions within it, it is required to identify the predominant building classes that, as previously mentioned, can be hazard-dependent. When housing census are available, they provide valuable data for this characterization. However, their completeness and availability are hardly the rule and therefore, the use of indicators provided by other initiatives such as the World Housing Encyclopedia (WHE 2023), or WAPMERR (Wyss et al. 2013), is needed. The latter repository provides the distribution of construction types by country considering three settlement categories although the changes to the building stock that have occurred during the past years may not be reflected. These categories are defined as a function of the number of people for which again the geographical distribution of the population is used as a proxy.

6.3.2.2 Estimating the Monetary Values at Risk

The lack of standardized and commonly accepted approaches to develop the exposure databases is not limited to the general methods, but also to the quantification of key attributes, being the estimation of the monetary values at risk a good example. In their most general form, risk assessments estimate direct damages and losses on the exposed assets (and even if secondary losses are sometimes accounted for, these are generally related to the direct losses), and as explained in Chapter 7, certain functions provide a relationship between a hazard intensity measure and the expected loss. The loss, described through vulnerability functions, is defined in proportions (meaning that 0 corresponds to a case without loss, and 1 to a case of total loss). To be able to obtain the risk results in monetary units, these relative values are multiplied by the value at risk of each asset, which is at the same time usually determined by estimating the replacement cost of each exposed asset, which corresponds to a stock measure.

The value at risk can be divided in four main categories, although not in all cases these are relevant, nor have been considered. The first category corresponds to the structural value of each asset, the second to the value of the non-structural elements, the third to the value of the contents, and the fourth one to the costs of business interruption, which are discussed next.

The replacement cost is an indicative measure of the monetary resources needed to repair or rebuild the damaged asset and bring it to the original condition, meaning that the costs of possible improvements are not accounted for. Although it depends on the characteristics of the asset, such as the construction material, structural system, and height, it does not account for any commercial aspect such as location, or land value. Despite being conceptually simple, large variations can exist in these costs for the same building class (for instance between urban and rural areas because of labour and material availability) and a commonly used approach for defining this cost is to make building class specific estimates of replacement cost by area. However, the replacement cost is not the only way that has been used to determine the monetary value at risk and in some cases, given the data availability, other indicators such as cadastral and commercial values have been used either directly, or as a reference (i.e., affected by a factor). In the case of national scale analyses where different types of assets are included in the exposure database, the capital stock value has been found to be representative of the monetary value at risk (De Bono and Chatenoux 2014).

In the case of exposure models for the built environment, particularly for commercial, industrial, and critical infrastructure mostly at the (re)insurance industry, the value at risk of the costs of the contents can be of relevance which in some cases may exceed the one of the structures only (e.g., hospitals). For this economic appraisal there is no standard method, and a common practice is that these values are estimated by the insurer to avoid the provision of wrong data by the owners. Also, of high relevance for the commercial assets is the definition of an estimate of a value (usually in annual terms), to be representative of the loss revenue as a consequence of the damage and disruption to the assets (i.e., business interruption). As in the case of the contents, there is no standard method for the quantification of these losses and typically have been related to the total value at risk (i.e., structure and contents), or by the insurer fixing a maximum amount.

6.4 Treatment of Uncertainties in the Exposure Models

In most probabilistic risk assessments (regardless of the hazard) the uncertainty of the hazard and vulnerability components is explicitly accounted for, with different methods available for their identification, quantification, and propagation. However, for the exposure component usually a deterministic approach is used and apart from sensitivity analyses to test the results under different assumptions, no additional methods have been formally established.

In professional practice, the uncertainties in the exposure models are rarely quantified, nor propagated throughout the risk assessments. This does not mean by any chance that they are non-existent and neither that they are negligible. Whereas most of the uncertainties related to the exposure databases can be classified as epistemic (i.e., they can be reduced by collecting new data and they exist because the models are based on data that were not created for that purpose, such as a national census), when using the bottom-up approach some of the attributes are treated as random variables and, therefore, their associated uncertainties can be classified as aleatory. This is the case for instance for the indicative replacement costs by built square meter, or the average constructed area per dwelling. Crowley et al. (2020) and Gómez-Zapata et al. (2022) discuss the different possibilities to integrate these uncertainties within a risk assessment for both, epistemic (e.g., the variability when defining the probabilities for each building class in the mapping schemes) and aleatory (e.g., treating as random variables some parameters to derive the exposure models, or when using random distributions to model the population density, (Mas et al. 2013)).

Population census data have significant spatio-temporal variability, and therefore, are prone to lack of data and consequently have associated large uncertainties. Regarding the temporal dimension, census data are static and are only updated in regular intervals of up to approximately 10 years or longer. Therefore, capturing the spatio-temporal distribution of the population exposed to a tsunami inherently involves variabilities related to the spatio-temporal population patterns. This still holds even if the census data are available, up-to-date, and provide full coverage of the study area. There are also uncertainties in demographic data as well as in the classification of vulnerable populations, which is reflected in the population exposure models. The uncertainty in almost all methods studied (census-based, dasymetric, and agent-based) stem from the baseline uncertainty in the census data, which is often well-documented. However, the dasymetric and agent-based methods are accompanied by other sources of uncertainties related to spatio-temporal downscaling of census data. These types of uncertainties can be (and should be) captured within a simulation-based framework (FW1B, Sect. 2.2). The uncertainties stemming from a lack of census data can be compensated, if available, by building data to be used as a proxy for population distribution. Cities or settlements with growing populations, typically expand in size and number of buildings.

The uncertainties associated with the exposure models can (and should) be directly propagated in the risk assessment analysis (FW1, Sect. 2.2). An approach that has been used to address the uncertainties is the development of exposure scenarios (i.e., daytime vs. nighttime, weekday vs. weekend) that within the probabilistic risk frameworks can have associated weights, as a function of their plausibility (equivalent to those of a logic-tree). Also, the variation in the exposure attributes can have consequences in the assignment of a given vulnerability function (i.e., the distribution of the number of potentially affected people). Ignoring the uncertainty in the exposure models can lead to under- or over-estimation of risk and therefore, it is one of the primary sources of uncertainty in PTRA when dealing with the loss estimation.

6.5 Available Models, Databases and Taxonomies for Population and Built Environment Exposure

This section includes a brief overview of the population and built environment exposure datasets and methods. For more details about the available public global datasets, Table 1 in Gunasekera et al. (2015) can also be consulted. In general, the population exposure models and datasets have a much wider coverage globally in comparison to those for the built environment, which tend to be limited to a geographic boundary and context.

6.5.1 Datasets for Population

6.5.1.1 Census Datasets

Census datasets are comprehensive collections of demographic, economic, and social information gathered through a systematic and periodic count of a population. There are different types of sources for this information, such as the national official census, usually made available through governmental statistical offices, and other datasets, as the ones described next, which may be available at varying levels of aggregation, say community, block, municipality, or province. Figure 6.5 shows the ISTAT census population counts for the Sicily region at two such granularity levels.

World Population Prospects

The *2022 Revision of World Population Prospects* is an official United Nations population estimate released by the United Nations Population Division (UNPD).

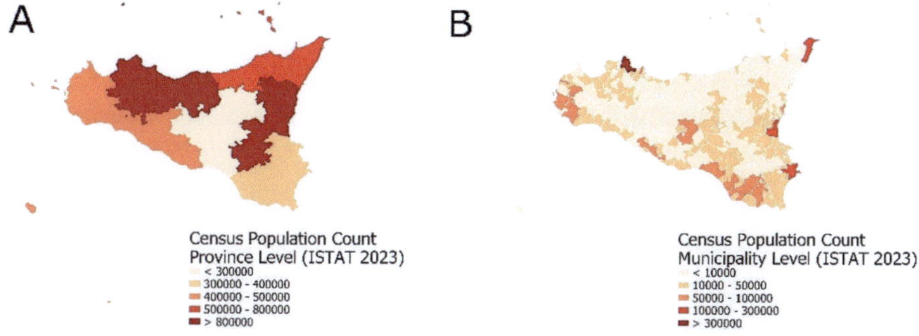

Fig. 6.5 Population count for Sicily at **a** province and **b** municipality level as of 2023 (population data shown are retrieved from the Italian national institute of statistics ISTAT, licensed under CC-BY 4.0)

It provides population projections from present up to 2100 considering a range of alternative scenarios at three levels: global, regional, and national. It is based on historical population trends collected from 1950 based on the national census results, vital registration systems, and national sample surveys. It also has aggregated results in annual intervals based on different age groups.

Eurostat

Eurostat is the statistical office of the European Union. This database includes statistics on a wide range of themes: general and regional statistics, economy and finance, **population** and social conditions, industry, international trade, transport, environment and energy, and science, technology, digital society. Includes data from 1990, 2001 and 2011 censuses.

6.5.1.2 Gridded Population Datasets

For more refined analyses, population distribution across a geographic area in a grid format are used, where each grid cell contains an estimate of the population count. Table 6.4 summarizes the open-access gridded population datasets with regional or global coverage. Some of these datasets are compared to each other in Fig. 6.6 highlighting their variation in terms of resolution and distribution.

Popgrid is a data cooperative that relies on various data providers such as Center for International Earth Science Information Network (CIESIN) at Columbia University, ESRI, the German Aerospace Center (DLR), and NASA. It uses satellite imagery and remote sensing to cover the gaps in population datasets that rely solely on Census data. The Popgrid mapper consists of layers from GPW, Landscan, WorldPop, and night-time lights (all described below).

Table 6.4 Some commonly available global and regional gridded population and census datasets

Dataset	Source	Population concept	Method	Spatial resolution	Year(s) represented	National-level population totals	Distribution policy	Reference link	
Global population grids									
Unmodeled	Gridded Population of the World (GPWv4.11)	Center for International Earth Science Information Network (CIESIN), Columbia University	Resident population (domicile)	GPW1: pycnophylactic; GPW2,3,4: areal weighting	v4 - 1 km (30 arcsec)	2000; 2005; 2010; 2015; 2020	2 versions: (1) official country census totals; (2) country totals adjusted to United Nations Population Division (UNPD) estimates and projections	Open access	http://sedac.ciesin.columbia.edu/data/collection/gpw-v4
Lightly modeled	Global rural urban mapping Project (GRUMPv1)	CIESIN, Columbia University; International Food Policy Research Institute, the World Bank, Centro International de Agricultural Tropical	Resident population (domicile)	Dasymetric	1 km (30 arcsec)	1990; 1995; 2000	UNPD estimates and projections	Open access	https://sedac.ciesin.columbia.edu/data/collection/grump-v1

(continued)

Table 6.4 (continued)

Dataset	Source	Population concept	Method	Spatial resolution	Year(s) represented	National-level population totals	Distribution policy	Reference link	
Highly modeled	Global human settlement layer – Population (GHS-POP-R2023A)	European commission joint research centre (JRC) and CIESIN, Columbia University	Resident population (domicile)	Dasymetric refinement, proportional to built-up density	100 m (3 arcsec)	1975–2030; in 5-year intervals	UNPD estimates and 2025 and 2030 projections from CIESIN GPWv4.11	Open access	https://ghsl.jrc.ec.europa.eu/ghs_pop2023.php
	Land scan global population database (Land Scan Global)	Oak Ridge National Laboratory (ORNL)	Ambient (24 h avg)	Multi-variable dasymetric modeling	1 km (30 arcsec)	2000–2021; annual releases	U.S. Census Bureau (normalized to CIA World Factbook)	Open access	https://landscan.ornl.gov/
	World pop	World pop, University of southampton	Resident population (domicile)	Statistical/ dasymetric	100 m (3 arcsec)	2000–2020	2 versions: (1) Country-official estimates, and (2) UNPD estimates and projections	Open access	https://www.worldpop.org/
	History Database of the Global Environment (HYDE) Population Grids (HYDEv3.2)	Utrecht University	Resident population (domicile) HYDE 3.3 (2023) in prep (to include base, high and low scenarios)	Dasymetric mapping using historical population, cropland, pasture data, satellite data	5 arcmin (ca. 10 km)	10 000 BC to CE 2015	United Nations World Population Prospects (2008 Revision) 1950 -2015	Open access	https://public.yoda.uu.nl/geo/UU01/MO2FF3.html

(continued)

Table 6.4 (continued)

Dataset	Source	Population concept	Method	Spatial resolution	Year(s) represented	National-level population totals	Distribution policy	Reference link	
Regional/continental population grids									
Lightly modeled	High Resolution Population Layer (HRSL) Available for 140 countries	Data for Good at Meta and CIESIN	Resident population (domicile)	Binary dasymetric	30 m (1 arcsec)	2015	Country-official estimates	Open access	https://ciesin.columbia.edu/data/hrsl/
									https://dataforgood.facebook.com/dfg/tools/high-resolution-population-density-maps
Highly modeled	ENACT-POP R2020A Available for EU-28 (2013)	European Commission Joint Research Centre(JRC)	Day and Night (resident) Population	Multi-layered and temporal dasymetric approach	1 km (30 arcsec)	2011. Monthly intervals (day and night)	Eurostat (statistical office of the European Union)	Open access	https://ghsl.jrc.ec.europa.eu/ENACT_POP.php

(continued)

Table 6.4 (continued)

Dataset	Source	Population concept	Method	Spatial resolution	Year(s) represented	National-level population totals	Distribution policy	Reference link	
	Gridded Population Mapping (Demobase) Available for select countries	U.S. Census Bureau	Resident population (domicile)	Statistical/ dasymetric	100 m	Depends on country (1998 to present)	U.S. Census Bureau International Database and national censuses	Open access	https://www.census.gov/geographies/mapping-files/time-series/demo/international-programs/demobase.html

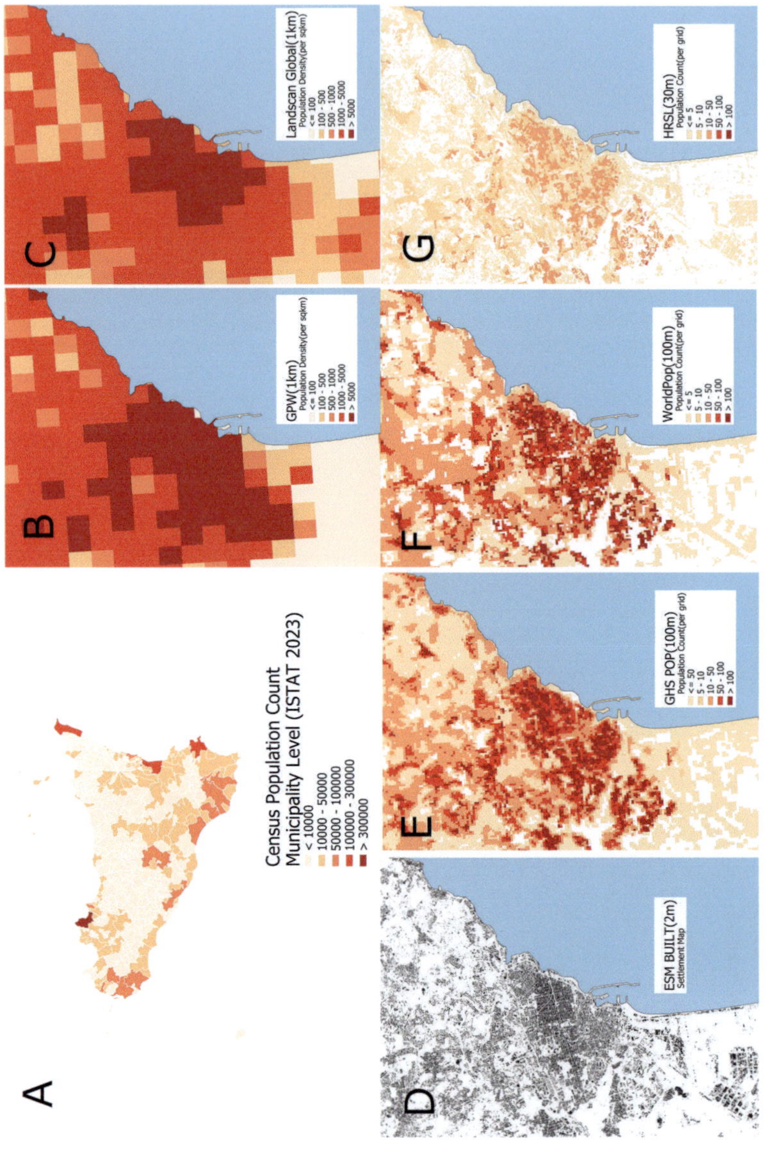

Fig. 6.6 Different gridded population for 2020 around catania, sicily. **a** Census dataset used as base information for generating such gridded population datasets. **b** GPW v4.11, **c** Landscan Global, **d** settlement map using ESM 2015, such datasets are used to constrain the dasymetric mapping to built-up area as in dataset, **e** GHS POP R2023, **f** WorldPop Constrained, **g** HRSL

Gridded population of the world (GPW), provided by CIESIN. The global (resident, counted at domicile) population grid is an open access dataset mainly based on (1) census data (2) total census population per country adjusted to account for the United Nation Population Division (UNDP) prospects (described above). It provides a spatial resolution of (30 arc-seconds) which is equivalent to 1 km at the Equator. Method: Areal weighted such that population is spread across all grid cells covering a census unit based on the proportion of the grid cell that falls in the census unit.

Global Human Settlement Layer—Population (GHS-POP) provided by the European Commission Joint Research Centre (JRC) and CIESIN. This open-access dataset provides Landsat, and proportional allocation to distribute population data from subnational census data to satellite-based built-up areas. The resolution is 3 arc-seconds (~100 m), 30 arc-seconds (~1 km) in the WGS84 standard. It uses the UNPD estimates and projections. See Schiavina et al. (2022) for more detail. The GHSL dataset is also widely used as a built-up mask for human settlements when developing exposure models for the built environment.

Global rural urban mapping project (GRUMP) The service providers are CIESIN, International Food Policy Research Institute (IFPRI), the World Bank, and the International Centre for Tropical Agriculture (CIAT for its acronym in Spanish). This open-access dataset provides resident population data counted at domicile. Method: it uses the binary dasymetric approach using nighttime lights and proportional allocation to distribute population data from subnational census data to the settlement extents. Resolution is 30 arc-seconds (1 km) and uses the UNPD estimates and projections.

WorldPop provided by WorldPop is an open-access dataset. It is a high-resolution, temporal data on human population and demographic distributions. As a method, it uses statistically based weighting layer combined with a dasymetric redistribution. It has 3-arc seconds (100 m) resolution. It has two versions (1) Country-official estimates, and (2) United Nations Population Division (UNPD) estimates and projections.

Landscan global population dataset provided by Oak Ridge National Laboratory (ORNL) is an open-access database providing ambient (24 h averaged) population within each cell. Method: Spatial data, imagery analysis, and multi-variable dasymetric modelling combined with geospatial science (remote sensing and ML technologies) and expert opinion to distribute population. Cells are weighted and tailored to match the data conditions and geographical nature of each country/region by using the most recent and highest resolution ancillary data. Resolution is 30 arc-seconds (1 km). Source: Individual country census normalized to the CIA World Factbook (2011).

History Database of the Global Environment (HYDE) Population Grids, version 3.2 provided by Utrecht University is an open-access database based on resident population. Method: Historical population, cropland and pasture statistics are combined with satellite information and specific allocation algorithms (which change over time) to create spatial population maps through dasymetry. Resolution:

5 arc-minutes (10 km). It covers 10,000 BC to 2000 AD. Population estimates are generally on the high end of the range of past estimates.

High Resolution Settlement Layer (HRSL) provided by CIESIN is an open-access dataset providing estimates of human population distribution at a resolution of 1 arc-second (approximately 30 m) for the year 2015. The population estimates are based on recent census data and high-resolution (0.5 m) satellite imagery from DigitalGlobe. The population grids provide detailed delineation of settlements in both urban and rural areas, particularly useful for many areas such as disaster response. The settlement's extent data is developed by the Connectivity Lab at Facebook using computer techniques to classify blocks of satellite imagery data as settled (with buildings) or not. CIESIN used proportional allocation to distribute population data from census data to the settlement extents.

ENACT POP (Batista e Silva et al. 2020), Schiavina et al. (2020)). This open-source spatial raster dataset depicts the seasonal nighttime and daytime population grids for 2011 produced in the ENACT project for European countries. Values represent the number of people per cell. The data is published at 1 km resolution (at Equator) which is equivalent to 30 arc-seconds in WGS-84 (12 monthly nighttime grids and 12 daytime grids).

6.5.2 Datasets for the Built Environment

6.5.2.1 Exposure Databases

UNDRR's Global Exposure Database is a global exposure database that accounts for individual taxonomies for earthquakes, hurricanes (wind and storm surge), floods and tsunamis population, distributing the exposed value into different categories among which the residential, education, health, and public buildings are included with a 5 $\times$ 5 km resolution level at inland locations and with a 1 $\times$ 1 km resolution level at coastal locations.

GEM Global Exposure Model is the exposure model used in the June 2023 GEM release of the Global Risk Model for earthquake, the national exposure models provide estimates on number of buildings, occupants, costs, and area aggregated at the first order administrative level and categories by building occupancy type and components (structural and non-structural). Additionally summary information at national, regional, and global level are available.

Exposure model for European seismic risk assessment is an exposure model for the European Seismic Risk Model 2020 (ESRM20) that describes the distribution of the main residential, industrial, and commercial building classes across all countries in Europe, as well as their occupants and replacement costs.

Seaside Testbed Data Inventory developed for the city of Seaside, Oregon provides an example of a local exposure model needed for detailed risk assessment. It provides information on the built environment and population exposure suitable for parcel-level analyses of damages, losses, and recovery for earthquakes and tsunamis on buildings, electric power networks, transportation and water and sanitation infrastructure.

6.5.2.2 Schemas, Taxonomies and Building Typologies

ATC-13 provides a comprehensive structural taxonomy for earthquakes based on structural and main-use classifications, which is representative for California only, even if some extrapolation of its use has been made for different locations in the world. Several assumptions made in the development of this model back in 1985 are not valid.

PAGER is one of the most complete global earthquake taxonomies that includes the most representative building types. It was developed with the objective of carrying out rapid earthquake human and physical risk estimations in the aftermath of large events, with the results made available through the USGS global earthquake monitoring system.

HAZUS is an example of a well-known multi-hazard taxonomy which is available for earthquakes, hurricanes, floods and tsunamis. However, it has been derived only for the United States and that is reflected in not having defined structural types that are common elsewhere, such as the case of stone and earthen buildings.

World Housing Encyclopedia Consists of a collection of resources related to earthquake resistant construction practices at seismically active areas worldwide. It is a global dataset consisting of 14 types and 45 subtypes.

EMS-98 is a simplified earthquake taxonomy developed for the European context consisting of 15 structural types. Although in some cases that simplification is an advantage, in some other cases the definitions are broad and do not allow a proper characterization. Another point to highlight of this taxonomy is that the structural types are somehow determined by the vulnerability characteristics, thus mixing the concepts of exposure and vulnerability, which can lead to misunderstandings in the conceptual bases of probabilistic risk assessments, regardless of the hazard.

WAPMERR Developed by the World Agency of Planetary Monitoring and Earthquake Risk Reduction, consists of a global database for which construction types have been defined for each country, considering 3 settlement categories. It was on this characterization that the distribution of building classes, at national level, was carried out for UNDRR's global exposure databases.

Open Exposure Data (OED) As part of Open Data Standards (ODS) provides the re/insurance and catastrophe modelling industry an open, and transparent data format. This set aims to improve efficiency and transparency for the cat modelling community, facilitating data transfer and analytics across models, model developing firms and open-source community. OED is a model agnostic data format for descriptions in terms of location, coverage, policy etc. relevant for loss modelling.

GED4ALL is an exposure data schema that is part of the open Risk Data Library Standard (RDLS) curated by GFDRR. The RDLS provides open standards for the description and sharing of risk modelling data including exposure, hazard, vulnerability, and losses (monetary and non-monetary).

6.5.2.3 Other Relevant Geospatial Datasets

Global Human Settlement Layer (GHSL) project is supported by European Commission, Joint Research Centre and Directorate-General for Regional and Urban Policy and apart from the GHS-POP population dataset also provide global spatial datasets (a) GHS-BUILT with built-up information on surface, height and volume, (b) city model GHS-SMOD with degree of urbanization. These are developed using a large amount of heterogeneous data including: global, fine-scale satellite image data streams, census data, and volunteering geographic information sources.

European Settlement Map (ESM R2019) is a spatial raster dataset that is mapping human settlements in Europe based on Copernicus Very High-Resolution optical coverage for reference year 2015 at resolution of 2 m.

OpenStreetMap (OSM) is a collaborative and open-source mapping project that aims to create a detailed and freely available map of the world. Unlike traditional maps produced by corporations, OSM allows anyone to contribute and edit the map data, resulting in a highly dynamic and constantly evolving map. The project utilizes a crowd-sourcing approach, where individuals and communities contribute information such as roads, buildings, landmarks, and points of interest.

EUBUCCO is a database of individual building footprints for 200 + million buildings across the 27 European Union countries and Switzerland, together with three main attributes—building type (only distinguishing between residential and non-residential buildings), height and construction year—included for respectively 45, 74, 24% of the buildings. EUBUCCO is composed of 50 open government datasets and Open Street Map that have been collected, harmonized, and partly validated.

6.6 Choosing What Method to Use

What method to use in the development of exposure databases for the population and the built environment for PTRA is conditioned by several aspects, such as data availability, expected use of the results, and available resources (i.e., time, computational, monetary). Some of these conflict with each other, and among the considerations described next, the balance between accuracy, detail, and resources (of all types) has a considerable weight in the final decision.

6.6.1 For the Development of Population Exposure Models

As far as the spatial resolution is concerned, most open-access human exposure databases with a wide coverage have a resolution of an order of 1 × 1 km, which is clearly inadequate and incompatible with the grid resolution for the tsunami hazard footprint. Therefore, top-down multivariate dasymetric approaches provide an acceptable spatio-temporal resolution, although noting that they are unaccompanied by a comprehensive characterization of the uncertainties. The population census-based methods, although inadequate in the spatio-temporal coverage, have the most reliable documentation of uncertainties.

In addition to population counts, demographic information which are typically available from census or other socio-economic indicators can be used to enhance the population exposure models. When such information is not available from census or other spatial records, it can be geo-statistically modelled from clustered surveys with full consideration of uncertainty. Different methods to account for the temporal dynamics can be used depending on the use case and the available resources, one such method is using mobile phone data where call densities can be empirically linked to population densities at daily to monthly time scales. Here, each mobile phone communication (voice call or text message) is used to identify the geolocation of an anonymous user by geolocating the transmitting cell phone tower and the cell device. Such a network-based positioning is simple, and the accuracy depends on the density of the cell phone towers, providing an alternative source of highly spatial and temporal detail for population mapping. Further, future population modelling techniques can be used for the long-term projection as time steps of the population in the exposure models.

6.6.2 For the Development of Built Environment Exposure Models

The purpose of use of the risk assessment defines in most cases the required resolution level. If the risk assessment is to inform local decision-making processes (e.g., those to be made by a city mayor), being able to obtain results at the household or maximum block level is of relevance. If on the contrary, the analysis is to be made for national or global scales, that granularity, despite being available, may be of little use if a compatible high-resolution level is not available for the hazard component.

The compatibility between the tsunami hazard and exposure resolution levels have also to do with the method selection since the spatial resolution of the exposure model must guarantee the representativeness of the data with respect to that one for the hazard component. Recent proposals about using variable resolution exposure databases (Pittore et al. 2020; Gómez-Zapata et al. 2021) when carrying out tsunami risk assessments have been made, with the objective of providing an appropriate balance between the needed (and scarce) resources, and the desired level of detail.

Bottom-up approaches are preferred for the local assessments, whereas top-down methods have been found as appropriate for large scale analyses, since using a high-resolution exposure database for global analyses has not been practical, besides posing computational challenges in the handling and storage of the datasets. Nevertheless, building-by-building exposure databases such as the Global Dynamic Exposure Model offers the possibility to easily handle and extract the data for the area of interest. An advantage with the aim of keeping manageable computational times when carrying out PTRA is that the exposure of interest is mostly concentrated at coastal areas, typically within 5 -10 km of the shoreline, which significantly reduces the required efforts and resources and with today's computational resources allow using building-by-building data.

6.7 Gaps in Exposure Data for Tsunami Risk Assessment

The most relevant gaps in the development of exposure models for population and the built environment to be used in PTRA have been subject of previous reviews and have been compiled by Behrens et al. (2021). These can be grouped into different categories, which are discussed next.

First, the lack of detail of the data, regarding the location of the population and assets, as well as about their main attributes and characteristics, which connects with another gap related to the lack of updated exposure data, applicable to areas where at some points these models have been developed, but are not anymore representative of the population and assets because of the major changes, expansion and, in some cases, damages inflicted by disasters.

The lack of a tsunami-specific taxonomy is also a gap, where the capture of key attributes such as the open ground floor, anchoring systems and type of foundations are missing. However, recent proposals for the development of multi-hazard taxonomies, where tsunamis have been included, can set the bases for further developments, and allow in the short- to mid-term, progresses in this matter. However, it must be mentioned that besides an exhaustive taxonomy for tsunamis, the necessary attributes of all buildings of interest need to be collected, and these data do not exist at the time of writing this book.

Also, there is the challenge to capture the spatio-temporal variability, for which proposals as that one by Pittore et al. (2017) have been made with the objective of capturing the dynamic aspects of interest in the exposure models, and to observe/monitor, at least during the lifespan of the exposed assets, these variations, which can occur either because of the normal use of the assets, or because they are inflicted by an external factor (e.g., a disaster).

Finally, the availability of global exposure data repositories is another gap which could support the development of other components of a tsunami risk model. For instance, in the case of hazard modelling, data on tsunami defence infrastructure (e.g., seawalls, dikes, breakwater and elevated roads) and the built environment (e.g., building footprint, land cover and their roughness) can facilitate their explicit representation in the tsunami inundation simulations and allow improving the accuracy of the estimated tsunami intensity measures (e.g., water depth and water velocity) at the sites of interest.

References

ADB – Asian Development Bank (2023). Multi-hazard disaster risk assessment, Togatapu. Interim Exposure Development Report. Regional: Pacific Disaster Resilience Program. Available at: https://www.adb.org/sites/default/files/project-documents/50028/50028-001-tacr-en_14.pdf Last accessed: August 7th, 2023.

ATC – Applied Technology Council (1985). ATC-13: earthquake damage evaluation data for California.

Aubrecht C, Özceylan D, Steinnocher K and Freire S (2013) Multi-level geospatial modelling of human exposure patterns and vulnerability indicators. NAT HAZ 68(147–163)

Batista e Silva F, Freire S, Schiavina M, Rosina K, Marín-Herrera MA, Ziemba L, Craglia M, Koomen E and Lavalle C, (2020) Uncovering temporal changes in Europe's population density patterns using a data fusion approach. NAT COMMUN 11(1):4631

Behrens J, Løvholt F, Jalayer F, Lorito S, Salgado-Gálvez MA, Sørensen M, Abadie S, Aguirre-Ayerbe I, Aniel-Quiroga I, Babeyko A, Baiguera M, Basili R, Belliazzi S, Grezio A, Johnson K, Murphy S, Paris R, Rafliana I, De Risi R, Rossetto T, Selva J, Taroni M, Del Zoppo M, Armigliato A, Bures V, Cech P, Cecioni C, Christodoulides P, Davies G, Dias F, Basak Bayraktar H, González M, Fritsevich M, Guillas S, Bonnevie Harbitz CB, Kanoglu U, Macías J, Papadopoulos GA, Polet J, Romano F, Salamon A, Scala A, Stepinac M, Tappin D, Thio HK, Tonini R, Triantafyllou I, Ulrich T, Varini E, Volpe M, Vyhmesiter E (2021) Probabilistic tsunami hazard and risk analysis – a review of research gaps. FRONT EARTH SCI 9:628772. https://doi.org/10.3389/feart.2021.628772

Bhaduri B, Bright E, Coleman P, Urban ML (2007) LandScan USA: a high-resolution geospatial and temporal modelling approach for population distribution and dynamics. GEOJ 69:103–117

Biljecki F, Chow Y, Lee K (2023) Quality of crowdsourced geospatial building information: a global assessment of open street Map attributes. BUILD ENVIRON 237:110295

Bright E, Coleman P, King A and Rose A (2008) LandScan Global 2007. Oak Ridge National Laboratory. https://doi.org/10.48690/1524203

Broughton V (2006) The need for a faceted classification as the basis of all methods of information retrieval. In: Aslib Proceedings, vol 58, no 1/2, pp 49–72

Brzev S, Scawthorn C, Charleson AW, Allen L, Greene M, Jaiswal K, and Silva V (2013) GEM building taxonomy version 2.0. GEM Technical Report 2013–02 v1.0.0

Central intelligence agency (2011) The World Factbook 2011. Central Intelligence Agency

Chen C, Mostafizi A, Wang H, Cox D, Chand C (2022) An integrative agent-based vertical evacuation risk assessment model for near-field tsunami hazards. RISK ANAL 42(12):2720–2734

Coccia G, Cereza P, Bussi G, Denaro S, Bazzurro P, Martina M, Fagà E, Avelar C, Ordaz M, Huerta B, Garay O, Raimbekova Z, Abdrakhmatov K, Mirzokhonova Ismailov V, Belikov V (2023) Large-scale flood risk assessment in data scarce areas: an application to Central Asia. Nat Hazard Earth Sys. https://doi.org/10.5194/nhess-2023-157

Cockings S, Harfoot A, Martin D, Hornby D (2011) Maintaining existing zoning systems using automated zone-design techniques: methods for creating the 2011 Census output geographies for England and Wales. Environ Plan A 43(10):2399–2418

Cremen G, Galasso C, McCloskey J, Barcena A, Creed M, Filippi ME, Gentile R, Jenkins LT, Kalaycioglu M, Mentese EY, Muthusamy M, Tarbali K, Sacic Groglic R (2023) A state-of-the-art decision-support environment for risk-sensitive and pro-poor urban planning and design in Tomorrow's cities. Int J Disast Risk Re. 85:103400

Crowley H, Despotaki V, Rodrigues D, Silva V, Toma-Danila D, Riga E, Karatzetzou A, Fotopoulou S, Zugic Z, Sousa L, Ozcebe S, Gamba P (2020). Exposure model for European Risk Assessment. Earthq Spectra. 36(1_suppl):252–273

Daniel CJ, Frid L, Sleeter BM, Fortin M-J (2016) State-and-transition simulation models: a framework for forecasting landscape change. Methods Ecol Evol 7(1413–1423):2016. https://doi.org/10.1111/2041-210X.12597

De Bono A (2013) Global Exposure Database for GAR 2013. Background paper prepared for the Global Assessment Report on Disaster Risk Reduction 2013
De Bono A, Chatenoux B (2014) A global exposure model for GAR 2015. Input paper prepared for the global assessment Report on Disaster Risk Reduction 2015
Deville P, Linard C, Martin S, Gilbert M, Stevens FR, Gaughan AE, Blondel VD, Tatem AJ (2014) Dynamic population mapping using mobile phone data. Proc Natl Acad Sci 111(45):5888–15893
Federal Emergency Management Agency - FEMA (2003) Multi-hazard loss estimation Methodology
Fraser SA, Wood NJ, Johnston DM, Leonard GS, Greening PD, Rossetto T (2014) Variable population exposure and distributed travel speeds in least-cost tsunami evacuation modelling. Nat Hazard 14(11):2975–2991
Freire S, Aubrecht C, Wegscheider S (2013) Advancing tsunami risk assessment by improving spatio-temporal population exposure and evacuation modelling. Nat Hazards 68:1311–1324
Gentile R, Cremen G, Galasso C, Jenkins LT, Manandhar V, Menteşe EY, Guragain M, McCloskey J (2022) Scoring, selecting, and developing physical impact models for multi-hazard risk assessment. Int J Disaster Risk Reduct. 82:103365
Goda K, De Risi R (2018) Multi-hazard loss estimation for shaking and tsunami using stochastic rupture sources. Int J Disaster Risk Reduct. 28:539–554
Gómez-Zapata JC, Brinckmann N, Harig S, Zafrir R, Pittore M, Cotton F, Babeyko A (2021) Variable-resolution building exposure modelling for earthquake and tsunami scenario-based risk assessment: an application case in Lima, Peru. Nat Hazard Earth Syst Sci. 21:3599–3628
Gómez-Zapata JC, Pittore M, Cotton F, Lilienkamp H, Shinde S, Aguirre P, Santa María H (2022) Epistemic uncertainty of probabilistic building exposure compositions in scenario-based earthquake loss models. Bull Earthq Eng 20:3401–2438
Greene M, Comartin C, Murty CVR and Brzev S (2008) The world housing encyclopedia: strategies for global information sharing and problem solving. In: Proceedings of the 14th World conference on earthquake engineering. Beijing, China
Grunthal G, Musson RMW, Schwarz J, Stucchi M (1998) European Macroseismic Scale 1998 (EMS-98). European Seismological Commission
Gunasekera R, Ishizawa O, Aubrecht C, Blakespoor B, Murray S, Pomonis A, Daniell J (2015) Developing an adaptive global exposure model to support the generation of country disaster risk profiles. Earth Sci Rev 150:594–608
Hawker L, Bates P and Neal J (2022) Future Pop-global gridded population projections at 90m resolution (No. EGU22–9606). Copernicus Meetings
Hoffmann EJ, Wang Y, Werner M, Kang J, Zhu XX (2019) Model fusion for building type classification from aerial and street view images. Remote Sensing 11(11):1259
Jaiswal KS, Wald DJ, Porter K (2010) A global building inventory for earthquake loss assessment and risk management. Earthq Spectra 26(3):731–748
Jin Y, Liu R, Fan H, Li P, Liu Y, Jia Y (2023) Multi-resolution population mapping based on a stepwise downscaling approach using multisource data. Remote Sensing 15(7):1947
Kang J, Körner M, Wang Y, Taubenböck H, Zhu XX (2018) Building instance classification using street view images. ISPRS J Photogramm Remote Sens 145:44–59
Kim K, Kaviari F, Pant P and Yamashita E (2022) An agent-based model of short-notice tsunami evacuation in Waikiki, Hawaii. Transportation research part D: transport and environment, 105:103239
Kobayashi T, Medina RM, Cova TJ (2011) Visualizing diurnal population change in urban areas for emergency management. Prof Geogr 63(1):113–130
Løvholt F, Glimsdal S, Harbitz CB, Horspool N, Smebye H, de Bono A, Nadim F (2014) Global tsunami hazard and exposure due to large co-seismic slip. Int J Disaster Risk Reduct. 10B:406–418
Løvholt F, Griffin J and Salgado-Gálvez, MA (2015). Tsunami Hazard and Risk Assessment on the Global Scale. In: Meyers, R. (eds) Encyclopedia of Complexity and Systems Science. Springer, Berlin, Heidelberg. https://doi.org/10.1007/978-3-642-27737-5_642-1

Lu X, Wetter E, Bharti N, Tatem AJ, Bengtsson L (2013) Approaching the limit of predictability in human mobility. Sci Rep 3(1):2923

Mas E, Adriano B, Koshimura S (2013) An integrated simulation of tsunami hazard and human evacuation in La Punta. Peru. Journal of Disaster Research 8(2):285–295

Mas E, Koshimura S, Imamura F, Suppasri A, Muhari A, Adriano B (2015) Recent advances in agent-based tsunami evacuation simulations: case studies in Indonesia, Thailand, Japan and Peru. Pure Appl Geophys 172(12):3409–3424

Muhammad A, De Risi R, De Luca F, Mori N, Yasuda T, Goda K (2021) Are current tsunami evacuation approaches safe enough? Stoch Env Res Risk Assess 35(4):759–779

Mori N, Satake K, Cox D, Goda K, Catalan PA, Ho TC, Imamura F, Tomiczek T, Lynett P, Miyashita T and Muhari A (2022) Giant tsunami monitoring, early warning and hazard assessment. Nature Reviews Earth & Environment, pp 1–16

Mostafizi A, Wang H, Dong S (2019) Understanding the multimodal evacuation behavior for a near-field tsunami. Transp Res Rec 2673(11):480–492

Murnane RJ, Allegri G, Bushi A, Dabbeek J, de Moel H, Duncan M, Fraser S, Galasso C, Giovando C, Henshaw P, Horsburgh K (2019) Data schemas for multiple hazards, exposure and vulnerability. Disaster Prev Manag: Int J 28(6):752–763

Nirandjan S, Koks EE, Ward PJ, Aerts JCJH (2022) A spatially-explicit harmonized global dataset of critical infrastructure. Scientfic Data. 9:150

OASIS Loss Modelling Framework (2022) Open Exposure Data (OED) Version 2.1.0. https://github.com/OasisLMF/ODS_OpenExposureData/blob/develop/OpenExposureData/Docs/Open_Exposure_Data_v2.1.0.pdf Accessed on 7 Aug 2023

Ordaz M, Salgado-Gálvez MA, Huerta B, Rodríguez JC, Avelar C (2019) Considering the impacts of simultaneous perils. Disaster Prev Manag 28(6):823–837

Paulik R, Craig H, Popovich B (2020) A national-scale assessment of population and built-environment exposure in tsunami evacuation zones. Geosciences 10(8):291

Paulik R, Horspool N, Griffiths WR, N, Beale T, Magill C, Wild A, Popovich B, Walbran G and Garlock R, (2022) Risk scape: a flexible multi-hazard risk modelling engine. Nat Hazards 119:1073–1090

Pittore M, Haas M, Silva V (2020) Variable resolution probabilistic modelling of residential exposure and vulnerability for risk applications. Earthq Spectra 36(1):321–344

Pittore M, Wieland M, Fleming K (2017) Perspectives on global dynamic exposure modelling for geo-risk assessment. Nat Hazards 86(S1):7–30

Pittore M, Haas M, Megalooikonomou KG (2018) Risk-oriented, bottom-up modelling of building portfolios with faceted taxonomies. Frontiers in Built Environment 4:41

Ruktanonchai NW, Ruktanonchai CW, Floyd JR, Tatem AJ (2018) Using google location history data to quantify fine-scale human mobility. Int J Health Geogr 17:28

Salgado-Gálvez MA, Momin S, Grajales S, Carreño ML (2024) Comparison of earthquake risk for masonry and reinforced concrete buildings in Barcelona using existing sets of vulnerability functions. Bull Earthq Eng 22:639–664

Salgado-Gálvez MA, Ordaz M, Huerta B, Garay O, Avelar C, Fagà E, Kohrangi M, Cereza P, Fasoulakis Z, Begaliev UT (2023) Development of a regionally consistent and fully probabilistic earthquake risk model for Central Asia. Nat Hazard. https://doi.org/10.5194/nhess-2023-137

Salgado-Gálvez MA, Zuloaga D, Henao S, Bernal GA, Cardona OD (2018) Probabilistic assessment of annual repair rates in pipelines and of direct economic losses in water and sewage networks. Application to Manizales, Colombia. Natural Hazards. 93(1):S5-S24

Salgado-Gálvez MA, Barbat AH, Cardona OD, Carreño ML (2016) Comparing observed damages and losses with modeled ones using a probabilistic approach. The Lorca 2011 case. Int J Disaster Risk Reduct. 19:355–365

Satake K et al (2006) Tsunami heights and damage along the Myanmar coast from the December 2004 Sumatra-Andaman earthquake Earth Planets Space. 58(2):243–252. https://doi.org/10.1186/BF03353384

Scaini C, Peresan A, Tamaro A, Poggi V, Barnaba C (2022) Can high-school students contribute to seismic risk mitigation? Lessons learned from the development of a crowd-sourced exposure database. Int J Disaster Risk Reduct. 69:102755

Scaini C, Tamaro A, Adilkhan B, Sarzhanov S, Ismailov V, Umaraliev R, Safarov M, Belikov V, Karayev J, Fagà E (2023a) A new regionally consistent exposure database for Central Asia: population and residential buildings. Natural Hazards and Earth System Sciences (preprint under Review). https://doi.org/10.5194/nhess-2023-94

Scaini C, Tamaro A, Adilkhan B, Sarzhanov S, Ismailov V, Umaraliev R, Safarov M, Belikov V, Karayev J, Fagà E (2023b) A regional scale approach to assess non-residential buildings, transportation and croplands exposure in central Asia. Natural Hazards and Earth System Sciences (preprint under Review). https://doi.org/10.5194/nhess-2023-95

Schiavina M, Freire S, Rosina K, Ziemba L, Herrera M, Craglia M, Lavalle C, Kemper T, Batista e Silva F (2020) ENACT 2011 Population grids

Schiavina M, Melchiorri M, Pesaresi M, Politis P, Freire S, Maffenini L, Florio P, Ehrlich D, Goch K, Tommasi P, Kemper T (2022) GHSL Data Package 2022, Publications Office of the European Union, Luxembourg. ISBN 978-92-76-53071-8, https://doi.org/10.2760/19817, JRC 129516

Silva V, Yepes-Estrada C, Dabbeek J, Martins L, Brzev S (2018) Global exposure database for multi-hazard risk analysis. D2 – Multi-hazard Exposure Taxonomy

Silva V, Brzev S, Scawthorn C, Yepes C, Dabbeek J, Crowley H (2022) A building classification system for multi-hazard risk assessment. Int J Disaster Risk Reduct 13(2):161–177

Sleeter BM, Wood NJ, Soulard CE, Wilson TS (2017) Projecting community changes in hazard exposure to support long-term risk reduction: a case study of tsunami hazards in the US Pacific Northwest. Int J Disaster Risk Reduct. 22:10–22

Sutton P (1997) Modelling population density with night-time satellite imagery and GIS. Comput Environ Urban Syst 21(3–4):227–244

Thomson DR, Gaughan AE, Stevens FR, Yetman G, Elias P, Chen R (2021) Evaluating the accuracy of gridded population estimates in slums: a case study in Nigeria and Kenya. Urban Science 5(2):48

United Nations Office for Disaster Risk Reduction – UNDRR (2017). Sendai Framework Terminology on Disaster Risk Reduction. Accessed on Jun 2 2023. https://www.undrr.org/terminology

United Nations Department of Economic and Social Affairs, Population Division (2022). World Population Prospects 2022: Summary of Results. UN DESA/POP/2022/TR/NO. 3

Wang Z, Jia G (2021) A novel agent-based model for tsunami evacuation simulation and risk assessment. Nat Hazards 105(2):2045–2071

Wesolowski A, Qureshi T, Boni MF, Sundsøy PR, Johansson MA, Rasheed SB, Engø-Monsen K, Buckee CO (2015) Impact of human mobility on the emergence of dengue epidemics in Pakistan. Proc Natl Acad Sci 112(38):11887–11892

WHE (2023) Online repository: www.world-housing.net

Wyss M, Tolis S, Rosset P, Pacchiani F (2013) Approximate model for worldwide building stock in three size categories of settlements. Background paper prepared for the 2013 Global Assessment Report on Disaster Risk Reduction

Yepes-Estrada C, Silva V, Valcárcel J, Acevedo AB, Tarque N, Hube MA, Coronel G, Santa María H (2017) Modeling the residential building inventory in South America for seismic risk assessment. Earthq Spectra 33(1):299–322

Chapter 7
Vulnerability Modelling

Marco Baiguera, Ignacio Aguirre Ayerbe, Raffaele De Risi, Marta Del Zoppo, Ufuk Hancilar, Rozana Himaz, Maria Papathoma-Köhle, Tiziana Rossetto, and Mislav Stepinac

7.1 Introduction

Vulnerability, understood as the conditions determined by physical, social, economic and environmental factors or processes that increase the potential for an individual, a community, assets or systems to be negatively affected by hazards (adapted from UNDRR, 2017), is a key component when assessing tsunami risk. Vulnerability can be affected by human interventions. Hence, the study of vulnerability is essential to

M. Baiguera (✉)
Risk Team, Global Earthquake Model (GEM) Foundation, Pavia, Italy
e-mail: marco.baiguera@globalquakemodel.org

I. Aguirre Ayerbe
Instituto de Hidráulica Ambiental (IH Cantabria), Universidad de Cantabria, Santander, Spain
e-mail: aguirrei@unican.es

R. De Risi
School of Civil, Aerospace and Design Engineering, University of Bristol, Bristol, UK
e-mail: raffaele.derisi@bristol.ac.uk

M. Del Zoppo
Università di Napoli Federico II, Naples, Italy
e-mail: marta.delzoppo@unina.it

U. Hancilar
Department of Earthquake Engineering, Kandilli Observatory and Earthquake Res. Inst., Boğaziçi University, İstanbul, Türkiye
e-mail: hancilar@bogazici.edu.tr

R. Himaz
Department of Risk and Disaster Reduction, University College London, London, UK
e-mail: r.himaz@ucl.ac.uk

M. Papathoma-Köhle
Institute for Mountain Risk Engineering (IAN), BOKU University, Vienna, Austria
e-mail: maria.papathoma-koehle@boku.ac.at

M. B. Sørensen et al. (eds.), *Probabilistic Tsunami Hazard and Risk Analysis*, Mathematics of Planet Earth 14, https://doi.org/10.1007/978-3-031-98115-9_7

support decision-making (e.g., land use planning and emergency planning) for more effective risk management.

A vulnerability assessment in a probabilistic analysis typically comprises the evaluation of potential losses that the natural process may cause to exposed assets (physical vulnerability) and people and communities (social vulnerability) in terms of monetary losses and casualties, respectively. Vulnerability is most commonly represented through a vulnerability function that expresses the likelihood of achieving a value of a loss parameter for increasing values of a hazard intensity measure. Vulnerability functions can be derived "directly" by fitting a statistical model to post-event loss data (empirical) or through the elicitation of expert opinion (heuristic). Vulnerability functions can be derived "indirectly" from the combination of fragility functions and a consequence model, also called damage-to-loss model (Rossetto and Ioannou 2018).

A fragility function expresses the likelihood of damage that will be sustained by a physical asset under increasing tsunami intensity values, and sets of fragility functions are commonly produced, with each individual fragility function representing the likelihood of exceedance of a specific damage state (DS), such as "collapse" or "heavy damage", conditioned on the expected intensity measure (IM), as given in Eq. (7.1) and illustrated in Fig. 7.1a:

$$\mathrm{P}(DS \geq ds_{\mathrm{i}}|IM) \tag{7.1}$$

To make the fragility analysis representative of the different structural responses to tsunami loading, buildings are typically classified according to structural properties and the fragility analysis is carried out for each class separately (Charvet et al. 2017). For example, recent research works have developed fragility curves for tsunami that consider structural type, material of the structure, construction year, and height, among other parameters (Nanayakkara and Dias, 2016; Macabuag et al. 2016a; Suppasri et al. 2014; Petrone et al. 2017; Valencia et al. 2011).

Vulnerability functions are defined as relationships providing the expected level of loss for a given IM, as expressed by Eq. (7.2):

$$P(L > l \mid IM) \tag{7.2}$$

T. Rossetto
Faculty of Civil Engineering and Geosciences (CEG), Delft University of Technology (TU Delft), Delft, The Netherlands
e-mail: T.Rossetto-1@tudelft.nl

M. Stepinac
Faculty of Civil Engineering, University of Zagreb, Zagreb, Croatia
e-mail: mstepinac@grad.hr

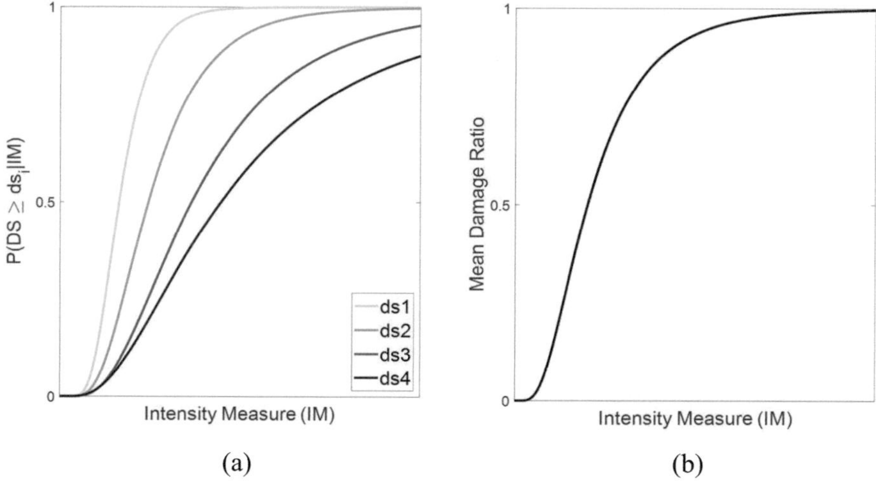

Fig. 7.1 Typical representation of **a** fragility functions and **b** vulnerability functions

where L is the loss parameter for a given asset (e.g., in terms of repair costs). Many vulnerability functions are given in terms of mean damage ratio (MDR) as a function of the IM - see Fig. 7.1b. MDR is defined as the expected value of the ratio of the cost of repair divided by the cost of replacement of the asset (repair-to-replacement cost ratio). Vulnerability functions can include the percentage of loss directly, where loss amount is used as the measure of impact when building the curve, e.g., when using insurance claims data.

Fragility functions can be transformed into "indirect" vulnerability functions using consequence models - see Fig. 7.2. For example, a typical tsunami consequence model used in the physical vulnerability assessment of buildings comprises a set of damage ratios (or loss ratios) for given damage states and takes values between 0.0 and 1.0 for no damage to collapse, respectively. In the case of human vulnerability, casualty rates are instead the most common form of consequence function (Rossetto et al. 2014). These express the proportion of occupants in a building that are likely to be injured/die conditioned on the building damage state. In both cases of consequence model, the damage ratios or casualty rates may have deterministic values for each damage state or present a mean value with standard deviation. Indirect vulnerability evaluation through use of fragility functions curves and consequence models is based on the fundamental assumption that the main cause of loss is due to asset damage (Rossetto et al. 2014). This assumption derives from earthquake risk assessment, and although it is realistic for the case of tsunami physical vulnerability evaluation, it does not hold true for tsunami human vulnerability evaluation, as the tsunami inundation can directly cause human fatalities.

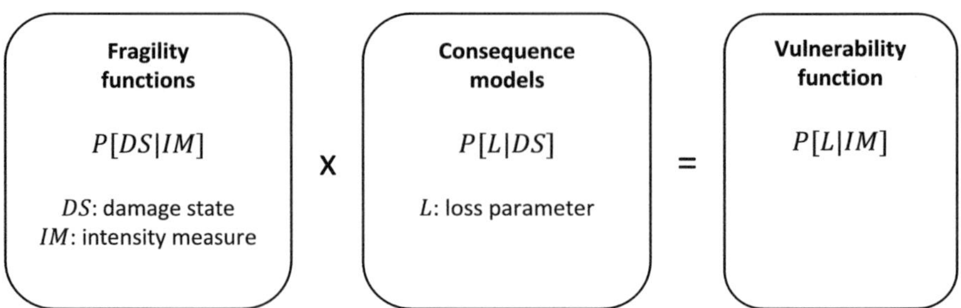

Fig. 7.2 Pseudo-formula for an "indirect" vulnerability function

Compared to earthquake vulnerability modelling, there are relatively few fragility functions for buildings and infrastructure under tsunami loading, and most of these are empirical in nature. Empirical fragility functions are developed by fitting a statistical model to observational data from real events. Charvet et al. (2017) present a review of available empirical fragility models and shows that these derive from observations made after a small number of recent tsunami events, given the rarity of this hazard. Empirical tsunami fragility functions are specific to the event and the location affected by the tsunami. Damage scales (e.g. MLIT, 2014, EEFIT, 2006, Fraser et al. 2013) are used to record levels of damage. These should then be used in the definition of each DS curve in the fragility function. It is also seen that most empirical fragility functions adopt inundation depth as an IM (e.g., Reese et al. 2011; Suppasri et al. 2011; Charvet et al. 2014a; De Risi et al. 2017a), since the inundation depth is measurable in the aftermath of a tsunami and can also be simulated numerically with relatively small errors. The values of the selected IMs are recorded through visual inspections after the event. For instance, if the inundation depth is the IM to use, then what is collected are the values of that measure (e.g., 1.5 m). However, it is often difficult to accurately determine the exact values of the IM. In the case of earthquake triggered tsunami, damage observations for buildings may reflect the cumulative action of both earthquake ground shaking and tsunami inundation, increasing the uncertainty.

Recently, advances in tsunami structural simulation approaches have led to the development of analytical fragility functions (e.g., De Risi et al. 2017b; Rossetto and Ioannou 2018). A review of existing analytical methods is provided by Rossetto et al. (2019). New analytical methods are based on the application of the tsunami load either in a dynamic regime according to tsunami inundation simulations or in a static load regime that mimics the effects of tsunami loads. While empirical fragility curves are typically obtained at community-scale for many buildings and assets, analytical fragility curves have been mostly developed at single-building level (e.g., Petrone et al. 2017; Alam et al. 2018; Karafagka et al. 2018). Only a few analytical studies have been developed at community level (e.g., Park et al. 2013; Aránguiz et al. 2018).

Approaches used for many years to understand physical vulnerability and model physical damages are now being applied to social vulnerability. When an approach is required encompassing several criteria, dimensions (e.g., social, physical, economic and environmental), and variables with different magnitudes and ranges of values, it may be necessary to apply an indicator-based approach (Aguirre-Ayerbe et al. 2018). There are some methods and operational tools available to facilitate the implementation of these approaches through the use of quantitative and qualitative indicators, e.g., the conceptual frameworks proposed by Aguirre-Ayerbe et al. (2018) and Strunz et al. (2011). The Papathoma Tsunami Vulnerability Assessment (PTVA) model (Papathoma et al. 2019) is one of the most widely used index-based methods for the assessment of physical vulnerability of buildings to tsunami. Index-based methods can feed into policies such as the Sendai Framework for Disaster Risk Reduction (SFDRR) 2015–2030 (UNISDR, 2015). There are emerging studies that aim to improve the way vulnerability is evaluated that are not necessarily constrained to physical vulnerability (Salmanidou et al. 2021; Himaz, 2022). Issues such as displacements and migration, pertaining to social conflicts, poverty, marginalisation and political vulnerabilities are areas that can significantly affect human and social vulnerability but tend to be avoided due to the challenging methods to quantify (Table 7.1).

7.2 Empirical Physical Vulnerability

7.2.1 Short Description

An empirical tsunami vulnerability function consists of a set of statistical models correlating some Explanatory Variables (EV) to some Response Variables (RV) based on observed tsunami events. A very comprehensive literature review of empirical models is provided by Charvet et al. (2017).

EVs are IMs and built environment characteristics. Examples of IMs are the inundation depth (h), tsunami velocity (v), moment flux (hv^2), moment of momentum flux (MMF), drag force (F), the quasi-steady tsunami force (FQS), and the debris impact. EVs describing the built environment include the building construction materials and typologies (e.g., Timber, Masonry, Reinforced Concrete, Moment resisting Frame, Shear Wall) and the coast type (e.g., Plain and Ria type).

For direct vulnerability models the RV is a measure of loss, such as loss ratios (cost of repair to replacement) for physical assets, or casualty rates. However, few direct empirical tsunami vulnerability functions exist. It is noted that the number of casualties (fatalities and injuries) depends on many variables, both physical and socio-economic ones, and is often expressed as a function of occupancy or of population (Marchand et al. 2009). Examples of empirical casualty models can be found in Berryman et al. (2005) and Reese et al. (2007). Most vulnerability functions are derived indirectly from fragility functions. Given this, the following text relates mainly to tsunami fragility functions.

Table 7.1 Summary table for tsunami vulnerability

What does the ingredient describe?	The assessment of the impact that tsunami may cause to exposed assets (physical vulnerability) and people and communities (social vulnerability)
Definitions of the ingredient	Tsunami vulnerability is understood as the conditions determined by physical, social, economic and environmental factors or processes that increase the potential for an individual, a community, assets or systems to be affected by tsunami
What is/are the metric(s) used for measuring/evaluating the ingredient?	Tsunami vulnerability is commonly represented through a vulnerability function that expresses the likelihood of achieving a value of a loss parameter for increasing values of a hazard intensity measure. Many vulnerability functions are given in terms of mean damage ratio (MDR) as a function of the intensity measure (IM)
Possible uses/applications of the ingredient	The study of tsunami vulnerability is a key part in a tsunami risk assessment, which is essential to support decision-making for effective tsunami risk management
Possible multi-hazard dimensions/applications?	Tsunami vulnerability models can be used to develop combined models that assess the impact of earthquake ground shaking and tsunami hazards simultaneously
Possible time-dependent/dynamic aspects	Tsunami vulnerability models can include accounting for the preceding earthquake damage due to near-field earthquake sources
List of the available methods to be described below	Empirical methods: a set of statistical models correlating some Explanatory Variables (EV) to some Response Variables (RV) based on observed tsunami events. Analytical methods: physical fragility functions are developed using a damage scale and transformed into vulnerability functions using consequence models. Indicator-based methods: vulnerability is evaluated qualitatively through indicators, which are operational representation of building exposure characteristics or the characteristics of the surroundings.

Fragility functions take a similar form to vulnerability functions but the RV used most typically is the DS, representing a damage condition experienced by the exposed physical asset to the tsunami. DS may vary from no damage to washed away. The descriptions of DS are not necessarily consistent among the reference publications, and the number of DS may vary from just 2 to more than 5. For instance, the Japanese Ministry of Land, Infrastructure, and Transportation (MLIT, 2014) adopts seven damage states, namely no damage (DS1), minor damage (DS2), moderate damage (DS3), major damage (DS4), complete damage (DS5), collapse (DS6), and wash-away (DS7). All the available empirical methods are based on tsunami damage databases that do not account for potential initial damage induced by the ground shaking caused by the triggering earthquake. However, it is evident that an initial earthquake effect/damage level is often present for structures damaged by a subsequent tsunami, especially in the case of near-field events. Small levels of ground shaking can cause damage in the physical assets that is exacerbated by the subsequent tsunami, and high levels of ground shaking could directly collapse assets before the tsunami hits. Both types of induced seismic damage can lead to potential overestimations and more in general bias on the final tsunami fragility functions. It is important to stress that the information on non-damaged assets (i.e., DS1) is as important as the information on damaged assets (i.e., DS > 1); therefore, it cannot be neglected.

Fragility functions can be mono-dimensional (i.e., a simple curve generally resembling a Cumulative Distribution Function) or multidimensional (i.e., a surface function of more than one IM). Examples of both typologies of fragility function are shown in Fig. 7.3. More advanced models (i.e., the ones using multiple parameters or machine-learning-based ones) are more challenging to represent, and in general, they are provided with coded functions ready to use (e.g., De Risi et al. 2017c).

Statistical methods used for the development of empirical fragility models are the Ordinary Least Squares Method (OLS), (e.g., Peiris and Pomonis, 2005; Dias et al. 2009; Gokon et al. 2009; Koshimura et al. 2009; Suppasri et al. 2009, 2011, 2012a, 2012b, 2014; Koshimura and Kayaba, 2010; Murao and Nakazato 2010; Valencia et al. 2011; Amakuni and Terazono, 2011; Koshimura and Gokon, 2012; Mas et al. 2012; Nihei et al. 2012; Hayashi et al. 2013; Maruyama et al. 2013; Narita and Koshimura, 2015; Tanaka and Kondo, 2015) and the Generalised Linear Model (GLM), (e.g. Reese et al. 2011; Charvet et al. 2014a, 2014b; Macabuag et al. 2016a,

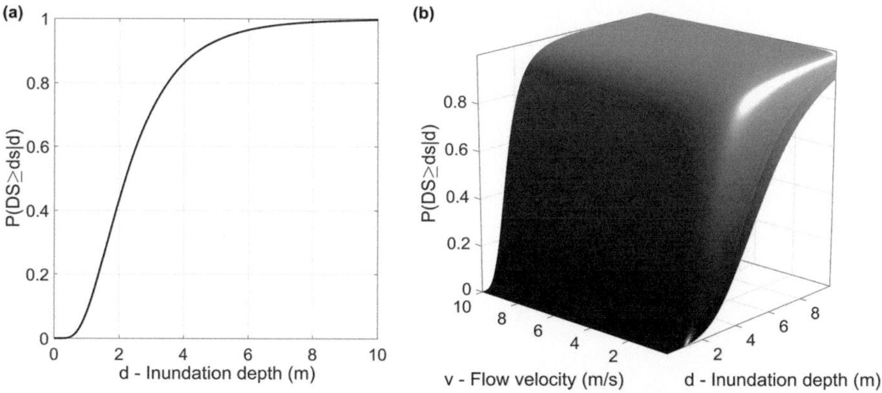

Fig. 7.3 Fragility of wooden structures: **a** monodimensional fragility function for collapse according to Suppasri:2013; and **b** multidimensional fragility for collapse/washed away (data from De Risi et al. (2017b))

2016b, 2018; Song et al. 2017; De Risi et al. 2017a, 2017b; Lahcene et al. 2021). More recently, Machine Learning (ML) algorithms have been used to develop such models (Mas et al. 2020; Rasheed et al. 2022).

OLS methods are the most traditional and used based on the probability-paper approach, where data for each damage class are binned for specific IM values and then fitted with a Normal or Log-Normal functional shape. These models are conventionally used to derive mono-dimensional models. Moreover, such models can be used when only a few data are available.

GLMs have recently become more popular because they allow for easier implementation of multivariate models and hence the use of multiple EVs at once. Specifically, GLM connects the damage probability with a linear predictor using simple formulations (e.g., probit, logit). Such a predictor is a simple linear combination of the explanatory variables since the coefficients of the linear combination are the object of the regression. However, GLMs need large databases to work correctly.

ML algorithms are used both for the automatic remote classification of damage and for the fitting of data. The simplest approach for the development of fragility models is the Artificial Neural Network (ANN), where the input layer consists of the EVs and the output layer provides the damage classification. The input layer and the damage classification are linked via a number of internal layers and nodes that may be the object of optimisation. Since such models do not need computational power as they are statistical tools, no cluster analyses or parallelisations are required. However, ML approaches may require more computational time with respect to OLS and GLMs as some model training is needed; however, this additional time is not comparable to the time needed, for example, for tsunami propagation and inundation analyses.

The abundance of potential methods and models requires the need for robust approaches for model diagnosis and selection. These may consist either in performing statistical tests (e.g., likelihood ratio test) or in looking at indicators, such as the Bayesian Information Criterion -BIC- (Schwarz, 1978), the Akaike Information Criterion -AIC- (Akaike, 1974), and the residual deviance. In general, the models presenting the smallest diagnostic indicator are the preferred ones. More sophisticated approaches also exist in the literature (Jalayer et al. 2022) consisting of simulation-based Bayesian method for parameter estimation and fragility model selection for mutually exclusive and collectively exhaustive (MECE) damage states making use of an adaptive Markov chain Monte Carlo simulation (MCMC) and based on likelihood estimation using point-wise intensity values.

7.2.2 Data and Technical Requirements

To develop the fragility models, databases of damage to the built environment collected after tsunami events are needed. Such databases must contain EVs and RVs. Most of the collected data and the developed models are event and country specific.

Among the EV, there is information about the hazard (i.e., the IMs), the affected asset (e.g., the building typology), and the local geomorphological conditions (i.e., Ria-type or Plain coast). Regarding the tsunami IM, a first attempt consists in retrieving them from the observation. This is generally feasible for the tsunami depth, even if some uncertainties exist due to judgement and spatial sampling (De Risi et al. 2017a). Simulation techniques are used when it is impossible to have a proper empirical estimation. For example, it is complicated to assess the tsunami flow velocity from video recordings of the events or sensors installed in buoys. It is also very complex to have such an estimation at urban or regional scales. Therefore, simulation values are often used as the IM in combination with other empirical values.

For what concerns the building typologies, a combination of governmental databases of the built environment and volunteered geographic information are typically used to characterise the affected assets. Such data are critical when buildings and constructions were entirely washed away (i.e., their posterior characterisation is impossible during the damage survey). Similarly, geomorphological features can quickly be investigated using observation of the topography.

Both building typology and topographic features can be used in two different ways. First, they may be used as a discriminant to define subsets of data to develop specific models. This first case is conventionally applied in OLS regression techniques. And second, they can become dummy variables; therefore, a single model is produced where all the data are used together as input. This is generally the case when GLMs are employed.

Regarding RVs, the damage states should have specific characteristics. First, they should be mutually exclusive and clearly distinguishable (e.g., no collapse, washed away). DS should be easily applied to populations of buildings. Finally, a division between structural and non-structural damage should be considered. As for some of the EVs, DSs can be used in two ways: i) creation of a subset of data, or ii) implementation in the model as dummy variables. As before, the first case is adopted for simple regression strategy. The second one is used for more sophisticated GLMs.

7.2.3 Treatment of Uncertainties

All the above-mentioned statistical regression methods provide a probabilistic interpretation of the observed data. The uncertainties associated with data are mainly epistemic and are associated with both the EVs and the RVs.

Regarding the IMs, uncertainties are associated with the data acquisition methods and/or with the computational models used to simulate missing data. For example, the tsunami inundation depth may be subjected to uncertain assessment associated with the survey method (e.g., wet marks on the walls or on the nearby still-standing infrastructures). Similarly, the tsunami velocity assessment can be affected by the data acquisition method (e.g., video interpretation) or by the simulation approach and grid resolution. Such uncertainties can be assessed and quantified either using multiple damage databases or using multiple tsunami inundation numerical models. De Risi et al. (2017a) provides an example of quantification and propagation of uncertainties on IMs. Regarding the DSs, an innovative method to propagate their uncertainties has been proposed by Jalayer et al. (2022). Both approaches above are based on Bayesian statistics; in fact, Bayesian methods constitute the ideal set of tools for such a task.

Conventional approaches to convey uncertainties consist of computing the variability of the model parameters (e.g., variance of the median and the logarithmic standard deviation). Such quantification of the uncertainty will allow to provide fragility curves and pre-defined confidence intervals (Table 7.2).

7.3 Analytical Physical Vulnerability

7.3.1 Short Description

All analytical physical vulnerability models are indirect and based on an evaluation of physical asset fragility. For the analytical assessment of fragility, a damage scale needs to be defined and mechanics-based damage states (DS) should be clearly described. Fragility functions can then be transformed into a vulnerability function using consequence models, which are derived based on reconstruction costs data from past tsunami events (if any). As an alternative, numerical simulations of reconstruction costs can be adopted. Up to now, there is a lack of available consequence functions for tsunami events, hence the following text focuses on tsunami fragility analyses.

Like the methods for deriving fragility curves in case of earthquake actions on structures, static or dynamic, linear or nonlinear structural analyses are performed for the assessment of structural response under tsunami loads and to define structural capacity and specific damage states (e.g., slight, moderate, extensive and complete). Once the structural responses and expected damage states under a range of tsunami intensity levels are analytically simulated, statistical processing similar to that used for empirical fragility function development is applied to this dataset to generate analytical fragility functions. The analytical simulation typically requires the development of structural models using finite element modelling (FEM). Examples of FEM packages that have been used for tsunami vulnerability simulations include OpenSees (McKenna et al. 2010) and SeismoStruct (SeismoStruct, 2022).

Tsunami forces are based on flow conditions and are mainly classified as hydrostatic forces under the quiescent flood states, buoyant forces resulting from vertical hydrostatic forces due to the volume of water displaced by a fully or partially submerged structure, hydrodynamic forces induced by steady flows, highly transient impulsive forces at the impact of the leading edge of the arriving water mass, debris impact forces caused by waterborne debris, and debris damming force due to the accumulation of debris across the structure (FEMAP-646, 2012).

Table 7.2 Summary table for empirical physical vulnerability

Short description of method	
Short outline of how the method works	The empirical vulnerability consists of a set of statistical models correlating some EVs to some RVs based on observed tsunami events. Explanatory Variables are both IMs and characteristics of the built environment. The typical RV is the DS, representing a damage condition experienced by the asset of interest after the tsunami. DS may vary from zero damage up to washed away (total damage)
How is/are the output(s) of the method(s) represented?	Empirical vulnerability models are usually presented as fragility curves, which are statistical relationships between the IMs and the probability of attaining a specific damage state conditioned on the IM
Do(es) the method(s) use parallelisation, machine learning, surrogate models, advanced simulation schemes, other computational reduction strategies (e.g., cluster analysis, etc.)?	ML algorithms have been used to develop such models. As such models do not need computational power as they are statistical tools, no cluster analysis or parallelisation is required. The machine learning approach may require more computational time as some model training is needed
What are the pros and cons and peculiarities of this specific method(s)? Are there known shortcomings / benefits of the method(s)? In what cases is this method(s) a good/poor choice?	Empirical tsunami vulnerability functions are, by definition, specific to the event and the location affected by the tsunami. Also, both IMs and building damage are typically recorded in visual inspections after the event. Hence, it is often difficult to accurately determine the actual IM. In addition, damage observations for buildings that are subject to both earthquake and tsunami may be due to both perils, increasing the uncertainty
Data and technical requirements for method	
What input data does the method(s) use?	Databases of damage to the built environment collected after tsunami events are needed. Such databases must contain EVs and RVs. Most of the collected data and the developed models are event-specific and country-specific
What are the technical requirements?	Typical statistical methods used for the development of empirical vulnerability models are the OLS, and the GLM
Treatment of uncertainties for the method	
Is the method(s) itself deterministic or uncertainty based?	The statistical regression methods provide a probabilistic interpretation of the data
Which are potential sources of uncertainties? Which parameters are used to characterize them?	The uncertainties associated with data are mainly epistemic and are associated with both the EVs and the RVs

(continued)

Table 7.2 (continued)

How are uncertainties quantified for the identified parameters? Are there significant correlations between different uncertain parameters?	Uncertainties in IMs are associated with the data acquisition methods and/or with the computational models used to simulate missing data. Novel approaches are based on Bayesian statistics; in fact, Bayesian methods constitute the ideal set of tools for such a task
Is the method(s) able to digest and propagate uncertainty in input (previous ingredient) and propagate it to the output (next ingredients)? Does the assessment of this ingredient require uncertainty propagation? If yes, how are uncertainties propagated?	Conventional approaches to convey uncertainties consists of computing the variability of the model parameters. Such quantification of the uncertainty will allow to provide fragility curves and pre-defined confidence intervals

Several experimental investigations have been conducted to evaluate the wave-structure interaction. Some of these focused on the generation of tsunami-like waves in hydraulic flumes (among others, Hamzah et al. 2001; Rossetto et al. 2011; Palermo et al. 2012; Chinnarasri et al. 2013; Park et al. 2013; Qi et al. 2014; Shafiei et al. 2016; Foster et al. 2017; Alam et al. 2020). The experimental measurements can provide valuable information on hydrodynamic flow conditions as well as lateral and vertical forces and pressures exerted on the structure.

Commonly used IMs for the development of tsunami analytical fragility curves are inundation depth, flow velocity, and specific momentum flux. Attary et al. (2017) proposed a methodology for the development of physics-based tsunami fragility curves using vector valued IMs, such as tsunami flow depth and flow velocity and several combinations thereof, relying on Monte Carlo Simulation to consider the material uncertainties and for epistemic uncertainties in the tsunami force calculation.

Different types of structural analysis can be selected for performing tsunami analytical fragility assessment of a facility, but these can be grouped into two main classes: (a) static analysis (pushover); and (b) dynamic time-history analysis. Nonlinear static analysis methods neglect the dynamic response of structures and apply tsunami loads in a static manner. Hydrodynamic forces are typically applied to each inundated story of the building along the seaward columns, assuming a constant tsunami inundation depth and increasing flow velocity, as illustrated in Fig. 7.4a. This approach is referred to as constant-depth pushover (CDPO). CDPO is similar to a seismic pushover analysis and can be implemented easily in commercial structural analysis software. Attary et al. (2017) and Alam et al. (2018) performed CDPO analyses in a fragility assessment of reinforced concrete structures, accounting for the local members' failure, e.g. columns.

Novel analysis methods developed by Petrone et al. (2017) are the tsunami time-history dynamic analysis (TDY) and the variable depth pushover (VDPO). TDY analyses are based on the same principle as a seismic time-history dynamic analysis, except for the input data. The latter typically consists of wave traces (i.e., tsunami inundation depth and flow velocity time-histories). These records can be generated

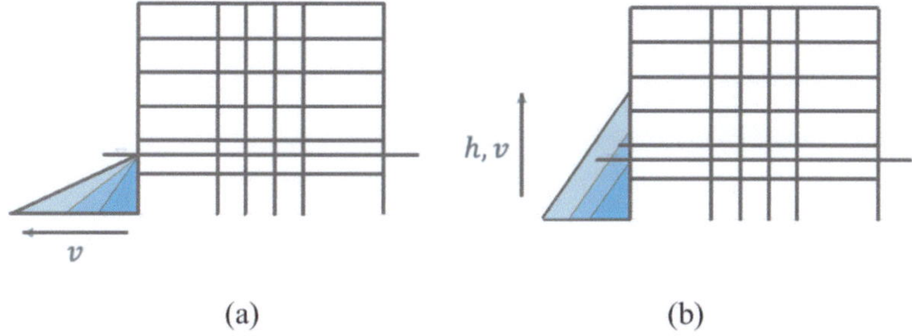

Fig. 7.4 Schematic illustration of recently proposed tsunami pushover analysis: **a** constant depth pushover; and **b** variable depth pushover

with numerical simulations (Goda et al. 2017). The VDPO analysis is a static analysis, where the tsunami force applied to the structure is increased incrementally by increasing the tsunami inundation depth monotonically, as illustrated in Fig. 7.4b. At each depth value, the corresponding flow velocity is calculated assuming a constant Froude number (*Fr*). Petrone et al. (2017) found that the VDPO analysis outperforms the CDPO analysis in terms of engineering demand parameter estimation, (i.e., inter-story drifts and column shear forces).

However, the VDPO as originally conceived in Petrone et al. (2017) is a load-control analysis, and therefore, it is not capable of capturing the degrading behaviour of the structure. This limitation is overcome by the so-called VDPO2 approach, developed by Baiguera et al. (2022). VDPO2 consists in a two-phase nonlinear static: in the first phase a load-control pushover analysis is carried out assuming that the inundation depth and flow velocity increase incrementally; in the second phase, the analysis switches to response-control pushover analysis, where the displacement is gradually increased, and the corresponding tsunami force is calculated. The switch from the first phase to the second one occurs either when a predefined load level is reached or when the analysis encounters a numerical convergence issue, whichever occurs first. The VDPO2 is referenced in ASCE7-22 (2022) as a recommended nonlinear static method for the tsunami design of buildings exposed to tsunami risk.

The VDPO method has been further enhanced for the analysis of reinforced concrete frames with breakaway infill walls. Del Zoppo et al. (2021) developed a bespoke method, referred to as VDPO-BI, in which the tsunami-induced lateral and vertical loads are monotonically applied to structural and non-structural components as a function of the tsunami inundation depth, up to the structural collapse (Del Zoppo et al. 2019). A mechanics-based method to develop fragility curves at community level has been proposed in Del Zoppo et al. (2022) with the application of the VDPO-BI analysis. In the case of masonry structures, Belliazzi et al. (2021) proposed methods to perform the damage assessment and develop fragility curves.

Most of the currently available analytical tsunami fragility curves have been developed at single-building level (e.g., Petrone et al. 2017; Alam et al. 2018). Although earthquakes are the primary trigger of tsunamis, few studies have investigated the impact of earthquake damage in deriving tsunami fragility curves (Park et al. 2012;

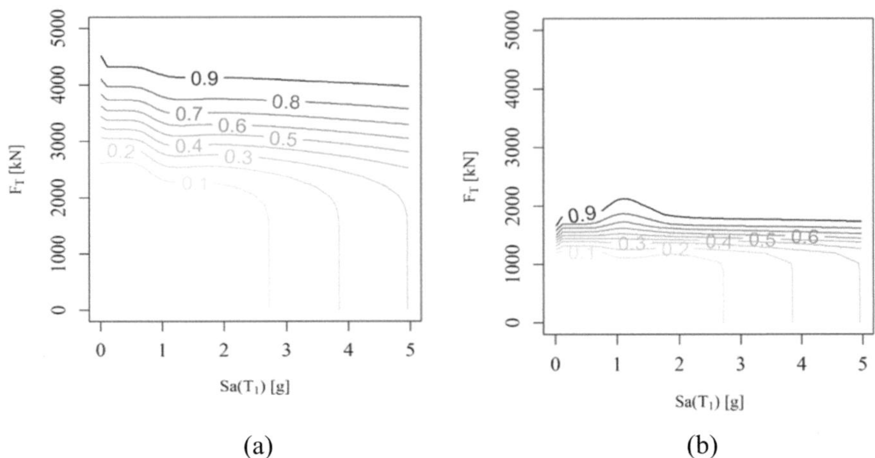

Fig. 7.5 Collapse fragility surface for a reinforced concrete building under sequential earthquake and tsunami: **a** global performance; and **b** local performance (Petrone et al. 2020)

Latcharote, 2015). Petrone et al. (2020) developed the combined earthquake-tsunami fragility curve for the first time for a reinforced concrete building, accounting for the global performance as well as the influence of local failure of structural components (see Fig. 7.5). A full nonlinear response history analysis of the building was used to simulate the earthquake and tsunami loading in sequence, since Rossetto et al. (2019) found that use of a static analysis for representing the earthquake damage, followed by a tsunami dynamic analysis, results in over-conservativism.

7.3.2 *Data and Technical Requirements*

To develop analytical fragility curves, input data related to the assets geometric and mechanical properties, and structural details must be known. For existing assets, structural details can also be derived from a simulated design process. Fragility curves are analytically derived for the assets under investigation performing a damage assessment through refined structural analyses in a probabilistic framework. Indeed, uncertainties related to load demand and structural capacity should be properly modelled.

The damage assessment should account for all the tsunami-induced loads and effects on structure and infrastructure. According to the ASCE7-22, (2022), a tsunami flow induces lateral and vertical (uplift) hydrostatic and hydrodynamic pressures on structure and infrastructure. Such flow can be considered as quasi-steady according to Qi et al. (2014), and the induced lateral pressure can be computed following the empirical equations proposed by Foster et al. (2017). Tsunami flows are unsteady; however, they can be considered quasi-steady when due to their extremely long wavelengths, the temporal variation of the flow is small, especially with respect to the lifetime of a building. There are no empirical models for induced uplift loads on structures, hence the equations provided by the ASCE 7–22 can be adopted to compute such load values. Besides the hydrostatic and hydrodynamic pressure, tsunami also induces impulsive loads, debris impact loads and scour. ASCE 7–22 provides models to compute such impulsive and impact loads on structure and infrastructure. About the scour, the American Standard provides information about the maximum scour depth induced by a tsunami, but research is ongoing to assess the evolution of the scour depth during the inundation (McGovern et al. 2019).

The most common IM adopted in both empirical and analytical tsunami vulnerability models is the inundation depth. Indeed, the inundation depth is easily measurable in the aftermath of a tsunami and can be simulated numerically with relatively small errors. However, recent analytical studies have observed that other parameters, such as the momentum flux, can be also adopted as IM given the good correlation with damages and losses (Park et al. 2017). The efficiency of vector-valued IMs consisting of scalar IMs such as inundation depth, flow velocity, specific momentum flux, kinematic moment of specific momentum flux, and their interactions were investigated by Alam et al. (2018) against the conventional scalar IMs. It was observed that the estimation efficiency of vector-valued IMs is higher than that of scalar IMs and the prediction of structural damage using vector-valued IMs can be further improved by considering the interaction among the IMs.

Different modelling approaches have been adopted for simulating physical assets and their response to tsunami loads. The asset modelling is of critical importance to the reliability of the resulting fragility function. In particular, the FEM model should be able to represent both global and local damage mechanisms under the horizontal and vertical loading imposed by tsunami inundations. Inevitably, this results in more sophisticated FEM modelling of buildings being required, and models based on single degree of freedom systems (e.g., Park et al. 2017) should be avoided. The complexity of the structural modelling needed is one of the reasons that analytical fragility functions have predominantly been derived for single buildings, or representative index buildings. Moreover, the modelling or assessment of structural element shear response is critical, as tsunami inundation loads impart high shear forces to structural elements (in particular, ground floor vertical elements in structures). It is noted that the local shear failure of vertical elements can lead to premature structural collapse, as shown in Petrone et al. (2017) and Petrone et al. (2020).

7.3.3 Treatment of Uncertainties

Uncertainties from demand parameters in tsunami loading (e.g., drag coefficient, among others) as well as in structural capacity modelling (e.g., material strength, structural details) should be accounted for and modelled during the fragility and vulnerability analysis. Capacity of structures, IM and tsunami-induced loads represent potential sources of uncertainties. Geometric and mechanical properties of materials are epistemic variables that should be considered with their probability distributions and statistics. Indeed, it is hard to achieve a full knowledge about the exact geometry, structural detailing and material properties of an existing building, and the level of knowledge is much lower when dealing with large scale fragility assessment. Epistemic uncertainties related to the modelling and structural analysis method should also be considered when assessing the structural capacity and damage. Only a few empirically validated models are available to simulate the tsunami loads and effects on structures (e.g., effect of openings, damming debris accumulation, debris impact, and scour), and available models can be affected by errors. Uncertainty in the tsunami load parameters (e.g., flow depth and velocity) need to be considered as well. Alam et al. (2018) observed that uncertainty in the IM is more sensitive to uncertainties in the tsunami demand parameters than those in the structural model idealisation. This implies that an improved understanding of the characteristic tsunami flow information, with a specific focus on flow velocities, will significantly reduce the uncertainty in the fragility curves.

Uncertainties related to the fragility of buildings can be quantified for the specific site and assets under investigation based on available historical data collected from literature about typical material types, material properties and structural design in the specific site, open data repositories on assets, such as OpenStreetMap, or in-situ surveys (Baiguera et al. 2021; Cels et al. 2023). One of the major challenges in incorporating uncertainty in fragility function development is the balance between computational demand and accuracy. Among the available methods, Monte Carlo-based simulation is considered as the most comprehensive one. However, to improve computational costs, importance sampling methods, such as Latin hypercube sampling, response surface, or artificial neural networks can also be used. Alam et al. (2018) used the First-order Sampling Method (FOSM) to propagate the uncertainty from demand parameters (e.g., tsunami loading) and structural capacity modelling to structural response estimation (Table 7.3).

7.4 Indicator-Based Vulnerability

7.4.1 Short Description of Method

Physical vulnerability has been defined as the degree of loss of a given element (e.g., buildings) that is affected by the impact of a natural process (UNDRO, 1984) and can vary from 0 (no loss) to 1 (total loss), when only direct losses are assessed. This definition suggests that vulnerability equals the consequences of this impact and is not related to a pre-existing condition, weakening in this way the predictive power of its assessment. By accepting the newer more general definition of vulnerability

Table 7.3 Summary table for analytical physical vulnerability

Short description of method	
Short outline of how the method works	The analytical vulnerability consists of a set of statistical models correlating some EVs to some RVs based on simulated tsunami scenarios. Explanatory Variables are both IMs and characteristics of the built environment. The typical RV is the DS, representing a damage condition experienced by the asset of interest after the tsunami. DS may vary from zero damage up to washed away (total damage)
How is/are the output(s) of the method(s) represented?	Fragility functions, consequence functions, and vulnerability functions
Do(es) the method(s) use parallelisation, machine learning, surrogate models, advanced simulation schemes, other computational reduction strategies (e.g. cluster analysis, etc.)?	The analytical simulation typically requires the development of structural models using FEM. Examples of commercial FEM packages that have been used for tsunami vulnerability simulations include OpenSees and SeismoStruct
What are the pros and cons and peculiarities of this specific method(s)? Are there known shortcomings / benefits of the method(s)? In what cases is this method(s) a good/poor choice?	One of the major challenges in incorporating uncertainty in fragility function development is the balance of computational demand and accuracy. Among available methods, Monte Carlo-based simulation is the most comprehensive method. However, to improve computational costs, importance sampling methods such as Latin hypercube sampling, response surface, or artificial neural networks can also be used
Data and technical requirements for method	
What input data does the method(s) use?	Input data related to the assets geometric and mechanical properties, and structural details should be known. For existing assets, structural details can also be derived from a simulated design process. Fragility functions are analytically derived for the assets under investigation performing a damage assessment through refined structural analyses in a probabilistic framework. The damage assessment should account for all the tsunami-induced loads and effects on structure and infrastructure
What are the technical requirements?	Different types of analysis can be selected for performing tsunami analytical fragility assessment of a facility. The structural analysis may be linear or nonlinear, depending on the nature of the materials constituting the building (e.g. confined masonry, steel frame, reinforced concrete), and either static or dynamic

(continued)

Table 7.3 (continued)

Treatment of uncertainties for the method	
Is the method(s) itself deterministic or uncertainty based?	The statistical regression methods provide a probabilistic interpretation of the data
Which are potential sources of uncertainties? Which parameters are used to characterize them?	The uncertainties associated with data are mainly epistemic and are associated with both the Explanatory Variables and the Response Variables. Geometric and mechanical properties of materials are aleatory variables that should be taken with their probability distributions and statistics. Only a few empirically validated models are available to simulate the tsunami loads and effects on structures (effect of openings, damming debris accumulation, debris impact, scour), and also available models can be affected by errors. Uncertainty in the tsunami load parameters (flow depth and velocity) need to be considered as well
How are uncertainties quantified for the identified parameters? Are there significant correlations between different uncertain parameters?	Uncertainties can be quantified for the specific site and assets under investigation based on available historical data collected from literature, open-source repositories, Open Street Maps, or in-situ surveys

(UNDRR, 2017) "*the conditions determined by physical, social, economic and environmental factors or processes, which increase the susceptibility of an individual, a community, assets or systems to the impact of hazards*", we accept that vulnerability is a pre-existing condition related to the characteristics of the elements at risk. Indicators can operationalise these characteristics (Birkmann, 2006) and form the basis for the development of an index that can express vulnerability qualitatively.

Indicators are widely used for the assessment of social vulnerability and may include information on age, health status, income, educational level, etc. In the case of physical vulnerability of buildings to tsunami the physical vulnerability indicators are operational representations of building exposure characteristics (e.g. building material, openings, etc.) or characteristics of their surroundings (e.g., surrounding vegetation or wall) that influence the overall vulnerability of the building to tsunamis. The theoretical background is described in detail in *Chapter 2*. One of the most-established indicator-based methods for tsunami focusing on physical vulnerability is the Papathoma Tsunami Vulnerability Assessment (PTVA). This method was first proposed in 2003 and has been applied to case-studies in Greece (Papathoma and Dominey-Howes, 2003). It was further developed following the 2004 Indian Ocean Tsunami (Dominey-Howes and Papathoma, 2007). The weighting method has been improved in further versions of the PTVA (DallOsso et al. 2009a; DallOsso et al. 2016) and has become a popular method as it has been validated and used in several countries. It has also been adapted for different hazards, including dynamic flooding

(Papathoma et al. 2019) and wildfire (Papathoma et al. 2022). A stepwise description of the method is illustrated in Fig. 7.6.

PTVA has been applied in Italy (DallOsso et al. 2010), Iran (Madani et al.2017), Chile (Izquierdo et al. 2018), Portugal (Santos et al. 2014), Australia (Dall'Osso et al. 2009b, 2016), in Indonesia for critical infrastructure (Husrin et al. 2013) and school buildings (Gentile et al. 2019), in Japan (Voulgaris and Murayama, 2014), in Malaysia (Ismail et al. 2012) and Sri Lanka (Sathiparan, 2020; Hasalanka et al. 2021). An overview of the indicator-based applications is presented in Table 7.4.

Indicator-based methods should account for the fact that vulnerability changes through space and time, therefore, they should be updated regularly to account for future changes. The GIS database containing vulnerability indicators for the buildings of the study area can be easily updated to describe the current situation and condition

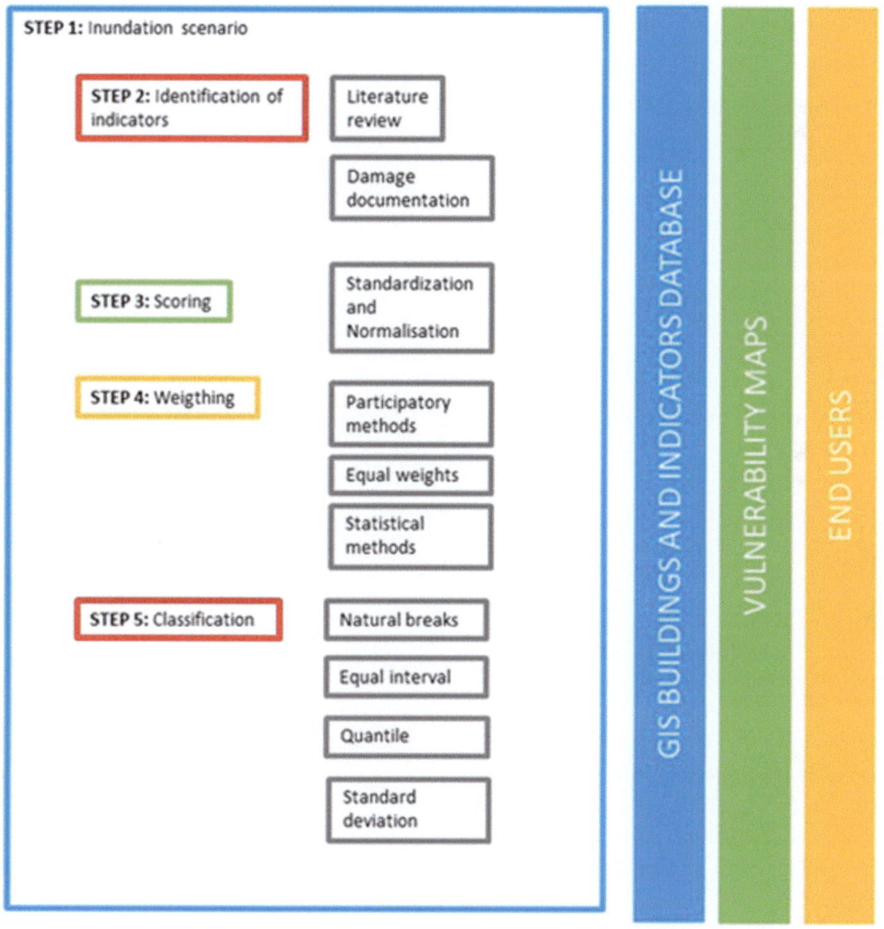

Fig. 7.6 Indicator-based vulnerability assessment workflow based on the PTVA method

Table 7.4 An overview of some of the studies that have applied the PTVA approach

Study	Country	Special focus	Application	Comments	Source of inundation map
Omira et al. (2010)	Morocco (Casablanca)	Buildings	Potentially for building modification and evacuation planning	Inspired from PTVA (only three indicators used)	Numerical modelling of the 1755 tsunami
Santoset al. (2014)	Portugal (Setubal city in Lisbon metropolitan area)	Buildings including services (police station, fire brigade, administrative buildings)	Validation of historical records from the 1755 tsunami. Potentially it could be used for urban planning	Use of PTVA but no comparison with another model (Ismail et al. 2012)	Numerical model of the 1755 tsunami
Voulgaris and Murayama (2014)	Japan	Buildings	None	PTVA Combined with population data	Imamura-Iida tsunami magnitude scale was used to calculate the maximum run-up for an 8.2 M earthquake
Alberico et al. (2015)	Italy (Naples)	Urban area (Census district)	Possibilities to reduce vulnerability (interventions), locations for vertical evacuation	PTVA applied to a different scale. The unit is not the building but the census district	Tsunami hydrodynamic modelling of a historic tsunami caused by a submarine debris avalanche in Ischia
Madani et al. (2017)	Iran (Chabahar Bay)	Buildings (residential area)	None	Use of PTVA	Numerical modelling of a Makran subduction worst-case scenario

(continued)

Table 7.4 (continued)

Study	Country	Special focus	Application	Comments	Source of inundation map
Izquierdo et al. (2018)	Chile	Buildings	Validation through a real event (2015), focus on historical buildings	PTVA material indicator was modified according to construction types in Chile	Reconstructed tsunami inundation map and flow depths from the data collected in the field after the 2015 tsunami
Batzakis et al. (2020)	Greece (Santorini)	Buildings (touristic area)	Potentially, Tsunami building codes and early warning systems	Use of PTVA	Static inundation method (run up based on previous studies)
Hasalanka et al. (2021)	Sri Lanka	Hospitals	Disaster risk management	Changes to PTVA are recommended to include hospital–relevant indicators	Potential tsunami scenarios based on available risk maps

of buildings. Moreover, the vulnerability index can be used during the reconstruction process to assess the quality of BBB (Build Back Better). Additionally, indicator-based methods could be expanded to include indicators related to the functionality of a building. For example, Baiguera et al. (2021) proposed a simple but quantitative approach for the assessment of the resilience of healthcare systems to tsunami. This is based on indicators that relate not only to hospital building fragility, but also to the maintenance of hospital functionality. The impact of tsunami on hospital critical units (e.g., Intensive Care Unit) in individual hospitals, as well as the impact on service provision across a network of hospitals, is measured through the Tsunami Relative Risk Index (TRRI). The ability of a hospital critical unit to function in the aftermath of a tsunami depends on three components: 1) the fragility of the structure where the hospital critical unit is located (i.e., the "building component"); 2) the integrity of non-structural elements relevant to the critical units, (e.g., the medical equipment that is required to ensure unit functionality); and 3) the functioning of the critical lifeline systems supporting unit functionality (e.g., electric power, water supply, telecommunications, etc.). It is noted that the main difference between the "building component" and the PTVA, is that the former one is based on a simplified assessment of the building failure and damage mechanisms, evaluated using physics and engineering-based formulations.

7.4.2 Data and Technical Requirements for Method

The main data required for the implementation of the indicator-based methods include geospatial data (e.g., building footprints), information on building characteristics and their surroundings and detailed information on past tsunami events or inundation modelling results for specific scenarios. This reflects the deterministic nature of the method. An expert panel should also develop a set of case study-related indicators. This expert panel should include local experts including researchers, engineers and environmental experts to develop an indicator dataset which is relevant to the local context, but also social scientists or economists if the vulnerability assessment is to be extended to include additional vulnerability dimensions (e.g., social). The main requirement involves detailed data on the buildings collected during extensive fieldwork. These data may include the number of floors, the building material, the use of the ground-floor, the building row facing the coast, the surrounding wall or vegetation and the presence of barriers between the building and the water. Alternative methods for data collection may include *Google Street View*, other remote sensing methods, existing inventories, *Youtube* videos or questionnaires completed by the house owners. Potential developments of the method could include using the *Boruta* algorithm (Kursa and Rudnicki 2010) for the weighting of indicators, using alternative methods for data collection and the inclusion of socioeconomic data to assess further vulnerability dimensions.

Indicator-based methods have many potential applications for disaster management and disaster risk reduction. Some of them are described in one of the first applications of the PTVA (Papathoma and Dominey-Howes, 2003). Applications may include the use of the vulnerability map as basis for decision-making (e.g., targeted building strengthening or relocation). Moreover, the combination of indicator-based methods with demographic data may support targeted evacuation and rescue planning, as well as public awareness and education programs. In any case, uncertainties related to the potential inundation zone, the quality of input data and the expert judgement may contribute to an overestimation or underestimation of physical vulnerability and must be treated before being used for decision-making. Nevertheless, as far as PTVA is concerned, uncertainties have not been addressed until now. In-depth information on the method can be found as supplementary material together with the latest publication (Dall'Osso et al. 2016). In more detail, the supplementary material contains an ARCGIS toolbox (PTVA4_Toolbox), a shapefile for the exposed buildings in the case study area and an Excel file with the questionnaire sent to the experts (Table 7.5).

Table 7.5 Summary table for indicator-based vulnerability

Short description of method	
Short outline of how the method works	Indicator-based methods make use of vulnerability indicators to assess the physical vulnerability of buildings to tsunamis. The vulnerability indicators are operational representations of building characteristics or characteristics of their surroundings that influence the overall vulnerability of the building to tsunamis
How is/are the output(s) of the method(s) represented?	The degree of loss of a given element (e.g., buildings) that is affected by the impact of the tsunami, which can vary from 0 (no loss) to 1 (total loss)
What are the pros and cons and peculiarities of this specific method(s)? Are there known shortcomings / benefits of the method(s)? In what cases is this method(s) a good/poor choice?	Indicator-based methods have many potential applications for disaster management and disaster risk reduction. For this reason, indicator-based methods have become a popular method: they have been validated through past tsunami events and expert elicitation. The indicator-based methods can be adapted to future changes. The GIS database containing vulnerability indicators for the buildings of the study area can be easily updated to describe the current situation and condition of buildings. The vulnerability index can also be used following a catastrophic and after the reconstruction process has been completed to demonstrate the BBB (Build Back Better)
Data and technical requirements for method	
What input data does the method(s) use?	Indicator-based methods require Geographic Information Systems (GIS), detailed information on past tsunami events or modelling possibilities and an expert panel by developing a set of case study-related indicators. Detailed data on the buildings collected during extensive fieldwork are needed. Alternative methods for data collection may include Google Streetview, other remote sensing methods, existing inventories, Youtube videos or questionnaires completed by the house owners
Treatment of uncertainties for the method	
Is the method(s) itself deterministic or uncertainty based?	The method has been mostly used for deterministic case studies

7.5 Social Vulnerability

7.5.1 *Short Description of Method*

Social vulnerability refers to the susceptibility of individuals and social groups to the adverse impacts of natural hazards, including disproportionate death, injury, financial and non-financial loss (e.g., loss to household assets, health) and disruptions to livelihood (Cutter et al. 2012; Birkmann et al. 2013). Social vulnerability has so far not been included in mainstream probabilistic tsunami risk assessment. However, the impact of tsunami on mortality, fertility, health, schooling and business continuity has been documented in the empirical economics literature, based on the analysis of large-scale household survey data before and after the hazard occurred (Cas et al. 2014; Frankenberg et al. 2011; Nobles et al. 2015; Himaz, 2022; Syukriyah and Himaz 2024). This literature is a subset of the wider, but still relatively young field of disaster impact evaluation in applied microeconomics.

A seminal effort to capture social vulnerability in tsunami risk assessment can be seen in Salmanidou et al. (2021). The empirical framework correlates EVs to RVs based on a observed tsunami event, the 2006 Java earthquake and tsunami with data coming from a survey of about 6,000 households. EVs are IMs and pre-tsunami impact characteristics at the household level. Although IMs such as the tsunami inundation depth can be used, they should be matched against each household in the sample. In the absence of such data, this study proposed to use a proxy measure of intensity that is constructed using field-level observations carried out by experts knowledgeable about the local area. These are calibrated for accuracy using before-after satellite imagery or remote sensing data. Considering vulnerabilities that may build up before the event, EVs are characteristics for the household just before the tsunami such as wealth, urban or rural location. The RV used in this study is loss ratio of the value of assets experienced by households, calculated as a percentage change to asset value using data just before and just after the tsunami. Assets include land, buildings, vehicles, furniture, livestock, jewellery, and hard-wood trees. The value of these RVs can vary from 0 to 100%. The OLS method is used to understand the impact of tsunami on damage to household assets. Multinomial logit regression modelling is used to estimate the distributions of damage by tsunami intensity.

Theoretically, RVs can capture negative impacts to various microeconomic indicators. At the level of the individual this can be, for example, the damage to earnings, mental and physical health, or schooling. At the household-level it can be the damage to assets or household standing in terms of poverty or wellbeing. At the community level, the indicator could be crime rates. Although the range is wide, there are two major econometric challenges to generating meaningful vulnerability functions

(Anttila-Hughes and Sharma, 2015; Porter and White, 2016). The first is the so-called 'internal validity', which refers to the establishment of a causal (or, at least, externally replicable) relationship between IMs and the microeconomic damage indicator. If the damages attributed to a tsunami are the result of the tsunami per se as well as other factors such as wealth and attitudes to risk, then the estimates may be biased. The second is the so-called 'external validity', or the extent to which assessed impacts can be generalised out of the sample, either to other contexts or locations, or to other time periods. Further discussion regarding internal and external validity can be found in Angrist and Pischke (2009).

7.5.2 Data and Techniques

Tsunami social vulnerability modelling requires large scale panel household survey data with detailed before-after information and objective IMs that can be linked to the unit of analysis (i.e., individual, household or community). A good example of such data is the Study of the Tsunami Aftermath and Recovery (STAR: http://stardata.org), a longitudinal survey carried out in Aceh and North Sumatra, Indonesia. The survey collected data on individuals, households, communities and facilities to provide evidence on the immediate and longer-term consequences of the 2004 Indian Ocean Tsunami. The first round of data was collected 5–14 months after the tsunami from over 6000 households from 487 communities. It also included retrospective information collected from households on pre-tsunami characteristics enabling before-after comparisons and accounting for vulnerabilities that built up before the event. The first round of data was followed by further annual rounds of data collected for four years from the same households with a 10-year follow-up survey. The data contains information on employment, income, spending, schooling, health, household and business assets, attitudes to risk, housing, migration and various other aspects. This enables the use of the data to create RVs capturing various dimensions of vulnerability. Data from this specific context, however, may not be applicable to other contexts across time and space. This is the issue of external validity discussed earlier. Tsunami-specific longitudinal surveys such as STAR are sparse, and an alternative source may be repeated cross section data collected from the same area before and after a tsunami. The latter data may be used to estimate community-level impacts. For example, Sri Lanka's Household Income and Expenditure Surveys conducted in 1990, 1995, 2002 and 2005 by the Department of Census and Statistics may be used to assess changes to consumption, income or spending at the level of an administrative unit such as a community or village, providing the geographical units can be linked over time. Apart from this, other data and techniques found in the wider economic and impact evaluation literature may be considered when estimating social vulnerability. These include constructing pseudo-panels with repeated cross sections (Verbeek, 2008), using administrative data (Wang et al. 2016; Karlan et al. 2016) and satellite and remotely sensed data (Heger and Neumayer, 2019). Even so, the relative rareness of tsunamis and the subsequent unavailability of historical data enabling the constructing of microeconomic indicators and IMs, as well as the small sample size in cases where data is available constrain the development of internal and externally valid estimates, and therefore meaningful tsunami social vulnerability functions.

7.5.3 Treatment of Uncertainties

In household survey data, systematic errors can arise from non-random biases in recollection data (such as respondents' recollection of pre-tsunami assets 12 months after the tsunami being systematically different to that reported 24 months after the event as noted in Himaz, (2022)), sample selection biases (arising from the people affected by the tsunamis being systematically different to those who were not affected) and non-random sample attrition in longitudinal surveys. Such systematic errors can bias the parameters estimated such that a causal interpretation cannot be given to an estimated parameter.

Various data design methods and statistical techniques can be used to address some of the issues especially in the context of impact evaluation (Angrist and Pischke, 2009). Usually, robustness checks using alternative specifications, data sources and statistical methods are used to validate results. Fundamentally, however, good quality, credible data is essential to measure social vulnerability accurately. About IMs, the discussion in Sect. 7.2.3 in the treatment of uncertainties in physical vulnerability models is relevant to social vulnerability.

Measures to convey uncertainty include measures that compute the variability of model parameters (standard errors, statistical significance, variance of the median, etc.), allowing to provide fragility curves with pre-defined confidence intervals (Table 7.6).

Table 7.6 Summary table for social vulnerability

Short description of method	
Short outline of how the method works	Social vulnerability is about the evaluation of the impact of tsunami on social outcomes, such as mortality, fertility, health, schooling and business continuity. This area of research is strongly linked to empirical economics. A social vulnerability framework aims to correlate EVs RVs based on an observed tsunami event. Examples of EVs are IMs (e.g. tsunami inundation depth) and pre-tsunami impact characteristics at the household level. RVs can vary from microeconomic indicators (e.g., damage to household assets) to community-level indicators (e.g., crime rates)
How is/are the output(s) of the method(s) represented?	The OLS method may be used to understand the impact of tsunami on damage to household assets. Multinomial logit regression modelling may be used to estimate the distributions of damage by tsunami intensity
What are the pros and cons and peculiarities of this specific method(s)? Are there known shortcomings / benefits of the method(s)? In what cases is this method(s) a good/poor choice?	There are two major econometric challenges to generating meaningful social vulnerability functions: (a) 'internal validity', which refers to the establishment of a causal (or, at least, externally replicable) relationship between IMs and the microeconomic damage indicator; (b) 'external validity', or the extent to which assessed impacts can be generalised out of the sample, either to other contexts or locations, or to other time periods
Data and technical requirements for method	
What input data does the method(s) use?	Tsunami social vulnerability modelling requires large scale panel household survey data with detailed before-after information and objective IMs that can be linked to the unit of analysis (i.e., individual, household or community). Such data are sparse and other techniques found in the wider economic and impact evaluation literature such as constructing pseudo-panels with repeated cross sections, using administrative data, satellite and remotely sensed data may be considered
Treatment of uncertainties for the method	
Is the method(s) itself deterministic or uncertainty based?	The statistical regression methods provide a framework for probabilistic interpretation of the data

References

Aguirre-Ayerbe I, Martínez Sánchez J, Aniel-Quiroga Í, González-Riancho P, Merino M, Al-Yahyai S, González M, Medina R (2018) From tsunami risk assessment to disaster risk reduction–the case of Oman. Nat Hazard 18(8):2241–2260

Akaike H (1974) A new look at the statistical model identification IEEE Trans. Autom Control 19(6):716–723

Alam MS, Barbosa AR, Scott MH, Cox DT, van de Lindt JW (2018) Development of physics-based tsunami fragility functions considering structural member failures. ASCE J. Struct. Eng. 144(3):04017221

Alam MS, Winter AO, GalantG, Shekhar K, Barbosa AR, Motley MR, Eberhard MO, Cox DT, Arduino P, Lomonaco P (2020) Tsunami-like wave-induced lateral and uplift pressures and forces on an elevated coastal structure. J Waterw Port Coast Ocean Eng 146(4)

Alberico I, Di Fiore V, Iavarone R, Petrosino P, Piemontese L, Tarallo D, Punzo M and Marsella E (2015) The tsunami vulnerability of urban environments through freely available datasets: the case study of Napoli City (Southern Italy). J Mar Sci Eng 3:981–1005

Amakuni K, and Terazono N (2011) "Basic analysis on building damages by tsunami due to the 2011 Great East Japan earthquake disaster. In: 15 world conference on earthquake engineering. http://www.iitk.ac.in/nicee/wcee/article/WCEE2012_3628.pdf

Angrist JD, Pischke JS (2009) Mostly harmless econometrics: An empiricist's companion. Princeton University Press

Anttila-Hughes J, Sharma M (2015) Linking risk models to microeconomic indicators. World Bank Policy Research Working Paper (7359)

Aránguiz R, Urra L, Okuwaki R, Yagi Y (2018) Development and application of a tsunami fragility curve of the 2015 tsunami in Coquimbo, Chile. Nat Hazard 18(8):2143–2160

ASCE (2022) Minimum Design Loads and Associated Criteria for Buildings and Other Structures. ASCE/SEI 7–22. Reston, VA, USA

Attary N, Unnikrishnan VU, van de Lindt JW, Cox DT, Barbosa AR (2017) Performance-based tsunami engineering methodology for risk assessment of structures. Eng Struct 141:676–686

Baiguera M, Rossetto T, Palomino JS, Dias P, Lopez-Querol S, Siriwardana C, Hashalanka H, Ioannou I, Robinson DI (2021) A New relative risk index for hospitals exposed to Tsunami. Front Earth Sci-Geohazards Georisks. https://doi.org/10.3389/feart.2021.626809

Baiguera M, Rossetto T, Robertson IN, Petrone C (2022) A procedure for performing nonlinear pushover analysis for Tsunami Loading to ASCE 7. ASCE J Struct Eng 148(2). https://doi.org/10.1061/(ASCE)ST.1943-541X.0003256

Batzakis DV, Misthos LM, Voulgaris G, Tsanakas K, Andreou M, Tsodoulos I and Karymbalis E (2020) Assessment of building vulnerability to tsunami hazard in Kamari (Santorini Island, Greece). J Mar Sci Eng 8

Belliazzi S, Lignola GP, Di Ludovico M, Prota A (2021) Preliminary tsunami analytical fragility functions proposal for Italian coastal residential masonry buildings. Structures 31:68–79

Berryman K (2005) Review of tsunami hazard and risk in New Zealand. Report no. CR2005–104, Institute of Geological & Nuclear Sciences, Lower Hutt

Birkmann J (2006) Indicators and criteria for measuring vulnerability: Theoretical bases and requirements. In: Birkmann J (ed.) Measuring vulnerability to natural hazards: towards disaster resilient societies. Tokyo: UNU Press

Birkmann J, Cardona OD, Carreño ML, Barbat AH, Pelling M, Schneiderbauer S, Kienberger S, Keiler M, Alexander D, Zeil P, Welle T (2013) Framing vulnerability, risk and societal responses: the MOVE framework. Nat Hazards 67:193–211

Cas AG, Frankenberg E, Suriastini W, Thomas D (2014) The impact of parental death on child well-being: evidence from the Indian Ocean Tsunami. Demography 51(2):437–457. https://doi.org/10.1007/s13524-014-0279-8

Cels J, Rossetto T, Dias P, Thamboo J, Wijesundara K, Baiguera M, Del Zoppo M (2023) "Engineering Surveys of Sri Lankan Schools Exposed to Tsunami. Frontiers Earth Sciences 11:1075290. https://doi.org/10.3389/feart.2023.1075290

Charvet I, Ioannou I, Rossetto T, Suppasri A, Imamura F (2014a) Empirical fragility assessment of buildings affected by the 2011 Great East Japan tsunami using improved statistical models. Nat Hazards 73:951–973. https://doi.org/10.1007/s11069-014-1118-3

Charvet I, Suppasri A, Imamura F (2014b) Empirical fragility analysis of building damage caused by the 2011 Great East Japan tsunami in Ishinomaki city using ordinal regression, and influence of key geographical features. Stoch. Environ. Res. Riskassess. 28:1853–1867. https://doi.org/10.1007/s00477-014-0850-2

Charvet I, Macabuag J, Rossetto T (2017) Estimating tsunami-induced building damage through fragility functions: critical review and research needs. Frontiers in Built Environment 3:36. https://doi.org/10.3389/fbuil.2017.00036

Chinnarasri C, Thanasisathit N, Ruangrassamee A, Weesakul S, Lukkunaprasit P (2013) The impact of tsunami-induced bores on buildings. Proc Inst Civ Eng Marit Eng 166(1):14–24

Cutter SL, Boruff BJ, and Shirley WL (2012) Social vulnerability to environmental hazards. In Hazards vulnerability and environmental justice. Routledge, pp 143–160

Dall'Osso F, Gonella M, Gabbianelli G, Withycombe G, Dominey-Howes D (2009a) A revised (PTVA) model for assessing the vulnerability of buildings to tsunami. Nat Hazard 9:1557–1565

Dall'Osso F, Gonella M, Gabbianelli G, Withycombe G, Dominey-Howes D (2009b) Assessing the vulnerability of buildings to tsunami in Sydney. Nat Hazard 9:2015–2026

Dall'Osso F, Maramai A, Graziani LBB, Cavalletti AMG, Tinti S (2010) Applying and validating the PTVA-3 Model at the Aeolian Islands, Italy: assessment of the vulnerability of buildings to tsunamis. Nat Hazard 10:1547–1562

Dall'Osso F, Dominey-Howes D, Tarbotton C, Summerhayes S, Withycombe G (2016) Revision and improvement of the PTVA-3 model for assessing tsunami building vulnerability using "international expert judgement": introducing the PTVA-4 model. Nat Hazards 83:1229–1256

Del Zoppo M, Rossetto T, Di Ludovico M, Prota A (2019) Assessing the effect of tsunami-induced vertical loads on RC frames. In: Proceedings of the 1st fib Italy YMG symposium on concrete and concrete structures (Vol. 49). FIB

Del Zoppo M, Di Ludovico M, Prota A (2021) Methodology for assessing the performance of RC structures with breakaway infill walls under tsunami inundation. J Struct Eng 147(2):04020330

Del Zoppo M, Di Ludovico M, Prota A (2022) A mechanics-based method towards risk assessment of RC buildings under tsunami and flow-type hazards. Eng Struct 264:114452

De Risi R, Goda K, Mori N, Yasuda T (2017a) Bayesian tsunami fragility modeling considering input data uncertainty. Stoch Environ Res Riskassess. 31:1253–1269. https://doi.org/10.1007/s00477-016-1230-x

De Risi R, Goda K, Yasuda T, and Mori N (2017b) Is flow velocity important in tsunami empirical fragility modeling? Earth Sci Rev. 166:64–82. 10.1016/j. earscirev.2016.12.015

De Risi R, Goda K, Yasuda,T., and Mori, N (2017c). MTEF. https://www.researchgate.net/publication/312627614_MTEF

Dias WPS, Yapa HD, Peiris LMN (2009) Tsunami vulnerability functions from field surveys and monte carlo simulation. Civ Eng Environ Syst 26:181–194. https://doi.org/10.1080/10286600802435918

Dominey-Howes D, Papathoma M (2007) "Validating the "Papathoma tsunami vulnerability assessment model (ptvam) using field data from the 2004 Indian Ocean tsunami. Nat Hazards 53:43–61. https://doi.org/10.1007/s11069-006-0007-9

EEFIT (2006) The Indian Ocean tsunami of 26 December 2004: Mission findings in Sri Lanka & Thailand. Pomonis A, Rossetto T, Peiris N, Wilkinson SM, Del Re D, Koo R, Gallocher S, Earthquake engineering field investigation team (EEFIT) Report, The Institution of Structural Engineers, London, UK. https://www.istructe.org/get-involved/supported-organisations/eefit/

FEMA-P646 (2012) Guidelines for Design of Structures for Vertical Evacuation from Tsunamis, Washington, DC: Federal Emergency Management Agency

Foster ASJ, Rossetto T, Allsop W (2017) An experimentally validated approach for evaluating tsunami inundation forces on rectangular buildings. Coast Eng J 128:44–57

Frankenberg E, Gillespie T, Preston S, Sikoki B, Thomas D (2011) Mortality, the family and the Indian Ocean tsunami. Econ J 121(554):F162–F182

Fraser S, Raby A, Pomonis A, Goda K, Chian SC, Macabuag J, Offord M, Saito K, Sammonds P (2013) Tsunami damage to coastal defences and buildings in the March 11th 2011 M w 9.0 Great East Japan earthquake and tsunami. Bull Earthq Eng 11:205–239

Gentile R, Galasso C, Idris Y, Rusydy I, Meilianda E (2019) From rapid visual survey to multi-hazard risk prioritisation and numerical fragility of school buildings. Nat Hazard 19:1365–1386

Goda K, Petrone C, De Risi R, Rossetto T (2017) Stochastic coupled simulation of strong motion and tsunami for the 2011 Tohoku, Japan earthquake. Stoch Env Res Risk Assess 31:2337–2355

Gokon H, Koshimura S, Matsuoka M (2009) Developing tsunami fragility curves for structural destruction in American Samoa. J Jpn Soc Civ Eng

Hamzah M, Mase H and Takayama T (2001) Simulation and experiment of hydrodynamic pressure on a tsunami barrier. In Proceedings 27th International Conference Coastal Engineering, 1501–1507. Reston, VA: ASCE

Hasalanka H, Siriwardana CSA, Kularatne D, Dias WPS (2021) Development of a structural robustness index against tsunamis for hospitals. Civ Eng Environ Syst 38(2):85–101

Hayashi S, Narita Y, Koshimura S (2013) Developing tsunami fragility curves from the surveyed data and numerical modeling of the 2011 Tohoku earthquake tsunami (inJapanese). J. Jpn. Soc. Civ. Eng. Coast. Eng. 69:1–5. https://doi.org/10.2208/kaigan.69.I_386

Heger MP, Neumayer E (2019) The impact of the Indian Ocean tsunami on Aceh's long-term economic growth. J Dev Econ 141:102365

Himaz R (2022) Business recovery in aceh and north sumatra following the indian ocean Tsunami. Int J Disaster Risk Reduct 73:102868

Husrin S, Kongko W and Putera A (2013) Tsunami vulnerability of critical insfrastructures in the city of Padang, West Sumatera. In: SIBE 2013 The 2nd international conference on sustainable infrastructure and built environment, 19–20 November 2013 2013 Bandung, Indonesia

Ismail H, Abd Wahab AK, Mohd Amin MF, Mhd Yunus MZ, Jaffar Sidek F, Esfandier BJ (2012) A 3-tier tsunami vulnerability assessment technique for the north-west coast of Peninsular Malaysia. Nat Hazards 63:549–573

Izquierdo T, Fritis E, Abad M (2018) Analysis and validation of the PTVA tsunami building vulnerability model using the 2015 Chile post-tsunami damage data in Coquimbo and La Serena cities. Nat Hazard 18:1703–1716

Jalayer F, Ebrahimian H, Trevlopoulos K, Bradley B (2022) Empirical tsunami fragility modelling for hierarchical damage levels: An application to damage data of the 2009 South Pacific tsunami. EGUsphere, pp 1–26

Karafagka S, Fotopoulou S, Pitilakis K (2018) Analytical tsunami fragility curves for seaport RC buildings and steel light frame warehouses. Soil Dyn Earthq Eng 112:118–137

Karlan D, McConnell M, Mullainathan S, Zinman J (2016) Getting to the top of mind: how reminders increase saving. Manage Sci 62(12):4–52

Koshimura S, Oie T, Yanagisawa H, Imamura F (2009) Developing fragility functions for tsunami damage estimation using numerical model and post-tsunami data from Banda Aceh. Indonesia. Coast Eng J. 51:243–273. https://doi.org/10.1142/S0578563409002004

Koshimura S, Kayaba S (2010) Tsunami fragility inferred from the 1993 Hokkaido Nansei-oki earthquake tsunami disaster. J. Jpn. Assoc. Earthq. Eng. 10:87–101. https://doi.org/10.5610/jaee.10.3_87

Koshimura S, Gokon H (2012) Structural vulnerability and tsunami fragility curves from the 2011 Tohoku earthquake tsunami disaster. J Jpn Soc Civ Eng Ser B2(Coast.Eng.) 68,I_336–I_340. https://doi.org/10.2208/kaigan.68.I_336

Kursa and Rudnicki, 2010Koshimura S, Gokon H (2012) Structural vulnerability and tsunami fragility curves from the 2011 Tohoku earthquake tsunami disaster. J Jpn Soc Civ Eng Ser B2(Coast.Eng.) 68,I_336–I_340. https://doi.org/10.2208/kaigan.68.I_336

Lahcene E, Ioannou I, Suppasri A, Pakoksung K, Paulik R, Syamsidik S, Bouchette F, Imamura F (2021) Characteristics of building fragility curves for seismic and non-seismic tsunamis: case studies of the 2018 Sunda Strait, 2018 Sulawesi-Palu, and 2004 Indian Ocean tsunamis. Nat Hazard 21(8):2313–2344

Latcharote P (2015) Development of nonlinear analysis method to simulate nonlinear behavior of rc walls in rc wall-frame buildings suffering damage from earthquake and subsequent tsunami

Macabuag J, Rossetto T, Ioannou I, Suppasri A, Sugawara D, Adriano B, Imamura F, Eames I, Koshimura S (2016a) A proposed methodology for deriving tsunami fragility functions for buildings using optimum intensity measures. Nat Hazards 84:1257–1285. https://doi.org/10.1007/s11069-016-2485-8

Macabuag J, Rossetto T, Ioannou I (2016a) Investigation of the effect of debris-induced damage for constructing tsunami fragility curves for buildings in 1st international conference on Natural Hazards & Infrastructure (Chania, Greece)

Macabuag J, Rossetto T, Ioannou I, Eames I (2018) Investigation of the effect of debris-induced damage for constructing tsunami fragility curves for buildings. Geosciences 8(4):117

Madani S, Khaleghi S, Jannat MRA (2017) Assessing building vulnerability to tsunami using the PTVA-3 model: a case study of Chabahar Bay. Iran. Natural Hazards 85:349–359

Marchand M, Buurman J, Pribadi A, Kurniawan A (2009) Damage and casualties modelling as part of a vulnerability assessment for tsunami hazards: a case study from Aceh. Indonesia. J Flood Risk Manag 2(2):120–131

Maruyama Y, Kitamura K, Yamazaki F (2013) Tsunami damage assessment of buildings in Chiba Prefecture, Japan using fragility function developed after the 2011 Tohoku-Oki earthquake" in Safety, Reliability, Risk and Life-Cycle Performance of Structures & Infrastructures (Taylor&Francis). http://ares.tu.chiba-u.jp/~papers/paper/2013/2013ICOSSAR_Maruyama.pdf

Mas E, Koshimura S, Suppasri A, Matsuoka M, Matsuyama M, Yoshii T, Jimenez C, Yamazaki F, Imamura F (2012) Developing Tsunami fragility curves using remote sensing and survey data of the 2010 Chilean Tsunami in Dichato. Nat Hazards Earth Syst Sci. 12:2689–2697. https://doi.org/10.5194/nhess-12-2689-2012

Mas E, Paulik R, Pakoksung K, Adriano B, Moya L, Suppasri A, Muhari A, Khomarudin R, Yokoya N, Matsuoka M, Koshimura S (2020) Characteristics of tsunami fragility functions developed using different sources of damage data from the 2018 Sulawesi earthquake and tsunami. Pure Appl Geophys 177(6):2437–2455

McGovern DJ, Todd D, Rossetto T, Whitehouse RJS, Monaghan J, Gomes E (2019) Experimental observations of tsunami induced scour at onshore structures. Coast Eng 152:103505

McKenna F, Scott MH, Fenves GL (2010) Nonlinear finite-element analysis software architecture using object composition. J Comput Civ Eng 24(1):95–107

MLIT (2014) Survey of tsunami damage condition. Japanese ministry of land, infrastructure, and transportation

Murao O, Nakazato H (2010) Vulnerability functions for buildings based on damage survey data in Sri Lanka after the 2004 Indian Ocean tsunami in international conference on Sustainable Built Environment (ICSBE-2010) Kandy, pp 371–378

Nanayakkara KIU, Dias WPS (2016) Fragility curves for structures under tsunami loading. Nat Hazards 80:471–486. https://doi.org/10.1007/s11069-015-1978-1

Narita Y, Koshimura S (2015) Classification of tsunami fragility curves based on regional characteristics of tsunami damage. J Jpn Soc Civ Eng Coast Eng. 71:331–336. https://doi.org/10.2208/kaigan.71.I_331

Nihei Y, Maekawa T, Ohshima R, Yanagisawa M (2012) Evaluation of fragility functions for tsunami damage in coastal district in Natori City, Miyagi Prefecture and mitigation effects of coastal dune. J Jpn Soc Civ Eng. 68, I_276–I_280. https://doi.org/10.2208/kaigan.68.I_276

Nobles J, Frankenberg E, Thomas D (2015) The effects of mortality on fertility: population dynamics after a natural disaster. Demography 52(1):15–38

Omira R, Baptista MA, Miranda JM, Toto E, Catita C, Catalao J (2010) Tsunami vulnerability assessment of Casablanca-Morocco using numerical modelling and GIS tools. Nat Hazards 54:75–95

Palermo DNistor I. Al-Faesly T, Cornett A (2012) Impact of tsunami forces on structures: The University of Ottawa experience. In: Proceedings 5th international tsunami symposium, ispra, Italy: Tsunami Society International

Papathoma M, Dominey-Howes D (2003) Tsunami vulnerability assessment and its implications for coastal hazard analysis and disaster management. Nat Hazard 6:733–744

Papathoma-Köhle M, Schlögl M Fuchs S (2019) Vulnerability indicators for natural hazards: an innovative selection and weighting approach. Scientific Reports, 9

Papathoma-Köhle M, Schlögl M, Garlichs C, Diakakis M, Mavroulis S, Fuchs S (2022) A wildfire vulnerability index for buildings. Scientific Reports, 12

Park S, van de Lindt JW, Cox D, Gupta R, Aguiniga F (2012) Successive earthquake-tsunami analysis to develop collapse fragilities. J Earthquake Eng 16(6):851–863

Park H, Cox DT, Lynett PJ, Wiebe DM, Shin S (2013) Tsunami inundation modeling in constructed environments: A physical and numerical comparison of free-surface elevation, velocity, and momentum flux. Coast Eng 79:9–21

Park H, Cox DT, Barbosa AR (2017) Comparison of inundation depth and momentum flux based fragilities for probabilistic tsunami damage assessment and uncertainty analysis. Coast Eng 122:10–26

Petrone C, Rossetto T, Goda K (2017) Fragility assessment of a RC structure under tsunami actions via nonlinear static and dynamic analyses. Eng Struct 136:36–53. https://doi.org/10.1016/j.engstruct.2017.01.013

Petrone C, Rossetto T, Baiguera M, De la Barra C, Ioannou I (2020) Fragility functions for a reinforced concrete structure subjected to earthquake and tsunami in sequence."Eng Struct, 205:1102120

Peiris N, Pomonis A (2005) December 26, 2004 Indian Ocean tsunami: vulnerability functions for loss estimation in Sri Lanka In: international conference on geotechnical engineering for disaster mitigation & rehabilitation (Singapore), pp 411–416

Porter C, White EJ (2016) Potential for application of a probabilistic catastrophe risk modelling framework to poverty outcomes: General form vulnerability functions relating household poverty outcomes to Hazard intensity in Ethiopia. World Bank Policy Research Working Paper (7717)

Qi ZX, Eames I, Johnson ER (2014) Force acting on a square cylinder fixed in a free-surface channel flow. J Fluid Mech 756:716–727

Rasheed A, Usman M, Zain M, Iqbal N (2022) Machine learning-based fragility assessment of reinforced concrete buildings. Computational Intelligence and Neuroscience

Reese S, Cousins WJ, Power WL, Palmer NG, Tejakusuma IG, Nugrahadi S (2007) Tsunami vulnerability of buildings and people in South Java–field observations after the July 2006 Java tsunami. Nat Hazard 7(5):573–589

Reese S, Bradley BA, Bind J, Smart G, Power W, Sturman J (2011) Empirical building fragilities from observed damage in the 2009 South Pacific tsunami. Earth Sci Rev. 107:156–173. https://doi.org/10.1016/j.earscirev.2011.01.009

Rossetto T, Allsop W, Charvet I, Robinson DI (2011) Physical modelling of tsunami using a new pneumatic wave generator. Coast Eng 58(6):517–527

Rossetto T, Ioannou, I, Grant, DN, Maqsood T (2014) Guidelines for empirical vulnerability assessment, Global Earthquake Model (GEM) Report, GEM Foundation, Pavia, Italy

Rossetto T, Ioannou I (2018) Empirical fragility and vulnerability assessment: not just a regression. In: Risk Modeling for Hazards and Disasters. Elsevier, pp 79–103

Rossetto T, De la Barra C, Petrone C, De la Llera JC, Vasquez J, Baiguera M (2019) Comparison of nonlinear static and dynamic analysis methods for assessing structural response under earthquake and tsunami in sequence. Earthquake Eng Struct Dynam 48:867–887. https://doi.org/10.1002/eqe.3167

Salmanidou DM, Ehara A, Himaz R, Heidarzadeh M, Guillas S (2021) Impact of future tsunamis from the Java trench on household welfare: Merging geophysics and economics through catastrophe modelling. Int J Disaster Risk Reduct 61:102291

Santos A, Tavares AO, Emidio A (2014) Comparative tsunami vulnerability assessment of an urban area: an analysis of Setubal city, Portugal. Appl Geogr 55:19–29

Sathiparan N (2020) An assessment of building vulnerability to a tsunami in the Galle coastal area, Sri Lanka. J Build Eng 27

Schwarz G (1978) Estimating the dimension of a model. Ann Stat 6(2):461–464

SeismoStruct (2022) A computer program for static and dynamic nonlinear analysis of framed structures. https://seismosoft.com/

Shafiei S, Melville BW, Shamseldin AY (2016) Experimental investigation of tsunami bore impact force and pressure on a square prism. Coastal Eng 110:1–16

Song J, De Risi R, Goda K (2017) Influence of flow velocity on tsunami loss estimation. Geosciences 7(4):114

Strunz G, Post J, Zosseder K, Wegscheider S, Mück M, Riedlinger T, Mehl H, Dech S, Birkmann J, Gebert N, Harjono H (2011) Tsunami risk assessment in Indonesia. Nat Hazard 11(1):67–82

Suppasri A, Koshimura S, Imamura F (2009) Tsunami fragility curves and structural performance of building along the Thailand coast. In: 8th international workshop on remote sensing for disaster management (Tokyo), 3–8

Suppasri A, Koshimura S, Imamura F (2011) Developing tsunami fragility curves based on the satellite remote sensing and the numerical modeling of the 2004 Indian Ocean tsunami in Thailand. Nat Hazards Earth Syst Sci. 11:173–189. https://doi.org/10.5194/nhess-11-173-2011

Suppasri A, Koshimura S, Imai K, Mas E, Gokon H, Muhari A (2012a) Damage characteristic and field survey of the 2011 Great East Japan tsunami in Miyagi Prefecture. Coast Eng J 54:30. https://doi.org/10.1142/S0578563412500052

Suppasri A, Mas E, Koshimura S, Imai K, Harada K, Imamura F (2012b) Developing tsunami fragility curves from the surveyed data of the 2011 Great East Japan tsunami in Sendai and Ishinomaki plains. Coast Eng J 54:16. https://doi.org/10.1142/S0578563412500088

Suppasri A, Mas E, Charvet I, Gunasekera R, Imai K, Fukutani Y, Abe Y, Imamura F (2013) Building damage characteristics based on surveyed data and fragility curves of the 2011 Great East Japan tsunami. Nat Hazards 66:319–341

Suppasri A, Charvet I, Imai K, Imamura F (2014) Fragility curves based on data from the 2011 Great East Japan tsunami in Ishinomaki city with discussion of parameters influencing building damage. Earthq Spectra 31:841–868. https://doi.org/10.1193/053013EQS138M

Syukriyah D, Himaz R (2024) Short and medium-run effects of the Indian Ocean tsunami on health costs in Indonesia. World Dev 180:106648. https://doi.org/10.1016/j.worlddev.2024.106648

Tanaka N, Kondo K (2015) Numerical analysis considering the effect of trapping the floatage by coastal forests and fragility curve of houses. J Jpn Soc. Civil Eng Ser B1(HydraulEng). 71,I_727–I_732. https://doi.org/10.2208/jscejhe.71.I_727

UNDRO (1984) Disaster Prevention and mitigation-a compendium of current knowledge, New York, UN

United Nations Office for Disaster Risk Reduction (UNDRR) (2017) The Sendai framework terminology on disaster risk reduction. "Vulnerability". https://www.undrr.org/terminology/vulnerability. Accessed 22 July 2025

UNISDR (2015) Sendai framework for disaster risk reduction 2015–2030. UNISDR, Geneva

Valencia N, Gardi A, Gauraz A, Leone F, Guillande R (2011) New tsunami damage functions developed in the framework of SCHEMA project: application to European-Mediterranean coasts. Nat Hazards Earth Syst Sci. 11:2835–2846. https://doi.org/10.5194/nhess-11-2835-2011

Verbeek M (2008) Pseudo-panels and repeated cross-sections. The Econometrics of Panel Data: Fundamentals and recent developments in theory and practice, pp 369–383

Voulgaris G, Murayama Y (2014) Tsunami vulnerability assessment in the Southern Boso Peninsula. Jpn Int J Disaster Risk Reduct 10:190–200

Wang P, Connor AL, Guo E, Nambao M, Chanda-Kapata P, Lambo N, Phiri C (2016) Measuring the impact of non-monetary incentives on facility delivery in rural Zambia: a clustered randomised controlled trial. Tropical Med Int Health 21(4):515–524

Chapter 8
Communicating Tsunami Risk and Uncertainties

Nikos Kalligeris, Irina Rafliana, Lorenzo Cugliari, Harald Spahn, Andrea Cerase, Jörn Lauterjung, Dimitra Salmanidou, Musavver Didem Cambaz, František Babič, Ignacio Aguirre Ayerbe, and Alessandro Amato

8.1 Introduction

8.1.1 Scope and Target Group of the Chapter

As communicating science and risks becomes a global concern, there is an increased demand for scientists' engagement with the public (Trench and Bucchi 2021). Hence, the purpose of this chapter is to provide scientists with guidance and background information on how to appropriately engage in, co-design, prepare, and deliver risk

N. Kalligeris (✉)
Institute of Geodynamics, National Observatory of Athens, Athens, Greece
e-mail: nkalligeris@noa.gr

I. Rafliana
Research Center for Area Studies, Badan Riset dan Inovasi Nasional (BRIN), Jakarta, Indonesia
e-mail: irin002@brin.go.id

Center for Life Ethics, Universität Bonn, Bonn, Germany

A. Amato
Centro Allerta Tsunami, Istituto Nationale di Geofisica e Vulcanologia, Roma, Italy
e-mail: alessandro.amato@ingv.it

L. Cugliari · A. Cerase
Istituto Nationale di Geofisica e Vulcanologia, Roma, Italy
e-mail: lorenzo.cugliari@ingv.it

A. Cerase
e-mail: andrea.cerase@uniroma1.it

Department of Communication and Social Research (CoRiS), Sapienza Università di Roma, Roma, Italy

M. B. Sørensen et al. (eds.), *Probabilistic Tsunami Hazard and Risk Analysis*, Mathematics of Planet Earth 14, https://doi.org/10.1007/978-3-031-98115-9_8

communication, with and to different audiences. It should be noted that risk communication is not only about classical, typically unidirectional science communication (from science to public or end users), but also about a range of different, multidirectional, interactive, and iterative communication approaches with all relevant actors related to tsunami risks. Figure 8.1 illustrates the three main pillars of tsunami risk communication and the process of building an effective communication strategy for the respective problem. First, we have to deal with understanding the context of what has to be communicated, by and to whom. This is described in Sects. 8.2 and 8.3. Then it comes to the choices of the appropriate methods and tools for risk communication, as described in Sect. 8.4. Last but not least, the evaluation of the effectiveness of the communication process will give indications for possible improvements (Sect. 8.5). The entire risk communication process is understood as a large feedback loop (blue arrows) to remain flexible and responsive to new challenges, with the arrows representing the process flow direction, and therefore not implying that the communication is necessarily in a singular or canonical direction. Given the limited space in this chapter to address the wide and complex scopes of tsunami risk communication, a reading list is provided for interested readers to further explore more resources that are not addressed here.

H. Spahn
Independent Consultant, Apen, Germany
e-mail: harald.spahn@web.de

J. Lauterjung
Geophysics, Helmholtz Centre Potsdam GFZ German Research Centre for Geociences, Potsdam, Germany
e-mail: lau@gfz.de

D. Salmanidou
Advanced Research Computing Centre, University College London, London, UK
e-mail: d.salmanidou.12@ucl.ac.uk

M. D. Cambaz
Kandilli Observatory and Earthquake Research Institute, Boğaziçi University, Istanbul, Türkiye
e-mail: didem.samut@bogazici.edu.tr

F. Babič
Faculty of Electrical Engineering and Informatics, Technical University of Kosice, Kosice-Sever, Slovakia
e-mail: frantisek.babic@tuke.sk

I. Aguirre Ayerbe
Instituto de Hidráulica Ambiental (IH Cantabria), Universidad de Cantabria, Santander, Spain
e-mail: aguirrei@unican.es

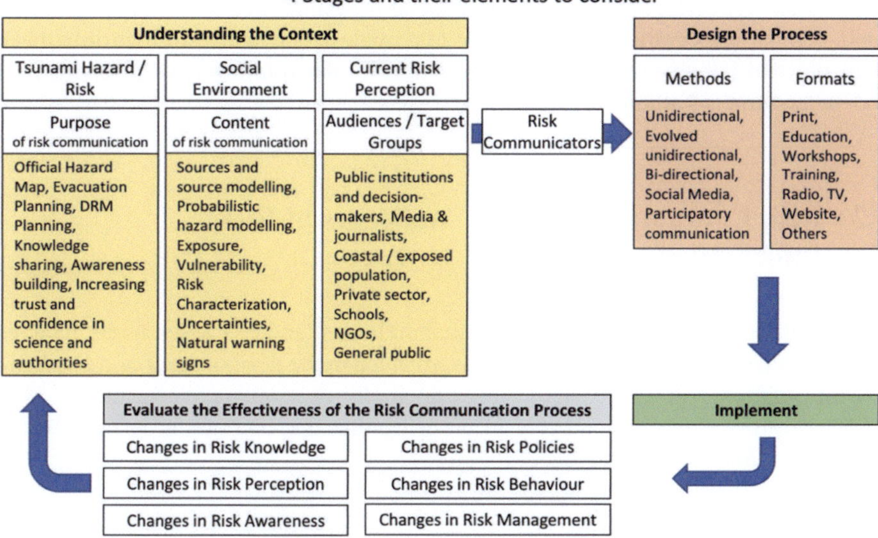

Fig. 8.1 Illustration of the process flow of tsunami risk communication, the three main pillars and the four stages to be considered

8.1.2 *Definition of Tsunami Risk Communication*

The purpose of risk communication is two-fold. First, it includes emphasizing both its informational and relational/ritual nature (Carey 1975): it must provide the information needed to understand and address "the nature, magnitude, significance, or control of a risk" (Covello 1992), thereby enhancing the ability of scientific knowledge to influence or socially construct risk perception. Second, risk communication is also required to create a climate of trust between authorities and experts, reciprocally, while engaging communities at risk to discuss these issues and make the necessary decisions (Renn and Levine 1991; Peters et al. 1997; Paton 2008). It is a huge task to achieve trustful communication exchanges, and Rohrmann (1992) rightfully argued that risk communication is a permanent and multidirectional activity among scientists, government, policy-makers, business and industrial sectors, private organizations, media and the public.

In this broad sense, risk communication should be understood as a "dynamic, holistic process extending beyond single products to the broader system within which we develop and exchange knowledge and awareness to make decisions individually and collectively about risk" (Doyle and Becker 2022). Given the general premises above, as well as the unique characteristics of tsunami risk, this definition of tsunami risk communication will be referred to below as a "two-way process" that occurs before and after an event and includes messages transmitted and relayed by scientists, government agencies, broadcasters, and social media, as well as by the population itself. Tsunami risk communication "aims to provide people with information to make decisions and influence the way tsunami risk and related information are perceived and understood" (Rafliana et al. 2022).

8.2 Risk Communicators and Audiences of Risk Information

Risk communication in the context of tsunami hazard and risk assessment involves interaction between scientists and experts involved in the development of hazard and risk information, and various audiences who are the users of such information. Risk communicators play a crucial role in ensuring that this interaction is fruitful and effective for all parties involved.

8.2.1 Risk Communicators

The role of risk communicators is often overlooked in the risk management realm, despite its paramount importance in disaster risk reduction efforts. Since their inception in the late 1980s, risk communication scholars have frequently argued that professionals, in some cases including scientists, engaged in this task require high-level training because of the high values and lives that are at stake, the volume and diversity of research findings, and the inherent technical complexity of the communication skills required of risk communication professionals, which are accompanied by high normative and ethical demands on both organizations and individual professionals (Covello et al. 1988; NRC 1989; Rowan 1991; Chess et al. 1995; Grabill and Simmons 1998; Johnson 1999; Reynolds 2002; Frewer 2004; Höppner et al. 2010; Renn 2010; Dickmann et al. 2016; Miller et al. 2017; Heath, and O'Hair 2010). In their systematic literature review of risk communication training needs, Miller et al. (2017) emphasize also on the importance of interagency collaboration and the training needed to work effectively with the media, including the ability to craft personalized messages for specific audiences and needs. Oftentimes, scientists, engineers or technologists that are used to moving across sectors and disciplines may have more aspirations to take part as risk communicators (Rubbia 2023).

Risk communicators are often called 'boundary crossers' of disciplines and of different epistemic groups. Their role is to bridge, narrate and make accessible risk information to different audiences and end users. Risk communication is a function that often is not assumed to be part of a scientific institution, but rather an additional, if not voluntary work of certain social groups. Risk communication requires capabilities from different disciplines. It is also necessary to consider that the effectiveness of communication has a large experiential component. Ideally, risk communication needs stronger involvement of social scientists to provide empirical baseline and evaluation on the changes of risk perceptions and how risk communication effectively takes place (Rohrmann 1992). Such inputs greatly contribute to the design of effective risk communication. A good example is the Istituto Nazionale di Geofisica e Vulcanologia (INGV) in Italy, where there is an attempt to provide spaces for social scientists to be directly involved in the day-to-day activities of tsunami hazard, risk analysis and risk communication.

The role of risk communication practitioners is still undermined in terms of job or career opportunities as well as training and higher education. Systematic analysis of global scientific publications on science communication reflects the asymmetries between high-, middle- and low-income countries in such capacities (Guenther and Joubert 2017). While literacy is key to shaping risk perceptions, the historical subjugations and colonization that lead to slow progress in science literacy in 'less advanced' countries compared to 'advanced' countries also play a critical role (du Plessis 2012; Regmi 2022). On the other hand, communicating risks also needs to take into account local knowledge, as well as avoiding judgements towards the general population as 'irrational', as different layers of communities have their own respective meanings of 'risk', and how these meanings are socially constructed through and in day-to-day lives. Local knowledge is essential to be included from the beginning of any initiative related to local disaster risk management, in order to ensure that the objectives pursued are truly fruitful.

8.2.2 Understanding the Audiences of Risk Information

When communicating scientific findings from tsunami hazard and risk analysis, scientists may face different audiences and users of risk information, hence there is a need to better understand how to engage in effective communication with these different target groups. It is important that all the actors involved in risk communication understand the different characteristics, needs, concerns and communication preferences of each group and take these into account in the communication process. Common audiences include public institutions, especially local authorities and decision-makers, public and private media, civil society organizations (NGOs, volunteer groups, neighbourhood associations, associations linked to economic activities such as tourism, recreation, fishing, etc.), the general public, including groups with special needs, and, to a greater extent particularly in the case for probabilistic approaches to risks, the private sectors such as businesses, critical infrastructures and insurance companies.

With the introduction of probabilistic tsunami hazard analysis, the already existing needs for risk communication to all these different target groups is further increased, as the results from this new approach are not easily understood by the non-experts and therefore require explanations.

Among public institutions, disaster management authorities from national to local levels have a strong interest in accessing and understanding scientific results regarding tsunami hazards and risk assessments as such information is of fundamental importance for all phases in the disaster risk management cycle. In particular, disaster management authorities rely on scientific knowledge for comprehensive hazard analysis, but may draw more on their own capacities in the case of vulnerability analysis. For example, the local disaster management authorities are better equipped to understand their own social contexts and vulnerabilities. The same applies in principle to local authorities and decision-makers, who play a critical role to manage tsunami preparedness and response at the local level. In this regard it is important to have in mind that probabilistic approaches may not match well with the common practice of disaster managers who often are basing preparedness and response planning on a single selected reference scenario.

It is particularly important to note that the institutional use of tsunami hazard and risk information for a practical application involves a strong regulatory and political component. Since tsunami hazard analysis can be based on different approaches and involves uncertainties and probabilities, it is always a political consideration to decide on which approach and scenarios local planning and preparedness should ultimately be based on. This often relates to the question of whether, for example, a scenario with a certain probability, a most probable, or the worst-case scenario will be used for planning purposes and hence what level of acceptable risk a community is willing to take. Experience has shown that such a political decision can be very challenging for the decision maker, especially when several hazard analyses by different actors are available for an area based on different approaches or methods. This might be different in cases where specific regulatory frameworks are in place providing indications on methods, parameters and data sources to be used when developing official tsunami hazard and risk maps. Another important issue to consider is what kind of tsunami source mechanisms are relevant and how they are taken into account, especially as non-seismically induced tsunamis, and particularly their recurrence times, are still poorly understood.

The communication of scientific findings and results, as well as explanations and accompanying advice to the relevant institutions and decision-makers at the national and local levels, is certainly one of the most important application fields for risk communication by tsunami scientists. On the other hand, it is always important to have the implications of the political dimension of hazard information in mind, when communicating with other audiences.

The general public is another primary target group for risk communication in the context of tsunami risk assessment. This group includes people who live, work, or visit the areas at risk of a tsunami. It is not always the case that scientists directly communicate with the public. Generally, this is mediated by risk communicators, relevant capacity development actors or the media. Nevertheless, when communicating risk information, it is critical that scientists understand the particular needs and characteristics of the population, inside and outside of the risk areas, as well as people with special needs. In particular children, the elderly, and people with disabilities, may have different needs and communication preferences. Another group to consider consists of tourists and visitors. They may not be familiar with the local geography or the tsunami risk and not speak the local language. Risk communication efforts should consider the unique needs of these special populations.

When preparing information for the public, various aspects have to be taken into account. First of all, information should be provided in a language that the majority of the population can understand and in a clear and simple way, avoiding technical jargon or complicated terminology. Information can be provided in different formats, including written materials, videos, and graphics, to reach people with different levels of literacy. Visualization and other forms of media can also help enhance understanding. Hazard information, especially when it comes to worst-case scenarios, should be communicated in such a way that it can be appropriately contextualized by the average citizen and with sensitivity to avoid causing unnecessary panic or anxiety.

Other aspects to have in mind refer to the cultural background of the audience as they may have different perceptions of risk. In this regard, it is essential to understand the respective cultural norms, beliefs, and values. In addition, it may play a role in whether people had previous experiences with tsunamis, as they may have different perceptions of the risk compared to those who have not.

Public and private media play an important role in disseminating information about a tsunami and raising awareness of the risks to different audiences. Effective communication with the media can help ensure that accurate information is provided to the public and that the message is disseminated widely. On the other hand, experiences show that the media often tend to sensationalize possible impacts of natural hazards and disasters, which can lead to misinformation and confusion (Walters and Mair 2019; Dootson et al. 2021; Rafliana and Arif 2022). Misinformation, due to problems in understanding concepts, terminology, and contexts is not uncommon, especially when treating complex or complicated issues. Root causes of vulnerability may sometimes be not revealed, thus affecting risk perceptions and communications. There are also cases where public and private media may be subject to political or economic influence or bias, which can impact the way information is presented and received by the public. Good preparation for interaction with the media can be very helpful in this respect. This may include doing some prior research on the media outlet and the persons who will be involved, preparing key messages and explanations on complicated issues which are easy to understand and being aware of critical aspects that might arise.

Businesses and industries located in areas at risk of a tsunami may have different needs and concerns than the general public. The assessment can help to make informed decisions about investments, site selection, and risk management strategies to ensure business continuity and protect assets, employees, and customers. Of special importance here are critical infrastructures, like harbours, airports, or industries located in tsunami-prone areas. By integrating a probabilistic tsunami hazard and risk assessment into their planning and design processes, operators of critical infrastructure can identify exposed assets, assess and estimate their social and economic vulnerabilities, design protective measures, establish back-up systems, and develop emergency response protocols to minimize the impact of tsunamis. Risk communication efforts should consider the unique needs of the private sector and provide information that is relevant to prepare for a tsunami and protect their employees and assets.

The results of probabilistic tsunami hazard assessments and derived guidelines for zoning and building codes are relevant for the insurance sector to develop appropriate insurance products, pricing strategies and incentives to reduce the vulnerability of insured properties to tsunamis, such as enforcing certain building standards or encouraging the installation of protective infrastructure.

8.3 Main Aspects and Challenges of Risk Communication Concerning Tsunami Hazard and Risk

The question of what to communicate is very important. Information can be obtained from nearly all the dimensions of tsunami risks: tsunami models, hazard analyses, vulnerability analyses, exposures, and risk analyses. Through Table 8.1, this chapter provides ideas on how the different ingredients of risk assessment can be conveyed, what main key messages to communicate, through which means, engaging which actors, and some good examples to benchmark. These sources of information can be communicated independently or combined, for example, through risk metrics and criteria that can be used by different user groups. This information could then be translated into hazard maps, flood and evacuation maps, preparedness messages, tsunami drills, dissemination material (e.g., brochures, smartphone apps, etc.), etc.

There are many sources of information related to tsunami risks that reach the public and stakeholders: hazard source information, hazard modelling, exposure data, vulnerabilities and risk metrics or assessments. Often, the information is not very easy to communicate. Sciences and studies on tsunamis need further translations to equip users with objective, well-founded decisions when dealing with tsunami threats. The challenges that sciences must deal with are often left as a remaining gap of information and translations, and thus also potentially create social problems such as misinterpretation, misunderstanding, misexpectations or even compromising trust. The challenges are even greater when addressing probabilities and uncertainties as part of the equations in the science of understanding risks. Two-way communication and stakeholder engagement are key aspects of strengthening decision-making under uncertainty. Participatory communication approaches can enhance inter-agency and stakeholder relationships to help identify decision-relevant uncertainties prior to an event and ensure that the right information is communicated

Table 8.1 Ingredients to tsunami hazard and risk assessment and their relations to risk communication

Hazard/risk information	Key point on probabilities and uncertainties to communicates	Possible use of ingredients	Actors involved	Methods and good examples
Ingredients	What	For What	Who/whom	Which/where
Tsunami source see Chap. 3	Tsunami sources comprise events which happened in the past, happen in the present and could happen in the future. It comes with two types of uncertainties: 1. Many variabilities of the sources both seismic or non-seismic are aleatoric uncertainties and related to natural or geological variabilities. 2. Epistemic uncertainties are related to the assignment and calculation of probabilities due to the limitations of the quality and availability of data	Tsunami warning system; Tsunami hazard maps; Evacuation planning	Tsunami scientists; social scientists; journalists; science communication practitioners; local government; NGOs; Civil society organizations (CSOs); social groups; Communities at risk	A merge or combination of classical and evolved unidirectional communication *Good examples*: GAR15, GAR19, New Zealand national PTHA maps, (MCDEM) and Italy (DPC) based on inundation maps for a design return period and prescribed level of epistemic uncertainty
Source modelling see Chap. 4	Source modelling attempts to address the important questions: how frequently a tsunami could be generated in a particular place? And how big? To do so, science needs to introduce varied probabilities. The difficulties in explaining probabilities to non-scientists or end users often lead to misunderstandings and eventually compromise public's trusts	Tsunami warning system; Tsunami hazard maps; Evacuation planning	Tsunami scientists; social scientists; journalists; science communication practitioners; local government; NGOs; CSOs; social groups; Communities at risk	

(continued)

Table 8.1 (continued)

Hazard/risk information	Key point on probabilities and uncertainties to communicates	Possible use of ingredients	Actors involved	Methods and good examples
Ingredients	What	For What	Who/whom	Which/where
Tsunami modelling see Chap. 5	Tsunami modelling aims to image a tsunami event, its propagation in the ocean, its height at the coastline and its inundation on land. Modelling is based on assumptions, mathematical calculations and references from historical events. As part of the probabilistic approach, tsunami modelling incorporates uncertainties in its calculations. It is important to be clear and honest with the limitations of tsunami model results, because often models are misunderstood by the public as fact or reality, rather than a suggestion	Tsunami warning system; Tsunami hazard maps; Evacuation planning; Contingency planning	Tsunami scientists; social scientists; journalists; science communication practitioners; local government; NGOs; CSOs; social groups; Communities at risk	
Exposure see Chap. 6	Exposure assessment encompasses humans, infrastructures, built environment and critical facilities in the tsunami inundation zone. Exposure assessment, combined with hazard assessment and vulnerability analysis is part of the risk analysis. Exposure assessment can provide a better tsunami evacuation simulation and planning, and estimation of loss and damage that is critical for mitigation policies	Exposure, inundation, land cover and urban morphology maps; Agent-based simulations; Evacuation planning; Contingency planning; Relocation planning; Rehabilitation and reconstruction planning	Scientists and engineers; science communication practitioners; journalists; local government; Community at risk	Can use a combination of all communication methods (Classical and evolved unidirectional methods, social media, bi-directional and participatory communication). *Good examples*: multi-methods for evacuation simulations and tsunami drills in the USA based on population exposed to tsunami risk to improve evacuation planning and identify demographic sensitivity (e.g., see example study 8 in Løvholt et al. 2019)

(continued)

Table 8.1 (continued)

Hazard/risk information	Key point on probabilities and uncertainties to communicates	Possible use of ingredients	Actors involved	Methods and good examples
Ingredients	What	For What	Who/whom	Which/where
Vulnerability see Chap. 7	Vulnerability is a key component for tsunami risk assessment. Vulnerability typically attempts to quantify the possible effects of a tsunami for exposed assets in terms of damage and/or loss. However, social vulnerability is also an important key aspect that includes subjective risk perceptions, behaviours, and decisions often even more uncertain and extremely challenging to measure quantitatively. This is also where local knowledge is important	Evacuation planning; Contingency planning; Relocation planning; Rehabilitation and reconstruction planning; Spatial planning, etc	Scientists and engineers; science communication practitioners; journalists; local government; relevant development sectors; social groups incl. schools; community leaders	Can use a combination of all communication methods *Good examples*: InaRISK Indonesia Apps, STEP-A Apps to measure school preparedness, "Community understanding of tsunami risk and warnings in Australia" Project
Aggregate risk assessment. See Chaps. 1 and 2	Risk assessment is based on hazard, exposure and vulnerability assessment and attempts to address the frequency of tsunami intensity and its consequences. The combination and the complexities of hazard, exposure, and vulnerability (often hindered by the limited knowledge and data availability) make risk analyses difficult and uncertain. On the other hand, it is important to opt for a comprehensive approach in analysing risks. The probabilities of hazard sources, the uncertainties of human and environment dimensions and the complex and cascading effects of a tsunami scenario are part of the challenges that need to be addressed effectively	Tsunami Hazard map; Evacuation planning, Contingency planning, relocation planning, Rehabilitation and reconstruction planning, spatial planning, etc	Tsunami scientists; journalists; science communication practitioners; national and local government; social groups; communities at risk; private sector; CSOs; relevant development sectors; community leaders	Can use a combination of all communication methods *Good examples*: 50-year resilience master plans for Seaside Oregon, Maximum Considered Tsunami, GAR15, Science for disaster risk management guidelines (European Commission)

at the time of crisis (Hudson-Doyle et al. 2018). Probabilistic tsunami hazard and risk results and their associated uncertainties can be communicated in a verbal, numeric or graphic format. Different presentations of statistical and visual data can greatly influence risk perception and understanding (Schneider et al. 2023).

A risk analysis provides information on the potential loss of life, injury, destroyed or damaged assets which could occur to a system, society or community in a specific period of time, determined as a function of hazard, exposure, vulnerability and capacity. It is important to consider the social and economic contexts in which disaster risks occur and that people do not necessarily share the same perceptions of risks and their underlying driving factors. Acceptable risk, or tolerable risk, is therefore an important concept (see Chap. 2: Frameworks for Assessing Tsunami Hazard and Risk) which describes the level of potential losses that a society or community considers acceptable given existing social, economic, political, cultural, technical and environmental conditions (see UNDRR 2023 terminologies on disaster risk reduction). These are considerations that risk communication practitioners should discuss thoroughly and at eye level with scientists, policy-makers, and end users of risk information.

8.3.1 Requirements for Risk Communication Before or During a Crisis

Timing is important. It makes a big difference for what purpose and at what moment information is to be provided. Risk communication is important in all segments and phases of both the traditional and simplistic disaster management cycle (Bosher et al. 2021) as well as in the whole risk governance process approach (IRGC 2009; Aven and Renn 2010). However, the main communication contents and messages might be different. In "peacetime", risk communication supports the prevention and mitigation phase as well as the preparedness and response phases, also building trust, confidence, credibility, and reputation for institutional and scientific sources, including local authorities and emergency managers. This is the most important part of risk communication and all measures, instruments, and contents which have been described above are relevant to this.

During times of crisis and especially immediately after disasters, local authorities, including emergency responders and local governments, play a critical role in managing the response to a tsunami, also relying on the credibility, trust, and reputation patiently built-up during peacetime. Effective communication with local authorities is essential to ensure that they have the information they need to make informed decisions and coordinate effective response efforts. Experiences from past events show also that scientists play an important role in communication with the media to explain the disaster and their consequences. This type of risk communication is also very sensitive because it tends to become political very quickly, since blame and responsibility are inherent dimensions of crisis and disasters (Seeger 2006), as in any disaster public attention and media are quite likely to focus on failures in the system and in finding responsible people or institutions to blame (Wachinger et al. 2013; Cerase 2021). This can lead to delicate situations where one misleading or failed statement can undermine all the efforts and confidence-building activities of the prevention and mitigation phase.

The dissemination and communication of early warnings immediately before or during a crisis is a very specific part of risk communication and has its own requirements, policies, and procedures. The provision and dissemination of tsunami early warning messages are the responsibility of public authorities worldwide, while science can contribute to monitoring information. For these reasons, this topic is not further explored in this chapter.

8.3.2 Considerations for Risk Communication Regarding Risk Perceptions

Since risk perception and risk communication are bonded together through a reciprocal relation, the imperative need that follows is to establish a strong connection between risk communication, strategic planning, risk perception theories and research findings. We refer to risk perception both as a cognitive phenomenon relating to the individual and as a social construct, that is the structure of socially shared meanings attributed to the phenomenon (Rafliana et al. 2022).

According to a still popular definition, risk perception is about "the judgments people make when they are asked to characterize and evaluate hazardous activities and technologies" (Slovic 1987). Researching risk perception is important for different reasons: (1) risk perception is both a logical and a statistical predictor of behavioural and social response to risks, as poor/inadequate risk perception is likely to result into poor mitigation actions; (2) consistent and solid results from risk perception research can improve scientists' and authorities' awareness on the way risk is perceived and understood by different individuals within different social contexts and should therefore be assumed as a fundamental basis to develop sound communication tools and strategies. Research literature has highlighted the importance of many relevant factors in understanding and predicting risk perception. Regarding tsunamis, we can identify three main factors that can influence the perception of tsunami risk: (1) the severity and extent of potential damage of a certain event (catastrophic potential); (2) the estimated probability of its occurrence (frequency/familiarity); and (3) the distance of household from the coastline (hazard proximity). However, with respect to other natural hazards, tsunamis appear as a particularly elusive and unfamiliar phenomenon, due to the relatively long return periods of larger events (particularly in the Mediterranean area), the difficulty to recognize and understand small tsunamis as an actual hazard, the variety of physical events that can trigger a tsunami, and the potential difficulty in addressing the relationship between causes and effects (Cerase and Cugliari 2023). Another factor that can influence tsunami risk perception, is potential conflicting social or economic interests. An example could be made for the tourism sector, of which narratives of risks often create inconvenience, and hence tend to neglect risks.

It is imperative to pay special attention to the impact of PTHA/PTRA approaches on risk communication, since the probabilistic assumptions and uncertainty estimates associated with these types of models can be unfamiliar and even misleading for lay people. In particular, low probability and long return periods of these events are likely to result in low familiarity, and thus in a tendency to underestimate or neglect tsunami risk, except in areas where a tsunami has actually occurred in relatively recent times (Cugliari et al. 2022). The greatest challenge is to communicate understandable, usable, and concretely viable complex concepts such as the distribution of probabilities and intensities and related uncertainties.

8.3.3 Considerations for Communicating Probabilistic Risk Information

Modern science emphasizes rigorous understanding of a broad variety of likely events, moving away from deterministic to probabilistic statements when analyzing risks (see Chap. 2: Frameworks for Assessing Tsunami Hazard and Risk). As knowledge on hazards and social uncertainties has improved, there is much better guidance for dealing with complex risks (Funtowicz and Ravetz 1990). However, decision-makers often prefer deterministic approaches that suggest a particular degree of certainty that can be easily translated into actions with appropriate investments to mitigate risks and implemented in a relatively short time. Visible, concrete, and tangible science results are what policy-makers prefer.

In the area of tsunami risk, the emerging role of the PTHA/PTRA approach challenges the role of scientific advice as well as the role of scientists as risk communicators in that they must provide the public with scientific information based on probabilistic concepts that are susceptible to ambiguous interpretations (Rafliana et al. 2022). The way lay people understand probabilistic information is relevant in addressing risk perception and can be very different from that of experts and scientists: psychological research from the heuristic-and-biases have dismantled the myth that people have a quasi-statistics reasoning following the rules of probability theory, showing how different presentations of statistical data can affect the way risk is perceived and understood (Meder and Gigerenzer 2014). Gerd Gigerenzer, a renowned social psychologist who advocates for teaching statistics in schools to improve individuals' and collective ability to make better decisions, has shown that even simple quantitative statements such as rain forecast, being assumed to be clear and unambiguous can result into very different interpretations from people who live in different cities (namely New York, Amsterdam, Berlin, Milan, and Athens) (Gigerenzer et al. 2005). Risk communication recipients may process probability and numerical concepts at diverse levels. This diversity can result in different ways of understanding probability information. Some numerical and graphical formats are more effective than others (e.g., graphic vs numerical, percentage vs frequencies) and some formats (e.g., comparisons) can be misleading for non-experts (Keller and Siegrist 2009). These biases are more likely to occur when it comes to thinking about risks as low-probability, high-damage events, fostering bad decisions and indirect damages due to poor statistical reasoning (Gigerenzer 2006). When it comes to talk of tsunami, scientists should carefully consider these issues before releasing public statements about probabilities, frequencies and return times, and should possibly involve experts from cognitive psychology to test better message formats to convey information to the different public involved in the communication process, including public officials that may have different levels of understanding of probabilistic data.

8.3.4 Considerations for Risk Communication Regarding Uncertainties

This chapter addresses risk communication that takes into account the different types of uncertainties. Two types of uncertainty are often discussed, the aleatory referring to the inherent randomness in the system (natural or geological variability), and the epistemic referring to the lack of knowledge of certain processes such as the structure or the parameters of the model. Both concur in the definition of the probability of a certain tsunami or damage intensity. Assigning probabilities in itself is often problematic and could lead to disagreement among scientists, and miscommunications between scientists and policy-makers as well as end users. A third category, described as ontological uncertainty or deep uncertainty, refers to the capacity of the entire hazard/risk estimation process for reflecting reality. This may for example depend on the abstract nature of the phenomenon or future behaviours/decisions not yet made, which may lead to several possible scenarios of undefined probabilities. Ontological uncertainty is more challenging to recognise and quantify and is evaluated through a qualitative assessment of the model (Spiegelhalter 2017).

Other forms of uncertainty are related to human and social dimensions. There are always certain social uncertainties regarding how individuals, social groups and institutions behave and decide or act upon threats, which are embedded into different social contexts. These forms of social uncertainty are related to uncertain information for decision making, uncertain moral and legal consequences of decision themselves, and the level of social cohesion/conflict in a given community (Funtowicz and Ravetz 1990). These forms of social uncertainty are also related to the confidence towards relevant institutions to foster communication, understanding and cooperation to manage challenges posed by risks, as well as accommodate individual and collective regulation constraints (Funtowicz and Ravetz 1990).

The way epistemic and aleatory uncertainties are managed within the tsunami risk management process is indeed affected by social uncertainty, which includes psychological and socio-cultural issues such as heuristics, biases, cultural, professional and ideological orientations, as well as political and economic goals pursued by different actors and groups (Donovan 2021). These uncertainties affect the way the public, scientists and other stakeholders address and decide on risk issues and to this effect, risk communication might be seen as an anti-entropic resource to reduce social uncertainty by producing, circulating, and legitimizing shared definitions of risk and its related consequences. Consequently, risk communication interventions should always take into account these different types of uncertainties.

An interesting case occurred in Indonesia, for example, where the media disseminated hazard information that was actually only discussed as a scenario in a closed group of experts. The case became viral on social media, causing widespread public anxieties (Rafliana and Arif 2022). Such cases demonstrate the unintended impact of a misleading tsunami risk communication on the reaction of the general public, police, law enforcement, parliament, and the media. There are more examples to learn from such unintended impact of risk and science communication related to uncertainties that occurred in L'Aquila, Italy (Cocco et al. 2015), from the 2010 Maule Earthquake and Tsunami in Chile (Soulé 2014), and the Fukushima Nuclear accident triggered by the 2011 North-East Japan tsunami (Svendsen et al. 2016). Further, the chapter describes the goal, the approach, and the currently available tools (approach, language, media) for effectively communicating tsunami risk. It is important to acknowledge, however, that the approaches need to adhere to the social and cultural contexts where risk communication is applied.

Although tsunami risk communication is steadily growing in many threatened areas, challenges remain cumbersome, notably when discussing risk criteria and acceptability (Guarascio et al. 2007). Once risk analysis results are available, further questions arise: what level of conservatism in risk assessment is acceptable? If so, acceptable by whom? And who should decide the acceptability? As such inquiries turn into political questions and relate directly with policy making, the dialogue approach to risk communication becomes a prerequisite. Last, one could not emphasize enough the importance of building trust towards motives of all parties involved and trust in competent experts and scientists (Twyman et al. 2008).

8.4 Available Methods and Approaches

Tsunami risk communication depends on different communication methods. The methods discussed in this section are applicable for addressing probabilities and uncertainties of tsunami risks. Each method requires different resources, level of engagements, and users. Table 8.2 is developed to provide an overview of five communication methods, listing for each of the methods their intended purpose, potential limitations, and valuable strengths.

Table 8.2 Characteristics of five risk communication methods

Method of communication	Purpose of communication	Examples of format and references	Strengths	Limitations
"Classical" unidirectional	Information transfer from A to B. Communication is hierarchical (Top-Down) No receiver feedback is provided for this method. Much of scientific communication follows this method	Website page, books, leaflets, signages, educational videos, TV news, documentaries, cinema, podcasts, etc Example from online video documentation: Repdeman Mentawai (2021); Website example: CAT-INGV (n.d.); In "itic.ioc-unesco" website there are several examples of brochures, signages and standard documents	Control the over-production process (content selection, proper use of concepts/ languages, unbiased, unmediated). Emphasis solely on information dissemination	No feedback from the public Receiver does not necessarily understand the message. It may generate manipulation or bias by the medium. To reach wide audiences, the cost is on average high. Less sensitive to local, social and cultural contexts
Evolved unidirectional	Information is generally mediated by professionals (journalists, media producers, bloggers, social media influencers, and science communicators) and not directly interacting with users. Information is also transformed according to the media organizer's choices (agenda setting, language, and relevance)	Photo, video and graphic contents, simulations, gamification, story maps, policy papers, with a limited degree of interaction with the audiences Examples from literature: Valenzuela et al. (2017); Sutton et al. (2020); Pignone et al. (2022); Doyle et al. (2022)	It slightly reduces the communication tensions from a hierarchical approach, through a more mediated communication by professionals	It is still difficult to tackle resistances or rejections on the notions of risks due to its lack of inclusivity towards the varied users

(continued)

Table 8.2 (continued)

Method of communication	Purpose of communication	Examples of format and references	Strengths	Limitations
Social-media communication	A multi-directional process by which social media platforms are considered both as channels to disseminate, and a source to collect relevant information. It can be produced by all those who populate the network (from experts to ordinary users). Platforms target information through algorithms that are based on users' tendency searches. It enables the production of data allowing reflections of personal opinions, thoughts, and behaviours	Twitter, Facebook, Instagram, TikTok, etc. Examples from literature: Kim (2021); Lacassin et al. (2020); Jayasekara et al. (2023) Examples from social networks: NEAMTIC (n.d.); Kodijat (n.d.); Research, to date, has not produced satisfactory results. No institutional/ official pages dealing with tsunami risk	The continuously increased number of mobile phone users worldwide have direct impacts on how social media play a role in risk communication. Information through social media is among the most accessible. Social channels can foster emergency communication and provide effective support to disaster-affected populations. It is possible that users interact directly with information providers	Often personal opinions have similar if not more power over official information or statements and co-shape the way risks are also perceived. In this sense, social media communication is prone to practices of hoax or misleading information. It is difficult to control information content widespread through social media networks. The platforms' algorithms generate echo chambers or content filtering bubbles that redirect users to similar content

(continued)

Table 8.2 (continued)

Method of communication	Purpose of communication	Examples of format and references	Strengths	Limitations
Targeted bi-directional communication	This model concerns the two-way exchange of information between two communication actors on the same (horizontal communication) or different levels (vertical communication) and involves both parties sharing knowledge, ideas, rights, and/or responsibilities to risk mitigation. Information is mostly orally transmitted as sharing scientific and field knowledge (community and field experiences, experienced in the past events)	In-person meetings, follow-up courses, seminars and scientific lectures, lectures with multimedia aids, field drills, knowledge, and risk perception test Meetings or courses with Civil Defence/ scientists; Journalists/ scientists; Civil Defence/students; Media producers/ scientists; Governors/ scientists Example from literature: Setiadi (2014)	This method allows a direct exchange between the two communication actors. There are no barriers and knowledge gaps about risks are directly bridged. Face-to-face communication reduces chances of misunderstandings and, at the same time, can generate innovative ideas. It is also possible to conduct learning or knowledge consolidation tests, including longitudinal tests	Usually, this communication takes place face-to-face, involves understanding between the parties and a planning of the subject of the meeting. It is recommended to have in-person meetings involving many people together to increase the exchange of views and facilitate communication. Communication facilitators familiar with the topic can be involved to stimulate debate
Participatory communication	Ideal option to a more symmetrical approach to risk communication where shared perspectives, needs and local concerns are part of the common motivation of communication. Local contexts are thus appreciated. Communication is built from intensive and continuous exchange, Interactions, and public engagements	'Tsunami science meets politics' workshops, co-designed risk reduction projects. Characterisation of open and usable data, co-creation of solutions, data democratization, interpretation of de-segregated data with community involvement Gamification Examples from literature: Rizzi and Porębska (2020); Repia and Bailey (2021)	Users likely to find the communication exchanges more transparent and all actors treated as equal partners. Through reciprocal processes, it is more possible to discuss the varied probabilities of risks with different risk information users. More conducive to build co-sharing responsibility upon the negotiated risks and what it means, and what aspects need to be valued	It is the least tangible outcome of risk communication methods. It takes time and longitudinal processes to ensure ideal participative exchanges. It highly depends on effective mediations/ facilitations. Although preferable and possible, it requires pre-emptive processes to take local social and cultural context into account

8.4.1 Choosing What Methods to Use

The choice of an appropriate risk communication method depends on a number of factors. First and foremost, it depends on the specific context, the purpose, who is to be involved, and the type of information to be conveyed. Secondly, other aspects such as energy, resources and costs cannot be neglected. Therefore, it is necessary to identify the actors who will be involved in the process to optimize the tools that will be used and to produce an initial estimate of the resources that will be needed. Considering socio-demographic data (e.g., age, educational degree) and the reference interests of the target audiences to which the communication is addressed makes it possible to quickly reach the communication target (if any) and to optimize costs and time. Other drivers for choosing the most appropriate method may be due to previous experiences or best practices with several methods or evidence that a method is effective in certain circumstances.

Figure 8.1 shows the process flow of tsunami risk communication, summarizing the options in choosing effective method(s) for tsunami risk communication, which are detailed in Table 8.2. The process flow, in its first phase, prompts the study of context-related factors, necessary for proper communication framing. This first stage, which is data-driven, involves the acquisition of territorial hazard data, aggregated socio-demographic data and data on tsunami risk perception. This information will facilitate the identification of the most effective method(s) for risk communication, message content and target audiences/groups to be addressed. The analysis of the collected data will allow the risk communicators to identify the most effective output for risk communication according to the chosen model(s). Following the process implementation, the effectiveness will be evaluated by testing the changes on the intervening variables, e.g., tsunami risk knowledge, policy knowledge, risk perception or situation-related behaviours.

Unidirectional risk communication (see Table 8.2 and Fig. 8.2 for examples) still remains the most widely used communication methodology in science. Books, documents, web pages, newspapers, brochures, podcasts, TV news documentaries, etc., base the transmission of information from the sender to the receiver.

This model is also called communication deficit model as it does not facilitate the opinion exchange and does not take into account many elements such as the target audience, cultural context, prior knowledge, personal experiences, beliefs or behaviour (Suldovsky 2017). In fact, the language used to convey the message depends on the medium (tool that conveys the message) and the communicator (Nakai and Nakano 2023). Among the relevant pros of this method is its ability to reach large audiences at low cost and to produce lasting risk information in tangible media. The message transmitted with this approach is not affected by external inputs that may modify it and there is no interaction between the communicator and the receiver. If the message is correctly oriented to the audience and the communicator (scientist or science communicator) uses the correct language, the result is a linear communication and the decoding of the information will be optimal (Simis et al. 2016). This category includes all printed communication on paper, which is needed in several situations

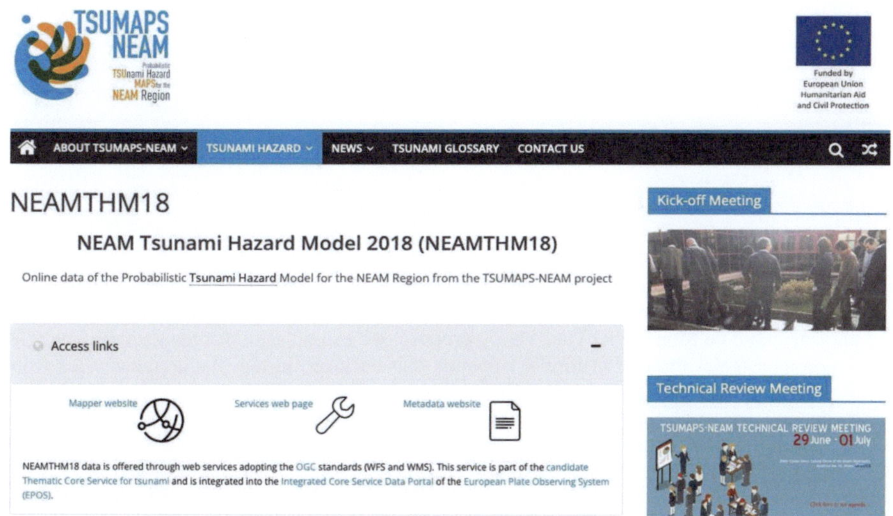

Fig. 8.2 The TSUMAPS NEAM website provides relevant information and tools for end users, including policy-makers, on probabilistic tsunami hazard assessments for the North-Eastern Atlantic, Mediterranean, and connected seas (NEAM) region. Users can access information on the state-of-the-art approaches towards tsunami risk analysis, how it was developed, and how it can be used. This website is an example of the classical unidirectional approach to tsunami risk communication. More use of the information provided by the website can be tailored and applied to other means of communication (from https://tsumaps-neam.eu/, licensed under CC-BY 4.0)

where no other information exchange would be possible. Moreover, this method, in some contexts, facilitates the communication of information where, for example, there is limited access to the internet. In recent years, there has been an increase in using hypertext tools (e.g., QR codes or OCR readers) that enhance the communication offer by creating interconnections between written or displayed information and other sources, even of different types. Among the cons of this method is the possible gap generated by the communicator's language, which cannot be adapted to all audiences and could be misled by personal experiences or misunderstandings.

The use of different devices in tsunami risk communication allows the transfer of information using multiple channels and languages (text, photos, videos, simulators, games, story maps, augmented reality, etc.; Dootson et al. 2023). The information conveyed by these tools is mediated by professionals (journalists, photographers, gamers, influencers or sometimes amateurs) and is part of the **evolved unidirectional method** (see Table 8.2 and Fig. 8.3 for examples). Non-interaction remains between communicator and end-user who, in this case, perceives a reduction in the tension generated by the hierarchy of communication roles. The communicator may in fact be a facilitator. This approach of tsunami risk communication leads to focusing the end-user's attention on the aspect or product of the risk to which the communicator wants to give more emphasis. Among the pros of using this methodology is the adaptation of the language to the target audience e.g., in a video game using virtual

Fig. 8.3 Story maps are interactive digital communication tools that combine maps, text, images, videos and other media for engaging and informative storytelling. The INGV Tsunami Alert Centre and the Université de Montpellier Paul Valéry propose two examples of different story maps. The first proposes a travel through historical tsunamis that occurred in the Mediterranean Sea from 365 B.C. to date with photographic details from historical and recent archives, videos, models, etc. (from CAT-INGV 2021, licensed under CC-BY 4.0). The second traces the French TASOMA project, aiming to contribute to understanding and improving evacuation processes in the event of a tsunami warning along the Mediterranean coast of France (Université de Montpellier Paul Valéry/LAGAM 2020). Both story maps integrate geographical (map-based) and narrative (text) elements to illustrate topics and information in a more engaging and comprehensible way. This tool is particularly useful for educating, raising awareness and involving audiences on geolocation-based topics who benefit from the visual presentation of information

reality visors, the language will be simple and usable for the age group at which the game is aimed, and the narration of the tsunami phenomenon will have various levels of complexity as the producer decides. The author of tsunami story maps will be able to decide the slant he/she wants to give the product to which language and video images will be commensurate. The journalist, depending on the agenda setting and context, which he or she will be clear about, will decide on the best approach.

When communicating with audiences in public meetings or workshops, **bi-directional risk communication methods** may be the choice (see Table 8.2 and Figs. 8.4, 8.5 and 8.6 for examples). The opportunity to ask questions, provide feedback and express concerns promotes understanding, trust, and engagement with the users of the risk information. If the purpose of risk communication is to enable authorities or people at risk to make informed decisions to prepare for or mitigate the effects of a threat, participatory communication, such as focus group discussions, may be more appropriate. Often, it would take more than one-time discussions to come across an agreed term, for example achieving a general agreement regarding 'acceptable' risks. In any case, it is important to identify the target audience in advance, as different groups of people may require different methods of communication. Local residents may be more familiar with the area and therefore require more detailed information about the risk and potential consequences, while non-residents may need more general information, so the provision of visual maps, for example through printouts or websites may be helpful.

When deciding to use **social media** to disseminate messages, it is necessary to analyse the user trends to which the message is addressed. Since social platforms evolve very quickly, the analysis should therefore be repeated each time a content is to be published. E.g., today's young people tend to use the social media that

Fig. 8.4 Activities in promoting better understanding of tsunami hazards and risks in Padang (West Sumatra) and Bali, Indonesia in dialogue formats. The meeting approaches in Padang were later known as the 'Padang Consensus' (see Taubenböck et al. 2009), to discuss and decide on the multi-scenario approach, as well as risk information produced by several scientific organizations. Similar approaches were adopted in Bali, whose outputs were technical documents later officially approved as the main reference for the decision-makers (for further info, visit www.gitews.org) (from www.gitews.org)

Fig. 8.5 Another form of risk communication was exemplified by one of the CoastWAVE project activities that took place in Turkiye—the IOC DG ECHO CoastWAVE project aims at improving understanding of tsunami risks and communication strategies to govern risks (UNESCO/IOC 2021). Risk communication effectiveness was discussed after testing the readiness and preparedness through tsunami drills in Buyukcekmece, Istanbul in 2022. Many methods were used, from tsunami animations to media engagements and post-evacuation meetings involving KOERI, local universities, national education directories, emergency responders and other related actors and institutions

convey information and which they use the most (e.g., TikTok, Instagram and Facebook); they prefer communication based on short, subtitled videos (reels) that contain attractive graphic elements that reinforce and focus the message. The message in this case should be concise and simple, preferably conveyed by people they know or who are popular in social media (influencers, youtubers, science communicators and professional communicators). Surveys and opinion polls can be used to gather data and opinions from a large sample of the target audience. They are useful for

Fig. 8.6 Examples of using targeted bi-directional communication methods in Minturno municipality, Italy. The left panel shows the application of a mixed method with the use of images and hazard maps to discuss several scenarios and the ensuing tsunami risks in the municipality, building interactive dialogue between national and local board members of Tsunami Ready—UNESCO program. The right panel shows the application of a bi-directional approach to communication, targeted at school children

assessing knowledge, attitudes and behaviours and can help to tailor risk communication messages to the needs of the target audience. Speaking the same language and being on the same communication level facilitates exchanges between actors. Trust, risk perception, hoaxes and viral information, misleading information, false expectations are important factors that can play a role, as well as the willingness of the parties to be directly involved in such processes.

Participatory communication for disaster risk reduction in general and specifically as a method of tsunami risk communication is still marginal. Active citizen engagement in scientific and public debate as a trigger for good practices in tsunami risk mitigation and resilience building in local communities and urban contexts may include several activities. Emerging participatory communication processes, for example, include Bottom-Up citizen Initiatives (BUIs) where citizens are involved in mediation processes for natural hazard risk management in their contexts (Thaler and Seebauer 2019). These processes involve local citizen committees proposing projects or initiatives discussed in the implementation of decision-making processes for the development of emergency plans or bottom-up mitigation interventions. Participatory communication also allows for more effective identification and targeting of tsunami risk uncertainties. Indeed, an open debate between scientists, stakeholders, communicators and citizens facilitates the identification of the uncertainty taxonomy related to tsunami risk and to co-develop methods, languages, models and outputs to facilitate interpretation even for events, such as tsunamis, that evolve over time and space. Furthermore, involving citizens to develop easily interpretable and audience-appropriate information on hazards, where uncertainty is embedded in the operational environment, follows the ethical principle of 'audience relevance' (Doyle et al. 2019). Participatory communication also includes the possibility of co-creation and co-design of hazard/evacuation maps based on the local and cultural context and the co-creation of governance through the free use of available data (socio-demographic, satellite, spatial). This last issue is particularly relevant in the contemporary context and, given the fast evolution and wide use of data-using technologies, will be increasingly so in the future (Kanbara and Shaw 2022).

8.4.2 Implications of Choosing the Different Methods

There is no 'one size fits all' category of communication methods. Therefore, it is important to understand each approach's strengths and limitations, and how the different communication methods can complement each other. In an ideal setting, a combination of different methods could complement each other and overcome some specific limitations as illustrated in Table 8.2. Targeted bi-directional communication, for example in a given context, could identify the target audience's information needs. This would not preclude the integration of other parallel communication methods that guide and complement the effectiveness of the transmission (such as stimulating discussions through the multimodality that social media provide through images, videos, navigable maps, and shared personal experiences). To make the communication of uncertainty more effective, the best choice is to adopt, in a complementary way, several communication methods so that one reinforces the other.

8.4.3 Relevance of Risk Communication Versus Crisis Communication to the Choice of Method

Risk communication is done in calm periods or peacetime. It draws from the five proposed methods and can take advantage of the potential and interconnection of different approaches by moving from a "classical" unidirectional communication plan to the plan of communication through social or participatory methods by involving different actors with the use of different tools. The two aspects move on different planes, at different times, and make use of different tools.

Crisis or emergency communication is done in case of an event (including "false" alarms) or during an exercise. Emergency systems are activated, and the communication flow is direct ("classical" communication, sometimes "evolved unidirectional" or via social media). Due to its higher sense of urgency for rapid decision-making, bi-directional targeted or participatory communication are often not the preferred options.

8.4.4 Risk Communication in a Multi-hazard Perspective

Risk communication in a multi-hazard perspective could influence the perception of a specific risk. This might lead to decreased ability to acquire specific information about a single risk. Festinger (1957) defines this behaviour as "cognitive dissonance": individuals select information by focusing more attention on information that reduces the distance between their actual behaviour and what they believe. Multi-risk communication would lead, in this case, to a "selective perception" through which individuals select risk information, depending on experiences and contextual variables to the point of distorting the information and making it consistent within their own value and belief system. Consequently, the most consistent method is chosen based on several communication variables such as the target audiences, the information to be conveyed, the tools through which to convey the message, and other context-related variables such as the occurrence frequency of a phenomenon, whether an event has happened more or less recently, the hazardousness of the area, population density, etc.

8.4.5 Treatment of Uncertainties Within a Chosen Method

In order to better approach the uncertainty aspects related to tsunami risk, it is necessary to consider the whole of the five methods of communication, assessing ex-ante the audiences to be addressed and considering the tools needed to achieve the purpose. The more canonical/one-directional the communication is, the less clarity it brings to the understanding of uncertainties and probabilities. The communication method differs as the target audience varies and, in order to optimize the message and reduce uncertainty, it is essential to involve communication experts, sociologists and social scientists, with a global vision of the entire process, who assess the context and applicability of that method before it is applied.

8.5 Evaluating the Effectiveness of Tsunami Risk Communication

The effectiveness of risk communication is certainly one of the most important and discussed topics in this field of research, making it possible to say that it is a predominantly evaluative discipline (feedback loop shown in Fig. 8.1). Albeit risk communication currently is still a small, interdisciplinary, and not yet fully mature science, it is quickly evolving from an under-theorized collection of practices based on good intuitions to a discipline grounded on well-researched principles, systematic hypothesis testing and formalized lesson learned (Bostrom and Löfstedt 2003).

The effectiveness of risk communication strategies, which can be defined as the "degree to which an initial situation is changed towards an intended state, as defined by the program goals", can be assessed through systematic empirical investigations considering the content, process, and outcomes of the methodology employed (Rohrmann 1992). These can be categorized into three areas of communication development: content, outcome, and process evaluations. The effectiveness criteria for evaluating the outcome of risk communication include improvement in risk knowledge, risk perception, risk awareness, improved risk-controlling behaviour, and changes in risk policies. Different methods and approaches can be used for evaluating the effectiveness of risk communication (e.g., Rohrmann 1992; Downs 2011), while the evaluation can benefit from a program/process model which describes the inter-connection between the risk communication activities, desired outcomes and intended impacts (Astbury and Leeuw 2010; Johnson et al. 2014).

Reasons for evaluating the effectiveness of risk communication include (Rohrmann 1992): (1) determination of the importance of risk communication, proving its effectiveness; (2) the evaluation process, which helps to identify not only if the risk communication program works, but also why. This information can serve to improve the effectiveness of the program in future activities; (3) examination which risk communication strategies work in which situations; (4) collection of information for choosing between alternative communication strategies depending on the objectives and target audiences.

Indicative successes of tsunami risk communication processes are proved by time, either in the occurrence of real events, or reflected in tsunami drills and exercises, as well as by re-assessment comparisons after initial risk communication interventions. The effectiveness of risk communication for real events can be judged in terms of risk communication as a whole, including individual behaviour, education, warning systems in place, risk reduction policies and political decisions. The lesson of the 2004 Indian Ocean tsunami highlighted how the inadequacy or absence of risk communication tools amplified the impact of the physical event, also and especially in terms of human lives lost (Rahman and Munadi 2019). In fact, the lesson from the event made clear how communication to the public is an essential part of any disaster warning system, not only for message dissemination but also to improve people's understanding of the nature and magnitude of the hazard, as well as its physical characteristics (Mayhorn and McLaughlin 2014). Moreover, a recent study conducted after the 2018 Palu earthquake and tsunami highlighted the importance of tsunami education to recall the essential hazard information and maximize the chances of survival (Wang et al. 2022). It is important to communicate the information clearly, on time and in a way that allows the end-users to consider options and act appropriately (Fakhruddin et al. 2020). At first, it is important to understand the social groups, how they think, feel and do about risk. Next, the technologies must be in place together with an early warning system. As Fakhruddin et al. (2020) suggest, "collective risk thinking needs to be the main objective where a catastrophic risk is modelled to a single citizen behaviour".

Tsunami drills and exercises help to build community resilience and offer the opportunity to evaluate tsunami preparedness during peacetime (UNESCO/IOC 2020). Tsunami drills, for example in schools, can also be evaluated based on the comprehension of protective actions practiced by the participants (e.g., Johnson 2013). Drills many times are part of wider tsunami educational programs which can be evaluated through empirical studies (e.g., Dengler 2005). The empirical studies used to evaluate the effectiveness of risk communication can be either cross-sectional or longitudinal, the latter considering the 'ante' condition in addition to the 'post' condition, i.e., the examined condition prior to the risk communication activity/intervention. The effectiveness of any tsunami risk communication activity or intervention can be evaluated through empirical studies. The results of the evaluations are used to update the risk communication context and if needed to adapt the process design accordingly in a continuous feedback loop (Fig. 8.1).

It is also important to evaluate the effectiveness of specific tsunami risk communication products such as hazard and evacuation maps, and educational material/products (brochures, websites, etc.) when possible, particularly in terms of the decision-making effectiveness of the products. To evaluate risk communication tools/products requires defining evaluation criteria based on the objectives. In this respect, Lindell et al. (2021) produced an evaluation rubric comprising eight criteria for emergency managers to assess the effectiveness of tsunami evacuation products, while Lindell et al. (2022) conducted a quasi-experimental evaluation of tsunami evacuation brochures distributed in the five Pacific Coast states of the US. An experimental example of evaluating the effectiveness of tsunami risk communication and knowledge raising was carried out in Italy by INGV's Tsunami Alert Centre, as part of a study on tsunami risk perception, involving students from a high school (Cugliari et al. 2023).

In general terms, it can be said that effectiveness cannot ever be taken for granted, and that it must be verified both before and after the deployment of tools and strategies (Morgan et al. 2001; Boholm 2019; Árvai and Rivers 2014). In this regard, it must be noted that in order to talk about communication effectiveness, first, it is necessary to clearly set objectives and limits (Rowan 1991; Leiss 2004; Fekete 2012), taking into account the human and economic resources that can be deployed and the peculiar characteristics of the audiences to which they are addressed (Bostrom 1997; Sellnow et al. 2009). Secondly, it is crucial to realize that no matter how clear, ethical and socially accepted the goals of communication are (Thompson 2012; Leiss and Larkin 2019), and what resources are available, it is unrealistic to expect messages to be equally effective for everyone, as demonstrated by a long tradition on the effects of communication. Third, it is important to remember that alongside the intended effects, a number of unintended or undesirable effects may occur, such as psychological reactance, source mistrust and stereotyping, and even boomerang effects (Salmon et al. 2014; Hart 2014; Rickard 2021), including the uncertainty of events that occurred as deviated or beyond what science had analysed and communicated.

More embedded impacts of effective risk communication can be traced through improvements of risk reduction policies at different societal levels. As a continuous feedback loop, effective risk communication will eventually attract more interests and sustainable support for risk communication practices. It would be a highly successful achievement also if exchanges and programs of risk communication are adopted as a permanent process, and through stronger interdisciplinary, intersectoral and intergenerational interactions.

8.6 Summary

This chapter focuses on risk communication in the context of tsunami hazard and risk assessment, highlighting the importance of scientists engaging with the general population and various stakeholders in order to effectively engage in conversations on risks and scientific findings with different users. This chapter emphasizes how risk communication goes beyond one-way communication and involves interactive and multidirectional approaches with different actors. It also underscores the dual purpose of risk communication: providing information to understand and address risks, and fostering trust between authorities, experts, and the public. The role of risk communicators is essential in bridging disciplinary gaps and making risk information accessible to diverse audiences. It also addresses challenges such as the involvement of social scientists, relational asymmetries, cultural and linguistic factors, and the role of media and private sectors. Additionally, it emphasizes the need to tailor risk communication to specific target groups, including public institutions, the general public, and private sectors, while emphasizing the importance of clear and contextualized communication to prevent misinformation.

The content and timing of the communication are some of the foremost important aspects of effective risk communication. Difficulties in conveying complex scientific information and the need for clear explanations to facilitate well-informed decision-making is emphasized. The importance of engagement in two-way communication and involving stakeholders in decision-making processes, particularly when dealing with uncertainties is also considered. Timing and purpose of risk communication, with different communication contents and messages during peacetime and crisis situations form part of the discussed issues together with the role of local authorities and scientists in risk communication.

Non-experts in hazard or risk and lay people may exhibit significant variations in understanding, perceiving and interpreting probabilistic information. Consequently, to be more effective, tsunami risk communication needs to take into account and adapt the language and communication tools catered to the target audience, while acknowledging concepts of risk perception. In order to achieve this, close collaboration between scientists from an interdisciplinary perspective is required.

Among the most challenging issues are the evaluation of uncertainty and the matter of conveying uncertainty in results. Throughout the chapter, the discussion has focused on two primary types of uncertainty: aleatory and epistemic. Aleatory uncertainty is rooted in the inherent randomness within the system, while epistemic uncertainty arises from a lack of knowledge regarding specific processes, such as the model's structure or parameters. The resulting assignment and calculation of probabilities can cause disagreements among scientists, miscommunications with policy-makers, and confusion for end users. Two other forms of uncertainty are discussed, namely ontological and social uncertainties, all of which should be taken into account for effective communication of probabilistic risk information.

A number of available methods each requiring different resources, level of engagements, and users are presented in the chapter, considering their potential limitations and valuable strengths. The discussion revolved around the main drivers (specific context, purpose, involved parties, identifying the involved actors, socio-demographic data, target audience interests, previous experience and evidence of effectiveness) in selecting the appropriate risk communication method. Recognizing different stakeholders, audiences, needs, knowledge levels and perspectives is important for effective risk communication. Effective risk communication also requires tailoring messages and delivery methods to specific audiences, such as local communities, policy-makers, scientists, or vulnerable populations. The effective translation of complex scientific information and probabilistic assessments into understandable language is an ongoing challenge. Bridging the gap between technical experts and the general public requires clear and concise communication strategies that convey probabilistic information in a way that is easily understandable, relevant and meaningful to different audiences.

The chapter also describes ways to evaluate the effectiveness of risk communication strategies, which can be assessed through systematic empirical investigations considering the content, process, and outcomes of the methodology employed. These are categorized into three areas of communication development: content, outcome, and process evaluations. The effectiveness criteria for evaluating the outcome of risk communication include improvement in risk knowledge, risk perception, risk awareness, improved risk-controlling behaviour, and changes in risk policies.

The role of emerging technologies is pivotal in enhancing risk communication endeavours by enabling more efficient and widespread dissemination of risk information. Technological advancements such as social media, data visualization tools, virtual reality, and mobile applications provide platforms for interactive and engaging communication, allowing for real-time updates, interactive maps, and immersive experiences. Moreover, these technologies facilitate two-way communication, empowering the public to actively participate in risk discussions and share their own perspectives and experiences. By integrating emerging technologies, risk communication becomes more accessible, reaching a wider audience and empowering individuals to make informed decisions and take appropriate actions in response to risks. As more research and implementations are needed related to the use of such emerging technologies, the chapter provides an outlook for further discussion in the following Sect. 8.7.

8.7 Outlook: The Role of Emerging Technologies in Risk Communication

Technological advancements and increased availability of computational power are expanding the available means and have the potential to improve the effectiveness of risk communication. The available risk communication methods outlined previously can be enhanced through emerging technologies such as Virtual and Augmented Reality. While in the Virtual Reality (VR) environment, the user is totally immersed in a synthetic world, where real-world physical laws may not apply, Augmented Reality (AR) includes virtualized and simulated content overlaid in the real environment and exists within the Mixed Reality (MR) environment where the real and virtual worlds are merging (Milgram and Kishino 1994; Anderson et al. 2021). VR, AR, and MR, also collectively called eXtended Reality (XR), allow the users of the technology to both immerse into an enhanced presentation of given virtualized and simulated contents, but also to interact with the content in both the real and virtual environment scenarios.

Literature reviews show that XR technology is applied extensively in disaster management, particularly in the areas of numerical simulations, interaction techniques, training, infrastructure assessment and reconnaissance, and public awareness (Khanal et al. 2022), and in emergency management in the built environments, including pre-emergency preparedness, response during an emergency, and post-emergency recovery (Zhu and Li 2021). Falsaperla et al. (2022) set up an AR exhibition through a "talking poster" for seismic preventive measures that enhance resilience. Rydvanskiy and Hedley (2021) developed single-user and collaborative flood visualization prototypes based on geospatial datasets and a state-of-the-art MR system, to explore the possibilities of MR technology in improving flood risk management and spatial planning. Erikson et al. (2018) developed VR and 360-degree 3D videos to communicate future coastal flood hazards for enhanced stakeholder engagement and outreach. Mol et al. (2022) used immersive VR technology to examine whether a simulated disaster can stimulate people to invest in flood risk-reducing measures. Another tool developed for science education and public outreach is the Augmented Reality Sandbox (e.g., Reed et al. 2016) which is a 3D visualization application allowing users to interact with the model and better understand the freshwater lake and watershed processes. Zhou and Lynett (2020) have developed an interactive AR tsunami simulation system that allows users to visualize a tsunami in a particular coastal location, aiming at raising coastal hazard education and awareness. Such applications rely on the increased availability of computational power (predominantly Graphical Processing Unit power), either on a local device or through cloud services. The aforementioned studies show promising applications of XR technologies. However, the use of this technology for disaster risk management is still in an early stage of development (see Khanal et al. 2022 for a review of existing gaps and future directions), and by extension also for tsunami risk communication.

Another approach is gamification (Hawthorn et al. 2021), which with further improvements in the near future will be able to contribute not only to enhanced tsunami awareness, but also to understanding risk and its complexities through vivid visuals. An interesting example is the application of tsunami survival game in Roblox (Playtad 2023), a widely recognized, engaging and accessible gaming platform accessible to children and younger generations in almost all parts of the world.

Currently, communication on tsunamis through different platforms is widely used. On the one hand, there are social information channels that operate in "peacetime", which disseminate information material (of various types), to the general public and maintain a continuity of information by interacting with the public to answer any questions or share material produced by users. On the other hand, social channels that operate "in an emergency", disseminate alerts and are uniquely (or more) operational if the need arises. Collecting and processing crowdsourced data can improve situational awareness and crisis risk communication.

Kankanamge et al. (2020) analysed the disaster-related Twitter posts between 2010 and 2011 in Southeast Queensland, Australia. The findings confirmed that social media information simply does not generate new information. The study by Kankanamge et al. (2020) shows that the value should be generated by "multilateral collaboration among disaster management authorities, the community, and community-based organizations". Brajawidagda and Chatfield (2012) performed a social network analysis of Twitter information flows among the central disaster warning agency's Twitter followers during the 2012 Indonesia earthquake. One of the most important implications of this study was a recommendation to develop government-to-citizen relationships with the most influential followers of the government agency's Twitter disaster warning network for timely information diffusion. Ai et al. (2016) designed and developed a prototype combining a geographical information system with a social media-based dynamic decision support system enabling self-organized information networks to support decision-making and collective actions in emergencies. It is deployed in Padang, Indonesia, to build a communication network between the government and the public, using the community leader as a middle actor in disseminating evacuation routes.

Tsunami risk communication can also benefit from advancements in Artificial Intelligence (AI). Ogie et al. (2018b, a) have found research activities in that field focusing on two broad areas of disaster risk management: (i) prediction and monitoring for early warning systems, and (ii) information extraction and classification for situational awareness. Mulia et al. (2020, 2021), Liu et al. (2021), and Makinoshima et al. (2021), amongst others, have explored the use of deep and machine learning algorithms to provide fast and high-resolution tsunami forecasts for tsunami early warning applications. The use of such methods can support risk communication by allowing more time for early warning during an emergency. In the field of information extraction and classification for situational awareness, AI technology can be used to extract information that is useful for emergency managers during disasters from crowdsourced social media content. Ogie et al. (2018b, a) provide a comprehensive review of such AI-based studies found in the literature.

Artificial intelligence and machine learning methods, such as natural language processing (NLP) and neural networks, are promising although not yet incorporated in emergency settings. NLP-based chatbots or smartphone personal assistants, for example, could be used for risk communication in emergency situations where human resources are limited. With the help of AI communication messages could be tailored to reach out to groups of diverse demographics and population minorities to increase the effectiveness of the message.

There is also a need in the future to turn towards sustainable urban development and adopt smart city initiatives. Consequently, emerging technologies such as intelligent communication systems and the Internet of Everything (IoE) that support continuous connectivity could be explored and employed for efficient risk communication.

8.7.1 What Does This Progress in Technologies Mean for More Effective Risk Communication?

New emerging technologies support more effective risk communication in several aspects. They can enable us to forward important information in a more personalized way to the right target groups, use prediction models to gain time for warning during emergencies, present prediction results in a more understandable way or train relevant safety and organization processes in more user-friendly ways. On the other hand, the use of new technologies will be only as effective as the quality of the inputs like data, domain knowledge or previous experiences represented in digital form is. Ultimately, new emerging technologies and communication methods are important to accommodate the complex information on tsunami risks and its layers of uncertainties, through more communicative and engaging approaches that can be provided through interactive visuals and images. To eliminate hesitance often felt by decision-makers in understanding and applying the tsunami risk probabilities in policy making, such new emerging technological supports could offer enhanced solutions and facilitate the current issues with regards to PTHA/PTRA approaches.

References

Ai F, Comfort LK, Dong Y et al (2016) A dynamic decision support system based on geographical information and mobile social networks: a model for tsunami risk mitigation in Padang, Indonesia. Saf Sci 90:62–74. https://doi.org/10.1016/j.ssci.2015.09.022

Anderson A, Boppana A, Wall R et al (2021) Framework for developing alternative reality environments to engineer large, complex systems. Virtual Real 25:147–163. https://doi.org/10.1007/s10055-020-00448-4

Árvai J, Rivers L III (2014) Introduction. In: Árvai J, Rivers L III (eds) Effective Risk Communication, 1st edn. Routledge, Oxon

Astbury B, Leeuw FL (2010) Unpacking black boxes: mechanisms and theory building in evaluation. Am J Eval 31(3):363–381. https://doi.org/10.1177/1098214010371972

Aven T, Renn O (2010) Risk management and governance: concepts, guidelines and applications. Springer Berlin, Heidelberg. https://doi.org/10.1007/978-3-642-13926-0

Boholm Å (2019) Lessons of success and failure: practicing risk communication at government agencies. Saf Sci 118:158–167. https://doi.org/10.1016/j.ssci.2019.05.025

Bosher L, Chmutina K, van Niekerk D (2021) Stop going around in circles: towards a reconceptualisation of disaster risk management phases. Disaster Prev Manag: Int J 30(4/5):525–537. https://doi.org/10.1108/DPM-03-2021-0071

Bostrom A, Löfstedt RE (2003) Communicating risk: wireless and hardwired. Risk Anal 23(2):241–248. https://doi.org/10.1111/1539-6924.00304

Bostrom A (1997) Vaccine risk communication: Lessons from risk perception, decision making and environmental risk communication research. RISK: Health Safety Environ 8(2):173–200

Brajawidagda U, Chatfield AT (2012) Twitter tsunami early warning network: a social network analysis of twitter information flows. In: 23rd Australasian conference on information systems, Geelong, Australia, 3–5 Dec 2012

Carey JW (1975) Review essay: communication and culture. Commun Res 2(2):173–181. https://doi.org/10.1177/009365027500200204

CAT-INGV (n.d.) Centro Alerta Tsunami—Istituto Nazionale di Geofisica e Vulcanologia. https://cat.ingv.it. Accessed May 2023

CAT-INGV (2021) Travelling through the Mediterranean Sea tsunamis. In: ArcGIS StoryMaps. https://storymaps.arcgis.com/stories/32091c82e42a4d30a2f24b1e7b5955b6. Accessed Jul 2023

Cerase A (2021) From 'good' intentions to principled practices and beyond: ethical issues in risk communication. Geol Soc London Spec Pub 508(1):245–264. https://doi.org/10.1144/SP508-2020-104

Cerase A, Cugliari L (2023) Something still remains: factors affecting tsunami risk perception on the coasts hit by the Reggio Calabria-Messina 1908 event (Italy). Sustainability 15(3):2787. https://doi.org/10.3390/su15032787

Chess C, Salomone KL, Hance BJ (1995) Improving risk communication in government: research priorities. Risk Anal 15(2):127–135. https://doi.org/10.1111/j.1539-6924.1995.tb00307.x

Cocco M, Cultrera G, Amato A et al (2015) The L'Aquila trial. Geol Soc Lond Sp Pub 419:43–55. https://doi.org/10.1144/SP419.13

Covello VT (1992) Risk communication: an emerging area of health communication research. Ann Int Commun Assoc 15(1):359–373. https://doi.org/10.1080/23808985.1992.11678816

Covello VT, Sandman PM, Slovic P (1988) Risk communication, risk statistics, and risk comparisons: a manual for plant managers. Chemical Manufacturers Association, Washington, DC, pp 1–57

Cugliari L, Cerase A, Amato A (2022) Tsunami risk perception, a state-of-the-art review with a focus in the NEAM region. Front Earth Sci 10:995536. https://doi.org/10.3389/feart.2022.995536

Cugliari L, Crescimbene M, Amato A (2023) Longitudinal assessment of tsunami knowledge in an Italian school. In: EGU general assembly, Vienna, Austria, 24–28 Apr 2023, EGU23–15923. https://doi.org/10.5194/egusphere-egu23-15923

Dengler L (2005) The role of education in the national tsunami hazard mitigation program. Nat Hazards 35:141–153. https://doi.org/10.1007/s11069-004-2409-x

Dickmann P, Abraham T, Sarkar S et al (2016) Risk communication as a core public health competence in infectious disease management: Development of the ECDC training curriculum and programme. Euro Surveill 21(14):30188. https://doi.org/10.2807/1560-7917.ES.2016.21.14.30188

Donovan A (2021) Experts in emergencies: a framework for understanding scientific advice in crisis contexts. Int J Disaster Risk Reduct 56:102064. https://doi.org/10.1016/j.ijdrr.2021.102064

Dootson P, Thomson T, Angus D et al (2021) Managing problematic visual media in natural hazard emergencies. Int J Disaster Risk Reduct 59:102249. https://doi.org/10.1016/j.ijdrr.2021.102249

Dootson P, Kuligowski E, Murray S (2023) Using videos in floods and bushfires to educate, signal risk, and promote protective action in the community. Saf Sci 164:106166. https://doi.org/10.1016/j.ssci.2023.106166

Downs JS (2011) Chapter 3: Evaluation. In: Fischhoff B, Brewer NT, Downs JS (eds) Communicating risks and benefits: an evidence-based user's guide. US Food and Drug Administration, Washington, DC, pp 11–18

Doyle EE, Johnston DM, Smith R et al (2019) Communicating model uncertainty for natural hazards: a qualitative systematic thematic review. Int J Disaster Risk Reduct 33:449–476. https://doi.org/10.1016/j.ijdrr.2018.10.023

Doyle EEH, Harrison SE, Hill SR et al (2022) Eliciting mental models of science and risk for disaster communication: a scoping review of methodologies. Int J Disaster Risk Reduct 77:103084. https://doi.org/10.1016/j.ijdrr.2022.103084

Doyle EEH, Becker JS (2022) Understanding the risk communication puzzle for natural hazards and disasters. In: Oxford Research Encyclopedia of Natural Hazard Science. https://doi.org/10.1093/acrefore/9780199389407.013.208. Accessed May 2023

Erikson L, Barnard P, O'Neill A et al (2018) Projected 21st century coastal flooding in the Southern California Bight. Part 2: Tools for assessing climate-change-driven coastal hazards and socio-economic impacts. J Mar Sci Eng 6(3):76. https://doi.org/10.3390/jmse6030076

Fakhruddin B, Clark H, Robinson L et al (2020) Should I stay or should I go now? Why risk communication is the critical component in disaster risk reduction. Prog Disaster Sci 8:100139. https://doi.org/10.1016/j.pdisas.2020.100139

Falsaperla S, Reitano D, Musacchio G (2022) Augmented reality in seismic risk management: a contribution to the reduction of non-structural damage. Geosciences 12(9):332. https://doi.org/10.3390/geosciences12090332

Fekete A (2012) Safety and security target levels: opportunities and challenges for risk management and risk communication. Int J Disaster Risk Reduct 2:67–76. https://doi.org/10.1016/j.ijdrr.2012.09.001

Festinger L (1957) Theory of cognitive dissonance, vol 2, Stanford University Press, Stanford, CA

Frewer L (2004) The public and effective risk communication. Toxicol Lett 149(1–3):391–397. https://doi.org/10.1016/j.toxlet.2003.12.049

Funtowicz SO, Ravetz JR (1990) Uncertainty and quality in science for policy. Kluwer Academic Publishers, Dordrecht

Gigerenzer G (2006) Out of the frying pan into the fire: behavioral reactions to terrorist attacks. Risk Anal 26(2):347–351. https://doi.org/10.1111/j.1539-6924.2006.00753.x

Gigerenzer G, Hertwig R, Van Den Broek E et al (2005) A 30% chance of rain tomorrow: how does the public understand probabilistic weather forecasts? Risk Anal 25(3):623–629. https://doi.org/10.1111/j.1539-6924.2005.00608.x

Grabill JT, Simmons WM (1998) Toward a critical rhetoric of risk communication: producing citizens and the role of technical communicators. Tech Commun Q 7(4):415–441. https://doi.org/10.1080/10572259809364640

Guarascio M, Lombardi M, Rossi G et al (2007) Risk analysis and acceptability criteria. WIT Trans Built Environ 94:131–138. https://doi.org/10.2495/SAFE070131

Guenther L, Joubert M (2017) Science communication as a field of research: identifying trends, challenges and gaps by analysing research papers. J Sci Commun 16(02):A02. https://doi.org/10.22323/2.16020202

Hart PS (2014) Boomerang effects in risk communication. In: Árvai J, Rivers L (eds) Effective Risk Communication, 1st edn. Routledge, Oxon, pp 304–318

Hawthorn S, Jesus R, Baptista MA (2021) A review of digital serious games for tsunami risk communication. Int J Serious Games 8(2):21–47. https://doi.org/10.17083/ijsg.v8i2.411

Heath RL, O'Hair HD (eds) (2010) Handbook of risk and crisis communication, 1st edn. Routledge, New York. https://doi.org/10.4324/9781003070726

Höppner C, Buchecker M, Bründl M et al (2010) Risk communication and natural hazards. CapHaz-Net WP5 Report. Swiss Federal Research Institute WSL: Birmendsdorf, Switzerland

https://www.roblox.com/games/8846144215/Tsunami-Disaster-Survival. Accessed May 2023

Hudson-Doyle EE, Paton D, Johnston D (2018) Reflections on the communication of uncertainty: developing decision-relevant information. In: Stock K, Bunker D (eds), Proceedings of the ISCRAM Asia Pacific conference, Wellington, NZ, 2018

IRGC (International Risk Governance Council) (2009). Risk Governance Deficits: An analysis and illustration of the most common deficits in risk governance. International Risk Governance Council, Geneva. https://irgc.org/wp-content/uploads/2018/09/IRGC_rgd_web_final1.pdf. Accessed June 2023

Jayasekara RU, Jayathilaka GS, Siriwardana C et al (2023) Identifying gaps in early warning mechanisms and evacuation procedures for tsunamis in Sri Lanka, with a special focus on the use of social media. Int J Disaster Resil Built Environ 14(1):1–20. https://doi.org/10.1108/IJDRBE-02-2021-0012

Johnson VA, Ronan KR, Johnston DM et al (2014) Evaluations of disaster education programs for children: a methodological review. Int J Disaster Risk Reduct 9:107–123. https://doi.org/10.1016/j.ijdrr.2014.04.001

Johnson BB (1999) Ethical issues in risk communication: continuing the discussion. Risk Anal 19(3):335–348. https://doi.org/10.1111/j.1539-6924.1999.tb00411.x

Johnson VA (2013) An impact evaluation of ShakeOut, an earthquake and tsunami drill in two coastal Washington state school districts. GNS Science Report 2013/19, Wellington, New Zealand

Kanbara S, Shaw R (2022) Disaster risk reduction regime in Japan: an analysis in the perspective of open data open Governance. Sustainability 14(1):19. https://doi.org/10.3390/su14010019

Kankanamge N, Yigitcanlar T, Goonetilleke A et al (2020) Determining disaster severity through social media analysis: testing the methodology with South East Queensland flood tweets. Int J Disaster Risk Reduct 42:101360. https://doi.org/10.1016/j.ijdrr.2019.101360

Keller C, Siegrist M (2009) Effect of risk communication formats on risk perception depending on numeracy. Med Decis Making 29(4):483–490. https://doi.org/10.1177/0272989X09333122

Khanal S, Medasetti US, Mashal M et al (2022) Virtual and augmented reality in the disaster management technology: a literature review of the past 11 years. Front Virtual Real 3. https://doi.org/10.3389/frvir.2022.843195

Kim J (2021) The meaning of numbers: effect of social media engagement metrics in risk communication. Commun Stud 72(2):195–213. https://doi.org/10.1080/10510974.2020.1819842

Kodijat AM (n.d.) Posts In: Instagram. https://instagram.com/iotsunami?igshid=YmMyMTA2M2. Accessed May 2023

Lacassin R, Devès M, Hicks SP et al (2020) Rapid collaborative knowledge building via twitter after significant geohazard events. Geosci Commun 3(1):129–146. https://doi.org/10.5194/gc-3-129-2020

Leiss W (2004) Effective risk communication practice. Toxicol Lett 149(1–3):399–404. https://doi.org/10.1016/j.toxlet.2003.12.050

Leiss W, Larkin P (2019) Risk communication and public engagement in CCS projects: the foundations of public acceptability. Int J Risk Assess Manag 22(3–4):384–403. https://doi.org/10.1504/IJRAM.2019.103339

Lindell MK, Bostrom A, Goltz JD et al (2021) Evaluating hazard awareness brochures: assessing the textual, graphical, and numerical features of tsunami evacuation products. Int J Disaster Risk Reduct 61:102361. https://doi.org/10.1016/j.ijdrr.2021.102361

Lindell MK, Jung MC, Prater CS et al (2022) Improving cascadia subduction zone residents' tsunami preparedness: quasi-experimental evaluation of an evacuation brochure. Nat Hazards 114:849–881. https://doi.org/10.1007/s11069-022-05415-7

Liu CM, Rim D, Baraldi R et al (2021) Comparison of machine learning approaches for tsunami forecasting from sparse observations. Pure Appl Geophys 178:5129–5153. https://doi.org/10.1007/s00024-021-02841-9

Løvholt F, Fraser S, Salgado-Gálvez M et al (2019) Global trends in advancing tsunami science for improved hazard and risk understanding. In: Contributing Paper to Global Assessment Report on Disaster Risk Reduction (GAR 2019). United Nations Office for Disaster Risk Reduction (UNDRR), Geneva

Makinoshima F, Oishi Y, Yamazaki T et al (2021) Early forecasting of tsunami inundation from tsunami and geodetic observation data with convolutional neural networks. Nat Commun 12:2253. https://doi.org/10.1038/s41467-021-22348-0

Mayhorn CB, McLaughlin AC (2014) Warning the world of extreme events: a global perspective on risk communication for natural and technological disaster. Saf Sci 61:43–50. https://doi.org/10.1016/j.ssci.2012.04.014

Meder B, Gigerenzer G (2014) Statistical Thinking: no one left behind. In: Chernoff EJ, Sriraman B (eds) Probabilistic thinking: Presenting plural perspectives. Springer Science + Business Media, Dordrecht, Netherlands, pp 127–148

Milgram P, Kishino FA (1994) Taxonomy of mixed reality visual displays. IEICE Trans Inf Syst 77(12):1321–1329

Miller AN, Sellnow T, Neuberger L et al (2017) A systematic review of literature on effectiveness of training in emergency risk communication. J Health Commun 22(7):612–629. https://doi.org/10.1080/10810730.2017.1338802

Mol JM, Botzen WW, Blasch JE (2022) After the virtual flood: risk perceptions and flood preparedness after virtual reality risk communication. Judgm Decis Mak 17(1):189–214. https://doi.org/10.1017/S1930297500009074

Morgan MG, Fischhoff B, Bostrom A et al (2001) Risk Communication: A Mental Models Approach. Cambridge University Press, Cambridge

Mulia IE, Ueda N, Miyoshi T et al (2021) Machine learning-based tsunami inundation prediction derived from offshore observations. Nat Commun 13(1):5489. https://doi.org/10.1038/s41467-022-33253-5

Mulia IE, Gusman AR, Satake K (2020) Applying a deep learning algorithm to tsunami inundation database of megathrust earthquakes. J Geophys Res Solid Earth 125(9):e2020JB019690. https://doi.org/10.1029/2020JB019690

Nakai F, Nakano G (2023) Community-mediated individual disaster preparedness practices: a case study in Kochi. Japan. Int J Disaster Risk Reduct 86:103532. https://doi.org/10.1016/j.ijdrr.2023.103532

NEAMTIC (n.d.) Tweets. In: Twitter. https://twitter.com/NEAMTIC1?t=353VzRDVpO3EGNdsVkoZYg&s=09. Accessed May 2023

NRC (National Research Council) (1989) Improving risk communication. The National Academies Press, Washington, DC. https://doi.org/10.17226/1189. Accessed May 2023

Ogie RI, Rho JC, Clarke RJ (2018) Artificial intelligence in disaster risk communication: a systematic literature review. In: 2018 5th international conference on information and communication technologies for disaster management (ICT-DM), Tohoku University, Sendai, 4–7 Dec 2018. https://doi.org/10.1109/ICT-DM.2018.8636380

Paton D (2008) Risk communication and natural hazard mitigation: how trust influences its effectiveness. Int J Glob Environ Issues 8(1–2):2–16. https://doi.org/10.1504/IJGENVI.2008.017256

Peters RG, Covello VT, McCallum DB (1997) The determinants of trust and credibility in environmental risk communication: an empirical study. Risk Anal 17(1):43–54. https://doi.org/10.1111/j.1539-6924.1997.tb00842.x

Pignone M, Amato A, Nostro C et al (2022) Public earthquake communication in Italy through a multi-source social media platform: the INGVterremoti experience (2010–2022). Front Earth Sci 10. https://doi.org/10.3389/feart.2022.1003867

Playtad (2023) Tsunami Disaster Survival. In: Roblox.

du Plessis H (2012) Whose science? What knowledge? Science, rationality and literacy in Africa. In: Schiele B, Claessens M, Shi S (eds) Science communication in the World. Springer, Dordrecht, pp 151–168. https://doi.org/10.1007/978-94-007-4279-6_10

Rafliana I, Jalayer F, Cerase A et al (2022) Tsunami risk communication and management: contemporary gaps and challenges. Int J Disaster Risk Reduct 70:102771. https://doi.org/10.1016/j.ijdrr.2021.102771

Rafliana I, Arif A (2022) Chapter 5 A tsunami of Anxiety: the 57-metre wave that shook Jakarta and Western Java. In: Siriwardane-de Zoysa R, Low K, Abdullah N, Hornidge A-K (eds) Coastal Urbanities: Mobilities, Meanings, Manoeuvrings. Leiden, Netherlands, pp 109–131. https://doi.org/10.1163/9789004523340_007

Rahman A, Munadi K (2019) Communicating risk in enhancing disaster preparedness: A pragmatic example of disaster risk communication approach from the case of Smong story. IOP Conf Ser: Earth Environ Sci 273(1):012040. https://doi.org/10.1088/1755-1315/273/1/012040

Reed S, Hsi S, Kreylos O et al (2016) Augmented reality turns a sandbox into a geoscience lesson. Eos 97:18–22. https://doi.org/10.1029/2016EO056135

Regmi KD (2022) Decolonizing epistemologies of the global south: perspectives from Canada. Multicult Educ Rev 14(4):258–276. https://doi.org/10.1080/2005615X.2023.2164977

Renn O (2010) The contribution of different types of knowledge towards understanding, sharing and communication risk concepts. Catalan J Commun Cult Stud 2(2):177–195. https://doi.org/10.1386/cjcs.2.2.177_1

Renn O, Levine D (1991) Credibility and trust in risk communication. In: Kasperson RE, Stallen PJM (eds) Communicating Risks to the Public. Technology, Risk, and Society, vol 4. Springer, Dordrecht, pp 175–217. https://doi.org/10.1007/978-94-009-1952-5_10

Repdeman Mentawai (2021) Repdeman English version. In: YouTube. https://youtu.be/GehtUWTTV38?si=GvMaW245pLcnNR9g&t=9. Accessed May 2023

Repia H, Bailey J (2021) Designing tsunami risk communication with communities: a site-specific case study from Tūanganui-a-Kiwa, Aotearoa New Zealand. Australas J Disaster Trauma Stud 25(1):3–16

Reynolds B (2002) Crisis and emergency risk communication. Centers for Disease Control and Prevention, Atlanta, GA. https://stacks.cdc.gov/view/cdc/6574. Accessed May 2023

Rickard LN (2021) Pragmatic and (or) constitutive? On the foundations of contemporary risk communication research. Risk Anal 41(3):466–479. https://doi.org/10.1111/risa.13415

Rizzi P, Poręebska A (2020) Towards a revised framework for participatory planning in the context of risk. Sustainability 12(14):5539. https://doi.org/10.3390/su12145539

Rohrmann B (1992) The evaluation of risk communication effectiveness. Acta Psychol 81(2):169–192. https://doi.org/10.1016/0001-6918(92)90004-W

Rowan KE (1991) Goals, obstacles, and strategies in risk communication: a problem-solving approach to improving communication about risks. Journal Appl Commun Res 19(4):300–329. https://doi.org/10.1080/00909889109365311

Rubbia GA (2023) Kaleidoscope of colours and an unexpected journey. In: Breviario D, Tuszynski JA (eds) Life in Science. Springer, Cham, pp 277–286. https://doi.org/10.1007/978-3-031-23717-1_19. Accessed May 2023

Rydvanskiy R, Hedley N (2021) Mixed reality flood visualizations: reflections on development and usability of current systems. ISPRS Int J Geo-Inf 10(2). https://doi.org/10.3390/ijgi10020082

Salmon CT, Byrne S, Fernandez L (2014) Exploring unintended consequences of risk communication messages. In: Árvai J, Rivers L (eds) Effective Risk Communication, 1st edn. Routledge, Oxon, pp 292–303

Schneider M, Cotton F, Schweizer P-J (2023) Criteria-based visualization design for hazard maps. Nat Hazards Earth Syst Sci 23(7):2505–2521. https://doi.org/10.5194/nhess-23-2505-2023

Seeger MW (2006) Best practices in crisis communication: an expert panel process. J Appl Commun Res 34(3):232–244. https://doi.org/10.1080/00909880600769944

Sellnow TL, Ulmer RR, Seeger MW et al (2009) Effective risk communication: a message-centered approach. Springer Science and Business Media, New York, NY. https://doi.org/10.1007/978-0-387-79727-4. Accessed May 2023

Setiadi NJ (2014) Understanding challenges at the "Last-Mile" in developing an effective risk communication to reduce people's vulnerability in context of tsunami early warning and evacuation. In: Kontar Y, Santiago-Fandiño V, Takahashi T (eds) Tsunami events and lessons learned. advances in natural and technological hazards research, vol 35. Springer, Dordrecht, pp 417–433. https://doi.org/10.1007/978-94-007-7269-4_24

Simis MJ, Madden H, Cacciatore MA et al (2016) The lure of rationality: why does the deficit model persist in science communication? Public Underst Sci 25(4):400–414. https://doi.org/10.1177/0963662516629749

Slovic P (1987) Perception of risk. Science 236(4799):280–285. https://doi.org/10.1126/science.3563507

Soulé B (2014) Post-crisis analysis of an ineffective tsunami alert: the 2010 earthquake in Maule. Chile. Disasters 38(2):375–397. https://doi.org/10.1111/disa.12045

Spiegelhalter D (2017) Risk and uncertainty communication. Annu Rev Stat Appl 4(1):31–60. https://doi.org/10.1146/annurev-statistics-010814-020148

Suldovsky B (2017) The information deficit model and climate change communication. In: Oxford Res Ency Clim Sci. https://doi.org/10.1093/acrefore/9780190228620.013.301

Sutton SA, Paton D, Buergelt P et al (2020) What's in a name? "Smong" and the sustaining of risk communication and DRR behaviours as evocation fades. Int J Disaster Risk Reduct 44:101408. https://doi.org/10.1016/j.ijdrr.2019.101408

Svendsen ER, Yamaguchi I, Tsuda et al (2016) Risk communication strategies: lessons learned from previous disasters with a focus on the Fukushima radiation accident. Curr Envir Health Rpt 3:348-359. https://doi.org/10.1007/s40572-016-0111-2

Taubenböck H, Goseberg N, Setiadi N et al (2009) Last-Mile" preparation for a potential disaster—interdisciplinary approach towards tsunami early warning and an evacuation information system for the coastal city of Padang, Indonesia. Nat Hazards Earth Syst Sci 9:1509–1528. https://doi.org/10.5194/nhess-9-1509-2009

Thaler T, Seebauer S (2019) Bottom-up citizen initiatives in natural hazard management: why they appear and what they can do? Environ Sci Policy 94:101–111. https://doi.org/10.1016/j.envsci.2018.12.012

Thompson PB (2012) Ethics and risk communication. Sci Commun 34(5):618–641. https://doi.org/10.1177/1075547012459177

Trench B, Bucchi M (2021) Global spread of science communication. In: Bucchi M, Trench B (eds) Routledge Handbook of Public Communication of Science and Technology, 3rd edn. Routledge, London

Twyman M, Harvey N, Harries C (2008) Trust in motives, trust in competence: separate factors determining the effectiveness of risk communication. Judgm Decis Mak 3(1):111–120. https://doi.org/10.1017/S1930297500000218

UNDRR (United Nations Office for Disaster Risk Reduction) (2023) Disaster risk terminology. https://www.undrr.org/terminology/disaster-risk. Accessed May 2023

UNESCO, IOC, (2020) Preparing for Community Tsunami Evacuations: From Inundation to Evacuation Maps, Response Plans and Exercises (IOC Manuals and Guides, 82). UNESCO, Paris

UNESCO/IOC (2021) CoastWAVE Project. http://www.ioc-tsunami.org/index.php?option=com_content&view=article&id=506&itemid=4188&lang=en. Accessed May 2023

Université de Montpellier Paul Valéry/LAGAM (2020) Se préparer face au risque tsunami en Méditerranée française. In: ArcGIS StoryMaps: https://arcg.is/0ifqmH. Accessed July 2023

Valenzuela S, Puente S, Flores PM (2017) Comparing disaster news on twitter and television: an intermedia agenda setting perspective. J Broadcast Electron Media 61(4):615–637. https://doi.org/10.1080/08838151.2017.1344673

Wachinger G, Renn O, Begg C et al (2013) The risk perception paradox—implications for governance and communication of natural hazards. Risk Anal 33(6):1049–1065. https://doi.org/10.1111/j.1539-6924.2012.01942.x

Walters G, Mair J (2019) Chapter 2 Theoretical foundations for the study of image and reputational management in tourism and hospitality. In: Walters, G, Mair J (eds) Reputation and image recovery for the tourism industry. Goodfellow Publishers, Oxford. https://doi.org/10.23912/9781911396673-4103

Wang H, Lindell MK, Siam MRK et al (2022) Local residents' immediate responses to the 2018 Indonesia earthquake and tsunami. Earthq Spectra 38(4):2835–2865. https://doi.org/10.1177/87552930221105104

Zhou Z, Lynett PJ (2020) Simulation system for coastal hazard education. Coast Eng Proc (36v):45. https://doi.org/10.9753/icce.v36v.management.45

Zhu Y, Li N (2021) Virtual and augmented reality technologies for emergency management in the built environments: a state-of-the-art review. J Saf Sci Resil 2(1):1–10. https://doi.org/10.1016/j.jnlssr.2020.11.004

Further Readings

Antoniou V, Nomikou P, Bardouli P et al (2019) The story map for Metaxa mine (Santorini, Greece): a unique site where history and volcanology meet each other. In: Proceedings of the 5th international conference on geographical information systems theory, applications and management (GISTAM 2019), vol 1, pp 212–219. https://doi.org/10.5220/0007715602120219

Boholm Å, Corvellec H (2011) A relational theory of risk. J Risk Res 14(2):175–190. https://doi.org/10.1080/13669877.2010.515313

Carlos-Arce RS, Onuki M, Esteban M et al (2017) Risk awareness and intended tsunami evacuation behaviour of international tourists in Kamakura City, Japan. Int J Disaster Risk Reduct 23:178–192. https://doi.org/10.1016/j.ijdrr.2017.04.005

Christgau S, Spazier J, Schnor B et al (2014) A comparison of CUDA and OpenACC: accelerating the tsunami simulation EasyWave. In: ARCS 2014; 2014 workshop proceedings on architecture of computing systems, pp 1–5

Cugliari L, Crescimbene M, La Longa F et al (2022) Tsunami risk perception in central and southern Italy. Nat Hazards Earth Syst Sci 22(12):4119–4138. https://doi.org/10.5194/nhess-22-4119-2022

Dhellemmes A, Leonard GS, Johnston DM et al (2021) Tsunami awareness and preparedness in Aotearoa New Zealand: the evolution of community understanding. Int J Disaster Risk Reduct 65:102576. https://doi.org/10.1016/j.ijdrr.2021.102576

Dominey-Howes D, Goff J (2013) Tsunami risk management in pacific island countries and territories (PICTs): some issues, challenges and ways forward. Pure Appl Geophys 170:1397–1413. https://doi.org/10.1007/s00024-012-0490-8

Donovan AR, Borie MAH, Blackburn SE (2019) Changing the paradigm for risk communication: integrating sciences to understand cultures. Contributing Paper to Global Assessment Report on Disaster Risk Reduction (GAR 2019). United Nations Office for Disaster Risk Reduction (UNDRR), Geneva

Fauzi A, Mizutani N (2020) Machine learning algorithms for real-time tsunami inundation forecasting: a case study in Nankai region. Pure Appl Geophys 177(3):1437–1450. https://doi.org/10.1007/s00024-019-02364-4

Fukutani Y, Moriguchi S, Terada K et al (2021) Time-dependent probabilistic tsunami inundation assessment using mode decomposition to assess uncertainty for an earthquake scenario. J Geophys Res Oceans 126(7):e2021JC017250. https://doi.org/10.1029/2021JC017250

GITEWS (German-Indonesian Tsunami Early Warning System) (2023) TSUNAMI*Kit*: Developing early warning and community preparedness in Indonesia. https://www.gitews.org/tsunami-kit/index_en.html. Accessed May 2023

Hawthorn S, Jesus R, Baptista MA (2022) Tsunami risk communication: assessing knowledge gaps and suitable serious games. Sci Tsunami Hazards 41(1):54–76

Howard A, Agllias K, Bevis M et al (2017) "They'll tell us when to evacuate": the experiences and expectations of disaster-related communication in vulnerable groups. Int J Disaster Risk Reduct 22:139–146. https://doi.org/10.1016/j.ijdrr.2017.03.002

Jacoby D, Coad Y, Dahl E et al (2019) VR tsunami. In: SIGGRAPH '19: ACM SIGGRAPH 2019 Appy Hour. SIGGRAPH '19: Special Interest Group on Computer Graphics and Interactive Techniques Conference, Los Angeles, CA. Association for Computing Machinery, New York, NY. https://doi.org/10.1145/3305365.3329728. Accessed May 2023

Kalampokis E, Tambouris E, Tarabanis K (2013) Understanding the predictive power of social media. Internet Res 23(5):544–559. https://doi.org/10.1108/IntR-06-2012-0114

Kasperson RE, Stallen PJM (eds) (1991) Communicating risks to the public: international perspectives, vol 4. Kluwer Academic Publishers, Dordrecht. https://doi.org/10.1007/978-94-009-1952-5. Accessed May 2023

Kasperson RE, Kasperson J (2005) Social contours of risk, vol 1. Earthscan, London

León J, Catalán PA, Gubler A (2021) Assessment of top-down design of tsunami evacuation strategies based on drill and modelled data. Front Earth Sci 9:744193. https://doi.org/10.3389/feart.2021.744193

Lundgren R, McMakin A (2018) Risk communication: a handbook for communicating environmental, safety, and health risks, 6th edn. IEEE Press, Wiley Inc, Hoboken, NJ

Mignan A, Mochel L, Ducos G (2022) Involving risk reduction practitioners and other experts in the management of super-catastrophes via an online interactive platform. Front Earth Sci 10:829145. https://doi.org/10.3389/feart.2022.829145

Neef A (2021) The contentious role of tourism in disaster response and recovery in Vanuatu. Front Earth Sci 9:771345. https://doi.org/10.3389/feart.2021.771345

Ogie R, Rho JC, Clarke RJ et al (2018) Disaster risk communication in culturally and linguistically diverse communities: the role of technology. Proceedings 2(19). https://doi.org/10.3390/proceedings2191256

Ongkrutraksa W (2015) International natural disaster communications: an exploratory study of signage for tsunami, earthquake and flood in Japan and Thailand. Journalism & Media, Nihon University, Japan (8):7–19. https://www.publication.law.nihon-u.ac.jp/pdf/journalism/journalism_8/each/04.pdf. Accessed May 2023

Renn O (1992) Risk communication: towards a rational discourse with the public. J Hazard Mater 29(3):465–519. https://doi.org/10.1016/0304-3894(92)85047-5

Rudloff A, Lauterjung J, Münch U (eds) (2010) The GITEWS Project (German-Indonesian Tsunami Early Warning System). Special Issue of Natural Hazards and Earth System Sciences. https://nhess.copernicus.org/articles/special_issue100.html. Accessed May 2023

Salvati P, Pernice U, Bianchi C et al (2016) Communication strategies to address Geohydrological risks: the POLARIS web initiative in Italy. Nat Hazards Earth Syst Sci 16:1487–1497. https://doi.org/10.5194/nhess-16-1487-2016

Spahn H, Hoppe M, Kodijat A et al (2014) Walking the last mile: contributions to the development of an end-to-end tsunami early warning system in Indonesia. In: Wenzel F, Zschau J (eds) Early Warning for Geological Disasters: Scientific Methods and Current Practice. Springer, Berlin, Heidelberg. https://doi.org/10.1007/978-3-642-12233-0_10

Sublime J (2021) The 2011 Tohoku tsunami from the sky: a review on the evolution of artificial intelligence methods for damage assessment. Geosciences 11(3). https://doi.org/10.3390/geosciences11030133

Tozato K, Moriguchi S, Takase S et al (2022) Optimal probabilistic placement of facilities using a surrogate model for 3D tsunami simulations. Nat Hazards Earth Syst Sci 23:1891–1909. https://doi.org/10.5194/nhess-2022-208

UNESCO, IOC, (2022) Standard guidelines for the tsunami ready recognition programme (IOC Manuals and Guides, 74). UNESCO, Paris

Wächter J, Babeyko A, Fleischer J et al (2012) Development of tsunami early warning systems and future challenges. Nat Hazards Earth Syst Sci 12(6):1923–1935. https://doi.org/10.5194/nhess-12-1923-2012

Wegscheider S, Post J, Zosseder K et al (2011) Generating tsunami risk knowledge at community level as a base for planning and implementation of risk reduction strategies. Nat Hazards Earth Syst Sci 11(2):249–258. https://doi.org/10.5194/nhess-11-249-2011

Wendling C, Radisch J, Jacobzone S (2013) The use of social media in risk and crisis communication. OECD Working Papers on Public Governance, 24. OECD Publishing. https://doi.org/10.1787/5k3v01fskp9s-en

Wogalter MS, Mayhorz CB (2006) The future of risk communication: Technology-based warning systems. In: Wogalter MS (ed) Handbook of Warnings. CRC Press. https://doi.org/10.1201/9781482289688

Xie X-F, Wang M, Zhang R-G et al (2011) The role of emotions in risk communication. Risk Anal 31(3):450–465. https://doi.org/10.1111/j.1539-6924.2010.01530.x

Chapter 9
Final Remarks

Jörn Behrens, Fatemeh Jalayer, Shane Murphy, Jacopo Selva, Mathilde B. Sørensen, Mario A. Salgado-Gálvez, Stefano Lorito, Finn Løvholt, and Irina Rafliana

If the reader has read through the book to finally reach these final remarks, they should be in the position to understand the individual steps in a PTHA or PTRA workflows. Of course, implementing and executing them may require some direct expert knowledge and skills. It should by now be clear that uncertainties are an integral part of the hazard and risk assessment processes, and that using probabilistic methods allows to quantify and constrain the uncertainties such that they become manageable and can be comprehended by the users of the results. The importance of open and transparent communication of methods, limitations and uncertainties should be evident. It should also be clear by now that probabilistic approaches are well established and yield useful and usable results.

Framing the topics in a cookbook style has allowed for making the material available to different target groups. Those who are mostly interested in getting inspiration for their specific situation should get information through the recipes. Those looking for a better understanding of the methods inside PTHA and PTRA should learn from the ingredients. Colleagues who are interested in more in-depth information should find references in the extensive lists included in the ingredients' sections.

J. Behrens (✉)
Department of Mathematics, Universität Hamburg, Hamburg, Germany
e-mail: joern.behrens@uni-hamburg.de

F. Jalayer
Department of Disaster Risk Reduction, University College London, London, UK
e-mail: f.jalayer@ucl.ac.uk

S. Murphy
Geo-Ocean, Ifremer, Plouzané, France
e-mail: shane.murphy@ifremer.fr

J. Selva
Dipartimento di Scienze della Terra, dell' Ambiente e delle Risorse (DiSTAR) , Università di Napoli Federico II, Naples, Italy
e-mail: jacopo.selva@unina.it

M. B. Sørensen et al. (eds.), *Probabilistic Tsunami Hazard and Risk Analysis*,
Mathematics of Planet Earth 14, https://doi.org/10.1007/978-3-031-98115-9_9

The management and all the choices related to PTHA and PTRA are rather complex. It is therefore advisable to seek expert advice, even though we have aimed to cover all topics in an exhaustive manner. Of course, there are still many open questions related to the workflows and individual components of PTHA and PTRA. In this section we address some of them, but first we review in brief where the community stands at the moment.

9.1 Current State of PTHA and PTRA

At the start of each PTHA or PTRA workflow stands the selection of a framework that is suitable for the intended purpose. We have formulated this in a rather simple way, distinguishing basically between two major framework types. The first framework (FW1) is related to quantitative probabilistic hazard and risk assessment based on the explicit simulation of tsunami sources and the quantification of their potential consequences. FW1 is based on the quantification of the hazard related to each individual source, typically considering a large number of scenarios. Sometimes only a few scenarios are evaluated. This case is here considered as a special case of PTHA/PTRA in which the various sources of uncertainties are poorly constrained, leading to a potential underestimation of uncertainty in the hazard and risk results. The second framework (FW2) is based on using compound indicators to evaluate vulnerability and risk. FW2 is a simplified approach to risk evaluation, but potentially lends itself well to participatory methods, aiming to engage the communities. It is

M. B. Sørensen
Department of Earth Science, University of Bergen, Bergen, Norway
e-mail: mathilde.sorensen@uib.no

M. A. Salgado-Gálvez
CIMNE, Barcelona, Spain

ERN International, Mexico City, Mexico

S. Lorito
Istituto Nazionale di Geofisica e Vulcanologia, Roma, Italy
e-mail: stefano.lorito@ingv.it

F. Løvholt
Offshore Energy, Norwegian Geotechnical Institute, Oslo, Norway
e-mail: finn.lovholt@ngi.no

I. Rafliana
Research Center for Area Studies, Badan Riset dan Inovasi Nasional (BRIN), Jakarta, Indonesia
e-mail: irin002@brin.go.id

Center for Life Ethics, Universität Bonn, Bonn, Germany

suitable also for evaluating non-physical vulnerabilities and risks that are related to a given societal context.

Within each of the frameworks, an appropriate tsunami source selection needs to be made. We have strived to give a comprehensive picture of relevant source types, deliberately excluding some low probability sources such as meteoroid impacts. We also give a comprehensive overview of available datasets on sources, at the base of probability quantification, acknowledging that there are often large gaps in historical and paleo-tsunami data. We discuss alternative ways to gather such data and how to derive quantifiable projections of probable future scenarios and how to quantify the related annual probability of occurrence. Since the quantification of source recurrence rates is the starting point of any tsunami hazard assessment, uncertainties in such rates and impact levels must be propagated through the whole analysis chain. Special care needs therefore be taken to capture a realistic and informing source parameter space. In most cases, past data on tsunamis are not sufficient to completely characterise the sources. Thus, a comprehensive PTHA/PTRA needs to go beyond actual observations and consider what is possible in the future. Even if not yet observed or properly recorded, sources with relatively low recurrence rates that are consequently often connected to a high epistemic uncertainty may be considered.

Once source types and probabilities are defined, there is a large diversity of approaches to model the sources such that they can be used in tsunami generation, propagation, and inundation modelling. The variety and sensitivity of available source modelling methods are large, with various degrees of sophistications and modelling for the different sources, rendering this ingredient complex to handle. For non-seismic sources there is also still the need for further research to close knowledge and data gaps, and monitoring the current literature in that area is certainly advisable.

Hydrodynamic modelling is the next step in the common workflow after defining sources. This is a much more mature and established field, but many subtle details may render plausible-looking results quantitatively useless. Therefore, a lot of care needs to be taken to select appropriate datasets, find robust simulation tools, validate the results, and extract the relevant pieces of information for the following steps of the analysis. Quantifying the uncertainty related to hydrodynamic modelling is an open research topic.

Exposure is an essential ingredient for any risk assessment regardless of the chosen framework. Exposure models describe population and assets at risk, for which relevant attributes such as their value can be difficult to capture (e.g. for historical and heritage buildings). In this cookbook, we focused on the population (human exposure) and buildings and infrastructure (physical exposure) without diminishing the relevance of other types of assets, such as environmental ones (e.g. mangroves), that are beyond the scope of this book. Although there are no exposure taxonomies specifically developed for tsunami hazards, the existing physical and human exposure models can still be used for tsunami risk assessments. However, it should be kept in mind that existing physical exposure taxonomies may lack some specific attributes important for characterisation of tsunami vulnerability. The main shortcoming of existing human exposure models, on the other hand, is related to an insufficient spatio-temporal resolution. For tsunami risk assessment, depending on the application, it may be crucial that the human exposure is modelled with a dynamic approach so that daily and/or seasonal variations are well captured. Although the uncertainty in the exposure models directly affects the risk assessment, there are limited examples of exposure models with explicit treatment of uncertainties, mostly related to physical exposure models.

Tsunami vulnerability assessment involves different dimensions (e.g., physical, socioeconomic, environmental), and in this cookbook we have discussed mostly the physical and social ones. We show that for physical vulnerability assessment in FW1, both analytical and empirical methods can be employed. Most physical tsunami vulnerability models are obtained empirically, and the analytical fragility models are generally more complex and challenging to develop. In fact, there are relatively fewer studies in the literature related to tsunami analytical fragility assessment. Nevertheless, empirical physical vulnerability models are available only for regions that have experienced damage due to tsunamis in the past and therefore they have a limited spatial coverage. This stresses the importance of tackling the challenges in developing analytical fragility models. We also show that social vulnerability models focus on the social and economic consequences of the tsunamis for the affected communities. We discuss different methods leading to social vulnerability models generally suitable for FW1. In the context of FW2, the physical vulnerability can be evaluated through weighted combinations of different indicators. In this case, the indicator-based vulnerability method can capture spatial and temporal variations in the vulnerability. Indicator-based vulnerability can be used to also evaluate other dimensions of vulnerability such as functional aspects.

An imperative part of the workflow is the evaluation and communication of risk, where interaction with stakeholders and society brings in a completely new level of complexity. We have compiled several best-practice experiences and suggestions. However, we stress that each situation requires a tailored solution. After all, risk communication is an always ongoing process, and it evolves with changes in society, economic development, and the progression of values and societal perception. While the ingredients' part of the cookbook ends with this section on risk communication, it can in fact be considered the beginning of a new analysis loop. For example, the definition of what an acceptable risk level is, and which are the risk metrics of interest to the communities, may initiate a new cycle of hazard and risk assessment for refined measures to be part of a comprehensive disaster risk management strategy.

9.2 Selecting the Best Workflow for a Specific Problem

With the variety of recipes that we collected from different application fields and from different parts of the world, we demonstrate that there are many suitable and successful ways to assess tsunami hazard and risk. The selection of methods heavily depends on the situation, available resources, the quality and availability of data, and societal situation. Practitioners and decision makers who enter a new probabilistic assessment cycle should therefore ask several fundamental questions, before designing their workflow:

- Purpose: What is the purpose of the PTHA or PTRA, and what key questions is the assessment expected to answer? These questions usually require consideration of the scale of the assessment, the expected use of the results, the type of hazard and/or risk metrics needed as well as the accuracy requirements.
- Funding: How much is the funding entity willing to pay for the assessment? Is it feasible to allocate new research, consulting, data acquisition efforts, or does the assessment have to rely on existing resources?
- Expertise: Which expertise is available and where can external expertise be found in case it is not available in the team of analysts? In other words, which parts of the assessment can be "self-catered", and which parts require external help?
- Resources: What kinds of data are available and where are the gaps? Can these gaps be easily filled or is there the need for further research and data acquisition efforts? Which computational resources are available? How can suitable databases be accessed?
- Scale: What is the spatial scale of the assessment? Does the assessment concern a whole region, a district, or just a city or critical infrastructure facility?
- Acceptable Risk: What is the acceptable risk, is the assessment meant for a critical infrastructure that may cause cascading hazards, or is it a general preparatory measure? Is there an existing regulating framework/legislation?
- Impact: It is wise to foresee the possible impact of the assessment to prevent undesired collateral damage. For example, assigning new inundation zones may have an effect on economic values, migration patterns, or social developments.

9.3 Future Challenges

While this book provides an overview of the current state of PTHA and PTRA, their evolution is ongoing as the research community strives to address outstanding challenges. Behrens et al. (2021) provide an extensive review of scientific gaps in PTHA and PTRA. Based on a poll of 50 researchers, they highlight gaps in quantifying landslide, earthquake and volcanic occurrence probabilities as well as properly accounting for multi- and cascading hazards to be of high priority and highly challenging. Other important future challenges for PTHA/PTRA practitioners and users are listed below:

Co-design: Improved communication is required between scientists developing the hazard and risk assessments and users of the assessments (i.e., reinsurance/finance industry, government and local policy makers, intergovernmental organisations, independent consultants, etc.). An organisation providing brokerage between the two groups and facilitating forums where the needs of the stakeholders and limitations of the assessments can be constructively discussed would be a welcome addition to the tsunami hazard and risk communities.

Standards: Setting standards for the input/output of both data used within the assessments and the final products is crucial. Construction of community-based standards for data input/output would greatly reduce the time spent wrangling data by researchers trying to implement their assessment workflows. Community benchmarking of codes and workflows is needed to understand how different codes/workflows may impact assessment. Additionally, tools are required to tailor the outputs from PTHA/PTRA for stakeholders, for instance in terms of maps at specific scales or metrics that are directly relatable to building codes (e.g. ASCE standards).

Multi-disciplinarity: Better interaction between different communities involved in hazard and risk assessment is required to produce assessments that are compatible with those produced in other, related disciplines (e.g. for earthquakes and volcanoes).

Scale: hazard and risk assessments are produced at a range of geographical scales from global to local. The compatibility of the assessments between these different scales remains an open topic of research.

Multi-hazard and multi-risk dimension: An open field is the exploration of the multi-hazard and multi-risk dimension of PTHA and PTRA. Tsunamis are typically secondary phenomena, and potential interaction with other phenomena (ground-shaking, volcanic products, etc.) may lead to significant variations in all the factors, from the hazard (changes in the probability of the sources as well as in the boundary conditions for propagation), to exposure (movements of population), and vulnerability (pre-damaged structures). This dimension is only partially explored in literature, and future research will surely open new methods to properly account for such interactions.

9.4 Future Opportunities

While the research community strives to close those research gaps discussed above, technological development opens completely new possibilities for improved PTHA and PTRA methods. Whereas we cannot predict where technological development will take us in the future, we already see the contours of new possibilities that are currently being implemented in tsunami science.

Monitoring tsunami generation and propagation is inherently difficult as it requires the deployment of sensors at sea which is challenging due to the hostile nature of the environment and the lack of infrastructure. To tackle this issue, the Science Monitoring and Reliable Telecommunications (SMART) Cables initiative aims to integrate temperature, pressure, and acceleration sensors on nodes along commercial telecommunication cables to provide trans-ocean information about the sea state. This has led to projects such as CAM (Continent - Azores - Madeira) in Portugal which will provide critical information about tsunamis for Portugal. New technologies are also being developed which can turn the fibre optic cables themselves into sensors which can record ambient variations (i.e. acoustic/seismic) along the cable. At present, the most promising technologies are DAS (Distributed Acoustic Sensing) which can convert kilometres of cable into dense arrays of virtual sensors and interferometry using ultra-stable lasers (Marra et al. 2022) which can operate at a trans-ocean scale. Cabled permanent observatories targeted at tsunami studies and providing warning of impending waves are now being developed in places like Japan (e.g. DONET 1 and 2, Dense Oceanfloor Network Systems for Earthquakes and Tsunamis, and S-Net, Seafloor observation network for earthquakes and tsunamis along the Japan Trench) and Canada (e.g., Neptune, NorthEast Pacific Time-series Undersea Networked Experiments). While these developments may clearly enhance real-time tsunami warning and monitoring capabilities, the denser and more complete data that they may record will help to better characterise tsunamis and hence the characterization of hazards and risks they pose.

In today's tsunami hazard modelling practise that this cookbook lends itself heavily on, we mostly explain procedures based on established methods. Yet, the use of Artificial Intelligence (AI) – or more precisely Machine Learning (ML) – to build emulators or surrogate models in tsunami science have recently accelerated. Tsunamis are infrequent events with limited data and their hazard assessment relies on intense use of numerical models. The ML techniques cannot rely on large resources of recorded data (which is perhaps more common in other disciplines), but they need to rely, to a large degree, also on synthetic model data produced by numerical models. Examples of the application of ML are trained emulators based on a large amount of ensemble simulations, which include using surrogate models with low resolution model output to model high resolution inundation (Mulia et al. 2022) or use of ML for providing new early warning technology (Makinoshima et al. 2021). However, with the rapid development in the field of ML applications, it is quite possible and even likely that we will see other types of applications beyond what is discussed above. Once such applications will become part of day-by-day routines, they will decrease substantially the computational cost related to the large amount of numerical simulations needed for thorough uncertainty quantification in PTHA/PTRA.

There are significant innovations also on the side of exposure modelling, which is a fundamental ingredient of PTRA. Taking advantage of the available earth observation data from satellites, other remote sensing capabilities (e.g., lidar, drones, aerial photos, traffic patterns), and citizen science initiatives (e.g., crowdsourcing), AI and ML techniques can be used to provide higher spatio-temporal resolution in exposure modelling. For instance, the merging of the ever-increasing computational and big-data analysis capabilities, with the crowd-sourcing approach, allowed the development of a Global Dynamic Exposure Model, which has the potential to significantly increase the quality of hazard and risk estimates.

We also note the effort to make the PTHA and PTRA results and tools openly available, embracing the principles of FAIR data provision. In this regard there is an important ongoing effort by the newly formed Thematic Core Service (TCS) Tsunami in the European Research Infrastructure Consortium EPOS. The TCS Tsunami provides open access to several fundamental data, models, and software for PTHA and PTRA such as event and source catalogues, modelling software, hazard, fragility, and vulnerability curves. The TCS Tsunami is also providing well-tested PTHA workflows by means of the competitive mechanism of the Transnational Access. Several institutions are currently offering the opportunity to use such workflows, providing initial support for new users for the usage of numerical simulation codes, computational facilities, PTHA aggregation and analysis tools.

9.5 A Vision for the Future

This work represents a snapshot of current knowledge and practices in PTHA and PTRA. As laid out before, new technologies, data, and scientific developments will continue to advance the field. Therefore, an ongoing reassessment of the state of the art will be necessary. It is our vision that the community of researchers contributing to PTHA and PTRA efforts will grow and that coastal communities will benefit from the growing knowledge base. To foster exchange and development of the field, additional efforts on a scientific as well as on a user level will be necessary. Such efforts will need to be founded in open and transparent international cooperation with easy exchange of methods and data, which is in the scope of the Global Tsunami Model (GTM). It is our dear hope that this book will contribute to bringing the communities involved in PTHA and PTRA together and help accelerate the development.

References

Behrens J, Løvholt F, Jalayer F, Lorito S, Salgado-Gálvez M, Sørensen M et al (2021) Probabilistic Tsunami Hazard and Risk Analysis: A Review of Research Gaps. Frontiers in Earth Science 9. https://doi.org/10.3389/feart.2021.628772

Makinoshima F, Oishi Y, Yamazaki T et al (2021) Early forecasting of tsunami inundation from tsunami and geodetic observation data with convolutional neural networks. Nat Commun 12:2253. https://doi.org/10.1038/s41467-021-22348-0

Marra G, Fairweather DM, Kamalov V, Gaynor P, Cantono M, Mulholland S, Baptie B, Castellanos JC, Vagenas G, Gaudron J-O, Kronjäger J, Hill IR, Schioppo M, Barbeito Edreira I, Burrows KA, Clivati C, Calonico D, Curtis A (2022) Optical interferometry-based array of seafloor environmental sensors using a transoceanic submarine cable. Science 376:874–879. https://doi.org/10.1126/science.abo1939

Mulia IE, Ueda N, Miyoshi T et al (2022) Machine learning-based tsunami inundation prediction derived from offshore observations. Nat Commun 13:5489. https://doi.org/10.1038/s41467-022-33253-5

Part II
Recipes

Chapter 10
R1–The NEAM Tsunami Hazard Model 2018 (NEAMTHM18)–Region-Wide Probabilistic Tsunami Hazard Based on a Large Number of Tsunami Scenarios

Roberto Basili and Stefano Lorito

10.1 Introduction

The purpose of this study was to provide the first homogeneous region-wide long-term probabilistic earthquake-induced tsunami hazard assessment (PTHA) for the coastlines of the North East Atlantic, the Mediterranean, and connected seas (NEAM) (Basili et al. 2021) to create a shared foundation for both localised and comprehensive hazard and risk evaluations, as well as for the development of tsunami early warning evacuation maps.

The hazard analysis considered only tsunamis generated by the co-seismic displacement of the seafloor due to fault dislocation, corresponding to earthquake sources as delineated in Sect. 3.2. The analysis used the following ingredients:

- This recipe follows framework FW1A (Fully simulation-based risk assessment; Sect. 2.2) but is concluded at the completion of the PTHA.
- Source selection and characterisation were based on geologic fault sources and seismicity from publicly available datasets (Sect. 3.2). These were used to model earthquake sources as a regularly spaced grid and individual three-dimensional faults.
- The process of tsunami source modelling (Sect. 4.2) involved two steps. The slip distribution for each earthquake scenario in the earthquake source model was determined in the first step. The seafloor deformation for each slip distribution was calculated in the second step.

R. Basili (✉) · S. Lorito
Istituto Nazionale di Geofisica e Vulcanologia, Roma, Italy
e-mail: roberto.basili@ingv.it

S. Lorito
e-mail: stefano.lorito@ingv.it

M. B. Sørensen et al. (eds.), *Probabilistic Tsunami Hazard and Risk Analysis*, Mathematics of Planet Earth 14, https://doi.org/10.1007/978-3-031-98115-9_10

- The tsunami modelling (Chapter 5) for each slip distribution was carried out using a realistic bathymetric model (Sect. 5.2) and by applying a filter to transform the seafloor displacement into a tsunami initial condition (Sect. 5.3).
- The tsunami propagation (Sect. 5.4) involved a domain extending from the open sea to a specified point of interest located in front of the coastline. This was accomplished through numerical simulation in the shallow water approximation and estimating the inundation distribution on the coast behind the point of interest using local stochastic amplification factors.

NEAMTHM18 provides the hazard curves, that is, the probability of exceedance in a 50-year exposure time of different Maximum Inundation Height (MIH) thresholds, at over 2000 Points of Interest (POI) approximately each 20 km along the coasts of the NEAM region. It also provides the uncertainty of the hazard curve estimates.

NEAMTHM18 was meant to promote a unified tsunami risk management approach within the region (e.g. IOC 2023). The approach described in this recipe is particularly valuable for the first screening of a region where a hazard model does not exist yet. The homogeneity of the approach and the scale over a relatively large region allows to inform, for example, national coastal long-term and evacuation planning (e.g. Tonini et al. 2021; see also recipe R24) and to prioritise local higher-resolution assessments (e.g. Gibbons et al. 2020; see also recipe R2).

The workflow followed for NEAMTH18 is summarised in Basili et al. (2021), and the extended project documentation is available in Basili et al. (2019). The hazard curves (Basili et al. 2018) are available in various formats and web services.

10.2 Framework Used in the Recipe

This recipe falls within the scopes of framework FW1A (Sect. 2.2), a probabilistic hazard model based on many earthquake scenarios with comprehensive uncertainty propagation. FW1A allows calculating hazard curves for any target point independently of past tsunami evidence preserved at those locations. The simulation-based approach is preferred because tsunami observations are too sparse in the NEAM region to constrain a hazard model sufficiently. Nonetheless, the obtained model does not consider tsunami sources different from earthquakes.

The recipe implementation follows the four-step workflow illustrated in Fig. 10.1, in which all variables are treated as aleatory. First, a probabilistic earthquake model is defined (Step-1), which is a collection of scenarios σ_K (the list of earthquake parameters for each scenario), accompanied by their respective occurrence rates $\lambda(\sigma_K)$. The tsunami corresponding to each earthquake scenario is numerically modelled to determine the coastal distribution of the MIH above sea level at each point of interest (Step-2,3). Adopting a time-independent (Poissonian) process, the earthquake rates are combined with the MIH distributions at each point of interest to obtain the hazard curves. An ensemble modelling approach is used to treat simultaneously aleatory and epistemic uncertainty. The latter is related to the different (weighted) alternative models considered (in practice, alternative workflow realisations).

A formalised decision-making process based on a multiple-expert protocol (Selva et al. 2024) was adopted to deal with epistemic uncertainty (Fig. 10.1). This protocol involved two elicitation experiments to identify the model alternatives to be implemented in each step and assign proper weights to the selected ones.

The advantage of the selected approach is the possibility of estimating the hazard at any location, independently of the local physical conditions, and of providing transparent and repeatable results. The time-independent hazard curves enable the user to explore the probability of exceeding any given tsunami hazard intensity threshold within different timeframes through interpolation and rescaling, providing the possibility to adapt the results for different applications and needs. A limiting factor in using this approach is finding a balance between the extent of the target coastlines, the size of the seismic source domain to be considered, and the accuracy of the results.

10.3 Source Selection and Activity Rate Characterisation Used in the Recipe

The basis for source selection is a tectonic regionalisation of the entire domain of potential sources that could generate a tsunami affecting the target coastlines. The regionalisation helped in defining two main seismicity types: predominant seismicity (PS) and background seismicity (BS) (Selva et al. 2016). The PS category encompasses significant earthquakes originating from well-known major faults, such as plate boundaries and subduction interfaces. The BS category encompasses all the scattered seismic activity within a region defined by tectonic considerations. This conceptual subdivision controls how the ingredients to derive the probabilistic earthquake model are blended based on the necessary and available data.

The entire BS domain is discretised into regularly spaced gridded sources. Crustal fault sources and focal mechanisms constrain the geometry and kinematics of earthquake sources in each grid node, and the depth dimension explores the entire crustal thickness. The BS Earthquake Rate Forecast (ERF) is based only on earthquake catalogues. The seismicity is first analysed for statistical completeness and declustering and then spatially smoothed (Hiemer et al. 2014) within the various tectonic regions. The PS ERF is based on both seismicity and fault-based seismic moment. The seismic moment rates of subduction interfaces are based on geodetic convergence rates and crustal fault slip rates.

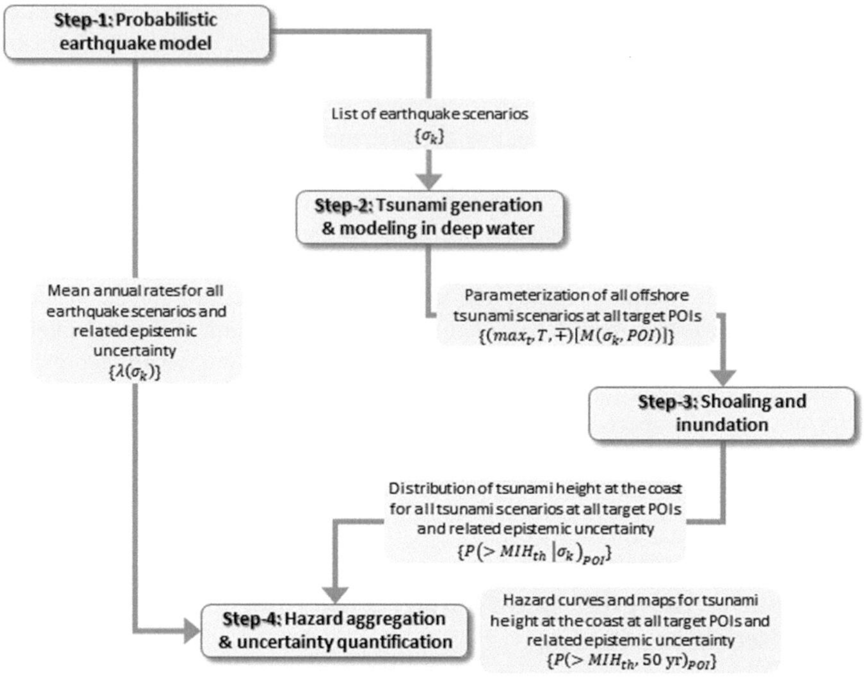

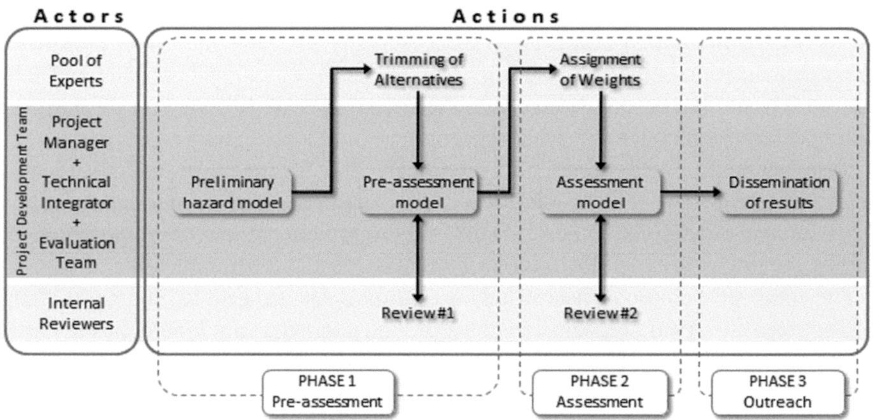

Fig. 10.1 NEAMTHM18 implementation (modified from Basili et al. 2021). (Top) Diagram showing the four-step workflow, including the information flux followed in building the hazard model. This procedure is repeated for each considered alternative model. (Bottom) Schematic illustration of the project's multiple-expert protocol involving different actors playing different roles during the development of the project's various phases. Phases 1 and 2 include an elicitation experiment and a revision stage each. Phase 3 includes external reviews where appropriate (e.g., peer review in journal articles)

The earthquake frequency magnitude distributions (FMD) and their associated epistemic uncertainties are quantified based on a set of Bayesian alternatives considering the BS/PS separation and different functional forms (truncated and tapered).

This implementation's main advantages are the possibility of modulating the discretisation of the entire parameter space on the relative importance of all possible earthquake sources and exploring several modelling alternatives without sacrificing resolution. A possible improvement of this approach would be the possibility of considering seismic-moment-based ERFs for the BS.

Individual earthquake rupture size and average slip are obtained with two alternative empirical scaling relations for the subduction sources (Strasser et al. 2010; Murotani et al. 2013) and by using specific versions of the same scaling relations (Leonard 2014) for both the tectonic setting (interplate and continental regions) and the faulting style (dip- and strike-slip).

The adopted fault plane representation is a rectangle with uniform slip for all BS-type and a selection of PS-type earthquake ruptures. A three-dimensional mesh made of triangular elements is adopted for the PS-type of subduction interfaces.

Information from both fault and moment tensor catalogues is used to predict future rupture direction and mechanisms for crustal faults. For subduction interfaces, the rake is derived from plate convergence. For the larger earthquakes on Mediterranean subduction interfaces, the approach considers two alternative slip probability models: uniform or controlled by depth-dependent rigidity and coupling variations, while preventing systematic slip excess at shallow depths over one or more seismic cycles (Scala et al., 2020).

10.4 Source Mechanics Modelling Used in Recipe

In each scenario, the earthquake sources are modelled as finite faults, using different configurations depending on their presumed importance based on the size of the larger earthquakes they could generate and their location relative to the target coastlines. For large subduction earthquakes in the Mediterranean Sea, different realisations of the heterogeneous slip distribution for the same rupture area are computed using a stochastic composite model (Ruiz et al. 2021). For all the others, a uniform slip distribution is used.

To model the co-seismic seafloor deformation field, Volterra's formulation of elastic dislocation theory was applied to faults in an infinite elastic half-space. The approach involved employing the analytical model by Okada (1992) for simplified fault ruptures, considering both single or multiple planar quadrangles, and the Meade (2007) code for more complex fault ruptures represented on 3D meshes of triangular elements.

10.5 Tsunami Modelling

The initial tsunami conditions at the surface are derived from the instantaneous seafloor vertical deformation. A two-dimensional filter (Kajiura 1963) is applied to depth-filtering the seafloor displacement to approximate the effect of lateral spreading of the water during the transmission to the sea surface through the water column.

In deep water (> 50 m depth), the tsunami propagation model simulates tsunami mareograms at the offshore points of interest. To save computational resources, such mareograms are obtained as linear combinations of Green's functions (Molinari et al. 2016) simulated with the Tsunami-HySEA NLSW GPU-optimized code (de la Asunción et al. 2013).

Starting from the synthetic mareograms and their offshore-derived parameters, the procedure calculates 1D amplification factors that transform the incident wave amplitude into a maximum inundation value conditional probability distribution based on local bathymetric profiles for each specific location as a function of the period and polarity. To address the two-dimensional nature of the inundation process, a stochastic component is superimposed to the amplified values, which is constrained with high-resolution inundation simulations (Glimsdal et al. 2019). Further uncertainties, like those related to the topo-bathymetric model the numerical simulations themselves, have been implicitly embedded in the amplification factors.

The adopted strategy has the advantage of saving computational resources to explore a large number of earthquake scenarios in a flexible and modular way while controlling the propagation of uncertainty.

10.6 Discussion

The main challenges faced in applying this recipe to the NEAMTHM18 were related to the diversity of tectonic environments hosting the potential seismic sources, the interconnections of large and small basins typical of the NEAM region, the necessity to treat both near and distant seismic sources appropriately, and the vastness of the tsunami propagation domain. Such an approach has pros and cons. On the one hand, consistently computed region-wide hazard enables us to compare the hazard of different and very distant places or even separate basins. Such comparisons will not depend on the differences in the input data or the adopted methods. On the other hand, regional-scale hazard models can be adopted as a common reference for performing local hazard quantifications in different areas (e.g. Bacchi et al. 2020) and enable using unsupervised filters of seismic sources for more detailed local analyses (Lorito et al. 2015; Volpe et al. 2019).

NEAMTHM18 used publicly available input data (bathymetry, geologic fault datasets, seismic catalogues), thereby conferring an element of transparency to the application. However, careful revision and harmonisation of these data were needed to ensure similar accuracy throughout the region. The heterogeneous provenance and the not necessarily purpose-oriented compilation of data required further consideration.

Since inundation could not be explicitly calculated, linear, relatively low-resolution and cheap numerical simulations were conducted. Nonetheless, significant computational resources were needed because of the spatial extent and the high number of individual scenarios needed to explore deeply enough aleatory and epistemic variability.

The above considerations and the application of a multiple-expert protocol imply that significant human resources need to be employed to address the subjective choices necessary to deal with and quantify the epistemic uncertainty, such as weighing the considered alternatives.

Links to Data, Models and Other Resources

- NEAMTHM18 website: https://tsumaps-neam.eu/neamthm18/
- NEAMTHM18 web services: https://tsumaps-neam.eu/web-services/
- Tsunami hazard and risk products available through the EPOS Thematic Core Service Tsunamis: https://tsunamidata.org/

References

Bacchi V, Jomard H, Scotti O, Antoshchenkova E, Bardet L, Duluc C-M et al (2020) Using metamodels for tsunami hazard analysis: an example of application for the French Atlantic coast. Front Earth Sci 8. https://doi.org/10.3389/feart.2020.00041

Basili R, Brizuela B, Herrero A, Iqbal S, Lorito S, Maesano FE et al (2018) NEAM tsunami hazard model 2018 (NEAMTHM18): online data of the probabilistic tsunami hazard model for the NEAM region from the TSUMAPS-NEAM project. Roma: Istituto Nazionale di Geofisica e Vulcanologia (INGV). https://doi.org/10.13127/tsunami/neamthm18

Basili R, Brizuela B, Herrero A, Iqbal S, Lorito S, Maesano FE et al (2019) NEAMTHM18 documentation: the making of the TSUMAPS-NEAM tsunami hazard model 2018. Roma: Istituto Nazionale di Geofisica e Vulcanologia (INGV), 352. https://doi.org/10.5281/zenodo.3406625

Basili R, Brizuela B, Herrero A, Iqbal S, Lorito S, Maesano FE et al (2021) The making of the NEAM tsunami hazard model 2018 (NEAMTHM18). Frontiers in Earth Science 8. https://doi.org/10.3389/feart.2020.616594

de la Asunción M, Castro MJ, Fernández-Nieto ED, Mantas JM, Acosta SO, González-Vida JM (2013) Efficient GPU implementation of a two waves TVD-WAF method for the two-dimensional one layer shallow water system on structured meshes. Comput Fluids 80:441–452. https://doi.org/10.1016/j.compfluid.2012.01.012

Gibbons SJ, Lorito S, Macías J, Løvholt F, Selva J, Volpe M et al (2020) Probabilistic tsunami hazard analysis: high performance computing for massive scale inundation simulations. Front Earth Sci. 8. https://doi.org/10.3389/feart.2020.591549

Glimsdal S, Løvholt F, Harbitz CB, Romano F, Lorito S, Orefice S et al (2019) A new approximate method for quantifying tsunami maximum inundation height probability. Pure Appl Geophys 176:3227–3246. https://doi.org/10.1007/s00024-019-02091-w

Hiemer S, Woessner J, Basili R, Danciu L, Giardini D, Wiemer S (2014) A smoothed stochastic earthquake rate model considering seismicity and fault moment release for Europe. Geophys J Int 198:1159–1172. https://doi.org/10.1093/gji/ggu186

Kajiura K (1963) The leading wave of a tsunami. Bull Earthq Res Inst Univ Tokyo 41:535–571
Leonard M (2014) Self-consistent earthquake fault-scaling relations: update and extension to stable continental strike-slip faults. Bull Seismol Soc Am 104:2953–2965. https://doi.org/10.1785/0120140087
Lorito S, Selva J, Basili R, Romano F, Piatanesi TMM, A, (2015) Probabilistic hazard for seismically induced tsunamis: accuracy and feasibility of inundation maps. Geophys J Int 200:574–588. https://doi.org/10.1093/gji/ggu408
Meade BJ (2007) Algorithms for the calculation of exact displacements, strains, and stresses for triangular dislocation elements in a uniform elastic half space. Comput Geosci 33:1064–1075. https://doi.org/10.1016/j.cageo.2006.12.003
Molinari I, Tonini R, Lorito S, Piatanesi A, Romano F, Melini D et al (2016) Fast evaluation of tsunami scenarios: uncertainty assessment for a Mediterranean Sea database. Nat Hazards Earth Syst Sci 16:2593–2602. https://doi.org/10.5194/nhess-16-2593-2016
Murotani S, Satake K, Fujii Y (2013) Scaling relations of seismic moment, rupture area, average slip, and asperity size for M ~9 subduction-zone earthquakes. Geophys Res Lett 40:5070–5074. https://doi.org/10.1002/grl.50976
IOC (2023) The North-eastern Atlantic, the Mediterranean and connected seas Tsunami Early Warning and Mitigation System (NEAMTWS) 2030 Strategy. Intergovernmental Oceanographic Commission of UNESCO IOC Technical Series IOC/2022/TS/171, Paris, pp 32
Okada Y (1992) Internal deformation due to shear and tensile faults in a half-space. Bull Seismol Soc Am 82(2):1018–1040
Ruiz JA, Baumont D, Bernard P, Berge-Thierry C (2011) Modelling directivity of strong ground motion with a fractal, k -2, kinematic source model. Geophys J Int 186:226–244. https://doi.org/10.1111/j.1365-246X.2011.05000.x
Scala A, Lorito S, Romano F, Murphy S, Selva J, Basili R et al (2020) Effect of shallow slip amplification uncertainty on probabilistic tsunami hazard analysis in subduction zones: use of long-term balanced stochastic slip models. Pure Appl Geophys 177:1497–1520. https://doi.org/10.1007/s00024-019-02260-x
Selva J, Tonini R, Molinari I, Tiberti MM, Romano F, Grezio A, Melini D, Piatanesi A, Basili R, Lorito S (2016) Quantification of source uncertainties in seismic probabilistic tsunami hazard analysis (SPTHA). Geophys J Int 205(3):1780–1803. https://doi.org/10.1093/gji/ggw107
Selva J, Argyroudis S, Cotton F, Esposito S, Iqbal SM, Lorito S, Stojadinovic B, Basili R, Hoechner A, Mignan A, Pitilakis K, Thio HK, Giardini D (2024) A novel multiple-expert protocol to manage uncertainty and subjective choices in probabilistic single and multi-hazard risk analyses. Int J Disaster Risk Reduct 110:104641. https://doi.org/10.1016/j.ijdrr.2024.104641
Strasser FO, Arango MC, Bommer JJ (2010) Scaling of the source dimensions of interface and intraslab subduction-zone earthquakes with moment magnitude. Seismol Res Lett 81:941–950. https://doi.org/10.1785/gssrl.81.6.941
Tonini R, Di Manna P, Lorito S, Selva J, Volpe M, Romano F et al (2021) Testing tsunami inundation maps for evacuation planning in Italy. Front Earth Sci 9. https://doi.org/10.3389/feart.2021.628061
Volpe M, Lorito S, Selva J, Tonini R, Romano F, Brizuela B (2019) From regional to local SPTHA: efficient computation of probabilistic tsunami inundation maps addressing near-field sources. Nat Hazards Earth Syst Sci 19:455–469. https://doi.org/10.5194/nhess-19-455-2019

BY

Chapter 11
R2–Local Probabilistic Tsunami Hazard Analysis

Manuela Volpe, Steven J. Gibbons, and Enrico Baglione

11.1 Introduction

This recipe describes a method for building simulation-based PTHA at local scale, by combining an adequate discretization of source scenarios with high-resolution inundation modelling. This leads to tens to hundreds of thousands of moderately intensive numerical simulations being required. In recent years, more efficient GPU-based High-Performance Computing (HPC) facilities, together with efficient GPU-optimized models for simulating tsunami inundation, have made local hazard assessment a feasible but still highly computationally demanding task. To reduce the computational cost of the analysis, different strategies can be implemented.

The local PTHA presented here includes the following ingredients:

- Fully simulation-based hazard assessment framework (FW1A, Sect. 2.2)
- Source selection for earthquakes sources—Sect. 3.2
- Source modelling for earthquake sources—Sect. 4.2
- Tsunami modelling with inundation–Chapter 5

The recipe is suitable for single-hazard analysis at local scale (i.e. covering a spatially limited study area with high resolution modelling) for seismically induced tsunamis, provided that a regional hazard model is available (e.g. NEAMTHM18, see recipe R1). It provides two different strategies for dealing with the computational

M. Volpe (✉)
Osservatorio Nazionale Terremoti, Istituto Nationale di Geofisica e Vulcanologia, Roma, Italy
e-mail: manuela.volpe@ingv.it

S. J. Gibbons
Natural Hazards, Norwegian Geotechnical Institute, Oslo, Norway
e-mail: steven.gibbons@ngi.no

E. Baglione
Istituto Nationale di Geofisica e Vulcanologia, Bologna, Italy
e-mail: enrico.baglione@ingv.it

M. B. Sørensen et al. (eds.), *Probabilistic Tsunami Hazard and Risk Analysis*, Mathematics of Planet Earth 14, https://doi.org/10.1007/978-3-031-98115-9_11

cost: a) in the absence of vast computational resources, a filtering procedure coupled with a cluster analysis can be applied for the identification of a subset of the most important sources, able to acceptably preserve the accuracy of results with a few hundreds of scenarios (Lorito et al. 2015; Volpe et al. 2019); b) given very large available HPC resources, a disaggregation procedure can be implemented to select the sources that contribute most to the hazard at the target site within one or more specific tsunami intensity ranges, while reproducing the entire hazard curve with a high level of accuracy (Gibbons et al. 2020).

The method can be suitable for use in local planning and/or local assessments in presence of strategic infrastructures.

11.2 Framework Used in Recipe

This recipe uses the framework FW1A (Fully simulation-based risk assessment; Sect. 2.2), a scenario-based framework that exploits a large number of scenarios to address uncertainty quantification using a probabilistic analysis. The characterization of the hazard implies large uncertainties related to the seismic sources as well as to the tsunami modelling. The probabilistic treatment of a large set of potential tsunami sources guarantees the full exploration of the aleatory uncertainty and the inclusion of alternative models needed for quantifying epistemic uncertainty and is essential in PTHA. The hazard curves are obtained by combining the annual rates of all the sources included in the analysis with the tsunami impact obtained from the numerical simulations, assuming a time-independent Poisson process. This recipe neglects the potential uncertainty stemming from the tsunami modelling and its propagation onto the hazard results.

The advantage of using this framework is the fact that it provides the most comprehensive treatment of the uncertainty in tsunami hazard analysis. The limitations are the need for high-resolution topo-bathymetric data and the extensive computational cost. The latter requires the introduction of some approximations to reduce the number of numerical simulations, in order to make it feasible.

11.3 Source Identification and Characterisation Used in Recipe

This recipe describes the estimation of tsunami hazard at local scale due to the earthquake sources described in Sect. 3.2. The local assessment uses as input the regional hazard model NEAMTHM18 described in recipe R1 and Basili et al. 2021. This is based on a tectonic regionalization and the definition, within each region, of two types of seismicity (predominant seismicity and background seismicity), with the earthquake parameters discretized through an event tree approach (Selva et al. 2016).

The full exploration of the source aleatory uncertainty is in principle achieved by including all earthquake sources as parameterized in NEAMTHM18. An ensemble modelling approach enables the epistemic uncertainty to be addressed, by including multiple alternative models weighted by their annual rates (Selva et al. 2016).

Treating the whole set of sources in local PTHA is nevertheless unfeasible, since it would require millions of high-resolution numerical simulations. To select a manageable set of sources, two different approaches can be implemented, depending on the availability of computational resources:

a) A two-step filtering procedure coupled with a cluster analysis can be implemented for a computationally efficient assessment given scarce computational resources. This approach helps to select a limited subset (a few hundreds) of representative scenarios, relevant for the target site. This method ensures that robustness of the assessment is preserved while still considering the practical constraints of computational resources (Lorito et al. 2015; Volpe et al. 2019). A key point of this approach is that separate treatments are applied for far-field and near-field sources, since the coseismic deformation can modify the actual local tsunami intensity at the nearby coast, due to coastal uplift or subsidence. For both far- and near-field sources, the two-step filtering algorithm is applied, aimed to i) discard all the sources giving a negligible contribution to the offshore tsunami height near the target area, assuming they would lead to negligible inundation; ii) discard scenarios whose cumulative mean annual rate (mean of the model epistemic uncertainty) is below a fixed threshold, possibly set at a negligible level in the framework of the overall analysis and/or with respect to other uncertainties. After these two filters, a *k*-medoids-clustering procedure is applied to the far-field sources, under the assumption that sources producing similar offshore profiles will also produce similar inundation patterns, so that each resulting source is representative of a cluster of sources producing comparable tsunamis. Near-field sources may challenge the general assumption that offshore tsunami amplitude profiles can be representative of the coastal inundation, due to the coseismic deformation, and for these sources the cluster analysis is carried out based on the similarity of the (vertical) coseismic fields, not of the offshore tsunami amplitudes.
b) A disaggregation analysis can be carried out on the regional hazard model given the availability of large HPC resources, to select all the sources (tens of thousands) which significantly contribute to the hazard at the target site, while reproducing the entire hazard curve from the regional model with a high level of accuracy (Gibbons et al. 2020). The disaggregation algorithm, once one or more offshore points close to the target site have been selected from the regional model, executes the following operations: i) ranking all the scenarios contributing to the hazard at those points for a given intensity, or intensity interval, according to their mean annual rates; ii) selection of a set of scenarios, starting from the most probable ones, able to reproduce the original offshore hazard curve, based on the desired "degree of accuracy" to which the hazard should be approximated. The mean annual rates of the selected scenarios are finally normalized by the degree of

accuracy chosen in the disaggregation procedure, to compensate for the removal of potentially impacting sources, under the assumption that the selected set of sources is an unbiased sample of all sources that may generate local tsunamis.

The advantage of the described approach is the full exploration and quantification of the uncertainty related to the adopted seismicity model. The limitations are due to the need to implement strategies to reduce the computational cost, which inevitably involves approximations that may lead to a lower accuracy, especially if a strong reduction is required.

11.4 Source Mechanics Modelling Used in Recipe

The earthquake sources included in the PTHA are modelled differently depending on the type of seismicity (see also Sect. 4.2):

- Background seismicity is treated through the Okada analytical model for rectangular finite faults (Okada 1992): this allows for fast modelling but excludes treatment of source complexities.
- Predominant seismicity is treated by discretizing the rupture area on a triangular mesh and computing the initial displacement as the superposition of the individual triangular patches (Scala et al. 2020): this allows a better constraint of the source geometry based on geological and seismological data and to model rupture complexities (i.e. heterogeneous slip) but requires a lot of preparatory work in building the meshes.

11.5 Tsunami Modelling Used in Recipe

The recipe uses the methods described in Chapter 5 for data preparation and tsunami generation, propagation, and inundation, as follows:

- Data preparation (see Sect. 5.2): high resolution and high accuracy models are required for the region where inundation is to be modelled (e.g. Griffin et al. 2015). A nested or telescopic grid system is applied, with increasing resolution as the target site is approached, from a few hundred meters for the offshore tsunami propagation to 5–10 m for simulating the inundation.
- Tsunami generation (see Sect. 5.3): the initial conditions are modelled as described above regarding the source mechanics modelling.
- Tsunami propagation and inundation (see Sects. 5.4–5.5): a model based on a GPU Faster-Than-Real-Time (FTRT) implementation is used, i.e. the nonlinear hydrostatic shallow-water code T-HySEA, based on a mixed finite-difference–finite-volume method (de la Asunción et al. 2013; Macías et al. 2016, 2017). Outputs (i.e. maximum wave height to be converted into flow depth) are saved at all the points within the inner grid.

Tsunami simulations are subject to various sources of uncertainty, including uncertainties in the topo-bathymetric data, the initial and boundary conditions, as well as model parameters and physics (see Sect. 5.7).

Using this modelling approach has the advantage of enabling relatively fast simulations. This is possible by simplifying the physics of the model on the basis of some assumptions, which however could be misleading in specific conditions (e.g. harbours or narrow channels). Moreover, great care must be taken in preparing the topo-bathymetric grids.

11.6 Discussion

A typical workflow for local PTHA is displayed in Fig. 11.1, divided into four main sections:

(1) User specifications: the topo-bathymetric grids as described previously; the estimate of the available HPC resources to allow for the required simulations; specific parameters required by the workflow (e.g. return periods and percentage of hazard captured in the disaggregation procedure or thresholds to be used in the filtering and the clustering procedures)

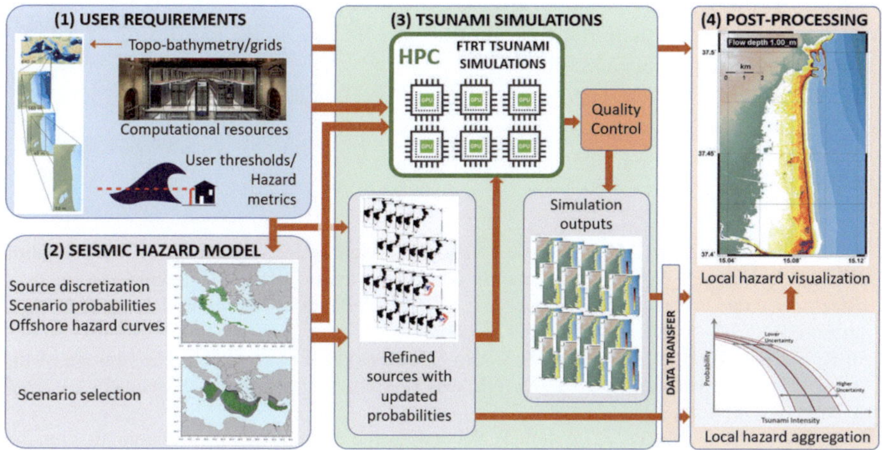

Fig. 11.1 Schematic representation of steps (1) to (4) in the local seismic PTHA. (1) The user specifies a high-resolution Digital Elevation Model for the region of interest and a set of topo-bathymetric grids covering all potential tsunami sources for the region, together with the required hazard metrics. A quantification of the computational resources available to the user is also needed. (2) A tsunami hazard model containing a discretization of tsunamigenic earthquakes with models of probabilities of occurrence is applied. (3) Faster Than Real Time (FTRT) numerical tsunami simulations are undertaken on High Performance Computing resources. (4) Hazard aggregation is performed on the outputs from the individual tsunami simulations. Visualization and reporting are provided as needed. Modified from Gibbons et al. (2020)

(2) Regional seismic hazard model: this must contain a discretization of possible earthquake sources, together with models for their probability of occurrence within a given time interval (see NEAMTHM18 in recipe R1).
(3) Inundation calculations: this is the HPC section where the tsunami inundation simulations are carried out, provided that codes optimized for rapid large-scale runs are available (e.g. Tsunami-HySEA, de la Asunción et al. 2013).
(4) Hazard aggregation: this is the stage at which the simulation outputs are combined with the source probabilities to produce local hazard curves and inundation maps.

In summary, the key elements for the local PTHA are: (1) robust and accurate digital elevation models, which must be carefully elaborated in order to build suitable simulation grids, (2) a reliable and rigorous regional hazard model, and (3) the availability of adequate HPC resources.

A robust PTHA is an important tool for civil protection purposes and local planning, and it has been already used for operational applications.

Both described approaches are based on approximations to reduce the computational cost of the analysis and make it feasible with present technologies. The uncertainties introduced by such approximations on the computed local hazard are not quantified. They are assumed to be negligible compared to those related to the seismic source process and occurrence rates. Moreover, the uncertainties related to the tsunami numerical modelling are not propagated onto the hazard results.

Future technologies, i.e., new faster GPUs or AI-based methods, could improve quantification of the epistemic uncertainties in PTHA while exploring to a large extent the natural aleatory uncertainty.

References

de la Asunción M, Castro MJ, Fernández-Nieto ED, Mantas JM, Acosta SO, González-Vida JM (2013) Efficient GPU implementation of a two waves TVD-WAF method for the two-dimensional one layer shallow water system on structured meshes. Comput Fluids 80:441–452. https://doi.org/10.1016/j.compfluid.2012.01.012

Basili R, Brizuela B, Herrero A, Iqbal S, Lorito S, Maesano FE et al. (2021) The Making of the NEAM Tsunami Hazard Model 2018 (NEAMTHM18). Frontiers in Earth Science 8. https://doi.org/10.3389/feart.2020.616594

Gibbons SJ, Lorito S, Macías J, Løvholt F, Selva J, Volpe M et al (2020) Probabilistic tsunami hazard analysis: high performance computing for massive scale inundation simulations. Front Earth Sci. 8. https://doi.org/10.3389/feart.2020.591549

Griffin J, Latief H, Kongko W, Harig S, Horspool N, Hanung R et al. (2015) An evaluation of onshore digital elevation models for modeling tsunami inundation zones. Front Earth Sci 3. https://doi.org/10.3389/feart.2015.00032

Lorito S, Selva J, Basili R, Romano F, Piatanesi TMM, A, (2015) Probabilistic hazard for seismically induced tsunamis: accuracy and feasibility of inundation maps. Geophys J Int 200:574–588. https://doi.org/10.1093/gji/ggu408

Macías J, Mercado A, Gonzalez-Vida JM, Ortega S, Castro M (2016) Comparison and computational performance of tsunami-HySEA and MOST Models for LANTEX 2013 scenario: impact assessment on puerto rico coasts. Pure Appl Geophys 173:3973–3997. https://doi.org/10.1007/s00024-016-1387-8

Macías J, Castro MJ, Ortega S, Escalante C, González-Vida JM (2017) Performance Benchmarking of Tsunami-HySEA Model for NTHMP's Inundation Mapping Activities Pure Appl Geophys 174:3147–3183. https://doi.org/10.1007/s00024-017-1583-1

Okada Y (1992) Internal deformation due to shear and tensile faults in a half-space. Bull Seismol Soc Am 82(2):1018–1040

Scala A, Lorito S, Romano F, Murphy S, Selva J, Basili R et al (2020) Effect of shallow slip amplification uncertainty on probabilistic tsunami hazard analysis in subduction zones: use of long-term balanced stochastic slip models. Pure Appl Geophys 177:1497–1520. https://doi.org/10.1007/s00024-019-02260-x

Selva J, Tonini R, Molinari I, Tiberti MM, Romano F, Grezio A, Melini D, Piatanesi A, Basili R, Lorito S (2016) Quantification of source uncertainties in seismic probabilistic tsunami hazard analysis (SPTHA). Geophys J Int 205(3):1780–1803. https://doi.org/10.1093/gji/ggw107

Tonini R, Di Manna P, Lorito S, Selva J, Volpe M, Romano F et al. (2021) Testing tsunami inundation maps for evacuation planning in Italy. Front Earth Sci 9. https://doi.org/10.3389/feart.2021.628061

Volpe M, Lorito S, Selva J, Tonini R, Romano F, Brizuela B (2019) From regional to local SPTHA: efficient computation of probabilistic tsunami inundation maps addressing near-field sources. Nat Hazards Earth Syst Sci 19:455–469. https://doi.org/10.5194/nhess-19-455-2019

Chapter 12
R3–Probabilistic Analysis for Tsunami Hazard in Macro Tidal Coasts

Íñigo Aniel-Quiroga, Ernesto Mauricio Gonzalez, Maria Ana Baptista, and Jorge Macías

12.1 Introduction

This recipe presents the application of a PTHA methodology that combines the statistical treatment of tsunami hazards and sea level variation (Gonzalez et al. 2021). The method improves on the existing approaches, where the treatment of sea level at the time of tsunami impact consists of adding or subtracting a constant sea level. The test site is the Bay of Cadiz in Spain characterized by a meso-tidal regime (higher than 2 m and lower than 4 m). The results are inundation maps expressing the likelihood of exceeding tsunami intensity measures (maximum wave elevation, maximum current speed and maximum wave forces).

This recipe is for earthquake-induced tsunamis only. The seismic study area is the SWIT (SW Iberian Transpressive Domain) in the NE Atlantic (Zitellini et al. 2009), home of the most damaging earthquake affecting Morocco, Portugal and Spain, namely the 1 November 1755 and the 28 February 1969 events, and the respective tsunamis (Baptista et al. 2019).

Í. Aniel-Quiroga (✉) · E. M. Gonzalez
Instituto de Hidráulica Ambiental (IH Cantabria), Universidad de Cantabria, Santander, Spain
e-mail: anieli@unican.es

E. M. Gonzalez
e-mail: mauricio.gonzalez@unican.es

M. A. Baptista
Instituto Superior de Engenharia de Lisboa, Instituto Politécnico de Lisboa, Lisboa, Portugal
e-mail: mavbaptista@gmail.com

J. Macías
Departamento de Análisis Mathemático, Estadística e Investigación Operativa y Mathemática Aplicada, Universidad de Málaga, Málaga, Spain
e-mail: jmacias@uma.es

M. B. Sørensen et al. (eds.), *Probabilistic Tsunami Hazard and Risk Analysis*, Mathematics of Planet Earth 14, https://doi.org/10.1007/978-3-031-98115-9_12

This PTHA includes the ingredients:

- Event-based probabilistic hazard analysis using discrete scenario modelling (FW1A, FW1B, Chap. 2)
- Source selection for earthquakes sources—Sect. 3.2
- Source modelling for earthquake sources—Sect. 4.2
- Tsunami hazard modelling, including inundation with numerical model – Chap. 5

The recipe is suitable for regional and local tsunami hazard analysis. This probabilistic method includes the application of high-resolution numerical simulations for tsunami propagation and inundation to derive inundation maps and hazard maps for a given return period and a given confidence interval. One of the unique characteristics of the present methodology based on Monte Carlo simulations is the consideration of the joint probability of the tsunamigenic earthquake and sea level occurrence, including both astronomical and meteorological effects. The PTHA incorporates the tide as a variable in the probabilistic framework.

12.2 The Framework Used in the Recipe

The applied PTHA methodology uses synthetic earthquake catalogues and incorporates the sea level variation into a Monte Carlo simulation (see Fig. 12.1). Therefore, this recipe uses Framework FW1A or FW1B in combination with FW1C (Sect. 2. 2), and its application follows these steps:

1. Build of a database of numerical simulations of tsunamis for different variables: fault geometry, earthquake magnitude, epicentre location and sea level;
2. Generation of a set of scenarios from synthetic seismic catalogues and sea levels based on the probabilistic distributions of the involved variables;
3. Map the maximum values of the selected tsunami intensity measures to assess the hazard for a given probability.

A total number of 619 scenarios were modelled. These scenarios are the result of the combination of focal mechanisms linked to magnitudes from 6.5 to 8.4, up to 4 different locations for each magnitude, and 4 sea level ranges. This methodology was developed in the European FP6 TRANSFER project (2009) and, in this recipe, it is applied to the Bay of Cádiz area in Spain (González et al. 2021).

The advantage of using this framework is the combined statistical treatment of both tsunami hazard and sea level by incorporating the tidal state into the PTHA. On the other hand, a limitation is that the tsunami-tidal current interaction is neglected and that epistemic uncertainties are overlooked (see Chapter 2).

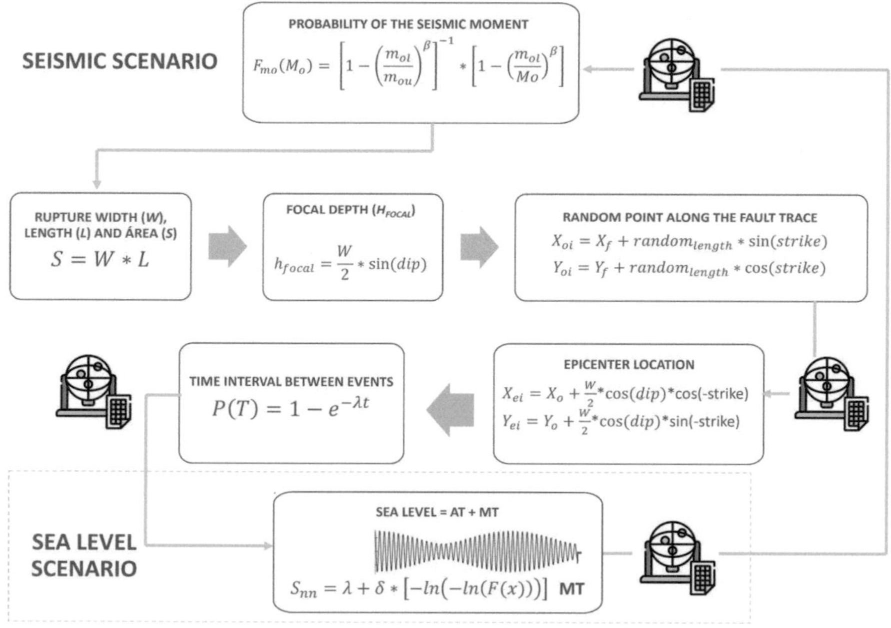

Fig. 12.1 Summary of the algorithm used to generate the synthetic catalogue of tsunamis (modified from Gonzalez et al. 2021)

12.3 Source Selection and Activity Rate Characterization Used in the Recipe

The tsunamigenic source characterization consists of the identification and characterization of the seismic faults in the area of interest, using the historical and instrumental catalogues, following these steps:

1. Seismic zonation with the identification of major geological features used to establish the maximum earthquake magnitude (Álvarez-Gómez et al. 2008 and Omira et al. 2009)
2. Use of Gutenberg-Richter Law to obtain the statistical parameters
3. Determination of geometrical parameters and scaling relationships between seismic moment and source parameters for the main fault ruptures

To compute the tsunami sources to be used as input in the propagation model, first, the earthquake scenarios in each fault are established, considering the relation between the seismic moment magnitude and the rupture area. The width of the fault (W) is constrained by the L/W constant ratio, where L represents the fault length, for each fault and momentum magnitude.

12.4 Sea Level Statistics

Based on the data measured by the tidal gauge closest to the study zone, the sea level probability distribution function is obtained and the tidal level is incorporated towards a joint probabilistic definition of scenarios. The net sea level variation is due to atmospheric processes, such as atmospheric pressure changes, the effect of wind and due to the astronomical effects. The former contribution has a random nature, while the astronomical tides are deterministic. Therefore, from the tide gauge data, conducting a harmonic analysis both contributions can be separated and treated as different processes. In meso-tidal and macro-tidal coastal areas, the inclusion of the tide level can make a relevant difference in terms of impact and helps to avoid underestimating flooded areas.

12.5 Monte Carlo Simulations

Based on the probabilistic characteristics of the sources, and by applying the Monte Carlo method, synthetic earthquake series are obtained for each of the considered faults. The time series corresponding to each fault are integrated to generate a global tsunamigenic earthquake and sea level catalogue for the study region. The length of the time series in the present study was selected to be 10,000 years. To derive confidence bands, for each thematic map, 1000 Monte Carlo simulations for each fault were performed.

12.6 Tsunami Modelling Used in the Recipe

In this recipe, tsunami modelling tools are used to generate a dataset of tsunami numerical simulations for different faults, earthquake magnitudes, epicentre locations and sea levels using the C3 numerical model (Olabarrieta et al. 2011). The model solves the conservative form of the Nonlinear Shallow Water equations (NLSW) in a Cartesian coordinate system. It integrates a finite difference scheme, in particular, leap-frog (COMCOT, Wang and Power 2011), for open ocean propagation and a finite volume model scheme (George 2008) in the higher resolution grids used for computing inundation. This numerical model has been validated following benchmarks proposed by Synolakis et al. (2008). Now we will split this recipe into the different steps described in Chapter 5, in particular, data preparation, interfacing data model, propagation and inundation step and validation.

- Data preparation (bathymetry and topography). As propagation and inundation simulations were performed, both bathymetric and topographic data were required. As the former, GEBCO data and SWIM (Zitellini et al. 2009) were used. To perform inundation in a finer resolution grid, IHM (Instituto Hidrográfico de la Marina) near-to-coast high-resolution bathymetric data and IGN (Instituto Geográfico Nacional) topographic data were used.

- Interfacing: 619 scenarios were simulated, covering the entire range of possible tsunami scenarios in each zone (potential sources, earthquake magnitudes, and epicentre locations), together with all the possible sea levels. They correspond to different initial conditions where different faults, earthquakes magnitudes, epicentre locations and sea levels are considered. The computation of this number of scenarios including inundation required HPC resources.
- Propagation: The NLSW model (Sect. 5.4.2.) C3 was used, and for propagation implements the leap-frog numerical scheme in finite difference staggered grids.
- Inundation: A nested grid approach was used to compute high-resolution inundation in the area of interest. Three different resolution grids were used with 864, 216 and 50 m resolution, respectively. As the leap-frog scheme is not suitable for inundation, a finite volume scheme was used in the lower level.
- C3 numerical model was validated (Sect. 5.6) previously to this work (Olabarrieta et al. 2011).

12.7 Tsunami Flooding Maps Elaboration

Once the 1000 time series of tsunamigenic earthquake events coinciding with a given sea level are created in the Monte Carlo computation, the impact of each tsunami on the coast is obtained using the scenarios database. The closest pair (sea level, earthquake) in the numerical database was identified for each scenario in the synthetic catalogue. For the interpolation of the events of the synthetic catalogue, we start from the basis of a 3D matrix of numerically modelled scenarios with three inputs: magnitude, earthquake epicentre location and sea level.

For a synthetic event of a given magnitude, the closest modelled magnitude is searched in the matrix, the closest modelled epicentre to the synthetic catalogue is searched among the modelled epicentres, and finally, the corresponding event with the closest sea level is searched in the matrix. Thus, the event of the synthetic series is associated with the closest simulated event in magnitude, location of the epicentre and sea level. The coastal variables corresponding to this numerical case are assigned to the tsunamigenic earthquake in the synthetic catalogue. The time series of the variables can be derived for the whole coastal region following this procedure. Therefore, these time series of the maximum value can be used to perform a statistical analysis of the selected variables and derive maps with an associated return period. For each return period and each synthetically generated catalogue, a value of the study variable is obtained in all the grid nodes. Therefore, for a given return period, the analysed variable can be mapped in the high-resolution grid. If several synthetically generated earthquake and sea level catalogues are used, the value of the analysed parameter for a given return period will present a mean value and a variance. Computing the distribution function of the values obtained for the given return period, the value for a given return period with given confidence can be obtained.

For a given return period it is possible in each grid point of the detailed mesh (the flooding area) to obtain the desired parameter (e.g., maximum water depth) using the grid point distribution function. The representation on a map of this parameter for all the grid points allows to produce thematic maps for specific return periods.

In the present study, return periods of 500, 1,000, 5,000 and 10,000 years with a confidence band of 99% were considered. This confidence band is representative of the mean extreme events.

12.8 Discussion

The recipe presents a PTHA methodology that treats aleatoric uncertainties on the seismic source and tidal stage through a Monte Carlo-type framework. Therefore, the present method does not assume that tsunamigenic earthquakes occur at a given tidal level condition and combines the tsunami hazard and sea level variation, including astronomical and meteorological effects.

The recipe allows for deriving probabilistic coastal hazard maps (wave heights, current speed, etc.) for different return periods. It deals only with tsunamis of tectonic origin, suitable for application to any tsunami-threatened coast where DEM and records of tidal variations are available. Increasing the number of scenarios in the database reduces the uncertainty of the results.

In order to consider the complexity of the bathymetric and topographic contours and properly represent the transition from offshore to onshore, those scenarios, following the presented methodology, must include the numerical modelling of the tsunami inundation.

In addition, applying this recipe allows obtaining a reduction of the uncertainties for probabilistic flooding maps, achieving a distribution function of different flooding variables (max. water depth, max. wave elevation, max. current sped, max. forces, etc.) in each nodal grid point inland. For each grid point, all the possible inundation scenarios are combined for a given return period to provide a more accurate result.

The inclusion of tide level in probabilistic tsunami hazard assessment is relevant because the flooded areas can be considerably larger in meso-tidal and macro-tidal areas if occurring at high-tide (and vice-versa for low-tide). For example, this inclusion should not be done by simply adding the maximum tide level, which may lead to a noticeable overestimate of the severity of the tsunami impact and the extension of the flooded area, since the tsunami may occur at different instants of the tidal cycle.

This recipe, however, overlooks the epistemic uncertainties such as those related to the source and the modelling tsunami generation, propagation, and inundation and those related to the accuracy of the bathymetric-topographic data.

References

Álvarez-Gómez JA, Gutiérrez OQG, Aniel-Quiroga Í, González M (2012) Tsunamigenic potential of outer-rise normal faults at the Middle America trench in Central America. Tectonophysics 574:133–143. https://doi.org/10.1016/j.tecto.2012.08.014

Baptista MA (2019) Tsunamis along the Azores Gibraltar plate boundary. Pure Appl Geophys 177:1713–1724. https://doi.org/10.1007/s00024-019-02344-8

George DL (2008) Augmented Riemann solvers for the shallow water equations over variable topography with steady states and inundation. J Comput Phys 227:3089–3113

TRANSFER FP6 project (2009) Tsunami risk and strategies for the European region, 6th Framework program https://cordis.europa.eu/project/id/37058 (last accessed 13.04.2023)

González, M., Álvarez-Gómez, J. A., Aniel-Quiroga, Í., Otero, L., Olabarrieta, M., Omira, R., Luceño, A., Jelinek, R., Krausmann, E., Birkman, J., Baptista, M. A., Miranda, M., Aguirre-Ayerbe, I. (2021) Probabilistic Tsunami Hazard Assessment in Meso and Macro Tidal Areas. Application to the Cádiz Bay, Spain. Frontiers in Earth Science, 9:328. https://doi.org/10.3389/feart.2021.591383

Olabarrieta M, Medina R, González M, Otero L (2011) C3: a finite volume- finite difference hybrid model for tsunami propagation and runup. Comput Geosci 37:1003–1014. https://doi.org/10.1016/j.cageo.2010.09.016

Omira R, Baptista MA, Matias L, Miranda JM, Catita C, Carrilho F et al (2009) Design of a sea-level tsunami detection network for the Gulf of Cadiz. Nat Hazards Earth Syst Sci 9:1327–1338. https://doi.org/10.5194/nhess-9-1327-2009

Synolakis CE, Bernard EN, Titov V, v., Kânoğlu, U., González, F. I. (2008) Validation and verification of tsunami numerical models. Pure Appl Geophys 165:11–12. https://doi.org/10.1007/s00024-004-0427-y

Zitellini, N., Gràcia, E., Matias, L., Terrinha, P., Abreu, M., DeAlteriis, G., et al. (2009) The quest for the Africa–Eurasia plate boundary west of the Strait of Gibraltar. Earth Planet. Sci. Lett. 280:13–50.

Wang, X., Power, W.L. (2011) COMCOT: a tsunami generation, propagation and run-up model. Lower Hutt, N.Z.: GNS Science. GNS Science report 2011/43, 121 p.

Chapter 13
R4–Event-Tree PTHA of Coastal Inundation with Tidal Uncertainty Treatment

Rachid Omira, Maria Ana Baptista, and Luis Manuel Matias

13.1 Introduction

This recipe describes the methodology used to derive probabilistic inundation maps for coastal regions threatened by tsunamis of tectonic origin. It includes the development of an event-tree probabilistic tsunami hazard assessment (PTHA) approach, accounting for both near and distant earthquake sources. The PTHA study's workflow includes the following ingredients:

- Event-tree PTHA to determine the likelihood of tsunami flood occurrence and exceedance for a given exposure time for source selection for earthquakes – Sect. 3. 2.
- Source modelling for earthquakes–Sect. 4.2.
- Tsunami modeling—Chapter 5.
- Tide integration—See also Recipe R3.

This recipe follows the framework FW1D (Working on a limited number of scenarios), and is limited to the quantification of the hazard. The recipe can be replicated for any coastal area exposed to the tsunami threat from earthquake sources. Decision-makers can use this study operationally to develop tsunami evacuation routes and define tsunami shelters. The PTHA workflow is described in detail by

R. Omira (✉)
Instituto Português do Mar e da Atmosfera–IPMA, Lisboa, Portugal
e-mail: rachid.omira@ipma.pt

M. A. Baptista
Instituto Superior de Engenharia de Lisboa, Instituto Politécnico de Lisboa, Lisboa, Portugal
e-mail: mavbaptista@gmail.com

L. M. Matias
Instituto Dom Luiz, Faculdade de Ciências da Universidade de Lisboa, Universidade de Lisboa, Lisboa, Portugal
e-mail: lmatias@fc.ul.pt

M. B. Sørensen et al. (eds.), *Probabilistic Tsunami Hazard and Risk Analysis*, Mathematics of Planet Earth 14, https://doi.org/10.1007/978-3-031-98115-9_13

Omira et al. (2016). It was first implemented in one of the NE Atlantic test-sites of the ASTARTE FP7 EU project, Sines, a port city in the SW of Portugal, 150 km south of Lisbon, belonging to a municipality with a total area of around 203.30 km^2 and a coastline extending for 35-km (Omira et al. 2016). Later, the PTHA workflow was applied to assess the probabilistic tsunami hazard in the Indian Ocean, namely along the coast of Oman exposed to tsunami hazard from the Makran Subduction Zone (El-Hussain et al. 2016) and specifically at the coastal city Sur, located about 203 km southeast of the Omani capital Muscat with approximately 20 km of coastline (El-Hussain et al. 2018).

13.2 Framework Used in the Recipe

This recipe addresses probabilistic tsunami coastal inundation (framework FW1D) using an event-tree framework (Fig. 13.1) that treats aleatoric uncertainties on the earthquake source location and tidal stage. It integrates seismic source probability models, high-resolution tsunami numerical modeling and statistical methods to derive the probability of tsunami occurrence and exceedance of a near-shore wave height /flow depth for a given exposure time. The generation of possible tsunami events requires the definition of the tsunamigenic source zones, attributing individual magnitudes for possible events and computing their recurrence rates and probability of occurrence, assuming a Poisson process.

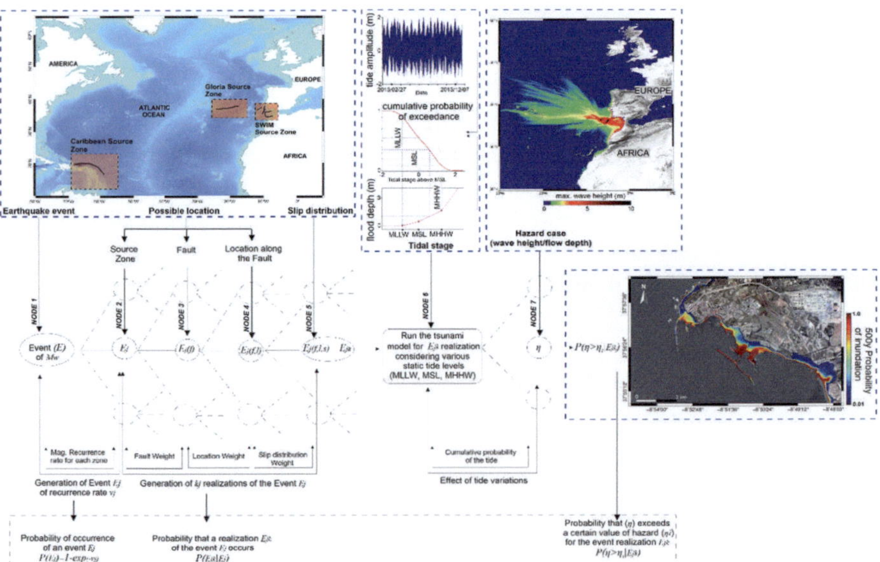

Fig. 13.1 Event-tree probabilistic tsunami hazard framework described in the recipe. Modified from Omira et al. 2016

The source-to-coast simulations of the whole set of tsunamigenic scenarios allow the generation of tsunami hazard curves that express the annual probability of exceeding intensity thresholds at selected locations on- or offshore. Considering a specified exceedance threshold, a vertical cut of these curves at each point of the grid domain enables deriving the tsunami exceedance probability maps.

Although the proposed PTHA model addresses the tsunami hazard locally (i.e., tsunami inundation), it is computationally less demanding as it only focuses on tectonically constrained tsunamigenic source zones. It also can be replicable for any tsunami-threatened coastal area where a high-resolution digital elevation model is available. Its disadvantages, however, concern using a limited number of tsunamigenic scenarios and the lack of treatment of epistemic uncertainties. If a logic tree approach is desired for the assessment of epistemic uncertainty, then the workflow described here would apply to any of the tree branches.

13.3 Source Selection and Activity Rate Characterization Used in the Recipe

This recipe accounts for both local and distant tsunami sources (i.e., when applied to the NE Atlantic three tsunami source zones were considered: the SW Iberia Margin (SWIM), the Gloria Fault (GF) and the Caribbean (Car)) and adopts the process described in Omira et al. (2015) to define the tsunamigenic scenarios and characterize their recurrence rates. The main steps of such a process are as follow (Fig. 13.1):

- Source zone: this step concerns the identification of source regions able to generate tsunamis that can impact / reach the target coast. Such an identification considers events from tsunami catalogs, seismic activity of the region and evidence of active tsunamigenic structures from geophysical and geological works.
- Magnitude range: to each source region the minimum and the maximum magnitudes of the tsunamigenic earthquakes need to be assigned. The minimum earthquake magnitude (MinEqMg) was selected on the basis of tsunami catalogues and tsunami numerical modeling tests. This led to attribute different MinEqMg for the considered source zones of the Atlantic, 7.5 for the SWIM and 8.0 for both GF and Car. The maximum possible earthquake magnitude (MaxEqMg) was, on the other hand, set from instrumental and historical catalogues or derived from fault size and earthquake scaling laws specific for each source region (i.e., Wells and Coppersmith 1994; Strasser et al. 2010), allowing to select the largest among the two magnitudes. Individual earthquake magnitudes are then defined by applying a regular interval (i.e., 0.1) to the magnitudes ranging from MinEqMg to MaxEqMg. Annual recurrence rate of each magnitude is then derived through the Bayesian method (Kijko 2004), which incorporates seismic information from historical and instrumental catalogs.

- Typical fault: for each earthquake magnitude, a typical fault is defined. According to Omira et al. (2009), typical faults must fulfill at least one of the following criteria: i) have been evidenced by seismo-stratigraphic research or at least deduced from morphostructural analysis of swath bathymetry data; ii) provide a clear indication of activity or iii) have been deduced from instrumental seismic data.
- Scenarios database: by variation of a typical fault along the fault traces of each source zone, a tsunamigenic scenario database is built. This database includes all earthquake fault parameters required for tsunami numerical modeling (Matias et al. 2013). It accounts for a total of 318 earthquake scenarios able to cause tsunamis that impact/reach the NE Atlantic coasts.

The advantage of this procedure is considering a limited number of tsunamigenic scenarios with simple source geometry and uniform slip distribution, which considerably reduces the computational cost of the tsunami simulations. Its disadvantage concerns the lack of considering complex earthquake ruptures and heterogeneous slip distributions as well as the potential under-estimation of the natural (aleatory uncertainty), such as the existence of sources not occurring in mapped faults and unexpectedly high magnitudes due to large slip or coalescence of faults.

13.4 Source Mechanisms Modeling Used in the Recipe

The sea-bed co-seismic deformation is computed using Okada's half-space elastic theory (Okada 1985), assuming an instantaneous rupture of the earthquake. The vertical sea bottom deformation is then transferred to the sea surface–with the assumption that the ocean water is incompressible (Kajiura 1970)—to form the tsunami generation.

13.5 Tsunami Modelling

The following set of steps were used to numerically model the source-to-coast tsunami:

- Data preparation: the first step concerned the preparation of bathymetric and topographic data sets required for the tsunami numerical simulations. It consisted of building a system of nested grids with increasing horizontal resolution towards the target coast. For the case-study of Sines in the NE Atlantic, two nested systems were considered. For the near-field sources, a system of four nested grids with a resolution of 640 m for the parent grid and applying a refinement factor of four to achieve the 10-m-resolution digital elevation model for the target coast (Sines).

For the far-field sources, five nested grids of resolutions ranging from 2560 m up to 10 m (refinement factor of four) were used. The bathymetry model for the larger grids was from the general bathymetric chart of the oceans (GEBCO) 30-arc sec gridded data (available at: http://www.gebco.net/). The high-resolution elevation model of Sines coast was built using three different types of datasets: 2-m-resolution LIDAR dataset, a 50-m-resolution bathymetric model and a 100-m-resolution nautical chart. This allowed an appropriate representation of the near-shore bathymetry and the coastal topography for tsunami inundation simulations.
- Deep water tsunami propagation: in the deep ocean, the linear approximation of the shallow water equations (SWEs), implemented in an explicit staggered leap-frog finite differences numerical code NSWING (Miranda et al., 2014), with Coriolis acceleration was used. This simple approximation consists of neglecting the non-linear convective inertia force as the tsunami propagates with wave heights much smaller than the water depth.
- Near-shore tsunami propagation and inundation: when tsunami-generated waves approach the shoreline and propagate in shallow water areas, the non-linear convective inertia force and bottom friction effects become increasingly important. In such a case, the linear approximation is no longer valid and the non-linear approximation of SWEs, as implemented in NSWING, was used to simulate the tsunami hazard in terms of wave height and inundation depth. To track the shoreline movement, allowing to compute tsunami inundation, the moving boundary treatment (Liu et al. 1995), based upon "wet" and "dry" cells, was used.

This tsunami simulation procedure presents the following advantages: (i) allowing to calculate the tsunami hazard locally, (ii) less computationally demanding as it uses linear approximation for the wider propagation domains. The disadvantages are: (i) lack of considering frequency dispersion, (ii) lack of any treatment of epistemic uncertainties such as those related with the accuracy of the bathymetric-topographic models and with the modeling of tsunami propagation and inundation.

13.6 Tidal Uncertainty Integration

To calculate the probability that a tsunami intensity (i.e., wave height above sea level, flow depth above ground level) exceeds a threshold for a given exposure time, the numerical code was run for all earthquake scenarios over multiple discrete tidal stages (i.e., mean lower low water: MLLW, mean sea level: MSL, and mean higher high water: MHHW). The tsunami-tide interaction was treated in a "static" way (the DEM was corrected to a given tidal level—MLLW, MSL or MHHW—before running the tsunami simulations), and the tsunami intensity was computed at each point of the grid. Then the curve of evolution of the tsunami intensity (vertical axis) was constructed as a function of the tidal stage (horizontal axis) for each grid point using a piecewise linear interpolation (see details in Omira et al. 2016). Also, from long time-series tide-gauge records the tide probability density function was computed, expressing the cumulative probability (vertical axis) of exceeding the tide levels (vertical axis). Selecting the exceedance threshold, a horizontal cut of the curve of evolution of the tsunami intensity in function of the tidal level gives the tidal stage (Ts) that leads to exceed the intensity threshold. Using this Ts value, a vertical cut of the tide probability density function curve gives the cumulative probability of exceeding the Ts. Finally, at that grid point, the probability to exceed a tsunami hazard intensity, considering the tidal variations, is equal to the cumulative probability that a tidal stage exceeds Ts.

The advantage of this procedure is the inclusion of tidal uncertainty in the PTHA, as the level of the tide at the tsunami arrival can significantly influence its coastal impact. Its disadvantage is disregarding the dynamic interaction between the tide and the tsunami (i.e., combined simulations of time-varying tide and tsunami).

13.7 Hazard Assessment

The PTHA presented in this recipe combined seismic sources probability models and high-resolution tsunami inundation modeling that accounts for the tidal variation. Tsunami hazard curves - expressing the annual probability of exceeding tsunami intensity thresholds at selected locations - were derived at each point of the grid of the coastal domain. A vertical cut of these curves, considering a selected exceedance threshold, allowed deriving the tsunami exceedance probability maps that express the likelihood of tsunami occurrence and exceedance of a near-shore wave height / flow depth for a given exposure time.

The advantage of this PTHA product is the provision of the hazard information at high spatial resolution at the coastal region of interest. Its disadvantage concerns the disregard of non-seismic tsunami sources, which can lead to an underestimation of the total tsunami hazard in the target coast.

13.8 Discussion

Overall, replication of the study presented in this recipe at other coastal sites requires:

1. Gathering the best knowledge on the tsunamigenic sources, of tectonic origin, threatening the target coast.
2. Building a high resolution digital elevation model of the target coastal area.
3. Collecting long time-series of tide-gauge records in the coastal region of interest.
4. Running tsunami inundation simulations using a benchmarked numerical code.

The limitations of the PTHA procedure presented in this recipe concern:

- The disregard of non-seismic tsunami sources, implying a possible underestimation of the tsunami hazard in some coastal regions.
- The selection of mapped faults within the potential seismogenic source zones, limiting the exploration of natural variability as large events can occur outside these mapped faults.
- The consideration of uniform slip distribution despite the recognized significant influence of this earthquake rupture feature in the resulting tsunami coastal impact, particularly for the local sources. Exploring a synthetic catalogue of heterogeneous slip distributions would be more appropriate to tackle the slip distribution problem.
- The lack of treatment of epistemic uncertainties, such as those related with the accuracy of the bathymetric-topographic models and with the modeling of tsunami generation, propagation and inundation.
- The consideration of a single value of the tidal stage during the whole tsunami propagation process; while in a real situation the tide level can be different for the first wave's arrival than for the later edge waves and a tide-tsunami interaction in a dynamic way would be more appropriate to account for the tidal variation in the numerical modeling. However, modeling tsunami-tide dynamic interaction would result in a higher computational cost as the PTHA involves several scenarios and the modeling is performed at local scale.

References

El-Hussain I, Omira R, Al-Habsi Z, Baptista MA, Deif A, Mohamed AME (2018) Probabilistic and deterministic estimates of near-field tsunami hazards in northeast Oman. Geoscience Lett 5(1):30. https://doi.org/10.1186/s40562-018-0129-4

El-Hussain I, Omira R, Deif A, Al-Habsi Z, Al-Rawas G, Mohamad A, Al-Jabri K, Baptista MA (2016) Probabilistic tsunami hazard assessment along Oman coast from submarine earthquakes in the Makran subduction zone. Arabian J of Geosc 9:668. https://doi.org/10.1007/s12517-016-2687-0

Kajiura K (1970) Tsunami source, energy and the directivity of wave radiation. Bull. Earthq Res Inst Tokyo Univ 48:835–869

Kijko A (2004) Estimation of the maximum earthquake magnitude, mmax. Pure and Appl Geophys 161:1655–1681

Liu PL-F, Cho Y-S, Brrigs MJ, Kanoglo U, Synolakis CE (1995) Runup of solitary waves on a circular island. J Fluid Mech 302:259–285

Matias LM, Cunha T, Annunziato A, Baptista MA, Carrilho F (2013) Tsunamigenic earthquakes in the Gulf of Cadiz: fault model and recurrence. Nat Hazards Earth Syst Sci 13:1–13

Miranda JM, Luis J, Reis C, Omira R, Baptista MA (2014) Validation of NSWING, a multi-core finite difference code for tsunami propagation and run-up. American Geophysical Union (AGU) Fall Meeting 2014, San Francisco – USA, 15 – 19 December, S21A-4390

Okada Y (1985) Surface deformation due to shear and tensile faults in a half-space. Bull Seismol Soc Am 75:1135–1154

Omira R, Baptista MA, Matias L, Miranda JM, Catita C, Carrilho F, Toto E (2009) Design of a sea-level tsunami detection network for the gulf of cadiz. Nat Hazards Earth Syst Sci 9:1327–1338

Omira R, Baptista MA, Matias L (2015) Probabilistic tsunami hazard in the northeast atlantic from near-and far-field tectonic sources. Pure and Appl Geophys 172(3–4):901–920

Omira R, Matias L, Baptista MA (2016) Developing an event-tree probabilistic tsunami inundation model for NE Atlantic coasts: application to a case study. Pure and Appl Geophys 173(12):3775–3794

Strasser FO, Arango MC, Bommer JJ (2010) Scaling of the source dimensions of interface and intraslab subduction-zone earthquakes with moment magnitude. Seismol Res Lett 81(6):941–950

Wells DL, Coppersmith KJ (1994) New empirical relationships among magnitude, rupture length, rupture width, rupture area, and surface displacement. Bull of Seismol Soc Am 84(4):974–1002

Chapter 14
R5–Combination of Multiple Tsunami Rate Estimates Weighted by Uncertainty with Likelihood Functions: An Example Runup Probability Map for the Caribbean Region

Tom Parsons and Eric L. Geist

14.1 Introduction

The purpose of this exercise is to develop methods for combining multiple tsunami rate estimates in the same region. These could, for example, be generated by multiple tsunami propagation models, source magnitude/frequency models, and/or empirical observations at sites. In this example after Parsons and Geist (2008) we discuss a spatial tsunami forecast for the Caribbean region made from numerical models of runup (Geist 2002; Geist and Parsons 2006), and a ~ 500-yr-long empirical catalogue (O'Loughlin and Lander 2003). Typical ingredients for calculating tsunami probability at a site are assessments of local and distant tsunamigenic sources, models of propagation, wave height and/or runup calculations, and aggregation of expected rates/probabilities. It can be difficult to account for every source, especially for the case of submarine landslides. Thus, independent empirical observations, such as from tide gauges, eyewitness accounts, or paleotsunami evidence are valuable. We present statistical methods for combining multiple results from models and empirical observation, assessing their uncertainties, and weighting those results accordingly. In this recipe we describe specific example steps towards calculating tsunami runup

T. Parsons (✉) · E. L. Geist
US Geological Survey, Moffett Field, USA
e-mail: tparsons@usgs.gov

E. L. Geist
e-mail: egeist@usgs.gov

M. B. Sørensen et al. (eds.), *Probabilistic Tsunami Hazard and Risk Analysis*,
Mathematics of Planet Earth 14, https://doi.org/10.1007/978-3-031-98115-9_14

($\geq$0.5 m) probability for all Caribbean region coastlines. However, these steps can readily vary depending on the application and development of new tools. The primary purpose here is to show how results from different models and data sources can be combined in a Bayesian framework that gives weight to different ingredients using likelihood functions.

14.2 Framework Used in Recipe

This recipe applies both a modular approach to catastrophe models (FW1B) and data-driven hazard assessment (see Sect. 2.3). Modules include source rate/location determination from 3D tectonic deformation models, numerical modelling of tsunami propagation, statistical modelling of calculated and observed runup rates, likelihood-function weighted combination of runup rates, and spatial time-independent probability calculations. We apply this recipe to the Caribbean region where, in addition to numerical modelling, there is a ~ 500-yr catalogue of tsunami events. Thus, there is an independent data-driven tsunami rate model that we combine with numerical rate estimates. This method can be applied to any set of rate or other parameters that we might want to combine into a single tsunami probability calculation.

14.3 Source Identification and Characterisation Used in Recipe

The example Caribbean region has not produced many large (M > 8) earthquakes in the modern catalogue era, which means we have little knowledge of the causative earthquake magnitude frequency distribution at these magnitudes and little information about fault slip distributions. For this reason, we model a wide array of possible earthquake scenarios constrained by the fault geometry and moment rate from plate motions. The empirical catalogue, while ~ 500 years long, is spatially incomplete because not all coastlines were populated over its duration. In any given 20 km $\times$ 20 km spatial cell there are less than five observations. We therefore apply Monte Carlo methods (e.g., Parsons 2008) to extrapolate recurrence parameters.

We apply/develop a finite element model of Caribbean and North American plate interaction (Parsons and Geist 2008) to calculate fault slip rates on the oblique-slip northern edge of the Caribbean plate and on the subduction zone interface. From that we calculate the expected rate of earthquakes greater than a given magnitude using a Gutenberg-Richter (G-R) magnitude-frequency distribution. Combinatorial optimization models can be used that fit the most efficient earthquake magnitude distribution to input data such as slip rates or stress constraints (e.g. Geist and ten Brink

2021; Geist and Parsons 2023). Some percentage of these events are expected to cause observable tsunamis. Source uncertainty was addressed with a Monte Carlo-type procedure in which synthetic earthquake catalogues of fixed duration were prepared from random samples of the earthquake rate model. Two different Monte Carlo analyses are conducted: (1) tsunami results are aggregated from 310 M $\geq$ 7 earthquakes representing a single catalogue of duration 4,442 years and (2) tsunami results are aggregated from 50 separate earthquake catalogues of duration 500 years each (approximately the duration of the observed earthquake and tsunami catalogues). Multiple catalogues are generated to account for variations in the distribution of near surface tsunamigenic earthquakes. The number of earthquakes sampled from the tapered G-R distribution varied for each source zone, depending on moment rate ($\dot{m}$). For a given period τ, the expected number of earthquakes of moment m_τ was derived from (cf., McCaffrey 1994): The location of each earthquake was randomly varied within each source zone according to a uniform distribution.

$$N(m > m_t) = \frac{(1-\beta)\, M_t\, \tau}{m_t^{\beta}\, m_c^{1-\beta}\, \Gamma(2-\beta) e^{m_t/m_c}} \tag{14.1}$$

We apply a third Monte Carlo analysis for empirical data. For a broad region with multiple sources, a time-independent process built around an exponential distribution (e.g., Geist and Parsons 2006) is most appropriate. In this example, observed tsunami runups are used to calculate rate parameters (λ) in the exponential distribution as

$$f(t) = \lambda e^{-\lambda t}, t > 0 \tag{14.2}$$

A series of distributions that covers all reasonable rates (1/10 to 1/3000 yrs) is considered, and rates are randomly drawn 100,000 times from each series and assembled into tsunami sequences. Monte Carlo sequences begin with events given freedom to happen any time before the start of the observation window (AD 1498 for the Caribbean region). The extra events yield starting points for possible observation periods. This is needed to account for the open interval prior to 1498, meaning that we do not know the interevent time before the first observations at each location. Each attempt that matches observed frequencies is tallied and a distribution of matches to the observed record is produced. The mode, median, or mean of that distribution can be taken to represent the rate parameter. This approach simultaneously estimates uncertainty in the rate parameter resulting from small sampling, and the influence of open intervals. This is achieved by building probability density functions that account for different possible parameter values.

14.4 Source Modelling

Heterogeneous, stochastic slip distributions are used for each earthquake, modelled using the *k-squared* model described in Chapter 4 of the ingredients. The plate boundary faults are discretized and vertical and horizontal coseismic seafloor displacements for the tsunami initial conditions (Tanioka and Satake 1996) are calculated using Okada's (1985) analytic functions. A kinematic component of tsunami generation is introduced by using a 20 s rise time applied uniformly along the rupture zone with no preferred rupture propagation direction.

14.5 Tsunami Modelling

We model the propagation of the tsunami wavefield using a finite-difference method to the linear long-wave equations (Satake 2002) as described in Sect. 5.4 of the ingredients. Runup (R_0) is approximated from a coarse 2 arc-minute grid model by finding the nearest model grid point to the coastline at the 250 m isobath and then multiplying the peak offshore tsunami amplitude by a factor of 3 that roughly accounts for shoaling amplification and the runup process itself—termed the "Bathtub" approach in Sect. 5.5. Recent methods described by Glimsdal et al (2019) can be used to define and reduce uncertainty associated with use of the amplification factor. Runup varies considerably throughout the model domain as a result of finite sampling of the G-R. Of the two Monte Carlo analyses described previously, using the fifty 500-year catalogues captured more variability in spatial runup distribution at the 0.5 m exceedance level than the single 4442-year catalogue. Multiple catalogues have more variety of earthquake locations since a few very large events can dominate the distribution of moment, and consequently regional tsunami runup distribution, due to the Gutenberg-Richter constraint.

14.6 Exposure Modelling

In this example we produce a map of time-independent tsunami probability of runup > 0.5 m for Caribbean coastlines within 20 × 20 km cells. The primary sources of epistemic uncertainty include: (1) expected tsunami runup rates, (2) tsunami sources not explicitly known or included in the model, (3) seismic coupling coefficient of the Caribbean plate boundary zones, and (4) degree of completeness in the empirical tsunami catalogue. To encompass these uncertainties into probability estimates, we create a Bayesian framework to build tsunami runup rate estimates within coastline cells. The cell scaling is set large enough to encompass uncertainty in historical locations and uncertainty in the nearshore bathymetry. The key advantage of this approach is that the empirical and numerical model results end up being weighted by their attendant uncertainties.

We combine rates that address the above-listed uncertainties as follows. (1) Monte Carlo rate-model matched to the empirical rates combined with results from 50 numerical model runs provide an ensemble of possible runup rate values at each cell. (2) Inclusion of the empirical rates add sources not accounted for in the numerical model; our forecast may suffer from incomplete knowledge if events not covered by our numerical models have also not occurred in the empirical catalogue over the past 500 years. (3) The historic earthquake catalogue implies a low seismic coupling of 0.32 (found by comparing seismic moment release to expected slip on Caribbean faults; Parsons and Geist 2008), whereas the numerical models have coupling coefficients of 1.0 and represent a maximum rate end member. (4) We account for low-rate plate-boundary events potentially not seen in the empirical catalogue with 50 numerical model runs.

We combine model-derived runup rates with empirical rates according to the following set of logic-tree branches: in cells where there are no empirical values, the numerical-model-derived rates are given full weight. Conversely, empirical rates are given full weight where numerical model rates are zero. Lastly, where there are empirical and model rate estimates within the same cells, we use Bayesian likelihood functions to weigh the two models. Distributions give the relative probability of different rates for a Poisson model that could have caused the empirical observations. Similarly, results from the 50 numerical model runs produce the relative probability of different rates in each model cell.

To rank different rate models for each cell where more than one estimate exists, we make a likelihood calculation to weigh the models. In the simplest, binomial case, likelihood is defined as proportional to the probability of obtaining results A given a fixed hypothesis H resulting from a set of fixed data. If $A_1, A_2, \ldots A_n$ are possible, mutually exclusive results, then

$$\mathrm{P(A_1 or\ A_2|H)} = \mathrm{P(A_1|H)} + \mathrm{P(A_2|H)} + \mathrm{P(A_n|H)} \tag{14.3}$$

and likelihood is thus

$$\mathrm{L(H|A)} = \mathrm{kP(A|H)}, \tag{14.4}$$

where k is an arbitrary constant.

The results from likelihood functions are used to obtain the final weights using Bayes' rule (Eq. 14.5), where the posterior distribution is proportional to the likelihood function multiplied by the prior. For this example, we use a uniform (noninformative) prior such that there is equal probability of all rates in each coastal cell. Further, since we update the prior twice, once for the empirical result, and again for the numerical result, the same result is achieved by simply multiplying the two likelihood functions. Thus, the likelihood of a given rate λ where there are empirical estimates ($e1$) and numerical-modeled estimates ($e2$) is

$$L(\lambda|e1, e2) = k[p_1(e1|\lambda)][p_2(e2|\lambda)] \tag{14.5}$$

where $p(e1|\lambda)$ is the probability of rate λ based on the Monte Carlo fits shown in Fig. 1a, b, and $p(e2|\lambda)$ is the probability of rate λ from the 50 numerical model runs. The constant k is used for normalizing the weights so that they add to 1. This can be expanded indefinitely if there are more information sources.

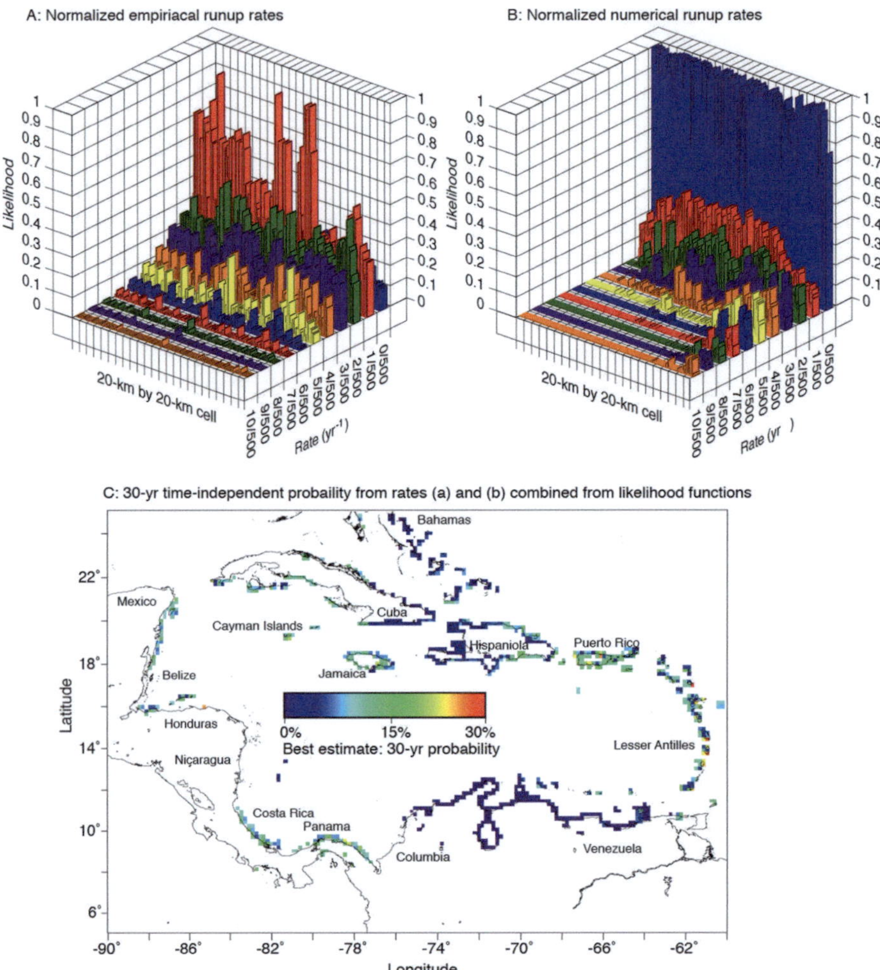

Fig. 14.1 (**a**) Normalized histogram (likelihood) of tsunami runup (R_0 0.5 m) rates in 214 20-km by 20-km cells defined using likelihood functions from empirical rates. (**b**) Normalized histogram of runup rates from numerical modelling in 685 cells where there are no empirical observations. Mean values from distributions of empirical and modelled tsunami runup rates are used in the best-estimate probability calculations mapped in (**c**), showing 30-year tsunami runup probability at coastal sites in the Caribbean region made from combined rate estimates from the two

We use likelihood functions to weight rate models over a range from 0 to 10 events in the 500-yr observation period assuming no further prior information. Final rates are found by weighted means of the posterior rates. To summarize the process, where model and empirical values are both absent for a given rate, the posterior distribution was zeroed. When one model provides rates, its likelihood function was used to update the priors, and when both empirical and numerical rate estimates are available, likelihood is developed through combination and renormalization using Eq. (14.5), which is then used to update the priors. Combining empirical and modeled rates makes up for some of the deficiencies in each approach; the empirical catalogue is likely not a complete record of all possible inter-plate tsunami sources, whereas the numerical model did not account for accommodating intra-plate faults and/or landslide sources that appear likely causes of tsunamis in the empirical record.

This recipe can be applied to any circumstance where there are multiple distributions of possible runup or wave height rates developed from any source where one might want to combine and weight different estimates based on their uncertainties.

References

Geist EL (2002) Complex earthquake rupture and local tsunamis. J Geophys Res 107:ESE2–1–ESE2–16

Geist EL, Parsons T (2006) Probabilistic analysis of tsunami hazards. Nat Hazards 37:277–314. https://doi.org/10.1007/s11069-005-4646-z

Geist EL, Parsons T (2023) Combinatorial optimization of earthquake spatial distributions under minimum cumulative stress constraints. Bull Seismol Soc Am 113. https://doi.org/10.1785/0120220175

Geist EL ten Brink US (2021) Earthquake magnitude distributions on northern Caribbean faults from combinatorial optimization models. J Geo Res: Solid Earth 126:e2021JB022050. https://doi.org/10.1029/2021JB022050

Glimsdal S, Løvholt F, Harbitz CB, Romano F, Lorito S, Orefice S et al (2019) A New approximate method for quantifying tsunami maximum inundation height probability. Pure Appl Geophys 176(7):3227–3246. https://doi.org/10.1007/s00024-019-02091-w

Okada Y (1985) Surface deformation due to shear and tensile faults in a half-space. Bull Seismol Soc Am 75:1135–1154

O'Loughlin KF, Lander JF (2003) Caribbean Tsunamis: A 500-Year History from 1498–1998. Kluwer Academic Publishers, Dordrecht, The Netherlands

McCaffrey R (1994) Dependence of earthquake size distributions on convergence rates at subduction zones. Geophys Res Lett 21:2327–2330

Parsons T (2008) Monte Carlo method for determining earthquake recurrence parameters from short paleoseismic catalogues: Example calculations for California. J Geophys Res 113. https://doi.org/10.1029/2007JB004998

Parsons T, Geist EL (2008) Tsunami probability in the caribbean region. Pure Appl Geophys 165:2089–2116. https://doi.org/10.1007/s00024-008-0416-7

Satake K (2002) Tsunamis. In: Lee WHK, Kanamori H, Jennings PC, Kisslinger C (eds) International Handbook of Earthquake and Engineering Seismology. Part A, Academic Press, San Diego, pp 437–451

Tanioka Y, Satake K (1996) Tsunami generation by horizontal displacement of ocean bottom. Geophys Res Lett 23:861–865

Chapter 15
R6–Worst-Case Tsunami Approach Applied to Catania (Eastern Sicily)

Filippo Zaniboni and Alberto Armigliato

15.1 Introduction

The Worst-case Credible Tsunami Scenario Analysis (WCTSA) is a methodology aiming at assessing the tsunami hazard for a target site based on a limited number of scenarios. These are defined starting from the events hitting the selected area in the past, and from potential events elaborated from the knowledge provided by geology, geomorphology, geodynamics and seismology of the studied zone. In this paper, the recipe related to the WCTSA is described through its application to the town of Catania (Sicily, south Italy), based on the paper by Tonini et al. (2011). The purpose here is not to discuss the details of the application to Catania itself and the results obtained for that specific site, but rather to outline the strategy in its key aspects, highlighting advantages and drawbacks.

In general, the steps constituting a WCTSA are the sources identification, the modelling of the generation, propagation and inundation phases of the tsunami for each source, the vulnerability and risk assessment. Coherently with the results illustrated in Tonini et al. (2011), here only the first three phases related to the hazard assessment will be discussed, while some results of the whole workflow including also vulnerability and risk assessment can be found in Tinti et al. (2010). The WCTSA approach should be regarded as an alternative to PTHA but not in opposition to it. First, it is particularly applicable to regions where seismic sources are not the predominant contributors to tsunami generation, or where they are negligible

F. Zaniboni (✉) · A. Armigliato
Dipartimento Di Fisica e Astronomia "A. Righi", Alma Mater Studiorum—Università Di Bologna, Bologna, Italy
e-mail: filippo.zaniboni@unibo.it

A. Armigliato
e-mail: alberto.armigliato@unibo.it

M. B. Sørensen et al. (eds.), *Probabilistic Tsunami Hazard and Risk Analysis*, Mathematics of Planet Earth 14, https://doi.org/10.1007/978-3-031-98115-9_15

or absent: it is well known that landslides, volcanoes and meteorological forcing can produce local relevant waves and that well-established PTHA approaches are presently lacking for these types of sources (for instance, because of the difficulties or even the impossibility to establish a magnitude frequency distribution). Secondly, especially local sources can produce large effects in limited coastal portions, often in conjunction with peculiar coastline morphologies, that can only be captured through very high-resolution grids, which perhaps can make the computational cost for a PTHA prohibitive.

15.2 Framework Used in Recipe

This recipe follows the framework FW1D (Working on a limited number of scenarios; Sect. 2.2). WCTSA can be considered a site-driven approach, with the concept of "worst-case" being declined in two respects, strictly connected to each other. The first regards the "size" of the tsunamigenic sources, whatever their nature; the second the effects in terms of hazard metrics. The specific target area determines the choice of the credible tsunami sources, both in the near and in the far field, that can involve any type of triggering phenomenon (earthquakes, landslides, volcanoes) relevant for the area. Following Tonini et al. (2011), "credible sources" are those that are compatible with the most up-to-date and reliable information on the seismotectonics and the geomorphology of the region under study, and they can coincide with known, well characterized tsunamigenic structures and/or with potential sources for which weaker constraints are available. In this second case, "representative sources" are selected. "Worst-case" means that the largest expected size (magnitude for earthquakes, volume and possibly rheology for landslides) is chosen with an uncertainty interval, which involves scientific knowledge and reasoning and an unavoidable portion of subjectiveness (which is anyway present at some stage in any type of approach). The idea of "worst-case" should not be regarded as "fixed for eternity": it is obviously affected by epistemic uncertainty and can hence evolve and be subject to modifications and updates as our knowledge progresses.

The chosen set of scenarios constitutes the input for the simulation of the ensuing tsunamis. The tsunami effects on the coast are measured through the maxima attained by different metrics, the most common of which are the wave amplitude at the coast, the run-up, the inundation line, the particle velocity (especially in the flooded areas), the Froude number, the momentum flux. The "aggregated scenario" is also part of the workflow, and it represents the "worst-case" as regards the tsunami effects in the target site: it reports the maximum of the maxima attained by each of the metrics for all the considered scenarios on each point of the computational grid. In principle, the final aggregated scenario can exhibit patterns varying from single-scenario dominated maps to very composite ones. In the first case, a posteriori further analyses should be performed to clarify whether this is the result of an actual characteristic of the site, or of an incomplete source selection process: if so, all missing sources must be included, and the overall workflow updated.

15.3 Source Identification and Characterisation Used in Recipe

The source selection processes are reviewed in Chapter 3, for both earthquake and landslide sources (Sects. 3.2 and 3.3, respectively). This phase is probably the most important and delicate for the WCTSA, since it can deeply affect the hazard assessment outcomes. It requires a detailed analysis of the past tsunami events in the target area, that must be complemented with the search of potential additional sources based on geophysical and geological studies. As to the former, one can rely on tsunami and paleo-tsunami catalogues, reporting the historical events and the observed effects; for the latter, the investigation ranges from the consultation of seismicity catalogues and of landslide databases, to the review of the available literature and to the knowledge of possible open debates.

An example of the analysis is summarised herein for the specific case of Catania (eastern Sicily), starting from the paper by Tonini et al. (2011), where a more extensive description can be found. Based on the past tsunami history of the Catania site, two source areas in the near field and one area in the far field are selected, corresponding to just as many past events:

- *the 1693 Eastern Sicily tsunami*, following a strong Mw = 7.4 earthquake that caused around 12,000 casualties in Catania. Tsunami catalogues (e.g., Tinti and Maramai 1996; Tinti et al. 2004) refer that a first negative polarity hit the whole eastern coast of Sicily, and the waves affected in particular the town of Catania and Augusta. The definition of the tsunami source is still an open issue (for instance Tinti et al. 1999; Gutscher et al. 2006; Argnani et al. 2012; Scicchitano et al. 2022) and the debate involves not only the specific tsunamigenic fault but also the possibility that an offshore landslide contributed to the generation. Two sources were introduced in the WCTSA, the first related to an offshore fault and the second to a huge offshore slump, both identified by marine geophysical surveys along the Hyblean-Malta escarpment (Argnani and Bonazzi 2005).
- the *1908 Messina Straits event*, originating from a Mw = 7.2 earthquake that devastated the towns of Messina and Reggio Calabria and several other villages on both sides of the Straits. A tsunami was observed soon after, and many run-up and inundation data collected and reported (e.g., Tinti et al. 2004). In Catania, waves with height up to 3 m were observed, with inundation reaching 700 m in the wide flat beach south of the town. Many studies have focused in the years on the definition of the tsunami seismic source, that is still an open issue, as testified for example

by the recent work of Barreca et al. (2021a) and the following debate (Pino et al. 2021; Barreca et al. 2021b; Argnani 2022; Barreca et al. 2022). Additionally, a source of different nature has been hypothesized, in order to fit the observations, i.e., an underwater landslide triggered by the seismic shaking whose evidence is still object of discussion too (Billi et al. 2008; Argnani et al. 2009; Schambach et al. 2020; Argnani 2021). However, the hypothesis of a combined source seems the most reasonable, even though there are not clear pieces of evidence yet (Tinti et al. 1999; Favalli et al. 2009). In the WCTSA for Catania, a seismic source scenario is defined basing on the fault proposed by Valensise et al. (2008), while a second composite-source scenario results from the addition to the previous fault of a static depression, resembling the initial condition related to a landslide-like source placed in correspondence with the landslide source area proposed by Billi et al. (2008).
- the *365 AD Crete tsunami*, originating from a Mw = 8.3–8.5 earthquake west of Crete (e.g., Papazachos 1996) and affecting the whole eastern Mediterranean basin, with coeval and later chronicles describing its effects in Greece, Egypt and south Italy (e.g., Shaw et al. 2008). Numerical simulations taking the western Hellenic trench as the most credible source area for the 365 event (e.g., Tinti et al. 2005; Lorito et al. 2008; Shaw et al. 2008; all sharing a thrust mechanism but varying in terms of geometry and position along the trench) demonstrated the effective possibility of such a source to generate tsunamis able to reach the coastal stretch of Catania. In order to account for the uncertainty in the source definition, that is still object of debate, three scenarios were adopted: one referring to the hypothesis by Shaw et al. (2008); the second indicating a possible new rupture located north of the first and representing a source with potential higher effects in Catania due to directivity; the third as a combination of the previous two.

The three described reference events originated in total 7 worst-case credible scenarios, representing the input for the numerical tsunami simulations. It is worth mentioning that in unpublished studies performed later (see Tinti and the ASTARTE Consortium 2015) many more sources were taken into account, there including geological structures not necessarily associated to historical events and other potential sources, such as for example small collapses along the Hyblaean-Malta escarpment (the steep slope running N-S offshore the Ionian coast of Sicily). These represent key additions to the recipe.

15.4 Source Mechanics Modelling Used in Recipe

The source modelling ingredients are reviewed in Chapter 4 (Sects. 4.2 and 4. 3 for earthquakes and landslides, respectively). The tsunami initial condition for earthquake-generated tsunamis was simulated by means of the "classical" Okada (1985) model, which allows to compute the displacement field of the Earth's free surface induced by a rectangular fault embedded in a perfectly elastic, homogeneous, and isotropic half-space. The vertical component of the coseismic displacement is copied identically and instantaneously to the initial perturbation of the water column.

For landslide-induced tsunamis the tsunamigenic impulse cannot be considered instantaneous and hence the sliding motion must be computed. Moreover, since the mass shape affects the wave generation, the related time history should be provided as well. The code UBO-BLOCK, developed by the Tsunami Research Team of the University of Bologna, meets all such requirements, solving the sliding motion equations by finite difference methods (further details on the code are found in Tinti et al.1997) and providing as output the complete time history of the sliding dynamics and geometry, necessary to compute the tsunamigenic dynamic impulse. This is obtained through another ancillary code, UBO-TSUIMP, mapping the seabed change in time due to the slide motion over the tsunami computational grid. Some more details about this code are reported in Tinti et al. (2006).

Landslide modelling is a particularly challenging task since it involves highly non-linear (when not chaotic) processes during the motion. It is then necessary to adopt strong simplifications to model and quantify the effects of a landslide, introducing various factors of uncertainty. These can be grouped into two main categories, the first concerning the modelling approach and the second the parameters adopted. Regarding the former, a full discussion of the adopted methodologies is out of the scopes of this paper: a comprehensive review can be found in Yavari-Ramshe and Ataie-Ashtiani (2016). As to the latter, a PTHA approach would require exploring all the possible values of the coefficient used for the parametrization of the motion, associating an arbitrary probability to each of them. The WCTSA allows to set a combination of such parameters representing the maximum credible scenario in terms of landslide motion, that limits the number of simulations to be run. Conversely, this approach adopts arbitrary assumptions, that can be reduced to some degree by running sensitivity analysis on the most important coefficients governing the motion.

15.5 Tsunami Modelling

The tsunami modelling "ingredient" is discussed in Chapter 5. Tsunamis are generally considered as "long-waves", meaning that their typical wavelength is much higher than the water depth where they are propagating. This means that the complex hydrodynamics equations can be simplified adopting the shallow water (SW) approximation. The UBO-TSUFD code (Tinti and Tonini 2013) implements a SW numerical solver, based on a Leap-Frog numerical scheme and on the staggered grids technique. The required input is a regular grid covering the computational domain. The boundaries can be treated as transparent for open sea border cells, while for water-land interface two possibilities can be set: in the linear version of the code, the higher-order terms of the SW equations are neglected, a uniform depth in shallow water is set and the coast can be seen as a reflecting wall, with no inland inundation; when also the non-linear terms of the SW equations are considered, the non-linear effects are considered together with flooding, that is accounted for through the moving boundary technique. Finally, the nested grids technique allows to focus on areas of particular interest with higher resolution computational grids. Recent applications of the whole numerical routine (from source to tsunami simulation) can be found in Zaniboni et al. (2021), Gasperini et al. (2022), Gallotti et al. (2023), and references therein.

Typically, the linear version of the code is adopted for simulations on a regional scale, providing hints for the coastal stretches affected by the most relevant effects that can be investigated, in the following phase, by finer grids and non-linear SW equations, accounting for coastal flooding and local hydrodynamic features (such as harbour basin resonance).

The SW are known to provide satisfactory results when dealing with seismic sources, while for landslide-generated tsunamis the shorter waves content can be relevant, making the SW approximation no longer valid, prompting for the need to account for hydrodynamic effects such as dispersion which in turn requires the inclusion of higher-order terms in the equations. The dispersive effects are known to produce a decrease of the wave amplitude at the coast (see for example Glimsdal et al. 2013). The adoption of SW approach produces then fair estimates of the tsunami impact for seismic sources and a slight overestimation for landslide-sources, so it can be considered adequate for the WCTSA approach, that considers the worst-case effects.

The application of the source modelling (Okada and UBO-BLOCK) and of the tsunami propagation (UBO-TSUFD) for the specific case of the target area of Catania produces 7 simulations, each providing a quantification of the respective tsunami effects on the coast. The results are reported and extensively commented in Tonini et al. (2011): here it is just worth underlining that the maximum tsunami effects for the area of Catania turn out to be associated to the more distant source, i.e., the 365 Crete scenario, and to the composite 1908 case (earthquake and landslide-like source).

15.6 Discussion

The described WCTSA approach has advantages and drawbacks. It allows to obtain a hazard assessment involving only a limited number of scenarios, hence requiring a lower effort in terms of data elaboration and computational power with respect to PTHA. Moreover, also non-seismic sources can be included in the analysis, especially landslide-generated events that are known to produce, at a local scale, non-negligible to relevant effects that presently PTHA cannot account for. On the other side, a WCTSA is deeply dependent on subjective choices, especially in the scenario selection, and requires experience and attention to avoid neglecting relevant sources of hazard for the target area. It is worth noting that PTHA is not free from subjective choices: while the process of source selection (currently limited to earthquakes) is by far more complete than in the WCTSA, the process to decide the weights to be assigned to each source (for example in a logic tree) has a non-negligible component of subjectiveness.

The WCTSA is a site-driven methodology, meaning that all the studies concentrate on the selected target area. This can be an advantage, since it limits the number of possible sources to explore and the simulations to run, but it represents a drawback too since the whole workflow must be run from scratch when the target area is changed.

The source selection phase is the most critical in the WCTSA: neglecting potential source areas or underestimating the tsunamigenic potential of some category of sources, are examples of typical mistakes that can be made in the scenario selection and that could deeply affect the final assessment. A significant degree of uncertainty characterizes local scenarios, especially with regards to small landslides close to the target area, that can produce relevant effects over limited coastal stretches and that are not usually mentioned nor observed in historical reports. The inclusion (or exclusion) of such sources among the investigated scenarios illustrates well the degree of subjectiveness of the WCTSA in this phase. As an example of application to the Catania target area, small underwater landslides along the steep Hyblaean-Malta escarpment, possibly triggered by seismic shaking or as a side effect of climate change, might produce some effects. On the other hand, no relevant piece of evidence of past or potentially unstable mass in shallow water in the Catania offshore is available: the respective hazard is considered then negligible, and for such reasons such type of event is not included among the investigated scenarios.

Another important source of uncertainty is connected to the source size. For earthquake sources, this is typically translated into the estimation of the largest expected magnitude, although this concept might be rather elusive and the way to estimate it is not univocal (see for instance the discussion in Shcherbakov et al. 2019). For the landslides, as a first guess it could be safely related to the volume of the sliding mass or to its thickness. The strategy introduced in the WCTSA approach presented in Tonini et al. (2011) foresees the adoption of the same "standard" scenarios illustrated before but with the source size incremented by a given percentage (20% in that study). More specifically, this implies increasing by the selected percentage the coseismic slip on the fault plane for earthquake sources and the tsunamigenic impulse in the case of landslides.

The modelling phases present less unknowns and indetermination: sensitivity analysis, comparison of different numerical methods, validation through benchmarks, allow to quantify the degree of acceptability of the results provided.

As a final remark, the WCTSA approach should be seen as complementary to and not as opposed to PTHA methodologies. Although it is intrinsically "less exhaustive" compared to PTHA, since it does not account for all possible sources and their recurrence time, still it can provide detailed scenarios with a resolution not at reach in some cases for probabilistic approaches. These scenarios, especially if characterized, at least approximately, by a rate of occurrence, can be a very valid tool for detailed risk assessments. Moreover, they can be used as benchmarks for PTHA at local scale.

References

Argnani A. (2021) Comment on "New simulations and understanding of the 1908 Messina tsunami for a dual seismic and deep submarine mass failure source" by L. Schambach, ST Grilli, DR Tappin, MD Gangemi, G. Barbaro [Marine Geology 421 (2020) 106093]. Mar Geol 442:106634

Argnani A. (2022) Comment on the paper by Barreca et al.:"The Strait of Messina: Seismotectonics and the source of the 1908 earthquake"[Earth-Science Reviews 218, 2021, 103685]. Earth-Science Reviews, 226:103961.

Argnani A. and Bonazzi C. (2005) Malta Escarpment fault zone offshore eastern Sicily: Pliocene-Quaternary tectonic evolution based on new multichannel seismic data, Tectonics, 24:TC4009. https://doi.org/10.1029/2004TC001656

Argnani A, Armigliato A, Pagnoni G, Zaniboni F, Tinti S, Bonazzi C (2012) Active tectonics along the submarine slope of south-eastern Sicily and the source of the 11 January 1693 earthquake and tsunami. Nat Hazard 12(5):1311–1319

Argnani A, Chiocci FL, Tinti S, Bosman A, Lodi MV, Pagnoni G, Zaniboni F.,Comment on "On the cause of the 1908 Messina tsunami, southern Italy" by Andrea Billi, et al (2009) Geophys Res Lett 36:L13307

Barreca G., Gross F., Scarfì L., Aloisi M., Monaco C., Krastel S. (2021a) The Strait of Messina: seismotectonics and the source of the 1908 earthquake. Earth-Sci. Rev.:103685, https://doi.org/10.1016/j.earscirev.2021.103685

Barreca G., Gross F., Scarfì L., Aloisi M., Monaco C., Krastel S. (2021b) Reply to: Comment on the paper by Barreca et al.:"The Strait of Messina: Seismotectonics and the source of the 1908 earthquake" by G. Barreca, F. Gross, L. Scarfì, M. Aloisi, C. Monaco, S. Krastel (Earth-Science Reviews 218, 2021, 103685). Earth-Science Reviews 223:103866.

Barreca G., Gross F., Scarfì L., Aloisi M., Monaco C., Krastel S., (2022) Reply to: Comment by A. Argnani on the paper:"The Strait of Messina: Seismotectonics and the source of the (1908) Earthquake"(Earth-Science Reviews 218, 2021, 103,685). Earth Sci Rev 226:103962

Billi A., Funiciello R., Minelli L., Faccenna C., Neri G., Orecchio B., Presti D. (2008) On the cause of the 1908 Messina tsunami, southern Italy. Geophysical Research Letters, 35(6).

Favalli M, Boschi E, Mazzarini F, Pareschi MT (2009) Seismic and landslide source of the 1908 Straits of Messina tsunami (Sicily, Italy). Geophys Res Lett 36:L16304. https://doi.org/10.1029/2009GL039135

Gallotti G., Zaniboni F., Arcangeli D., Angeli C., Armigliato A., Cocchi L., Muccini F., Zanetti M., Tinti S., Ventura, G. (2023) The tsunamigenic potential of landslide-generated tsunamis on the Vavilov seamount. Journal of Volcanology and Geothermal Research, 107745. https://doi.org/10.1016/j.jvolgeores.2023.107745

Gasperini L, Zaniboni F, Armigliato A, Tinti S, Pagnoni G, Özeren MS, Ligi M, Natali F, Polonia A (2022) Tsunami potential source in the eastern Sea of Marmara (NW Turkey), along the North Anatolian Fault system. Landslides 19(10):2295–2310. https://doi.org/10.1007/s10346-022-01929-0

Glimsdal S, Pedersen G, Harbitz CB, Løvholt F (2013) Dispersion of tsunamis: does it really matter? Nat Hazards Earth Syst Sci 13:1507–1526. https://doi.org/10.5194/nhess-13-1507-2013

Gutscher M-A, Roger J, Baptista M-A, Miranda JM, Tinti S (2006) Source of the 1693 Catania earthquake and tsunami (southern Italy): New evidence from tsunami modeling of a locked subduction fault plane. Geophys Res Lett 33:L08309

Lorito S, Tiberti MM, Basili R, Piatanesi A, Valensise G (2008) Earthquake-generated tsunamis in the Mediterranean Sea: Scenarios of potential threats to southern Italy. J Geophys Res 113:B01301. https://doi.org/10.1029/2007JB004943

Okada Y (1985) Surface deformation to shear and tensile faults in a halfspace. Bull Seismol Soc Am 75:1135–1154

Papazachos BC (1996) Large Seismic Faults in the Hellenic Arc. Ann Geofis 39:891–903

Pino N. A., Palano M., Ventura G. (2021) Comment on the paper by Barreca et al.:"The Strait of Messina: Seismotectonics and the source of the 1908 earthquake"(Earth-Science Reviews 218, 2021, 103685). Earth-Science Reviews, 223:103865.

Schambach L, Grilli ST, Tappin D, Gangemi MD, Barbaro G (2020) New simulations and understanding of the 1908 Messina tsunami for a dual seismic and deep submarine mass failure source. Mar Geol 421:106093

Scicchitano G, Gambino S, Scardino G, Barreca G, Gross F, Mastronuzzi G, Monaco C (2022) The enigmatic 1693 AD tsunami in the eastern Mediterranean Sea: new insights on the triggering mechanisms and propagation dynamics. Sci Rep 12:9573. https://doi.org/10.1038/s41598-022-13538-x

Shaw B, Ambraseys NN, England PC, Floyd MA, Gorman GJ, Higham TFG, Jackson JA, Nocquet J-M, Pain CC, Piggott MD (2008) Eastern Mediterranean tectonics and tsunami hazard inferred from the AD 365 earthquake. Nat Geosci 1:268–276. https://doi.org/10.1038/ngeo151

Shcherbakov R, Zhuang J, Zöller G, Ogata Y (2019) Forecasting the magnitude of the largest expected earthquake. Nat Comm 10:4051. https://doi.org/10.1038/s41467-019-11958-4

Tinti S, Armigliato A (2003) The use of scenarios to evaluate the tsunami impact in southern Italy. Mar Geol 199:221–243

Tinti S, Armigliato A, Bortolucci E, Piatanesi A (1999) Identification of the source fault of the 1908 Messina earthquake through tsunami modelling. Is it a possible task?, Phys. Chem. Earth, Part b: Hydrology, Oceans and Atmosphere 24:417–421

Tinti S., Armigliato A., Pagnoni G., Tonini R., Zaniboni F. (2010) Atlante delle zone esposte al rischio di maremoto nell'area di Catania. © 2010 Editrice Compositori, Bologna. ISBN 978–88–7794–727–7 (in italian).

Tinti S., Armigliato A., Pagnoni G., Zaniboni F. (2005) Scenarios of giant tsunamis of tectonic origin in the Mediterranean, ISET Journal of Earthquake Technology, Paper No. 464, 42(4):171–188

Tinti S. and the ASTARTE Consortium (2015) ASTARTE DELIVERABLE D8.8 "Tsunami Hazard Assessment Methods: application in the NEAM region and in the ASTARTE test sites". Available online at http://194.117.20.221/files/astarte/documents/deliverables/d8-8/ASTARTE%20D8%208%20-%20v0.9.4.pdf (last accessed 22/02/2024).

Tinti S, Bortolucci E, Vannini C (1997) A block-based theoretical model suited to gravitational sliding. Nat Hazards 16:1–28

Tinti S, Maramai A (1996) Catalogue of tsunamis generated in Italy and in Côte d'Azur, France: a step towards a unified catalogue of tsunamis in Europe. Ann Geofis 39:1253–1299

Tinti S, Maramai A, Graziani L (2004) The new catalogue of Italian tsunamis. Nat Hazards 33:439–465

Tinti S, Pagnoni G, Zaniboni F (2006) The landslides and tsunamis of 30th December 2002 in Stromboli analysed through numerical simulations. Bull Volcanol 68:462–479. https://doi.org/10.1007/s00445-005-0022-9

Tinti S, Tonini R (2013) The UBO-TSUFD tsunami inundation model: validation and applicatio case study focused on the city of Catania, Italy. Nat Hazards Earth Syst Sci 13:1795–1816. https://doi.org/10.5194/nhess-13-1795-2013

Tonini R, Armigliato A, Pagnoni G, Zaniboni F, Tinti S (2011) Tsunami hazard for the city of Catania, eastern Sicily, Italy, assessed by means of Worst-case Credible Tsunami Scenario Analysis (WCTSA). Nat Hazards Earth Syst Sci 11:1217–1232

Valensise G., Basili R., Burrato P. (2008) La sorgente del terremoto del 1908 nel quadro sismotettonico dello Stretto di Messina, in "Il terremoto e il maremoto del 28 dicembre 1908: analisi sismologica, impatto, prospettive", edited by: Bertolaso G., Boschi E., Guidoboni E. and Valensise G., 161–182 (in italian).

Yavari-Ramshe S, Ataie-Ashtiani B (2016) Numerical modeling of subaerial and submarine landslide-generated tsunami waves—recent advances and future challenges. Landslides 13:1325–1368. https://doi.org/10.1007/s10346-016-0734-2

Zaniboni F, Pagnoni G, Paparo MA, Gauchery T, Rovere M, Argnani A, Armigliato A, Tinti S (2021) Tsunamis From Submarine Collapses Along the Eastern Slope of the Gela Basin Strait of Sicily). Front Earth Sci 8:602171. https://doi.org/10.3389/feart.2020.602171

Chapter 16
R7—ASCETE Coupled Earthquake-Tsunami Modelling

Jörn Behrens

16.1 Introduction

This recipe covers only a small part of the probabilistic tsunami hazard analysis workflow, namely the seismic induced earthquake-tsunami modelling. While most approaches in modelling earthquake-induced tsunami events use simplified source models, in this recipe, developed in a 6-year long project called Advanced Simulation of Coupled Earthquake Tsunami Events (ASCETE), a chain of complex earthquake cycle—earthquake rupture—tsunami models allow for realistic case studies of seismic tsunamis.

The coupled seismic-tsunami framework, presented in this recipe, is generally useful for detailed hazard assessment and event analysis (or hindcast). It is predominantly deterministic but can reveal the uncertainty in complex earthquake source mechanics. It may help to explain so far surprising tsunami events such as tsunami amplification through splay-faulting, strike-slip induced tsunami sources, outer rise sources, etc. (see e.g. Ulrich et al. 2019a, b).

16.2 Framework Used in Recipe

The ASCETE coupled earthquake-tsunami simulation workflow follows FW1D described in Chap. 2: Usually, only one or few scenarios, based on geologic or geodynamic evidence are simulated and a dynamic behaviour is considered.

J. Behrens (✉)
Department of Mathematics, Universität Hamburg, Hamburg, Germany
e-mail: joern.behrens@uni-hamburg.de

M. B. Sørensen et al. (eds.), *Probabilistic Tsunami Hazard and Risk Analysis*, Mathematics of Planet Earth 14, https://doi.org/10.1007/978-3-031-98115-9_16

This simulation chain strives to simulate the event based on best possible physics-based numerical algorithms and thus assumes that uncertainty lies mainly in the basic assumptions on the geological layout of the scenario. Once the very basic assumptions on geometry and material properties are stated, the cycle model creates a subduction area and seismicity cycles, which may serve as input to the rupture model (van Zelst et al. 2019). The rupture model then takes the output of the cycle model, such as stresses and nucleation location as input for a physics-based rupture simulation. The three-dimensional rupture model generates a realistic earthquake scenario with seismic waves up to frequencies of several Hz, horizontal and vertical time-dependent displacements, and complex slip behaviour including re-activation, supershear rupture velocity, fault jumps, etc. (Pelties et al. 2014, Ulrich et al. 2019b).

The displacement information is then modified (filtered) to be used in an adaptive tsunami model, capable of simulating propagation and inundation seamlessly, adhering to conservation principles (Vater et al. 2019). The output of the tsunami model can then be used for tsunami hazard assessment studies, sensitivity analysis or hindcasting efforts. The whole modelling chain is visualized in Fig. 16.1.

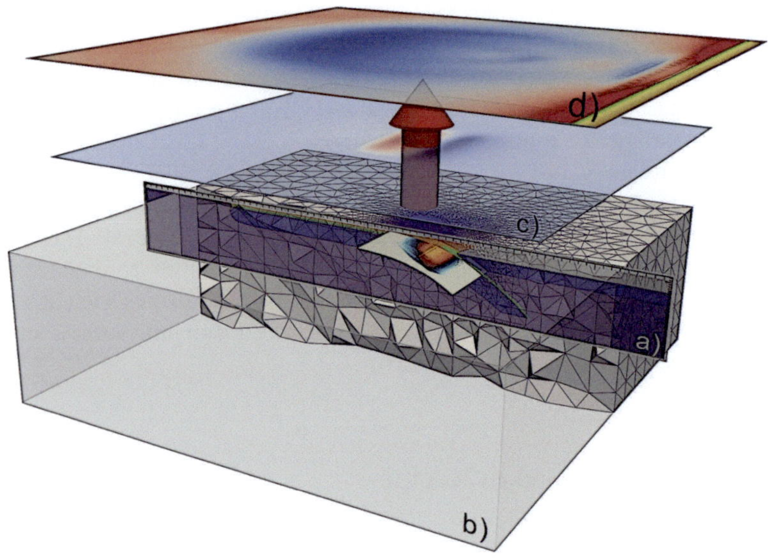

Fig. 16.1 ASCETE coupled seismic-tsunami framework: **a** a seismic cycle model generates realistic (but synthetic) subduction geometry and stress distribution. **b** Physics-based three-dimensional rupture model computes realistic earthquake scenario, yielding **c** sea floor displacement, which is filtered to be used in **d** the adaptive nonlinear tsunami wave model including propagation and inundation

16.3 Source Identification and Characterisation Used in Recipe

In this recipe only seismic sources are considered. The method can be seen as a complement to the source identification in Sect. 3.2, although in this case no specific source characterization or identification is performed. Rather, a modelling chain is used to determine source parameters. This recipe can either be used in a setting where the geologic situation is known from historic seismicity, stress analysis, and geological records regarding the material properties of the location in focus. Or it can be used in a synthetic situation, where the geologic setting is computed by a seismo-thermo-mechanical model simulating long term subduction dynamics and seismic cycles (van Dinther et al. 2013; van Zelst et al. 2019). Such models allow for studying the contributions of rheology, temperature, subduction dynamics and geometry, loading and evolution of faulting to the uncertainty in constraining inputs to the rupture model that can act as important preconditioning factors.

The dynamics, such as nucleation, propagation and arrest, of earthquake ruptures are controlled by material properties, fault stress, strength and geometry, which can be deducted from the seismic cycle model.

16.4 Source Mechanics Modelling Used in Recipe

The earthquake source is computed by a physics-based rupture model—SeisSol (Pelties et al. 2014). This recipe implements source modelling by the dynamic rupture modelling method as described in Sect. 4.2. Here, physics-based means that the elastodynamic equations for spontaneous dynamic rupture and seismic wave propagation are used together with advanced friction laws such that fault slips can evolve as frictional shear failure without pre-determination of the slip. SeisSol is based on the three-dimensional seismic wave equation in velocity-stress formulation. Numerically it implements an ADER-DG (Arbitrary high-order DERivative Discontinuous Galerkin) method with advanced subcycling time-stepping, highly parallelized and capable of running on large parallel computing architectures (Dumbser and Käser 2006; Breuer et al. 2016). Three-dimensional non-uniform tetrahedral meshes with millions of elements have been used in landmark applications.

SeisSol computes a three-dimensional displacement field, which is sampled with 1 Hz in a domain that is large enough to hide boundary effects of the earthquake wave field (here approximately one fault length in each direction of the fault). For the two-dimensional sea floor displacement, the three-dimensional displacement field is mapped by the formula of Tanioka and Satake (1996) taking into account both vertical and horizontal displacements. Additionally, a low-pass spectral filter is applied to the seismic (Rayleigh-) waves, which are incompatible with wave propagation speeds in the hydrostatic shallow water tsunami model (Madden et al. 2021). This time-dependent sea floor displacement is added to the bathymetry, which initialises and sources the tsunami model (see Sect. 5.3—Interface).

16.5 Tsunami Modelling

Tsunami modelling according to Sect. 5.4 (Propagation) and (Inundation) is performed with an adaptive nonlinear shallow water model, based on a nodal Runge–Kutta Discontinuous Galerkin (RKDG) numerical scheme as described in Vater et al. (2014, 2019). An alternative implementation exists for high-performance computing environments, using the same numerical approximation but different algorithmic implementation based on the sam(oa)^2 framework (Meister et al. 2016).

The tsunami model uses adaptive mesh refinement to allow high-resolution in areas of small-scale wave activity and gains efficiency by using coarse mesh resolution in areas of little interest. An inundation scheme, based on well-balanced flux computations from wet to dry cells is incorporated and seamlessly activated, where a wet-dry-interface is detected. Sensitivity studies have been performed to study the dependence of deep or shallow earthquake ruptures on the inundation in Madden et al. (2021).

A study confirming that the tsunami magnitude is explainable by purely seismic activity has been performed for the Palu 2018 tsunami event (Ulrich et al. 2019a, b). In this study the coupled earthquake-tsunami modelling framework shows that not only seismic parameters, but also seismic wave forms, propagation velocities, GNSS-based displacement measurements, tsunami wave arrival times, tide gauge measurements and inundation data can be incorporated into the analysis. These metrics are computable by the complex modelling approach and can therefore be used for constraining the model's parameters and by comparing with measurements for validating hypotheses on processes and uncertainties.

16.6 Discussion

The three-dimensional rupture model is extremely computationally intensive (Breuer et al. 2014). All components (except for the cycle model) are open source, many parts optimised for high performance architectures.

The coupling is one-way, that means the cycle model computes input for the rupture model, which in turn derives input for the tsunami model. Feedback has not been considered due to the largely differing time-scales, on which each modelling component operates—the cycle model on millions of years, the rupture model on seconds, the tsunami model on minutes to hours.

Due to these considerations the models employ different dimensionalities. The cycle model's runtime would not have been feasible in a three-dimensional version, so the assumption of symmetry in one spatial dimension was made for deriving initial conditions for the rupture model. The tsunami model also uses a two-dimensional approximation and integrates over depth due to geophysical approximations. Since the tsunami model employs a hydrostatic approximation (i.e. a uniform density in the whole water column), no realistic wave dispersion of seismic waves is possible, requiring the low-pass filtering. Uncertainties due to such approximations are not yet fully determined nor constrained.

While these insufficiencies call for further development, the existing tool set is already very useful and allows for unprecedented abilities to study complex earthquake-tsunami dynamics, such as the surprising magnitude of the Palu 2018 tsunami event (Ulrich et al. 2019a, b).

16.7 Links to Data, Models and Other Resources

- SeisSol Homepage: http://www.seissol.org
- SeisSol Github Page: https://github.com/SeisSol/SeisSol
- TsunaFlash Homepage (TsunaFlash is part of the amatos software suite): https://www.amatos.info
- sam(oa)^2 Gitlab Page: https://gitlab.lrz.de/samoa/samoa
- sam(oa)^2 Benchmarks: https://doi.org/10.5281/zenodo.3803664

References

Breuer A, Heinecke A, Rettenberger S, Bader M, Gabriel A-A, Pelties C (2014) Sustained petascale performance of seismic simulations with SeisSol on SuperMUC. In: PRACE ISC Award 2014, international supercomputing conference (ISC) proceedings 8488:1–18. https://doi.org/10.1007/978-3-319-07518-1_1

Breuer A, Heinecke A, Bader M (2016) Petascale local time stepping for the ADER-DG finite element method. In: 2016 IEEE international parallel and distributed processing symposium (IPDPS), Chicago, IL, USA, pp 854–863. https://doi.org/10.1109/IPDPS.2016.109

Dumbser M, Käser M (2006) An arbitrary high-order discontinuous Galerkin method for elastic waves on unstructured meshes—II. The three-dimensional isotropic case. J Geophys Int 167(1):319–336. https://doi.org/10.1111/j.1365-246X.2006.03120.x

Madden E, Bader M, Behrens J, Van Dinther Y, Gabriel A-A, Rannabauer L, Ulrich T, Uphoff C, Vater S, Wollherr S, Van Zelst I (2021) Linked 3-D modelling of megathrust earthquake-tsunami events: from subduction to tsunami run up. Geophys J Int 224(1):487–516. https://doi.org/10.1093/gji/ggaa484

Meister O, Rahnema K, Bader M (2016) Parallel memory-efficient adaptive mesh refinement on structured triangular meshes with billions of grid cells. ACM Trans Mat Softw 43(3):27 Article 19. https://doi.org/10.1145/2947668.

Pelties C, Gabriel A-A, Ampuero J-P (2014) Verification of an ADER-DG method for complex dynamic rupture problems. Geosci Model Dev 7:847–866. https://doi.org/10.5194/gmd-7-847-2014

Tanioka Y, Satake K (1996) Tsunami generation by horizontal displacement of ocean bottom. Geophys Res Lett 23(8):861–864. https://doi.org/10.1029/96GL00736

Ulrich T, Vater S, Madden EH, Behrens J, Van Dinther Y, Van Zelst I, Fielding E, Liang C, Gabriel A-A (2019a) Coupled, physics-based modeling reveals earthquake displacements are critical to the 2018 Palu, Sulawesi Tsunami. Pure Appl Geophys 176(10):4069–4109. https://doi.org/10.1007/s00024-019-02290-5

Ulrich T, Gabriel A-A, Ampuero JP et al. (2019b) Dynamic viability of the 2016 Mw 7.8 Kaikōura earthquake cascade on weak crustal faults. Nature Comm 10:1213. https://doi.org/10.1038/s41467-019-09125-w

van Dinther Y, Gery TV, Dalguer LA, Mai PM, Morra G, Giardini D (2013) The seismic cycle at subduction thrusts: insights from seismo-thermo-mechanical models. J Geophys Res Solid Earth 118(12). https://doi.org/10.1002/2013JB010380

Van Zelst I, Wollherr S, Gabriel A-A, Madden E, Van Dinther E (2019) Modelling megathrust earthquakes across scales: from geodynamics and seismic cycles to dynamic rupture. J Geophys Res Solid Earth 124(11):11414–11446. https://doi.org/10.1029/2019JB017539

Vater S, Beisiegel N, Behrens J (2019) A limiter-based well-balanced discontinuous Galerkin method for shallow-water flows with wetting and drying: triangular grids. Int J Num Meth Fluids 91(8):395–418. https://doi.org/10.1002/fld.4762

Vater S, Behrens J (2014) Well-balanced inundation modeling for shallow-water flows with discontinuous galerkin schemes. In: Fuhrmann et al (eds) Finite volumes for complex applications vii—elliptic, parabolic, and hyperbolic problems, Springer, Berlin, proceedings in mathematics and statistics vol 78. https://doi.org/10.1007/978-3-319-05591-6_98

Chapter 17
R8—Probabilistic Tsunami Hazard Analysis for Landslide Sources—The Lyngen Case

Finn Løvholt and Sylfest Glimsdal

17.1 Introduction

The purpose of this study is to carry out a landslide probabilistic tsunami hazard analysis (LPTHA) within a fjord system. The study is semi-local, as it aggregates many probabilistic inundation maps within a small region covering a fjord system in Norway. The LPTHA combines ensembles of landslide, tsunami propagation, and inundation models which are aggregated into hazard maps. The tsunami propagation is modelled across the fjord system, and individual local probabilistic hazard results using high resolution grids in 31 different coastal sites are compiled. This LPTHA includes the following ingredients:

- Event-based probabilistic hazard analysis using discrete scenario modelling and probabilities (FW1A)—Sect. 2.2
- Source selection for landslide sources—Sect. 3.3
- Source modelling for landslide sources—Sect. 4.3
- Tsunami hazard modelling including tsunami propagation and inundation modelling—Chap. 5.

This recipe is suitable for local or small-scale regional hazard analysis. Because it is reliant on local tsunami inundation simulations, it implies that regional coverage must be limited. The model can be used for local hazard zonation and helps to link the tsunami hazard analysis to land use planning, which is defined in terms of the annual exceedance probability of inundation. Full details of this study are given by Løvholt et al. (2020).

F. Løvholt (✉)
Offshore Energy, Norwegian Geotechnical Institute, Oslo, Norway
e-mail: finn.lovholt@ngi.no

S. Glimsdal
Natural Hazards, Norwegian Geotechnical Institute, Oslo, Norway
e-mail: sylfest.glimsdal@ngi.no

M. B. Sørensen et al. (eds.), *Probabilistic Tsunami Hazard and Risk Analysis*, Mathematics of Planet Earth 14, https://doi.org/10.1007/978-3-031-98115-9_17

17.2 Framework Used in Recipe

This recipe uses a subset of FW1A (Sect. 2.2), an event-tree framework to propagate uncertainties taking as input landslide probabilities for identified unstable landslide volumes. The event-tree combines the landslide rates with conditional probabilities for each different realisation of the landslide dynamics for each volume. The probabilistic quantification is attributed only to the sources, but tsunami propagation and inundation models propagate the sources' probabilities to inundation maps. As the method is scenario based, each scenario and related inundation map is attributed with a given discrete probability. The total probabilities are estimated by summing these different onshore probability maps for the flooding by assuming that each scenario probability is small. Each cell that is inundated for a given scenario is given the scenario probability, and the corresponding non-inundated cells are given zero probability. FW1A is not exploited fully, as only mean values of the probabilities are considered, and hence not quantifying the hazard uncertainty. The output of the analysis are 1000-year and 5000-year return period hazard maps, which are extracted by a contouring algorithm applied to the probability map providing exceedance probabilities for the given return periods. For the temporal probabilities representing the hazard, a Poissonian process (time independent) is assumed.

The advantage of using the framework is the possibility of linking the hazard analysis to the national legislation in Norway (e.g., land use planning) that is only possible through annual exceedance probabilities. Also, the possibility of dissecting the analysis and understanding which scenarios contributing most to the hazard makes the analysis transparent and flexible. It also offers a platform that can be potentially expanded to be used also for acute situations (i.e. when the local hazard changes because of increased slope movements) reminiscent of probabilistic tsunami forecasting (PTF) (Selva et al. 2021), although this PTF technology is presently not developed for landslide-triggered tsunamis. The limitation of this method is mainly that the activity rates, which are key inputs to the analysis, are mainly set using expert judgement, although the rates are partly based on regional landslide statistics. Another limitation is the lack of uncertainty quantification in the inundation maps, as well as in the volume likelihood quantification, although the uncertain source (landslide dynamics) uncertainties are aggregated to the mean hazard values.

17.3 Source Selection and Activity Rate Characterization Used in the Recipe

This recipe represents probabilistic tsunami hazard from subaerial landslides described in Sect. 3.3, using the following steps:

Step 1—Basis for source selection: Landslide sources and rates based on slope stability evaluations—source determination for unstable rock slopes (for more details see Sect. 3.3). The basis for the analysis were four different rock slopes (Jettan, Indre Nordnes, Revdalsfjell 1B and 2), which volumes were identified through extensive field instrumentation and remote sensing monitoring. These volumes were used directly in the numerical analysis.

Step 2—Source parameterisation: Each landslide volume was represented by four different uncertain parameters representing the landslide kinematics; the landslide impact velocity, the landslide run-out distance, the frontal area of impact, and the shape of the landslide (i.e., short block or an infinitely long block). The parameters were set based on calibrated run-out distances using statistics of past events, while the impact velocity and frontal areas were set based on sensitivity analysis using a numerical model to simulate the landslide dynamics.

Step 3—Probabilistic treatment: The method by Herrmanns et al. (2012) was used to quantify the rock slope rates based on rock slope instrumentation and remote measurements, and the resulting failure rates were given as input to the method of Løvholt et al. (2020). Based on their analysis, mean values of annual rates were identified for each of the four rock slope volumes. Conditional probabilities for each selected landslide dynamics parameter under step 2 (representing the full ensemble) were set through group discussions using expert judgement. All the probabilities were implemented in an event-tree framework.

The advantage of this source representation is the possibility of having results of the hazard analysis in terms of annual probabilities. As discussed above, the disadvantage is the lack of local empirical basis for estimating the occurrence rates for each volume. Yet, the rockslide rates are expected to be similar to national rates compared to past documented events in the relevant rockslide regions in Norway, but the slope failure rates are still not directly linked to source statistics. In other words, in some areas with past landslides, the method gives synthetic Magnitude Frequency Distributions (MFDs) that are similar to rate functions from past events nationally. This way of establishing the rates represents a nationally established procedure in Norway, and is necessary to provide hazard results compatible with national legislation. Because of this, it cannot be used directly in other countries or dissimilar settings.

17.4 Source Mechanics Modelling Used in the Recipe

A landslide block model with fixed shape and prescribed kinematics is used for modelling each landslide scenario in the ensemble, (see description in Sect. 4.3). A ¼ period part of a sinusoidal shaped velocity profile as a function of time is used to represent the landslide kinematics (for the representation of such slide shapes see e.g. Løvholt et al. 2005). Supplementary depth averaged landslide dynamics simulations using a Voellmy model (Kim 2014) are carried out to span the uncertainty space and calibrate the input to the block model to comply with the expected landslide mechanics.

The strength of using the block model is the offered robustness for a probabilistic analysis, although some physics is lost as the shape deformations of the landslide are not taken into account. This was deemed as of secondary importance, since the tsunami generation is believed to be mainly controlled by the impact. The smoothing of the blocks implied that interface modelling of tsunami generation was not needed (see below).

17.5 Tsunami Modelling Used in Recipe

The model uses the following methods and steps, as described in Chap. 5, for data preparation, tsunami generation, tsunami modelling, and inundation modelling.

- For the first step on data preparation (see Sect. 5.2), a local moderately resolved (25 m) bathymetry for the Lyngen fjord is prepared as input to provide the bathymetric data to the tsunami model. For the local inundation analysis, high resolution grids for the onshore were sampled to 10 m resolution for the local inundation sites, and merged by interpolation with the local bathymetry to the same spatial resolution.
- For the second step, the tsunami generation (see Sect. 5.3), the landslide induced sea bottom displacement is copied directly to the sea surface at given time steps (2 s intervals) as flux sources. The direct use of the seabed deformation to the sea surface is possible due to the smoothing of the block sources.
- For the third step, the tsunami propagation modelling, a simple (linear) weakly dispersive wave model based on a Boussinesq approximation (Sect. 5.4.2.1) is used (Løvholt et al. 2008). A staggered C-grid finite difference scheme is used for solving the tsunami propagation numerically. A set of propagation files (Løvholt et al. 2010), which constitute a down-sampled (in space and time) time dependent set of surface elevations and velocity fields are stored for each scenario, as input to drive the inundation analysis simulations.
- For the fourth step, simulating the tsunami inundation, the MOST inundation model (Titov and Synolakis 1998; Titov et al. 2011) is used, driven by the propagation files stored for each scenario simulation (Løvholt et al. 2010). Inundation simulations for about 600 scenarios (i.e. probabilistic landslide volumes and related landslide dynamics realisations) are computed for each of the 31 sites, which yields 18,600 inundation simulations. They are carried out on a workstation, but with slightly larger numbers High Performance Computing (HPC) resources would be required.

The main advantage of this simulation procedure is that it is relatively simple (no smoothing needed, linear model), which makes it possible to integrate it into PTHA even without HPC resources. Some disadvantages are that some physics is lost by the linear modelling, as well as the simplification of the landslide dynamics in the modelling (wave propagation can be highly non-linear in generation and propagation including cratering).

17.6 Risk Communication

The results are provided in the form of technical reports that are conveyed to the national authorities, and further conveyed to local stakeholders through a series of meetings, to ensure local ownership and implementation of relevant mitigation measures.

17.7 Discussion

LPTHA is generally less developed than for earthquake sources. This is partly due to the infrequency of these tsunami sources, which often prevents using past source statistics as basis for determining annual sources rates as input to the analysis. In this study, a schema based on slope stability evaluations is used to estimate the source rates, on which slopes with high movements or other evidence of instabilities are given higher rates. These rates strongly determine the hazard outputs. This implies that there is a strong factor of judgement that enters into the LPTHA, and this is considered the main factor of uncertainty. On the other hand, rates are expected to comply with landslide statistics. An important point to consider when adapting this method is that the slope stability method is linked to the Norwegian practice and past landslide data related to rock slope hazard analysis, and includes expert judgement treatment of probabilities that is calibrated towards regional statistics of past landslides. Therefore, the determination of rates cannot necessarily be transferred and used directly in other countries.

References

Hermanns RL, Oppikofer T, Anda E, Blikra LH, Böhme M, Bunkholt H et al (2012) Recommended hazard and risk classification system for large unstable rock slopes in Norway. NGU report, 0800–3416

Kim J (2014). Finite volume methods for tsunamis generated by submarine landslides (Doctoral dissertation). Univ Washington

Løvholt F, Harbitz CB, Haugen KB (2005) A parametric study of tsunamis generated by submarine slides in the Ormen Lange/Storegga area off western Norway. Mar Pet Geol 22(1–2):219–231

Løvholt F, Pedersen G, Gisler G (2008) Modeling of a potential landslide generated tsunami at La Palma Island. J Geophys Res Oceans 113:C09026

Løvholt F, Glimsdal S, Harbitz CB (2020) On the landslide tsunami uncertainty and hazard. Landslides 17(10):2301–2315. https://doi.org/10.1007/s10346-020-01429-z

Løvholt F, Pedersen G, Glimsdal S (2010). Coupling of dispersive tsunami propagation and shallow water coastal response. Open Ocean J 4(1)

Selva J, Lorito S, Volpe M, Romano F, Tonini R, Perfetti P et al (2021) Probabilistic tsunami forecasting for early warning. Nat Commun 12(1):5677

Titov VV, Synolakis CE (1998) Numerical modeling of tidal wave runup. J Waterw Port Coast Ocean Eng 124(4):157–171

Titov VV, Moore CW, Greenslade DJM, Pattiaratchi C, Badal R, Synolakis CE, Kanoğlu U (2011) A new tool for inundation modeling: community modeling interface for tsunamis (ComMIT). Pure Appl Geophys 168:2121–2131

Chapter 18
R9—Probabilistic Hazard Analysis for Tsunamis Generated by Subaqueous Volcanic Explosions in the Campi Flegrei Caldera, Italy

Anita Grezio, Raphaël Paris, and Jacopo Selva

18.1 Introduction

An eruption event tree is developed to explore the natural variability of potential submarine volcanic explosions on the basis of past volcanic activity and to quantify the tsunami hazard conditional upon the occurrence of an underwater eruption at Campi Flegrei, Italy.

Very few events of this kind are documented, but both instrumental and observational data are available for two case-studies (Myojin-sho 1952 and Karymsky Lake 1996, in Chap. 3).

This recipe requires a good knowledge on past eruptions (sizes of the vents and energy of explosions, inferred from geological data), as well as a good bathymetry of the study area (a minimum resolution of 50 m is recommended). The initialization of the waves is static, i.e. instantaneous, water surface deformation, as explained in Chap. 4. It is recommended to use a depth-averaged weakly dispersive tsunami propagation model.

A. Grezio (✉)
Istituto Nazionale di Geofisica e Vulcanologia, Bologna, Italy
e-mail: anita.grezio@ingv.it

R. Paris
CNRS, IRD, OPGC, Laboratoire Magmas et Volcans, Université Clermont Auvergne, Aubière, France
e-mail: raphael.paris@uca.fr

J. Selva
Dipartimento di Scienze della Terra, dell'Ambiente e delle Risorse (DiSTAR), Università di Napoli Federico II, Naples, Italy
e-mail: jacopo.selva@unina.it

M. B. Sørensen et al. (eds.), *Probabilistic Tsunami Hazard and Risk Analysis*, Mathematics of Planet Earth 14, https://doi.org/10.1007/978-3-031-98115-9_18

This recipe is presented in Paris et al. (2019). Considering that there are no other tsunami studies using this workflow, we may refer to studies of volcanic hazard with a similar approach but related to tephra and pyroclastic density currents (Tierz et al. 2018; Grezio et al. 2020).

The present recipe includes the following ingredients:

- A fully simulation-based hazard framework with a large number of scenarios and comprehensive uncertainty treatment—FW1A or FW1B of Sect. 2.2
- Source selection for volcanic sources—Sect. 3.4
- Source modelling for volcanic sources—Sect. 4.4
- Tsunami modelling—Chap. 5.

18.2 Framework Used in Recipe

The framework is fully probabilistic based on a sufficiently large set of scenarios and defined through a Bayesian Event Tree, as commonly adopted for volcanic hazards (Marzocchi et al. 2008, 2010). The analysis is limited to hazard quantification, and provides both individual scenario results and hazard curves. Thus, this recipe produces a hazard quantification compatible to both frameworks FW1A (Fully simulation-based risk assessment) or FW1B (Simulation-based hazard convoluted with vulnerability and loss using numerical integration). The method applied is summarized in Fig. 18.1. Nodes 1–2–3 of the event tree (Fig. 18.1d) can be neglected, since for tsunami-generating explosive eruptions, probabilities are conditioned upon the occurrence of an eruption (Orsi et al. 2009). Node 4 for the vent location assumes that the relative probability distribution is not uniform, eruptions in the middle of the gulf being most probable due to the presence in the main NE-SW trending fault system of the caldera. Node 5 for the eruption size (here, the energy of the explosion) presents the distribution parameters with a decreasing probability for increasing explosion size and links the tsunamigenic capability of explosions with the size of emerged past craters for which estimation of eruption magnitudes and volumes were available. Node 6 for the tsunami generation assumes that effusive eruptions certainly do not produce tsunamis (conditional probability equal to 0 and no epistemic uncertainty) and explosive eruptions certainly produce tsunamis if the vent is located at a depth larger than 10 m (conditional probability equal to 1). Nodes 7–8 for the tsunami intensity and hazard curves are respectively defined by the maximum wave amplitude (Hmax) at 1 m depth in the numerical simulations and by the probability to overcome the tsunami intensity measures in the 421 sites, from Cuma to Positano.

To account for both the aleatory variability of the source space and the intrinsic uncertainty in the tsunami generation and propagation models, we assume that each simulation result is associated with a conditional hazard curve (conditional upon a scenario in that vent location and that size class) described by a log-normal distribution with a sigma equal to 1 (Davies et al. 2017; Selva et al. 2022).

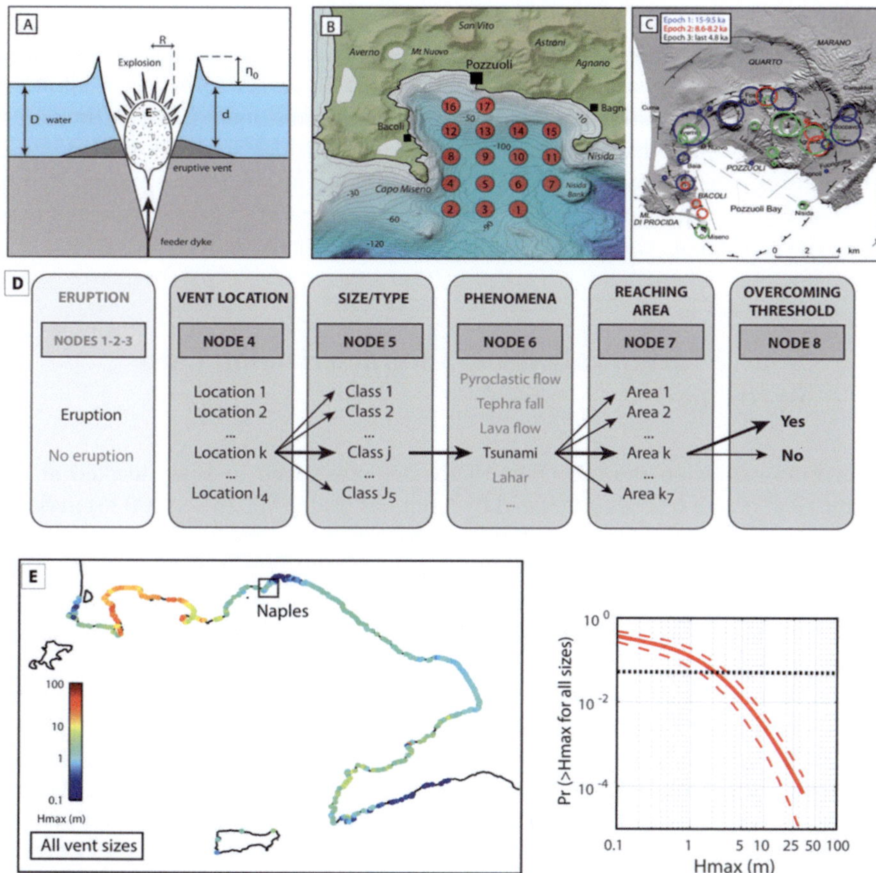

Fig. 18.1 **a** Physical parameters controlling the initial displacement of the water surface and subsequent tsunami (E: explosion energy; d: explosion depth; D: total water depth around the volcanic cone; R: eruptive vent radius; η_0: initial water surface displacement). **b**: Locations of the 17 vents for tsunami simulations. **c**: Craters of the past eruptions at Campi Flegrei (the size of the craters -circles- determine the 3 classes of radii of the eruptive vents). **d**: Bayesian Event Tree for Volcanic Hazard (light gray: nodes that are not quantified, due to the choice of conditioning the probabilities upon the occurrence of subaqueous explosions). **e**: Conditional mean hazard maps for the probability threshold = 0.05 (map on the left) and conditional hazard curve (mean and 10th–90th percentiles confidence interval) in one point (black square) within the city of Naples (red lines on the right)

Tsunami generation is controlled by a range of physical parameters: water depth, size of eruptive vent, energy and depth of explosion, and complex magma-water interactions which determine the explosion itself. The Bayesian Event Tree quantifies and propagates this natural source variability uncertainty through the conditional probabilities and provides the epistemic uncertainty in each conditional probability.

The main advantage of this event tree method is the fully probabilistic hazard quantification, decoupling the uncertainties in volcanic phenomena and tsunami generation and propagation issues. Moreover, the Bayesian formulation allows a simultaneous quantification of both aleatory and epistemic uncertainty. A limitation is represented by the fact that the results do not consider absolute unconditional probabilities, which may be obtained by multiplying the conditional hazard by the mean annual rate of eruptions, even if the quantification of this mean annual rate is not trivial as volcanic eruptive sequences are not stationary in time.

18.3 Source Identification and Characterization Used in Recipe

Submarine explosion sources (Fig. 18.1b) are positioned in vents located at the geometrical centre of cells generated by a regular grid (1500 m × 1500 m) in order to decrease the computational cost of the subsequent tsunami simulations. Cells are limited to the areas with non-zero conditional probability and water depth capable of generating tsunamis (D > 10 m). Thus, the centres of 17 cells are identified as potential vent locations, with the explosion depth (d) similar to water depth (D) in the range 23–111 m.

Four representative eruption classes are estimated by the past explosive caldera activity in three different periods: 34 explosive eruptions occurred 15–9.5 ka before present (BP), 6 explosive eruptions in 8.6–8.2 ka BP, and 22 explosive and 3 effusive eruptions in 4.8–3.8 ka BP. The four classes are related to different explosion energies and are linked to standard eruptive sizes for which specific probabilistic studies were available. Also, on the basis of the corresponding crater characteristics, we selected 3 crater radii (200 m, 650 m, and 900 m) assuming that both submerged and emerged areas have similar explosive processes.

Important sources of epistemic uncertainty related to the eruption areas exist, due to several major faults in the submerged part of the caldera. In fact, the last eruption of Campi Flegrei (Monte Nuovo, 1538 CE) occurred along these structures. Due to computational cost, we only explored the aleatory variability of the vent location and its crater radius. This limitation was compensated by introducing a rather large uncertainty into simulation results.

We considered only tsunamis generated by explosive eruptions (Fig. 18.1a), specifically at Campi Flegrei, fully accounting for the corresponding source variability. We do not include other volcanic processes that may generate tsunami waves in the same area, such as flank collapses or pyroclastic density currents entering the sea. Given the characteristics of Campi Flegrei volcanism, the contribution of the other sources at Campi Flegrei is probably negligible. However, the contributions of other volcanoes (e.g. Mt. Vesuvius and Ischia) are surely not negligible for the local hazard but are not considered here. By following the "representative scenario" approach, each combination of the 17 vent locations and 3 crater radii (Fig. 18.1b,c) represents a single scenario with representative parameters. In this way, the intra-class variability is assumed negligible with respect to the inter-class variability, as a reasonable first order approximation (Sandri et al. 2016).

18.4 Source Mechanics Modelling Used in Recipe

The volcanic explosions are located in the water column with $D > 10$ m (potential cavities) and an instantaneous upward displacement is formed in each water column, shaping a crater and a steep cone of water in the centre which collapses and turns to a second cylindrical bore (as demonstrated by experimental explosions and numerical simulations).

The energy of the volcanic explosion is estimated following the empirical relationship $E = 3.56\ 10^7\ R^3$ given by Sato and Taniguchi (1997) where E is the explosion energy in Joules and R the radius of the eruptive vent in meters. Using the empirical formula of Le Méhauté and Wang (1996), the initial surface displacement generated by the explosion is a parabolic cavity with a vertical steep water rim. The size of the water cavity (R_c, in meters) is estimated based on explosion energy by the relationship $R_c = 0.0361\ E^{0.25}$. The initial maximum vertical water surface displacement η_0 of the cavity rim is a function of the explosion energy at a given depth and is based on the empirical formula: $\eta_0 = c\ E^{0.24}$ where $c = 0.0143$ if $6.15 \times 10^{-4} < d/E^{1/3} < 1.85 \times 10^{-2}$ or $c = 0.0291$ if $d/E^{1/3} < 6.15 \times 10^{-4}$.

The vent radii R of 200 m, 650 m and 900 m corresponding respectively to explosion energies E of 2.9×10^{14} J, 9.8×10^{15} J and 2.6×10^{16} J create initial displacements of the water surface η_0 of 85 m, 200 m and 250 m, respectively.

Also, we assumed that the explosion occurs at the sea floor or at the rim of a pre-existing eruptive vent. However, the evolution of the vent geometry during the eruption (crater excavation, growth of the ejecta ring) and location of the explosion inside the vent might have an influence on tsunami generation.

A limitation comes from the explosion model. In fact, the size of the crater reflects the energy of the most powerful explosion at shallow depth, regardless of the total number of explosions and post-eruptive erosion. In fact, the morphology of the crater might result from multiple vertically- and horizontally-migrating explosions. Thus, assuming that the size of the crater corresponds to the size of the explosion might lead to an overestimation of the explosion energy in some cases.

18.5 Tsunami Modelling

The tsunami numerical simulations were performed using the FUNWAVE-TVD code (Shi et al. 2012) solving the fully nonlinear Boussinesq-type wave equations in spherical coordinates. The Boussinesq model includes the effects of amplitude non-linearity and frequency dispersion in the case of intermediate-to-deep water waves neglecting the effect of bottom friction. The third-order Runge–Kutta method is used to advance the model in time and an adaptive time step is chosen following the Courant-Friedrichs-Lewy criterion (=0.5) to ensure the stability of the scheme for all simulations. A fourth-order spatial discretization is employed.

In order to prevent reflections from the end-walls at the open sea, we impose a 3 km wide sponge layer at the limit of the computational domain. The spherical grid covers a bathymetric area of about 3000 km^2 with a resolution of 39 m in longitude and 50 m in latitude and is obtained by integrating printed nautical charts with the available digital elevation and bathymetric models. Wave heights at the coast were computed at virtual tide gauge stations at the coast. The propagation of the tsunami in the entire Gulf of Naples takes only 15 min and Pozzuoli is affected in less than 2 min.

The uncertainty in the tsunami propagation model is treated with a log-normal distribution centred on simulation results, following a common approach adopted in literature (Choi et al. 2002; Grezio et al. 2017).

Tsunami waves generated by the explosion propagate radially away from the explosion centre, all over the computational area, with a significant influence of the bathymetry (e.g. higher wave heights due to shallow banks) and coastal morphology (e.g. effects of wave refraction around the islands and capes). So, a high horizontal and vertical resolution of the bathymetry and a well-defined position of the vent is essential ($\leq$50 m). As the distance from the explosion increases, the arrival of longer-period waves might predate the short-period waves because they travel faster. This is a direct effect of frequency dispersion. Wave refraction creates local amplification that is well reproduced by the numerical models. Thus, the inclusion of tsunami dispersion is also essential.

One of the main advantages of the adopted framework is that it allows a reasonable exploration of vent positions, accounting for its uncertainty, and to couple potential vent position with dispersive tsunami models. However, tsunami wave behaviour at the coast (e.g. wave breaking, interaction with coastal structures) and inundation inland were not simulated, because it would require high-resolution terrain models (typically 1 m). Future simulations could be conducted at local scale, i.e. for areas of interest (e.g. harbours) using simulations at the larger scale as input.

18.6 Discussion

This analysis represents a pioneering study on full PTHA for volcanic sources. The main results of the PTHA are the probabilities (and relative uncertainties) of exceeding given tsunami intensity thresholds (i.e., wave amplitudes at the coast), given the occurrence of a subaqueous eruption. Final conditional hazard curves and the relative epistemic uncertainty are provided in each of the coastal sites of the Gulf of Naples.

The main limitation is the estimation of the conditioning probabilities at eruptions, which are highly non-stationary in time. The quantification of long-term conditional hazard is common in volcanic hazard studies and its extension to absolute (unconditional) PTHA may be obtained by multiplying the conditional hazard by the mean annual rate or the probability of occurrence of volcanic eruptions (Selva et al. 2022).

Another limitation is related to the exclusion of other nearby sources of volcanic tsunamis, like pyroclastic flows at Mt. Vesuvius (Grezio et al. 2020) and flank collapses at Ischia.

The adopted approach, being based on standard volcanic hazard quantification methods, reached a quite advanced level of maturity. However, the parametrization of tsunami-related sources is still suffering a significant lack of local data and/or global analogues, increasing the epistemic uncertainty in hazard estimations.

References

Choi BH, Pelinovsky E, Ryabov I et al (2002) Distribution functions of tsunami wave heights. Nat Hazards 25:1–21. https://doi.org/10.1023/A:1013379705323

Davies G, Griffin J, Løvholt F, Glimsdal S, Harbitz CB, Thio HK, Lorito S, Basili R, Selva J, Geist E, Baptista MA (2017) A global probabilistic tsunami hazard assessment from earthquake sources. Geol Soc Lond Spec Publ 456. https://doi.org/10.1144/SP456.5

Grezio A, Babeyko A, Baptista MA, Behrens J, Costa A, Davies G, Geist EL, Glimsdal S, Gonzàlez FI, Griffin J, Harbitz CB, LeVeque RJ, Lorito S, Løvholt, F, Omira R, Mueller C, Paris R, Parsons T, Polet J, Power W, Selva J, Sørensen MB, Thio HK (2017) Probabilistic tsunami hazard analysis: multiple sources and global applications. Rev Geophys 55. https://doi.org/10.1002/2017RG000579

Grezio A, Cinti FR, Costa A, Faenza L, Perfetti P, Pierdominici S, Pondrelli S, Sandri L, Tierz P, Tonini R, Selva J (2020) Multisource Bayesian probabilistic tsunami hazard analysis for the Gulf of Naples (Italy). J Geophys Res Ocean 125. https://doi.org/10.1029/2019JC015373

Marzocchi W, Sandri L, Selva J (2008) BET_EF: a probabilistic tool for long- and short-term eruption forecasting. Bull Volcanol 70:623–632. https://doi.org/10.1007/s00445-007-0157-y

Marzocchi W, Sandri L, Selva J (2010) BET_VH: a probabilistic tool for long-time volcanic hazard assessment. Bull Volcanol 72:705–716. https://doi.org/10.1007/s00445-010-0357-8

Le Méhauté BL, Wang S (1996) Water waves generated by underwater explosion. Adv Ser Ocean Eng World Sci 10 (New Jersey). https://apps.dtic.mil/sti/pdfs/ADA304244.pdf

Orsi G, Di Vito MA, Selva J, Marzocchi W (2009) Long-term forecast of eruption style and size at Campi Flegrei caldera (Italy). Earth Planet Sci Lett 287:265–276. https://doi.org/10.1016/j.epsl.2009.08.013

Paris R, Ulvrova M, Selva J, Brizuela B, Costa A, Grezio A et al (2019) Probabilistic tsunami hazard analysis for tsunamis generated by subacqueos volcanic explosions in the Campi Flegrei caldera (Italy). J Volcanol Geoth Res 379:106–116. https://doi.org/10.1016/j.jvolgeores.2019.05.010

Sandri L, Costa A, Selva J, Tonini R, Macedonio G, Folch A, Sulpizio R (2016) Beyond eruptive scenarios: assessing tephra fallout hazard from Neapolitan volcanoes. Sci Rep 6:24271. https://doi.org/10.1038/srep24271

Sato H, Taniguchi H (1997) Relationship between crater size and ejecta volume of recent magmatic and phreato-magmatic eruptions: implications for energy partitioning. Geophys Res Lett 24:205–208. https://doi.org/10.1029/96GL04004

Selva J, Sandri L, Taroni M, Sulpizio R, Tierz P, Costa A (2022) A simple two-state model interprets temporal modulations in eruptive activity and enhances multivolcano hazard quantification. Sci Adv 8:44. https://doi.org/10.1126/sciadv.abq4415

Shi F, Kirby JT, Harris JC, Geiman JD, Grilli ST (2012) A high-order adaptive time-stepping TVD solver for Boussinesq modeling of breaking waves and coastal inundation. Ocean Model 43–44:36–51. https://doi.org/10.1016/j.ocemod.2011.12.004

Tierz P, Stefanescu ER, Sandri L, Sulpizio R, Valentine GA, Marzocchi W, Patra AK (2018) Towards quantitative volcanic risk of pyroclastic density currents: probabilistic hazard curves and maps around Somma-Vesuvius (Italy). J Geophys Res Solid Earth 123:6299–6317. https://doi.org/10.1029/2017JB015383

Chapter 19
R10—Adriatic APTHA

Cléa Denamiel

19.1 Introduction

The Adriatic Sea, and more precisely some coastal cities along different Croatian islands, are well-known locations of interest for weather related meteotsunami (or atmospherically driven tsunami) events (Sect. 3.5). However, in contrast to Ciutadella (Balearic Islands, Spain)—another location of interest, where operational meteotsunami forecasts have been derived since 1983 (Ramis and Jansà 1983)—it is only very recently that a meteotsunami early warning system prototype has been implemented in this region. The reasons might be that, in the Adriatic Sea, (1) the observational network providing 1-min sea-level and pressure measurements was underdeveloped until 2017 and (2) the atmospheric mesoscale gravity waves (MGWs) driving most of the meteotsunami events in the area are highly variable in space and shape and thus difficult to forecast with numerical models.

The presented recipe is based on the work of Denamiel et al. (2019a, b) and Tojčić et al. (2021) who developed and evaluated the Croatian Meteotsunami Early Warning System (CMeEWS) prototype following the framework presented in Vilibić et al. (2016). In operational mode, this system applies an observational-deterministic-stochastic approach: (1) the MGWs are forecasted with deterministic models, at least a day in advance, and/or detected through measurements at least 2 h before the meteotsunami reaches locations of interest and (2) the meteotsunami surge hazard assessment (or Atmospheric PTHA—APTHA) at those locations are derived with a surrogate model (Surrogate-based PTHA recipe, R11).

C. Denamiel (✉)
Division of Marine and Environmental Research, Ruđer Bošković Institute, Zagreb, Croatia
e-mail: cdenami@irb.hr

M. B. Sørensen et al. (eds.), *Probabilistic Tsunami Hazard and Risk Analysis*, Mathematics of Planet Earth 14, https://doi.org/10.1007/978-3-031-98115-9_19

19.2 Framework

The CMeEWS prototype is mostly linked to Frameworks FW1A, FW1D, and FW2. Practically, the CMeEWS receives two different kinds of data: (1) high-resolution atmospheric and ocean model results provided by the Adriatic Sea and Coast (AdriSC) modelling suite (Denamiel et al. 2019a), and (2) measurements from the observational network along the Adriatic coast. However, the evaluation of the AdriSC Meteotsunami Forecast system (Denamiel et al. 2019a,b; Tojčić et al. 2021) has shown that the capacity of the tsunami models to deterministically reproduce or forecast the observed floods occurring during meteotsunami events is extremely sensitive to the capacity of the kilometre-scale atmospheric models to reproduce the MGWs. As implementing dense observational networks, ensemble of kilometre-scale models or data assimilation is extremely expensive, a cheaper solution had to be envisioned to account for the uncertainty linked to the deterministic forecast of MGWs. This is why a surrogate model approach, based on the "Synthetic Catalogue" method and using uncertainty quantification (Sect. 4.5), was implemented within the CMeEWS prototype following Denamiel et al. (2019b, 2020, 2021).

19.3 Source Identification and Characterization

In the Adriatic Sea (as also seen in the Surrogate-based PTHA recipe), most meteotsunami events are driven by MGWs propagating from the Italian coastline towards the eastern Adriatic and most particularly towards the Croatian coasts. The generation of the MGWs in the Adriatic region is not well understood but is most probably linked to unbalanced jet streaks through the so-called spontaneous balance adjustment (Sect. 3.5). The MGWs can cause rapid surface pressure oscillations > 5 hPa in a few minutes and, while traveling over the Adriatic Sea, can transfer their energy via the Proudman resonance process to generate meteotsunami waves (Sect. 3.5). In Croatia, known locations of interest—where the meteotsunami waves are amplified by the local geomorphology and therefore can cause extreme sea-levels and flooding (hereafter meteotsunami surges)—are, for example, the harbours' of Vela Luka, Stari Grad and Vrboska in the islands of Hvar and Korčula where tourism is quite intensive.

19.4 Source Mechanics and Tsunami Modelling

The deterministic and stochastic atmosphere–ocean models (providing source mechanism and tsunami modelling simultaneously) as well as the operational use of the CMeEWS prototype are described below.

Models. Two different approaches are used for the source mechanics and tsunami modelling: (1) a deterministic forecast and (2) a stochastic hazard assessment.

The deterministic approach is based on the AdriSC model (with no data assimilation performed) composed of a basic module providing high-resolution regional atmospheric and ocean results for the entire Adriatic Sea and a dedicated meteotsunami module. The basic module uses a modified version of the Coupled Ocean-Atmosphere-Wave-Sediment Transport (COAWST) modelling system developed by Warner et al. (2010) which couples (online—i.e., models are run in parallel) the Regional Ocean Modelling System (ROMS; Shchepetkin and McWilliams 2005) at up to 1 km resolution and the Weather Research and Forecasting (WRF) model (Skamarock et al. 2005) at up to 3 km resolution. The dedicated meteotsunami module is forced by the hourly fields produced by the basic module (i.e., WRF 3 km and ROMS 1 km) and couples (offline—i.e., models are run one after the other) the WRF model at 1.5-km resolution with the ADvanced CIRCulation (ADCIRC) model (Dietrich et al. 2012) using a mesh of up to 10-m resolution at locations of interest. Every day at midnight, the 1-min forecast results from WRF 1.5 km and ADCIRC simulations for the next day are provided. The stochastic hazard assessment is based on a surrogate model propagating the uncertainties from the atmospheric input (i.e., the MGWs responsible for the meteotsunami generation) to the meteotsunami surges at different locations along the Croatian coastline by using synthetic pressure disturbances (no wind, no tide) as described in Denamiel et al. (2018, 2019a) and the surrogate-based PTHA recipe (R11). The synthetic forcing depends on 6 stochastic input variables: start location, direction of propagation, amplitude, period, propagation speed and attenuation factor.

Operational Mode. As seen in Fig. 19.1 and presented in Denamiel et al. (2019b), 5 steps are involved in providing the daily meteotsunami surge hazards derived with the surrogate model approach from both deterministic model results (Steps 1–4) and observations (Step 5).

Step 1: AdriSC Forecast. At least 30 h before any potential meteotsunami event (M) can occur, the mean sea-level pressure is high-pass filtered in order to potentially reveal the high-frequency oscillations characteristic of the MGWs and is automatically extracted from the WRF 1.5-km results.

Step 2: Automatic Detection. No later than 28 h before any meteotsunami event, the maximum filtered pressure amplitude per time unit is derived for each WRF 1.5 km model grid point. If the number of grid points exposing more than 20 Pa/4-min (a threshold derived from observational studies; Vilibić et al. 2016), or spatial coverage, is larger than 5% then a potential meteotsunami is detected (event mode of the warning) and an automatic email is sent to the AdriSC team (red box, Fig. 19.1). Otherwise (silent mode of the warning), at this stage, it is assumed that no meteotsunami will occur in the next forecasted day.

Step 3: Manual Extraction. No later than 27 h before any meteotsunami event, in case of automatic meteotsunami detection, the filtered pressure field is visualised and analysed by the AdriSC team. If the detected pressure disturbance is recognized

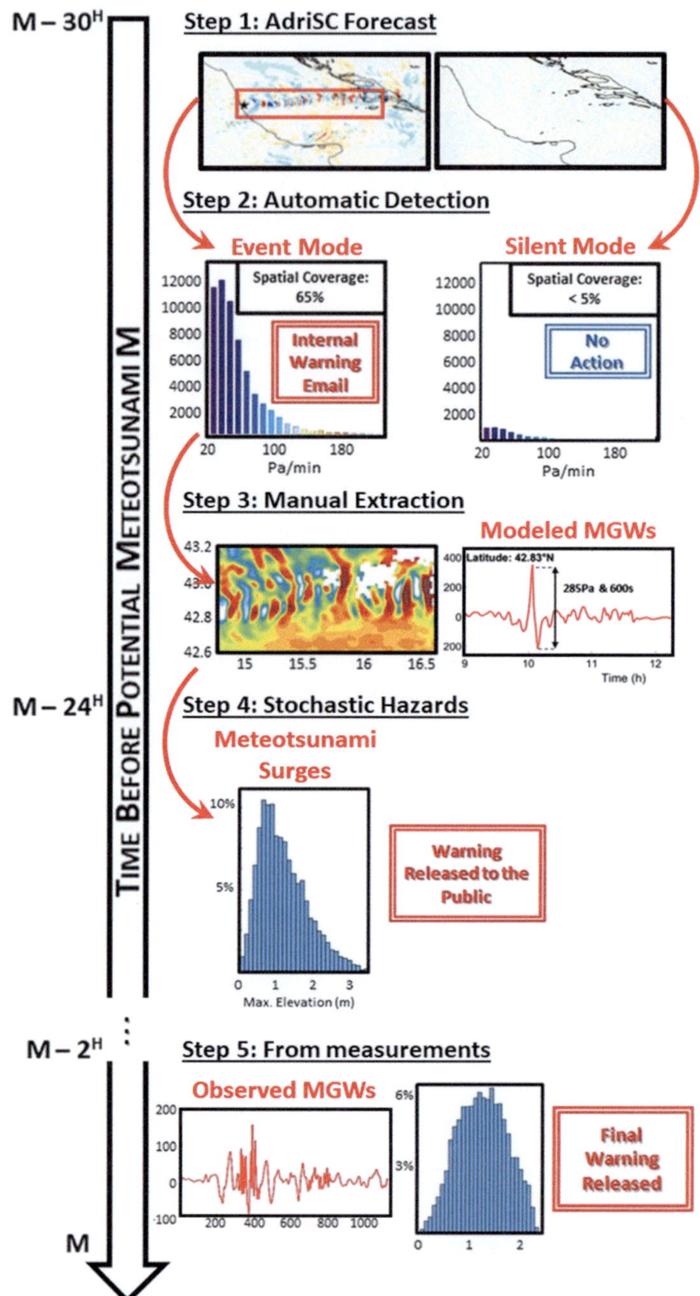

Fig. 19.1 Operational stochastic meteotsunami (M) surge hazard forecast within the croatian meteotsunami early warning system (CMeEWS) based on mesoscale gravity wave (MGW) input from both (1) the deterministic model results and (2) the measurements. Adapted from Denamiel et al. (2019b)

as an MGW, the ranges of variation of the forecasted MGW parameters (e.g., location, speed, period, amplitude) are manually estimated from the WRF 1.5-km 1-min filtered air pressure results.

Step 4: Stochastic Hazards. At least 24 h before the forecasted meteotsunami event, the meteotsunami stochastic surrogate model is used to produce the meteotsunami surge distributions at the locations of interest. These distributions are the output of the surrogate model, derived from 20,000 random combinations of the input variables (e.g., location, speed, period, amplitude) selected in the range of the MGW parameters extracted from the WRF 1.5 km filtered air pressure results. At each location of interest, a first warning provides (1) the probability of the expected meteotsunami surge derived from the surrogate model (including maximum daily tidal elevation) and (2) the deterministic meteotsunami surge from the ADCIRC model run, which is also considered during the decision process for timely managing the hazard.

Step 5: From Measurements. In the 2-h period before the forecasted meteotsunami event (i.e., the estimated time for the MGWs to cross the Adriatic Sea from the Italian cost), the 1-min air pressure measurements from Ancona, Ortona, and Vieste along the Italian coast and from Svetac and Vis in the middle of the Adriatic (Fig. 19.1) are analysed by the AdriSC team, and if any MGW is detected, amplitude and period are extracted from the observations. These parameters are then used as constant values in the surrogate model and updated meteotsunami surge distributions are produced with 20,000 random combinations of the other input variables selected in the range of the parameters extracted from the WRF 1.5-km results. The final meteotsunami warning using the updated distributions (including maximum daily tidal elevation) is then ready to be published and accessible to users at the locations of interest.

19.5 Discussion

The Adriatic APTHA approach presents two main advantages. First, it takes into account the uncertainties associated with the MGWs particularly during storm events when the deterministic model lacks accuracy. Second, it updates the final warnings using meteotsunami hazards based on input parameters extracted from the measured MGWs. However, in operational mode, the CMeEWS prototype has three major weaknesses: (1) due to its extreme numerical cost, the AdriSC modelling suite is run operationally only once per day, the warnings derived from the deterministic models are thus always based on conditions forecasted from a 72-hour old assimilation cycle which can lack accuracy; (2) human intervention is required to extract the MGW parameters from the deterministic forecast as, till now, no algorithm is capable to do it automatically; and (3) in event mode, the microbarograph data should be analysed in a timely manner with efficient operational tools which are not currently available within the CMeEWS.

In terms of the maturity, the CMeEWS prototype, which combines scarce atmospheric observations, computationally expensive kilometre-scale deterministic models and not fully tested surrogate models, has been successfully evaluated against a dozen of well-observed historical meteotsunami events in the Adriatic Sea (Denamiel et al. 2019b) and has proven to provide meteotsunami surge hazard assessments accurate enough to be used to warn the coastal communities of Vela Luka, Stari Grad and Vrboska at least a day in advance (Tojčić et al. 2021).

As also discussed in the Surrogate-based recipe, alternative methods to produce APTHA include increasing the density and quality of the observational networks, implementing ensemble modelling combined with data assimilation and using machine learning (e.g., neural networks).

References

Denamiel C, Šepić J, Vilibić I (2018) Impact of geomorphological changes to harbor resonance during meteotsunamis: the Vela Luka Bay test case. Pure Appl Geophys 175:3839–3859. https://doi.org/10.1007/s00024-018-1862-5

Denamiel C, Šepić J, Ivanković D, Vilibić I (2019a) The Adriatic Sea and coast modelling suite: evaluation of the meteotsunami forecast component. Ocean Model 135:71–93. https://doi.org/10.1016/j.ocemod.2019.02.003

Denamiel C, Šepić J, Huan X, Bolzer C, Vilibić I (2019b) Stochastic surrogate model for meteotsunami early warning system in the eastern Adriatic Sea. J Geophys Res Oceans 124:8485–8499. https://doi.org/10.1029/2019JC015574

Denamiel C, Huan X, Šepić J, Vilibić I (2020) Uncertainty propagation using polynomial chaos expansions for extreme sea level hazard assessment: the case of the eastern Adriatic meteotsunamis. J Phys Oceanogr 50(4):1005–1021. https://doi.org/10.1175/JPO-D-19-0147.1D

Denamiel C, Huan X, Vilibić I (2021) Conceptual design of extreme sea-level early warning systems based on uncertainty quantification and engineering optimization methods. Front Mar Sci 8:650279. https://doi.org/10.3389/fmars.2021.650279

Dietrich JC, Tanaka S, Westerink JJ, Dawson CN, Luettich RA Jr, Zijlema M et al (2012) Performance of the unstructured-mesh, SWAN+ADCIRC model in computing hurricane waves and surge. J Sci Comput 52:468–497. https://doi.org/10.1007/s10915-011-9555-6

Ramis C, Jansà A (1983) Condiciones meteorológicas simultáneas a la aparición de oscilaciones del nivel del mar de amplitud extraordinaria en el Mediterráneo occidental. Rev Geofisica 39:35–42

Shchepetkin AF, McWilliams JC (2005) The regional oceanic modeling system: a split-explicit, free-surface, topography-following-coordinate ocean model. Ocean Model 9(4):347–404. https://doi.org/10.1016/j.ocemod.2004.08.002

Skamarock WC, Klemp JP, Dudhia J, Gill DO, Barker DM (2005) A description of the advanced research WRF Version 2. NCAR Technical Note NCAR/TN-468+STR. https://doi.org/10.5065/D6DZ069T

Tojčić I, Denamiel C, Vilibić I (2021) Performance of the Adriatic early warning system during the multi-meteotsunami event of 11–19 May 2020: an assessment using energy banners. Nat Hazards Earth Syst Sci 21:2427–2446. https://doi.org/10.5194/nhess-21-2427-2021

Vilibić I, Šepić J, Rabinovich AB, Monserrat S (2016) Modern approaches in meteotsunami research and early warning. Frontiers Mar Sci 3(57). https://doi.org/10.3389/fmars.2016.00057

Warner JC, Armstrong B, He R, Zambon JB (2010) Development of a coupled ocean-atmosphere-wave-sediment transport (COAWST) modeling system. Ocean Model 35(3):230–244. https://doi.org/10.1016/j.ocemod.2010.07.010

Chapter 20
R11—Surrogate-Based PTHA

Cléa Lumina Denamiel and Dimitra Salmanidou

20.1 Introduction

In the last decade, surrogate model approaches—providing fast-running approximation of complex dynamical models—have been gaining traction in the geoscience community (e.g., Formaggia et al. 2013; Wang et al. 2016; Sraj et al. 2014; Giraldi et al. 2017). Compared to sampling approaches (e.g., Monte Carlo simulations), surrogate models are highly efficient for uncertainty quantification, and, once trained for a given location, can be used without additional numerical resources (Knio and Le Maître 2006; Najm et al. 2009). Further, as surrogate models are driven by information needs and as their training is relatively cheap, they require limited computational resources. For example, surrogate models can be built with a limited knowledge of the uncertain parameter prior distributions (e.g., uniform distributions can be used when observations are scarce), and a scarce sampling of these distributions compared to Monte Carlo simulations (e.g., sparse grids) can be used to minimize the training dataset.

Consequently, despite producing higher uncertainty for PTHA, surrogate models can be seen as a natural solution for PTHA in regions of the world where observations and modelling capacities are lacking. In this context, surrogate models propagate the uncertainties associated with the tsunami sources (see Chaps. 3 and 4) in order to assess the probability of scenario distributions to exceed defined tsunami hazard

C. L. Denamiel (✉)
Division of Marine and Environmental Research, Ruđer Bošković Institute, Zagreb, Croatia
e-mail: cdenami@irb.hr

D. Salmanidou
Advanced Research Computing Centre, University College London, London, UK
e-mail: d.salmanidou.12@ucl.ac.uk

M. B. Sørensen et al. (eds.), *Probabilistic Tsunami Hazard and Risk Analysis*, Mathematics of Planet Earth 14, https://doi.org/10.1007/978-3-031-98115-9_20

metrics. Two examples are described hereafter to demonstrate the applicability of surrogate models for tsunami science: (1) tsunami hazards in the northern Cascadia (Salmanidou et al. 2021) and (2) meteotsunami hazards in the Adriatic Sea (Denamiel et al. 2020). In this recipe, a common Framework (Chap. 2) is first described for both applications. Then Identification and Characterization and Modelling of the sources (Chaps. 3 and 4) and Tsunami Modelling (Chap. 5) are presented separately for each application. Finally, the advantages and limitations of surrogate-based methods for PTHA are discussed.

20.2 Framework

Surrogate-based PTHA falls within a framework where a large number of scenarios is considered (FW1A or FW1B; Sect. 2.2), even though the epistemic uncertainty (for example, the one linked to the characterization of the seismic and atmospheric sources) is not considered in the two presented applications. Statistical surrogates are largely used for uncertainty quantification purposes thanks to the large number of scenarios that can be explored when building a surrogate and the added benefit of providing a credibility interval of their predictions. The general methodology can be summarized in five stages, some of which are similar to deterministic studies: (1) the set-up of the experiments, defined by the model and data choices; (2) the experimental design, which relates to the selection of experiments for the training, common approaches include one-shot designs (Latin Hypercube Sampling) and sequential design algorithms; (3) the training runs, which relate to the numerical modelling of the selected scenarios; (4) the surrogate modelling, that is the process where the surrogates are built based on the predictor and response variables, kernel functions and hyperparameter choices; evaluation of the surrogates in this step can circle back to (1), and (5) the prediction stage, where the surrogates can be used effectively for the predictions of numerous scenarios.

20.3 Source Identification and Characterization

The approach selected for the source identification and characterization can be driven by the research question and available information. The identification of the most influential input parameters is an important step of the surrogate modelling and can be implemented in various ways such as expert knowledge, sensitivity analysis, model calibration or statistical methods. The scarcity of the observations can render the quantity of the data insufficient for the study, thus synthetic data from source modelling are often used in such cases.

Tsunami hazards in northern Cascadia. The last largest earthquake in the Cascadia subduction zone occurred in 1700 and few historical observations have transcended the passage of time. However, evidence of an orphan tsunami in the Japanese shorelines around this period led to the source inversion and first characterization (Satake et al. 2003). The megathrust earthquake was estimated to range between a moment magnitude, Mw, of 8.7–9.2 and most likely around Mw 9, rupturing the full zone. Nowadays, a potential subduction earthquake in Cascadia could have catastrophic consequences for the populated shorelines of the north-western Pacific Ocean. Scientific studies examine the potential of full or partial zone rupture estimating high tsunami waves that will reach the shores in short times, giving little room for warning (Witter et al. 2013).

Meteotsunami hazards in the Adriatic Sea. In the Adriatic Sea (as also seen in the Adriatic APTHA recipe R10), most meteotsunami events are driven by Mesoscale Gravity Waves (MGWs) propagating from the Italian coastline towards the eastern Adriatic and most particularly towards the Croatian coasts. The generation of the MGWs in the Adriatic region is not well understood, but is most probably linked to unbalanced jet streaks through the so-called spontaneous balance adjustment (Sect. 3.5). The MGWs can cause rapid surface pressure oscillations > 5 hPa in a few minutes and, while traveling over the Adriatic Sea, can transfer their energy via the Proudman resonance process to generate meteotsunami waves (Sect. 3.5 for details). In Croatia, known locations of interest—where the meteotsunami waves are amplified by the local geomorphology and can cause extreme sea-levels and flooding (hereafter meteotsunami surges)—are, for example, the harbours of Vela Luka, Stari Grad and Vrboska in the islands of Hvar and Korčula where tourism is quite intensive. However, at these locations, the capacity of the tsunami models to deterministically reproduce or forecast the observed meteotsunami surges is extremely sensitive to the capacity of the atmospheric models to reproduce the parameters of the MGWs (e.g., location, speed, period, amplitude).

20.4 Source Mechanics Modelling

In the two cases described, two different approaches are followed to model the source as the source mechanism of the events is distinct (seabed deformation vs atmospheric forcing).

Tsunami hazards in northern Cascadia. Salmanidou et al. (2021) modelled a full rupture using an idealized geometry of the seabed deformation. This approach allowed for a simple and straightforward representation of the source, however did not incorporate complex rupture mechanics that can influence wave generation. One of the most critical steps when training the surrogates is the source characterization as the most influential parameters need to be identified and adequately sampled to lead to meaningful predictions. Six source parameters were selected as the input parameters, or predictor variables, to modify and assess their effect on the tsunami magnitude. These were the shape parameters and the displacement time. In later iterations, the displacement time was omitted as its variance did not significantly influence the response.

For the experimental design, the sequential design algorithm MICE (Mutual Information for Computer Experiments, Beck and Guillas 2016) was adopted. This active learning algorithm uses a mutual information-based design criterion, to determine the optimal setting of design of experiments with Gaussian process emulation. The objective is to maximize the information gain for training and building the emulators with as little information as possible. 60 simulations were used in this setting to predict thousands of scenarios.

Meteotsunami hazards in the Adriatic Sea. The surrogate-based Atmospheric PTHA (APTHA) was implemented for meteotsunami hazards in the Adriatic Sea following Denamiel et al. (2019, 2020, 2021) with the purpose to account for and to propagate the MGW uncertainty to the meteotsunami surges. It relies on a series of synthetic pressure disturbances (i.e., idealized MGW fields) following the "Synthetic Catalogue" method (Sect. 4.5). The synthetic pressure disturbances can be analytically written in a reference frame (xx, yy) rotated in the direction of propagation (θ) of the MGWs (Fig. 20.1) as follows: $P_{MGW}(xx, yy, t, \omega) = P_A(\omega)e^{-5\left(\frac{yy}{d(\omega)}\right)^2} A_t(t)sin\left(\frac{2\pi}{T(\omega)}\left(t - \frac{xx}{c(\omega)}\right)\right)$. Here, the 6 stochastic variables depending on (ω) and defining the MGWs are: $(f(x_0(\omega)), y_0(\omega))$ the start location in the (x, y) frame with f the function defining the longitude depending of the latitude, $\theta(\omega)$ the direction of propagation, $P_A(\omega)$ the amplitude, $T(\omega)$ the period, $c(\omega)$ the propagation speed, $d(\omega)$ the attenuation factor in the yy-direction and A_t the time attenuation function (see Denamiel et al. 2018). The range of variation of these six stochastic input parameters has been defined based on both pressure measurements and numerical simulations (Denamiel et al., 2018, 2019). However, given the sparsity of the available measurements and the rarity of strong meteotsunami events, it is assumed that all stochastic variables follow a uniform distribution such as: $y_0(\omega) \sim U([41.25°, 43.65°])$, $\theta(\omega) \sim U([0°\text{N}, 150°\text{N}])$, $P_A(\omega) \sim U([50\,Pa, 400\,Pa])$, $T(\omega) \sim U([300\,\text{s}, 1800\,\text{s}])$, $c(\omega) \sim U([15\,\text{m/s}, 40\,\text{m/s}])$ and $d(\omega) \sim U([30\,\text{km}, 150\,\text{km}])$.

20.5 Tsunami Modelling

The choice of the numerical model for the tsunami simulations is source-dependent. For example, Nonlinear Shallow Water Equation (NSWE) solvers are commonly used in the modelling of earthquake-generated tsunami propagation and inundation. Different approaches can be adopted for the surrogate modelling, such as Gaussian Processes or Polynomial Chaos Expansion as described in the examples below.

Tsunami hazards in northern Cascadia. The source displacement was superimposed over the current bathymetry and used as an input in the NSWE code VOLNA (Reguly et al. 2018) to model the tsunami generation, propagation and inundation. A High-Performance workflow was developed to automate the design process, where

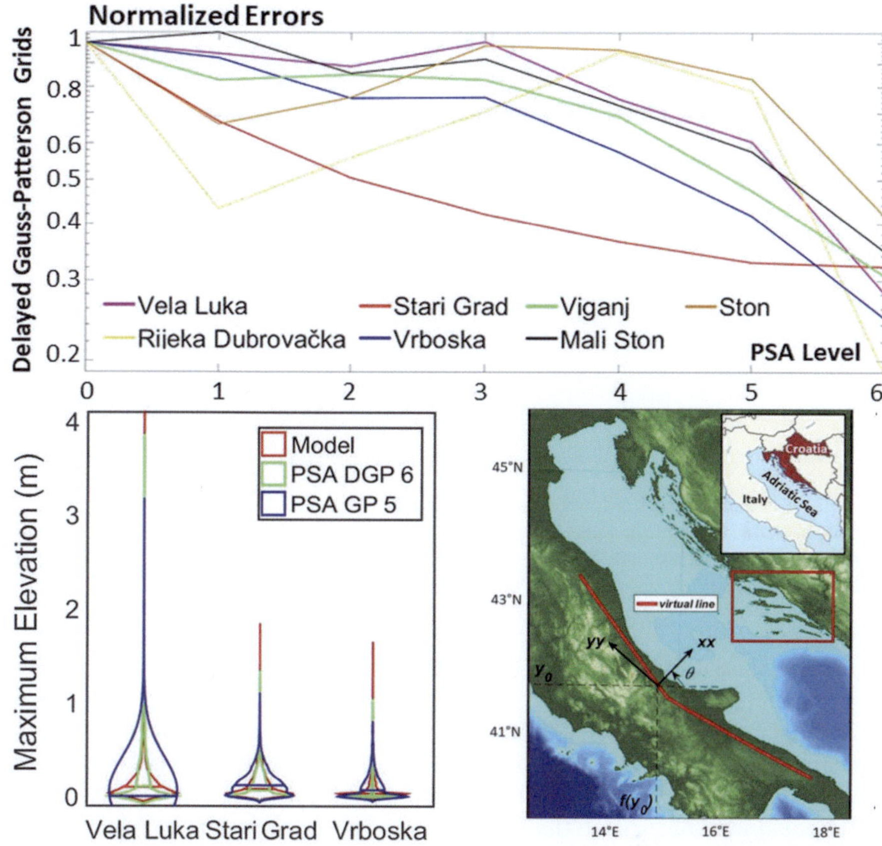

Fig. 20.1 Top panel: Evolution of the normalized errors derived at locations of interest (e.g., Vela Luka, Stari Grad, Vrboska) depending of the order p of the PSA. Bottom left panel: Violin plot distributions of the deterministic tsunami model results (in red), the PSA of order 5 with a Gauss–Patterson rule (GP 5; in blue) and the PSA of order 6 with the delayed Gauss–Patterson rules (DGP 6; in green) depending on the location of interest. Bottom right panel: Rotated coordinate system (xx, yy) depending on start location $(f(y_0), y_0)$ and direction θ of each synthetic atmospheric MGW

the outputs of the sequential design were used as input values in the source modelling, the displacement was then coupled with the tsunami model, the outputs of which were used as inputs in the design algorithm starting the next iteration (Fig. 20.2, Experimental Design). The outputs of the tsunami model (flow depths at the shoreline level in ca. 5 K locations) were used as the response variable for training the surrogates. The hazard metrics were the maximum onshore run-up heights at each grid cell. One surrogate was built in each location using the Multi-Output Gaussian Process (MOGP) method. Gaussian processes form a reliable tool in surrogate modelling for their usability and flexibility and provide an estimate of their uncertainty. The MOGP code allowed for parallel building of the surrogates without considering correlation between them. Time-dependency and nonlinear effects were not considered in this

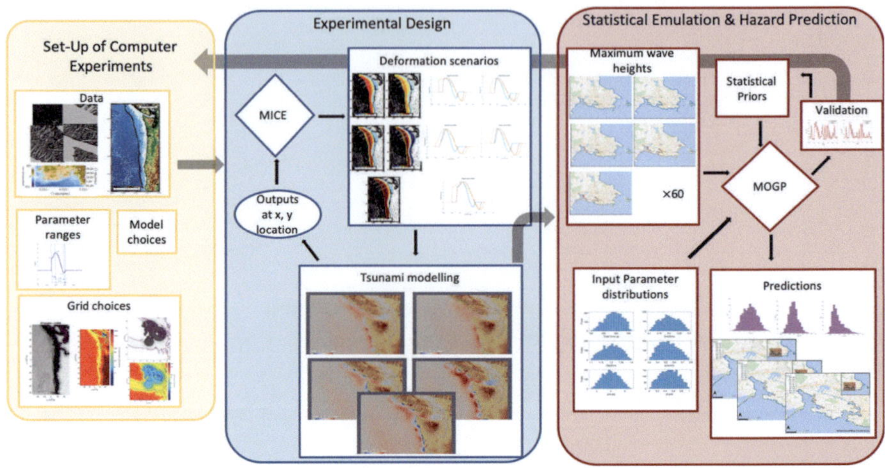

Fig. 20.2 The surrogate workflow characterized by three stages: stage 01 (yellow panel) study specification and set-up of the experiments, stage 02 (blue panel) study design and conduction of the numerical experiments and stage 03 (red panel) building of the emulators and their use for prediction. Image source: Salmanidou et al. (2021), licensed under CC-BY 4.0

study. To analyse the likelihoods of the events occurring over time, the distribution of the input parameters was linked to the magnitude-frequency distribution of the events. A Tapered Gutenberg-Richter (TGR) distribution was used to associate frequency of events with earthquake magnitudes, which led to a probabilistic assessment of the surrogate predictions within specific time frameworks.

Meteotsunami hazards in the Adriatic Sea. The mathematical framework of the surrogate-based APTHA is fully described in Denamiel et al. (2020). It is based on polynomial expansions that decompose into deterministic coefficients and random orthogonal bases. The coefficients, which are the projection of the meteotsunami surges onto each polynomial basis, are derived from a quadrature-based approximation using numerical simulations undertaken with hydrodynamic model forced only by the synthetic pressure disturbances (no wind, no tide). For the surrogate-based APTHA in the Adriatic Sea, a nonintrusive quadrature-based Pseudo-Spectral Approximation (PSA; Gerstner and Griebel 1998; Constantine et al. 2012) of the meteotsunami surge generalized Polynomial Chaos Expansions (gPCE; Soize and Ghanem 2004; Xiu and Karniadakis 2002) is used. The number and values of the synthetic atmospheric pressure conditions forcing the deterministic tsunami model only depend on the total order p of the polynomial decomposition and the chosen quadrature rule. In the Adriatic Sea, the meteotsunami surrogate model is built using (1) the delayed Gauss–Patterson nested rules (Patterson 1968; Holtz 2011) which keep a high precision but dramatically reduce the number of stochastic variable values needed for the quadrature compared to other rules, and (2) the barotropic version of the unstructured ADvance CIRCirculation model (ADCIRC; Dietrich et al. 2012). Practically, the equation of the synthetic MGW is directly coded within the ADCIRC model and the tsunami simulations are only performed for a 12-h period. The resolution of the ADCIRC unstructured mesh ranges from about 5 km in the deepest part of the Adriatic Sea to 10 m at the locations of interest. The ADCIRC model is used in parallel with 100 CPUs (i.e., with the mesh decomposed in 100 subdomains). The computation time can reach a maximum of 5 min for the strongest meteotsunami surges but is, on average, about 3 min. At the end, a good convergence of the PSA method and a good fit with the tsunami model are obtained for a total order $p = 6$ of the polynomial decomposition using 4161 synthetic atmospheric MGW conditions (Fig. 20.1).

20.6 Discussion

Three main advantages of surrogate-based PTHA are identified. First, surrogate models achieve fast and accurate uncertainty quantification analysis and provide rigorous accounting of the tsunami source uncertainty to the tsunami surges. Second, they can be developed and updated offline ahead of any emergency situation, in a data-driven manner using tsunami simulations that form a training database over time. Once trained, surrogate models can run online when the severe events are (about to) taking place at very high speeds and with nearly no computational cost. Third, surrogate models can be built with high-fidelity physics-based tsunami models too computationally demanding to be used in operational mode.

However, some limitations of surrogate-based PTHA are also identified. First, the models cannot extrapolate, the surrogate predictions are limited to the boundaries of the input parameter spaces used for the training. Outside those boundaries, predictions tend to be unreliable to be considered in the assessment. In addition, the surrogate-based PTHA fully relies on the quality of the models used in the deterministic simulations. These models generally have numerous limitations, including the bathymetry used to generate their grids/meshes and the oversimplifications made by neglecting the non-linear interactions with tidal and storm surge forcing as well as with sea-level rise due to on-going climate warming. Finally, the use of synthetic forcing over-simplifying the tsunami sources, can be another major limitation: the more realistic the synthetic forcing the more likely the number of stochastic variables to increase and, consequently, the more expansive the surrogate model.

In terms of maturity, both presented applications have demonstrated the interest of surrogate models for PTHA despite not accounting for the epistemic uncertainty. In the Adriatic Sea, the meteotsunami surrogate model is implemented within the prototype Croatian Meteotsunami Early Warning System (CMeEWS; see Adriatic APTHA recipe, R10) and was fully evaluated both in hindcast and operational modes (Denamiel et al. 2019b; Tojčić et al. 2021). Overall, it has proven to be extremely useful and reliable during recent events spanning between 2014 and 2020. In particular, during the 2020 multi-meteotsunami event, if the prototype CMeEWS had been fully operational, the coastal communities of Vela Luka, Stari Grad and Vrboska would have been warned, at least a day in advance, of the meteotsunami surges forecasted with the surrogate model (Tojčić et al. 2021). However, the meteotsunami surrogate model has not yet been used in the context of long-term hazard assessments as described in the "Climate Change" method (Sect. 4.5). The Cascadia model was not built for early warning; the scope was to provide a high-resolution hazard model that can be used as a step towards loss estimation in British Columbia. To support this scope, the model has been expanded in later iterations to predict tsunami hazard in more areas in the region.

Surrogate models are the poor man's solution for PTHA and other solutions can be envisioned as better alternatives. This includes but is not restricted to increasing the density and quality of observational networks, implementing ensemble modelling combined with data assimilation for atmospheric sources, using machine learning techniques. Another application can be found in catastrophe modelling where the surrogate predictions can form a part of the hazard module which forms an element towards a holistic risk assessment for loss predictions. Successful PTHA may thus combine deterministic models, observational networks and surrogate model approaches.

References

Beck J, Guillas S (2016) Sequential design with mutual information for computer experiments (mice): emulation of a tsunami model. SIAM/ASA J Uncertain Quantif 4:739–766. https://doi.org/10.1137/140989613

Constantine PG, Eldred MS, Phipps ET (2012) Sparse pseudospectral approximation method. Comput Methods Appl Mech Eng 229–232:1–12. https://doi.org/10.1016/j.cma.2012.03.019

Denamiel C, Šepić J, Vilibić I (2018) Impact of geomorphological changes to harbor resonance during meteotsunamis: the Vela Luka Bay test case. Pure Appl Geophys 175:3839–3859. https://doi.org/10.1007/s00024-018-1862-5

Denamiel C, Šepić J, Huan X, Bolzer C, Vilibić I (2019) Stochastic surrogate model for meteotsunami early warning system in the eastern Adriatic sea. J Geophys Res Oceans 124:8485–8499. https://doi.org/10.1029/2019JC015574

Denamiel C, Huan X, Šepić J, Vilibić I (2020) Uncertainty propagation using polynomial chaos expansions for extreme sea level hazard assessment: the case of the eastern Adriatic meteotsunamis. J Phys Oceanogr 50(4):1005–1021. https://doi.org/10.1175/JPO-D-19-0147.1D

Denamiel C, Huan X, Vilibić I (2021) Conceptual design of extreme sea-level early warning systems based on uncertainty quantification and engineering optimization methods. Front Mar Sci 8:650279. https://doi.org/10.3389/fmars.2021.650279

Dietrich JC, Tanaka S, Westerink JJ, Dawson CN, Luettich RA Jr, Zijlema M et al (2012) Performance of the unstructured-mesh, SWAN+ADCIRC model in computing hurricane waves and surge. J Sci Comput 52:468–497. https://doi.org/10.1007/s10915-011-9555-6

Formaggia L, Guadagnini A, Imperiali I, Lever V, Porta G, Riva M et al (2013) Global sensitivity analysis through polynomial chaos expansion of a basin-scale geochemical compaction model. Comput Geosci 17(1):25–42. https://doi.org/10.1007/s10596-012-9311-5

Gerstner T, Griebel M (1998) Numerical integration using sparse grids. Numer Algorithms 18:209–232. https://doi.org/10.1023/A:1019129717644

Giraldi L, Le Maître OP, Mandli KT, Dawson CN, Hoteit I, Knio OM (2017) Bayesian inference of earthquake parameters from buoy data using a polynomial chaos-based surrogate. Comput Geosci 21(4):683–699. https://doi.org/10.1007/s10596-017-9646-z

Holtz M (2011) Sparse grid quadrature. In: Sparse grid quadrature in high dimensions with applications in finance and insurance, 1st ed. lecture notes in computational science and engineering 77:51–76

Knio OM, Le Maître OP (2006) Uncertainty propagation in CFD using polynomial chaos decomposition. Fluid Dyn Res 38(9):616–640. https://doi.org/10.1016/j.fluiddyn.2005.12.003

Najm HN, Debusschere BJ, Marzouk YM, Widmer S, Le Maître OP (2009) Uncertainty quantification in chemical systems. Int J Numer Meth Eng 80:789–814. https://doi.org/10.1002/nme.2551

Patterson TNL (1968) The optimum addition of points to quadrature formulae. Math Comput 22:847–856. https://doi.org/10.1090/S0025-5718-68-99866-9

Reguly I, Giles D. Volna, Github. https://github.com/reguly/volna

Salmanidou DM, Beck J, Pazak P, Guillas S (2021) Probabilistic, high-resolution tsunami predictions in northern Cascadia by exploiting sequential design for efficient emulation. Nat Hazard 21(12):3789–3807. https://doi.org/10.5194/nhess-21-3789-2021

Satake K, Wang K, Atwater BF (2003) Fault slip and seismic moment of the 1700 Cascadia earthquake inferred from Japanese tsunami descriptions. J Geophys Res 108:2535. https://doi.org/10.1029/2003JB002521

Soize C, Ghanem RG (2004) Physical systems with random uncertainties: Chaos representations with arbitrary probability measure. SIAM J Sci Comput 26(2):395–410. https://doi.org/10.1137/S1064827503424505

Sraj I, Mandli K, Knio O, Dawson CN, Hoteit I (2014) Uncertainty quantification and inference of Manning's friction coefficients using DART buoy during the Tohoku tsunami. Ocean Model 83:82–97. https://doi.org/10.1016/j.ocemod.2014.09.001

Tojčić I, Denamiel C, Vilibić I (2021) Performance of the Adriatic early warning system during the multi-meteotsunami event of 11–19 May 2020: an assessment using energy banners. Nat Hazards Earth Syst Sci 21:2427–2446. https://doi.org/10.5194/nhess-21-2427-2021

Wang M, Wan Z., Huang Q (2016) A new uncertain analysis method for the prediction of acoustic field with random and interval parameters. Shock Vib 16. https://doi.org/10.1155/2016/3693262

Witter R, Zhang Y, Wang K-K, Priest G, Goldfinger C, Stimely L, English J, Ferro P (2013) Simulated tsunami inundation for a range of Cascadia megathrust earthquake scenarios at Bandon, Oregon, USA. Geosphere 9:1783–1803. https://doi.org/10.1130/GES00899.1

Xiu D, Karniadakis GE (2002) The Wiener-Askey polynomial chaos for stochastic differential equations. SIAM J Sci Comput 24(2):619–644. https://doi.org/10.1137/S1064827501387826

Chapter 21
R12—Multi-Source (Bayesian) PTHA in the Gulf of Naples, Italy

Anita Grezio and Jacopo Selva

21.1 Introduction

Tsunamis can be generated by multiple sources impacting the same area, but if only a single source is considered the hazard may be underestimated and/or biased. This study deals simultaneously with different types of tsunamigenic sources in a coherent and multidisciplinary probabilistic framework for the PTHA.

A comprehensive methodology takes into account PTHAs for submarine seismic sources (*SSSs*), submarine mass failures (*SMFs*), and pyroclastic density currents (*PDCs*). Eventually a Bayesian procedure merges the prior probabilities with historical data, mainly taken from catalogs, to provide a simultaneous evaluation of both aleatory and epistemic uncertainties.

This multi-source PTHA approach is relevant in all those sites where more than one tsunamigenic source exists and potentially interacts with others. This is the case of the Campania region in Italy, although the method is applicable to many other areas worldwide, where near-coast or submarine landslides are possible in addition to earthquake triggered tsunamis.

This recipe in full is presented in Grezio et al. (2020).

The present recipe includes the following ingredients:

- Framework FW1A (Probabilistic tsunami risk assessment; Sect. 2.2)
- Source selection for earthquakes, landslide and volcanic sources—Chap. 3
- Source modelling for earthquakes, landslide and volcanic sources—Chap. 4
- Tsunami modelling—Chap. 5

A. Grezio (✉)
Istituto Nazionale di Geofisica e Vulcanologia, Bologna, Italy
e-mail: anita.grezio@ingv.it

J. Selva
Dipartimento di Scienze della Terra, dell' Ambiente e delle Risorse (DiSTAR) , Università di Napoli Federico II, Naples, Italy
e-mail: jacopo.selva@unina.it

M. B. Sørensen et al. (eds.), *Probabilistic Tsunami Hazard and Risk Analysis*, Mathematics of Planet Earth 14, https://doi.org/10.1007/978-3-031-98115-9_21

21.2 Framework Used in Recipe

The framework for the multi-source PTHA is fully probabilistic (Fig. 21.1). The procedure evaluates separately the PTHAs for different source types (*SSSs, SMFs*, and *PDCs*) and assumes that these are independent. In each PTHA both source probabilities and conditional probabilities of inundation are quantified at the coastal sites of interest.

For the generic i-th source type, the source probabilities Pij represent the occurrence probability of the j-th individual source scenario. The natural variability is covered by multiple source scenarios j, with $j \in [1,\ldots, \text{NSSS}]$, $j \in [1,\ldots, \text{NSMF}]$ and $j \in [1,\ldots, \text{NPDC}]$ belonging to the different source types (i.e., $i = SSS, SMF, PDC$). Source probabilities are evaluated by multiplying the three independent probabilities: temporal, spatial, and the magnitude/size. So, the following event-tree strategy is implemented for the different tsunamigenic sources to reduce the number of simulations:

SSSs:

- *Node 1:* submarine earthquake occurrence. A Poissonian model is assumed for the next 50 years with mean annual rates calculated from the European-Mediterranean Earthquake Catalog (EMEC) for the last millennium using a Gaussian filter with correlation distance equal to 20 km for the spatial distribution of the simulated events.

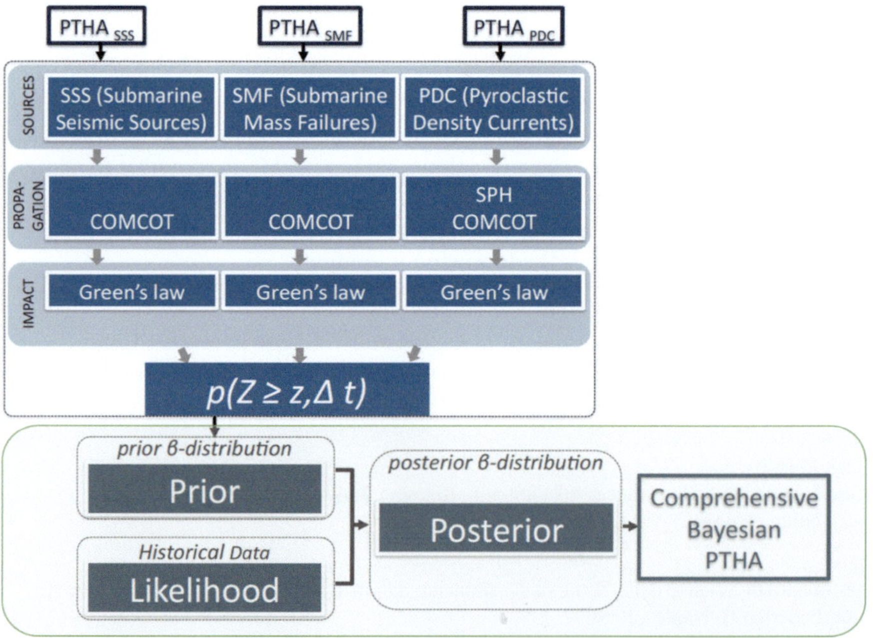

Fig. 21.1 Schema for a multi-source PTHA, green box represents the Bayesian analysis

- *Node 2:* magnitudes. The truncated Gutenberg-Richter relationship (Cornell and Vanmarcke 1969) is assumed.
- *Node 3:* fault depth. The geometrical centers of the faults are assumed equally probable at three possible depths (5, 10, and 15 km), although some depths are physically not possible for certain magnitudes and are discarded.
- *Node 4:* fault geometry (i.e., strike, dip, and rake angles). In each cell we assume only one combination of the angles with probability set to 1, based on historical seismicity.

SMFs:

- *Node 1:* submarine landslide occurrence. The probability of SMF in the next 50 years follows a Poissonian model with a mean annual rate of 0.0004/years (i.e., 2500 yr return period) estimated by geological investigations and instrumental data in the area.
- *Node 2:* spatial probabilities based on a susceptibility analysis. Spatial weights are introduced to represent the failure probability of the submarine masses. Some predominant geomorphological features are considered as controlling factors (e.g., slope angles and depths of the center of mass, which affect the propensity of failure; scars of past mass failures, which left unstable margins; and earthquake occurrences, which increase slide instability). Those controlling factors are assigned to source locations as weights when the detailed geotechnical data are not available for an adequate slope stability analysis. Because the earthquake load increases instability and the consequent SMFs probability of occurrence, an additional stability factor related to the horizontal peak ground acceleration is introduced.
- *Node 3:* slide volume. A frequency-size power law is calculated in order to weight 5 selected SMFs classes with a *b*-value based on the background studies in the area.

PDCs:

- *Node 1:* eruption event. The probability of at least one eruption at Somma-Vesuvio in the next 50 years is calculated using a Poissonian occurrence model with a mean annual rate of 8×10^{-3}/yr, estimated from the literature.
- *Node 2:* eruption vent. All eruptions are assumed to be located at the summit of the Somma-Vesuvio volcano.
- *Node 3:* eruption size. The probability of the eruption size was taken from literature, based on categorization into small, medium, and large events. The probability density function for the total erupted mass is based on a frequency-size power law, continuous on the whole spectrum of possible erupted mass values. All simulations within a given size class are assumed equiprobable. The probability distribution among the three size classes is instead set according to a power law based on the frequency of such classes in the geological record of the Somma-Vesuvius.
- *Node 4:* dense flow. The probability of PDC generation and the frequency of the arrival at the coast by PDCs is based on the statistics and simulations provided by previous studies.

Finally, the source probabilities for each source scenario are combined with the corresponding conditional probabilities of the simulated tsunami heights at different thresholds $z \in$ [0.1 m,..., 5 m] at the coastal sites. For simplicity, in this study no uncertainty was modeled for the generation, propagation and impact at coastal sites. Consequently, the exceedance probabilities are evaluated by stepwise Heaviside functions (sometimes called identity functions, Grezio et al. 2017).

The Bayesian Analysis (green box in Figure 21.1) combines all source types for the final and comprehensive PTHA. The prior β-distribution is computed using the probability $p(Z \geq z, \Delta t)$ from the multi-source analysis; the likelihood is computed on the basis of the historical tsunamis from the Italian tsunami catalog, counting time windows with or without tsunami observations to input a Binomial model; the posterior is a β-distribution analytically evaluated given the conjugacy of the prior and the likelihood.

The largest uncertainties are related to the unknown locations of the submarine sources (SSSs and SMFs), their magnitudes and occurrence frequencies. Also, a rather limited coverage of the natural variability is used to reduce the computational effort and does not effectively explore the aleatory uncertainty of the phenomena. These limitations are somehow compensated for by an increase of epistemic uncertainty in the Bayesian Analysis.

The use of the event-tree strategy represents an advantage, although the large number of scenarios required yields a computational limitation. Also, we did not account for other potential volcanic sources (e.g., the underwater explosions at Campi Flegrei, see Recipe R9/Chap. 18), nor did we model potential interaction among the different types of sources (e.g., earthquake-triggered landslides).

21.3 Source Identification and Characterisation Used in the Recipe

The potential tsunamigenic sources are identified in the region on the basis of the regional seismotectonics and the geologic features. The source identification and the relative magnitudes and characteristic parameters were located in the centre of a gridded domain.

SSSs are located at the centre of a 0.5° × 0.5° grid covering the entire Tyrrhenian Sea, with a finer inner 0.25° × 0.25° grid at the Gulf of Naples. Magnitudes were grouped in four classes in the [6.0–8.0] interval. The fault parameters (length, width, and slip) and related variability for each magnitude were calculated by the empirical relationships by Wells and Coppersmith (1994) and were randomly sampled twice to introduce a certain variability in the source data set. The fault geometry (i.e., strike, dip, and rake angles) was predetermined for each grid element from cumulative moment tensors obtained with a summation of source parameters of the earthquakes that occurred in the grid element during the past 100 years.

SMFs are located at the centre of 1' × 1' cells, due to the limited amount of available data in the Tyrrhenian Sea. Five volume classes in the range [5 × 10^5–5 × 10^{10}] m^3 were chosen on the basis of the past regional events and were assigned to each cell. The lower class is set according to a known subaerial mass failure (at Stromboli volcano in December 2002). The upper class represents the sizes of the largest historical slides that occurred in the area. A simplified block describes the mass failures deducing the relative geometry parameters (length, thickness, and width) from the volume classes. SMFs moving over plane slopes were schematized in two idealized types: translational failures, which are underwater slides, and rotational failures, which are slumps.

PDCs entering the sea were only considered from the summit crater of the Somma-Vesuvio. A set of 64 TITAN2D simulations of Tierz et al. (2018) for each size class, accounting for the bed-friction angle, indicates that small eruptive size classes do not generate dense flow that reaches the sea, whereas medium and large-size classes do reach the coastal sites.

A rather large number of tsunamigenic events was considered to explore aleatory uncertainty, despite the scarce information on the submarine earthquake locations and faults in the region and no past records of pyroclastic flows. No alternative modeling was developed, and the significant epistemic uncertainty was quantified using a Bayesian Analysis and assuming the largest possible uncertainty in the prior model.

This approach allows the use of larger cells of the gridded domain for the far sources in the Tyrrhenian Sea and a finer grid for the near-coast tsunamigenic sources to differentiate the directivity of the initial tsunami waves. The limited time coverage of the instrumental database represents the major limitation for the setting of the source datasets. Volcanic sources are limited to the *PDCs* from Somma-Vesuvio.

21.4 Source Mechanics Modelling Used in Recipe

The seafloor deformations of *SSSs* are modeled using an Okada (1992) model assuming a rectangular fault in a semi-infinite elastic half plane.

The seafloor deformation of the *SMFs* is represented by an initial block with a transient seafloor motion in the time of the temporal variation of the water depth.

The potentially tsunamigenic sliding wedges entering the sea of the *PDCs* have the flow thickness ≥ 1 m and the specific weight set by a generic background density value (1,250 kg/m^3) on an average coastal slope. This input at the coastal sites is provided to the numerical model SPHysics (Smoothed Particle Hydrodynamics theory). During a transient time, each sliding wedge moves the sea water particles producing surface local peaks which reconstruct 2-D waveform by the use of smoothing routines. Finally, the width of pyroclastic flow along the coast sets the 3D initial sea surface disturbance to be propagated.

For all sources, the simplest possible description was adopted to limit computational times. This implies that a potentially large number of sources of uncertainty may be neglected with the selected procedures. As discussed above, this significant epistemic uncertainty is here somehow compensated by assuming the largest possible uncertainty in the prior model in the Bayesian Analysis.

21.5 Tsunami Modelling

The initial tsunami waves are propagated near the coast by the COMCOT (Cornell Multi-grid Coupled Tsunami) numerical model (Wang 2009), which uses an explicit leap-frog finite difference scheme. Multi-size nested grids were not used because of the limited extension of the domain. The free surface elevation and the volume flux are staggered both in time and in space. The model runs in two modes that resolve in spherical coordinates: the Boussinesq Shallow Water Equations (SWEs) and the weakly nonlinear SWEs, when the tsunami waves propagate from offshore to near-shore areas, and the nonlinear effects significantly increase. COMCOT uses spherical coordinates and GEBCO (General Bathymetric Chart of the Oceans) bathymetry with 30-arc second intervals of latitude and longitude for the *SSS* and *SMF* simulations, but a reduced domain and a higher bathymetry resolution (grid size of about 50 m) for the *PDC* simulations.

For the *SSSs*, COMCOT computes the seafloor deformation imparted by the fault parameters (latitude and longitude of the epicenter, focal depth, length and width of the fault plane, fault slip, strike, dip, and rake angles). The model assumes a rectangular fault in a semi-infinite elastic half plane with an idealized representation of the interface between two colliding crustal blocks during the earthquake event. The relative motion of the blocks is simulated as an instant seafloor deformation neglecting the duration of the rupturing process. The seafloor disturbance generates the initial tsunami wave to be propagated.

For the *SMFs*, COMCOT simulates the seafloor deformation using the slide parameters (starting and stopping coordinates of the center of mass, typical slope angle along the sliding path, length, width, and thickness of the sliding volume) similarly to the seismic case, but with the additional input of the duration of the transient seafloor motion (time of the temporal variation of the water depth). The longwave assumption is satisfied in the numerical model implementation.

For *PDCs*, the 3D initial sea surface disturbances previously computed by the numerical model SPHysics are provided as input to COMCOT to be propagated.

The wave amplitude is computed near the coast using Green's law $h_c = (d_{iso}/d_c)^{1/4}\, h_{iso}$, where h and d the tsunami amplitude (maximum water surface elevation) and the sea depth, respectively; whereas the subscripts *iso* and *c* indicate the 50 m and 1 m isoline depths, respectively.

The h_c values are used to estimate the tsunami parameter Z overcoming the thresholds z at the coastal sites.

Uncertainty in tsunami modelling for relatively short wavelengths, like the one generated by non-seismic sources, is challenging. Also the Green's law empirical relationship assumes the wave front is almost parallel to the coastline where the wave refraction is toward the directions of the down-gradient depth and the convergence and/or divergence of the rays is neglected. With this assumption the wave amplitude variability due to the wave refraction is considered a second-order problem in the probabilistic analysis. Also, this energy conservation law estimates the evolution of the initial tsunami wave at the coastal line from an appropriate isoline where the non-linearities are negligible.

Here we selected the simplest method, which is uniformly applied to all tsunami sources, instead of adopting more sophisticated amplification models that account for more specific coastal characteristics and allow estimating the uncertainty. Also, we note that for local hazard quantifications it would be preferable to consider simulations that include inundation, eventually coupled with strategies to reduce the computational effort.

The resolution of the domain allows the use of (a) the first level model grid, without the sub-level grids, and (b) the nonlinear shallow water mode, with the Manning's roughness coefficient set equal to 0.013. This model setting provides the further advantage of reaching a good compromise between the computer running time and the number of simulations required for the statistical analysis.

21.6 Discussion

The multi-source PTHA uses a procedure that traces and weights the different tsunami sources with varying probability of occurrence, with the scope of reducing possible biases and underestimations that may arise by focusing on a single tsunamigenic source.

The procedure provides a first-order prioritization of the different sources in the long-term. Moreover, the Bayesian Analysis merges computational hazard quantification (based on source-tsunami simulations) and past data, including appropriately the quantification of the epistemic uncertainty.

To our knowledge, this analysis represents a unique example of multi-source PTHA within a comprehensive framework.

References

Cornell CA, Vanmarcke EH (1969) The major influences on seismic risk. In: Proceedings of the 4th world conference on earthquake engineering, Santiago, Chile

Grezio A, Babeyko A, Baptista MA, Behrens J, Costa A, Davies G et al (2017) Probabilistic tsunami hazard analysis: multiple sources and global applications. Rev Geophys 55:1158–1198. https://doi.org/10.1002/2017RG000579

Grezio A, Cinti FR, Costa A, Faenza L, Perfetti P, Pierdominici S, Pondrelli S, Sandri L, Tierz P, Tonini R, Selva J (2020) Multisource Bayesian probabilistic tsunami hazard analysis for the Gulf of Naples (Italy). JGR. https://doi.org/10.1029/2019JC015373

Okada Y (1992) Internal deformation due to shear and tensile faults in a half-space. Bull Seismol Soc Am 82:1018–1040

Paris R, Ulvrova M, Selva J, Brizuela B, Costa A, Grezio A et al (2019) Probabilistic tsunami hazard analysis for tsunamis generated by subacqueos volcanic explosions in the Campi Flegrei caldera (Italy). J Volcanol Geoth Res 379:106–116. https://doi.org/10.1016/j.jvolgeores.2019.05.010

Tierz P, Stefanescu ER, Sandri L, Sulpizio R, Valentine GA, Marzocchi W, Patra AK (2018) Towards quantitative volcanic risk of pyroclastic density currents: probabilistic hazard curves and maps around Somma-Vesuvius (Italy). J Geophys Res: Solid Earth 123:6299–6317. https://doi.org/10.1029/2017JB015383

Wang X (2009) User manual for COMCOT version 1.7 (First Draft). In: Liu PL-F, Woo S-B, Cho Y-S (eds) Computer programs for tsunami propagation and inundation, edited by (1998), Cornell University

Wells DL, Coppersmith KJ (1994) New empirical relationships among magnitude, rupture length, rupture, width, rupture area and surface displacement. Bull Seismol Soc Am 84:974–1002

Chapter 22
R13—Global Risk Model 2015—Tsunami Component—Global Risk and Loss Computation

Finn Løvholt, Mario A. Salgado-Gálvez, and Gareth Davies

22.1 Introduction

The purpose of this study is to provide a global tsunami risk analysis quantifying direct economic losses from building damage caused by earthquake-triggered tsunamis. This necessitates an event-based PTHA because of the need to link modelled losses to scenario probabilities. The PTRA includes the following ingredients:

- Event-based probabilistic risk analysis using discrete scenario modelling and probabilities aggregated to losses (FW1A), Sect. 2.2
- Source selection for earthquakes sources—Sect. 3.2
- Source modelling for earthquake sources—Sect. 4.2
- Tsunami propagation modelling and amplification—Chap. 5
- Exposure databases—Chap. 6
- Vulnerability models—Chap. 7

F. Løvholt (✉)
Offshore Energy, Norwegian Geotechnical Institute, Oslo, Norway
e-mail: finn.lovholt@ngi.no

M. A. Salgado-Gálvez
CIMNE, Barcelona, Spain
e-mail: mario.sal.gal@gmail.com

ERN International, Mexico City, Mexico

G. Davies
Geoscience Australia, Symonston, Australia
e-mail: gareth.davies@ga.gov.au

M. B. Sørensen et al. (eds.), *Probabilistic Tsunami Hazard and Risk Analysis*, Mathematics of Planet Earth 14, https://doi.org/10.1007/978-3-031-98115-9_22

The recipe is suitable for global or regional single hazard analysis, (i.e. with need for covering large study areas when the level of model granularity is low). As the analysis produced the same loss metrics as for other natural hazards, the tsunami risk metrics can directly compared and integrated with those from other perils. Such models are normally used by high-level stakeholders such as UN agencies, DG-ECHO in the EU, and potentially at country level, for instance for regional statistics, interregional comparisons, inter-hazard comparison and political prioritizations. This type of study is not intended to be used in local decision-making processes. The PTHA is described in detail by Davies et al. (2018), whereas the full details of PTRA (hazard analysis with less rigour) are presented in greater detail by Løvholt et al. (2015).

22.2 Framework Used in Recipe

This recipe uses framework FW1A (Fully simulation-based risk assessment; Sect. 2.2), a full probabilistic event tree framework, to propagate uncertainties. The recipe takes as input scenario earthquake probabilities aggregated to compute Loss Exceedance Curves (LEC) and related metrics (e.g., Average Annual Loss—AAL, and return period specific losses such as Probable Maximum Losses—PML). It includes three different Moment Magnitude (M_w) weights in a logic-tree for the source branches. For the temporal probabilities representing the hazard, a Poisson process (time independent) is assumed. The hazard output is the mean value and uncertainties of the Maximum Tsunami Inundation Height (MIH) for given return periods, as well as flooded areas for all the scenarios in the study, later used to estimate tsunami risk for the mean scenario return periods. The outcome of the probabilistic risk analysis is a LEC country-by-country, from which different risk metrics such as the AAL and PML are derived.

The advantage of using the framework is both the transparency a probabilistic study gives, but also the possibility of quantifying losses based on probability estimates that allows a full PTRA with results that can be directly compared with other hazards (e.g., earthquakes, riverine floods, hurricane wind and storm surge, in the framework of the Global Assessment Report 2015—GAR15). The limitation of the framework used here is the lack of detailed local analysis.

22.3 Source Selection and Activity Rate Characterization Used in Recipe

This recipe represents the global tsunami hazard due to earthquake sources (Sect. 3.2), using the following source selection models:

- Fault sources are used. The Slab1.0 model (Hayes et al. 2012) is used for representing irregular rectangles as unit sources where available, with some local adaptation. A constant shear modulus, assuming an average rigidity of 30 GPa combined with the scaling relationship of Strasser et al. (2010) is used for setting up the source geometries in a synthetic catalogue of earthquake events. All scenarios used in this framework have uniform slip and all subduction zones globally are included. In addition, local faults are included in certain areas to supplement the global model.
- The GEM faulted earth model (Berryman et al. 2015) for activity rates is used to define the rates for the individual earthquake sources. Sources down to M_w 7.5 are simulated, whereas smaller magnitude sources are not covered. Simulation results for different sources with different rates are aggregated to probabilistic hazard.
- The probabilistic treatment and parameter representation is implemented in R.

The advantage of this source representation is among others the simplicity of using unit sources and uniform slip. The use of unit sources also reduces simulation times. Disadvantage is lack of inclusion of heterogeneous slip that can give some bias (underestimation) in the probability distributions (e.g. Davies 2019). Also, not including smaller magnitude earthquakes or other sources than earthquakes limits the analysis.

22.4 Source Mechanics Modelling Used in Recipe

An Okada type analytical model is used to simulate the seafloor response for each unit source (layered approach). Aggregated synthetic sources are created by adding several unit sources and adjusting the slip. This is the simplest type of seafloor deformation model among those described in Sect. 4.2. The strength and weaknesses of this method are shared with those described for the source selection procedure above.

22.5 Tsunami Modelling Used in Recipe

The model uses the following methods and steps as described in Chap. 5 following all the different steps from data preparation, tsunami generation, tsunami modelling, and inundation modelling.

- For the first step on data preparation (see Sect. 5.2), a global GEBCO bathymetry is prepared as input to provide the bathymetric data to the tsunami model.

- For the second step (see Sect. 5.3, interface model between the earthquake unit source and tsunami model) a Kajiura filter is used to smooth the co-seismic displacement provided by the Okada solution described above. This is used as the initial condition for tsunami modelling.
- For the third step, for the tsunami propagation, a simple (linear) shallow water model (Subsection 5.4.2.1) is used for modelling the tsunami propagation, taking the smoothed unit sources as input (Thio et al. 2007). A staggered finite difference scheme is used for modeling the tsunami propagation numerically (see Subsection 5.4.4). Time series of the tsunami surface elevations are extracted at synthetic gauge locations near shore (at approximately 100 m depth) as outputs. HPC resources are needed to run all the source simulations and aggregate the results from the different scenarios.
- For the fourth step, simulating the tsunami inundation, the approximative amplification factor method is used (Løvholt et al. 2012). This method relates the offshore tsunami surface elevations from the time series gauges to Maximum Tsunami Inundation Heights (MIHs). Amplification factors are found through look up tables using wave periods and polarity as input. As time series gauges are located at different water depths than the 50 m depth used in generating the amp factors, the derived maximum heights at the gauge points are corrected to 50 m depth using Green's Law before the factors are applied. The uncertainty in the inundation height is also quantified assuming a lognormal distribution of the MIHs (Davies et al. 2018), but this uncertainty information is only used in the hazard model and not propagated to the exposure and loss modelling. For the tsunami inundation modelling in this recipe, the MIHs are projected from the offshore points to the shoreline through nearest neighbourhood interpolation. Then, open-source topography data (in this case SRTM was used) is combined with the shoreline MIH to estimate the inundated area for each scenario. A quadratic friction law is used as an analytical model to reduce the tsunami inundation height landward (Løvholt et al. 2014). The topography intersection with the MIH then defines the tsunami trim line. The inundation procedure is implemented in ArcGIS.

Advantages of this simulation procedure: The simplicity (linear shallow water tsunami model and amplification factors) is necessary to make a global risk analysis feasible. Inclusion of an uncertainty-based amplification factor gives a statistically based MIH representation globally, despite not being locally accurate. Disadvantages of this simulation procedure: Lack of inclusion of local details that may be important for the hazard. The statistical methods have very large prediction errors (low accuracy), which in principle could be improved via higher resolution tsunami modelling combined with a representation of tsunami source heterogeneity. Use of open-source topo-bathymetric data can restrict accuracy in particular in the inundation modelling, and artificial effects of vegetation and buildings can block inundation in the model due to imperfections in the global open-source data. Higher order hydrodynamics such as frequency dispersion could not be included, but this is likely less important than the factors discussed above.

22.6 Exposure Modelling

The inundation model is superimposed on the SRTM 30 m using ArcGIS (https://www.earthdata.nasa.gov/sensors/srtm). The inundated area is taken as input to calculate exposure for various assets at risk. Different sets of exposure data are used, including the exposure data for the global economy stock from UNEP Grid representing the replacement cost of buildings (de Bono et al. 2014), population (LandScan 2012), and critical facilities (UNEP-GRID 2011). The exposure integration is carried out by combining inundated areas from the tsunami modelling with exposure datasets in ArcGIS. The final exposure database is determined in a scenario-by-scenario manner.

The advantage of this procedure is that it can cover the globe and does not need high resolution or proprietary data. By integrating exposure on a scenario-by-scenario basis, the procedure allows using the exposure integration into losses in quantitative risk analysis. The disadvantage is that the SRTM 30 m data applied at the time of the implementation is very inaccurate, in particular where vegetation and buildings are present and may obstruct tsunami inundation in an artificial manner. Since data quality since then has increased a lot, it may be possible to achieve greater accuracy in future studies by updating the topography data, but also through updating the exposure database.

22.7 Vulnerability Modelling

Tsunami hazard and exposure are convoluted with vulnerability models to calculate direct losses due to building damage (Løvholt et al. 2015). Although exposure of critical facilities and human population is calculated, related losses are not calculated in the GAR as reliable global vulnerability models for these two elements at risk are not available. Two different vulnerability datasets are used, one based on an analytical vulnerability model (Sect. 7.3) and the other based on an empirical fragility model (Sect. 7.2). Each vulnerability dataset includs building type specific vulnerability functions, providing a relationship between flow depth and the expected loss.

The analytical model is based on input from flood risk management and is established in an expert workshop facilitated by UNISDR (now UNDRR) and including regional experts from Southeast Asia. The expert workshop determined which models to use and where to apply them. The expert-based models are employed for the Southeast Asia region only. The empirical model is based on the fragility observations from Suppasri et al. (2013) and applied for the rest of the globe.

CAPRA (Cardona et al. 2012) is the risk assessment tool and allows the estimation of LEC, AAL and PML for all tsunami exposed countries on the globe. Vulnerability functions described herein are the best available at the time they are employed. Shortcomings of the study are the heterogeneity in the vulnerability models, the lack of site-specific models, and the lack of using the uncertainties associated with the vulnerability functions in the loss calculations (only mean values of the curves are used).

22.8 Risk Communication

A compact and aggregated presentation of the results is presented in several of the biannually issued Global Assessment Reports by UNDRR, where the most important was GAR15, which integrated the tsunami model into a global multi-risk analysis together with other natural hazards in terms of comparable loss metrics. The presentation of the hazard is performed by the UNDRR and integrated with other hazards to focus on the integral risk rather than for individual hazards. In addition, the tsunami risk metrics are included for creating country profiles and results of the analysis are made available online through the GAR Atlas repository, although not all services have been maintained by UNDRR since 2019.

22.9 Discussion

The tsunami risk model for the GAR15 naturally necessitates a highly elaborate study, which requires:

1. A global probabilistic analysis of earthquake sources.
2. Integration of loss metrics allowing comparison with risk from other hazards. Still the low probability—high impact nature of the tsunami hazard necessitates use of certain loss metrics that are not used for more frequent hazards (e.g., AAL and PML).
3. Need to carry out inundation estimates, exposure analysis, and risk integration.
4. Use of open-source data owing to the global character of the analysis.

Due to the need to simplify the analysis and scope, in order to make the study feasible, the following limitations are noted:

- Heterogeneous (random) slip is not included for the earthquake sources.
- Only earthquake sources are considered. In areas where non-seismic sources are significant this implies that the tsunami hazard and risk may be underestimated.
- The amplification factors are not accurate for estimating MIH, and therefore have large uncertainties associated. Local analyses such as NLSW modelling would provide more accurate estimates, but their application is generally infeasible for global PTHA.
- Use of open-source topography for inundation and exposure estimation is likely one of the largest source of uncertainty. In such datasets artificial obstacles can reduce inundation and lead to underestimation of the exposure and losses.

- Only direct economic losses are considered in the study. The metric Urban Reduced Capital is used as the loss metric, which is particular for the GAR study. Mortality losses are difficult to include as they are likely to strongly vary from country to country, and sufficiently elaborate data for this is not available.

References

Berryman K, Wallace L, Hayes GP, Bird P, Wang K, Basili R et al (2015) The GEM faulted earth subduction interface characterisation project, version 2.0. http://itic.ioc-unesco.org/index.php?option=com_oe&task=viewDocumentRecord&docID=15688

Cardona OD, Ordaz M, Reinoso E, Yamín LE, Barbat AH (2012) CAPRA—comprehensive approach to probabilistic risk assessment: international initiative for risk management effectiveness. In: Proceedings of the 15th world conference on earthquake engineering. Lisbon, Portugal

Davies G, Griffin J, Løvholt F, Glimsdal S, Harbitz CB, Thio HK et al (2018) A global probabilistic tsunami hazard assessment from earthquake sources. Geol Soc Lond Spec Publ 456:219–244. https://doi.org/10.1144/sp456.5

Davies G (2019) Tsunami variability from uncalibrated stochastic earthquake models: tests against deep ocean observations 2006–2016. Geophys J Int 218(3):1939–1960. https://doi.org/10.1093/gji/ggz260

De Bono A, Chatenoux B (2014) A global exposure model for GAR 2015. Input paper for the global assessment report on disaster risk reduction 2015. http://www.preventionweb.net/english/hyogo/gar/2015/en/home/documents.html#contributing_papers

Hayes GP, Wald DJ, Johnson RL (2012) Slab1,0: a three-dimensional model of global subduction zone geometries. J Geophys Res 117. https://doi.org/10.1029/2011JB008524

LandScan (2012) High resolution global population data set # UT-Battelle, LLC, Operator of Oak Ridge National Laboratory, USA. Dataset is available upon demand to ONRL

Løvholt F, Glimsdal S, Harbitz CB, Zamora N, Nadim F, Peduzzi P et al (2012) Tsunami hazard and exposure on the global scale. Earth Sci Rev 110(1–4):58–73

Løvholt F, Glimsdal S, Harbitz CB, Horspool N, Smebye H, De Bono A, Nadim F (2014) Global tsunami hazard and exposure due to large co-seismic slip. Int J Disaster Risk Reduct 10:406–418

Løvholt F, Griffin J, Salgado-Gálvez MA (2015) Tsunami hazard and risk assessment on the global scale. Encyclopedia of complexity and systems science, complexity in tsunamis, volcanoes, and their hazards, pp 213–246

Strasser FO, Arango MC, Bommer JJ (2010) Scaling of the source dimensions of interface and intraslab subduction-zone earthquakes with moment magnitude. Seismol Res Lett 81(6):941–950

Suppasri A, Mas E, Charvet I, Gunasekera R, Imai K, Fukutani Y et al (2013) Building damage characteristics based on surveyed data and fragility curves of the 2011 Great East Japan tsunami. Nat Hazards 66:319–341

Thio HK, Somerville P, Ichinose G (2007) Probabilistic analysis of strong ground motion and tsunami hazards in Southeast Asia. J Earthq Tsunami 1(02):119–137

UNEP/GRID (2011) Nuclear power stations of the world, geodata-set generated from various sources, situation as of December 2010

BY

Chapter 23
R14—Event-Based Probabilistic Earthquake and Tsunami Risk Assessment—Application in Acapulco (Mexico) and Callao (Peru)

Mario A. Salgado-Gálvez, Mario Ordaz, and Benjamín Huerta

23.1 Introduction

The purpose of these two separate risk assessments is to obtain prospective and probabilistic estimates of direct economic losses at urban scale in two study areas along the Pacific Coast, Acapulco (Mexico) and Callao (Peru), caused by simultaneous losses triggered by earthquakes and tsunamis. This study requires the development of an event-based PSHA to account for the future earthquake occurrence, from which a PTHA is derived allowing the obtention of a stochastic tsunami event-set. With the two event-sets, and making use of a loss combination rule, the simultaneous direct economic losses caused by the ground motion by earthquakes and the waves by the tsunamis are estimated. The PTRA described in this recipe includes the following ingredients:

- Simulation-based hazard convoluted with vulnerability and loss using numerical integration FW1B (Sect. 2.2)
- Source selection for earthquake sources (Sect. 3.2)
- Source modelling for earthquake sources (Sect. 4.2)
- Tsunami propagation modelling and amplification (Chap. 5)

M. A. Salgado-Gálvez (✉)
CIMNE, Barcelona, Spain
e-mail: mario.sal.gal@gmail.com

M. Ordaz
Instituto de Ingeniería, Universidad Nacional Autónoma de México (UNAM), Mexico City, Mexico
e-mail: MOrdazS@iingen.unam.mx

M. A. Salgado-Gálvez · B. Huerta
ERN International, Mexico City, Mexico
e-mail: benjamin.huerta@ern.com.mx

M. B. Sørensen et al. (eds.), *Probabilistic Tsunami Hazard and Risk Analysis*, Mathematics of Planet Earth 14, https://doi.org/10.1007/978-3-031-98115-9_23

- Exposure database (Chap. 6)
- Physical vulnerability (Chap. 7)

This recipe is suitable for local and regional tsunami hazard and risk assessments, which are typically used by local institutions and governments because of the high resolution of the different components (e.g., hazard and exposure). Results of these studies can be used in local decision-making processes, including risk-informed land use planning and the design of disaster risk financing instruments. Only tsunamis triggered by earthquakes are considered with this approach. The PSHA and PTHA frameworks are described in detail by Ordaz et al. (2021) and Jaimes et al. (2016), respectively. Details of the simultaneous loss combination rule can be found in Ordaz (2015) and the PTRA methodology in Ordaz et al. (2019). The same methodology can be used at coarser scale, as described by Løvholt et al. (2022).

23.2 Framework Used in Recipe

This recipe uses simulation-based hazard convoluted with vulnerability and loss using numerical integration (FW1B), in which different uncertainties related to the hazard, physical vulnerability and loss modelling for both earthquakes and tsunamis are identified, quantified and propagated through the assessment. Losses are estimated using R-CAPRA (ERN 2021) which implements the risk assessment framework by Ordaz (2000). Since simultaneous losses caused by the ground motion and tsunami waves are estimated, the combination rules by Ordaz (2015) are used to account for the simultaneous losses within a probabilistic framework. For this, the development of two stochastic event-sets, one for earthquakes and one for tsunamis, is required. The main results of these risk assessments are the Event Loss Tables (ELT) and the loss exceedance curve (LEC), from which other probabilistic risk metrics such as the average annual loss (AAL) and return-period specific losses (PML) can be derived.

A classic PSHA (Esteva 1967; Cornell 1968) has been developed for the study areas and it has been assumed that only subduction interface earthquakes can trigger tsunamis. The selected event-based approach (Ordaz 2000; Salgado-Gálvez et al. 2014) provides a direct relationship between the two stochastic event-sets generated, one for the PSHA and another for the PTHA. The chosen hazard intensity measures (IM) are spectral accelerations for the ground motion and flood depth for the tsunami.

23.3 Source Selection and Activity Rate Characterization Used in Recipe

This recipe allows describing the tsunami hazard because of interface earthquake sources, as described in Sect. 3.2. For this, the following data sources and parameters are used:

- 3D representation of the interface sources along the Pacific Coast using the Slab1.0 model (Hayes et al. 2012), as described in Salgado-Gálvez et al. (2023).
- A declustered and homogenised earthquake catalogue for the study area considering historical (1520–1899) and instrumental (1900 onwards) seismicity.
- Characterization of future earthquake occurrence through Poissonian occurrence models (i.e., modified Gutenberg-Richter, and characteristic earthquake).
- The treatment of the uncertainties on the seismicity parameters at each source is described in detail in Ordaz and Salgado-Gálvez (2020) and Ordaz et al. (2021).

23.4 Source Mechanics Modelling Used in Recipe

For each modelled interface earthquake included in the stochastic event-set derived from the PSHA, the sea floor displacement is calculated. Using the relationships by Hanks and Kanamori (1979) and Singh et al. (2012) for seismic moment and average fault slip, respectively, the rupture area for each event is calculated. Also, using the dip for each modelled earthquake, the vertical component of the slip, Uz, is computed as $Uz = U^{*}Sin(\delta)$, which is required for the tsunami propagation analysis. In this relation, U is the slip along the fault plane and δ is the dip.

23.5 Tsunami Modelling Used in Recipe

This recipe uses the following methods and steps, as described in Chap. 5, following all the different steps from data preparation, tsunami generation and tsunami modelling.

- For the first step on data preparation (see Sect. 5.2), global topographic and bathymetric datasets are initially used, adjusted and corrected using local models where available, for instance in Mexico.
- For the second step (see Sect. 5.3), for each synthetic interface earthquake in the event-set, the vertical component of the slip is estimated as the initial condition for tsunami modelling.
- For the third step, GeoClaw's numerical model for tsunami is used (Berger et al. 2011). For each synthetic tsunami, the wave heights are computed at different reference points along the shore. Flood depth is calculated as the difference between the wave height (at the closest reference point), and the terrain elevation (indicated by the topographic model).

23.6 Exposure Modelling

For the two study areas, and making use of updated and official cadastral databases, high-resolution building-by-building exposure models are developed. The exposure models include public and private buildings, for which a comprehensive characterization in terms of the most relevant structural parameters (i.e., construction material, construction year, number of stories, openings) is carried out. The economic appraisal of the exposed assets is developed in terms of the replacement cost, using typology specific reference values per constructed area.

The exposure resolution level must be consistent with that of the hazard modelling. Whereas for the case of earthquakes there are usually not abrupt changes in the hazard intensity values (at rock conditions) within distances between 100 to 300 m, for the case of tsunamis large variations in the hazard intensity values can be found within very short distances (1 to 10 m).

An advantage of the two exposure models developed is the use of official data which can be updated when new information is available, and the high-resolution that allows considering the exact location of each asset, both in plan and height conditions. More details about the development of the exposure database can be found in Ordaz et al. (2019).

23.7 Vulnerability Modelling

By using a simulation-based hazard convoluted with vulnerability and loss using numerical integration (FW1B), a continuous and probabilistic representation of the physical vulnerability of the assets in the study area is required. This is achieved by using vulnerability functions that are specific for each building typology. Because the losses caused by both earthquakes and tsunamis are quantified in this study, separate sets of vulnerability functions are needed for each peril.

In each case, the vulnerability functions provide a relationship between the IM, the expected loss, and an uncertainty measure. For earthquake losses, the chosen IM is the spectral acceleration at the fundamental period of each typology, whereas for tsunami losses, the IM is the flood depth. Earthquake vulnerability functions are derived from an analytical procedure and are subject to validations and calibrations given the relatively large earthquake loss databases for the study areas. Tsunami vulnerability functions are based on empirical models (Suppasri et al. 2013) and some adjustments based on expert criteria are implemented to reflect local conditions.

23.8 Discussion

The combined earthquake and tsunami risk models for Acapulco and Callao require:

- A regional probabilistic analysis of earthquake sources within a PSHA framework.
- The generation of two separate but related stochastic event-sets.
- The development of high-resolution exposure databases for buildings.
- Use of an open-source risk assessment tool and a simple simultaneous loss combination rule.
- Use of an event-based risk assessment framework that allows for obtaining typical catastrophe risk assessment outputs such as the ELTs, LEC, AAL and return period specific losses.

The following limitations are disclosed:

- Only earthquake-triggered tsunamis are accounted for.
- Only direct economic losses are estimated in this study.
- The PTHA method does not allow a comprehensive quantification of the tsunami hazard IM for each modelled event.
- Little validation and calibration of vulnerability fragility functions exist due to the lack of historical tsunami events in the study areas.
- No human impact or consequences (e.g., fatalities, injured, internally displaced people) are estimated.
- High dependence of the PTHA method on the quality and reliability of topographic and bathymetric data.
- Use of open-source topographic and bathymetric data is likely a large source of uncertainty, even if corrected using local data when available for these studies.

References

Berger MJ, George DL, LeVeque RJ, Mandli KT (2011) The GeoClaw software for depth-averaged flows with adaptive refinement. Adv Water Resour 34(9):1195–1206

Cornell CA (1968) Engineering seismic risk analysis. Bull Seismol Soc Am 58(5):1583–1606

ERN-Evaluación de Riesgos Naturales (2021) R-CAPRA. Program for probabilistic and multi-hazard risk assessment. Mexico City, Mexico

Esteva L (1967) Criterios para la construcción de espectros para diseño sísmico. In: Proceedings of the 3er Simposio Panamericano de Estructuras. Caracas, Venezuela

Hanks TC, Kanamori H (1979) A moment magnitude scale. J Geophys Res 84(5):2348–2350

Hayes GP, Wald DJ, Johnson RL (2012) Slab1.0: a three-dimensional model of global subduction zone geometries. J Geophys Res 113:B01302

Jaimes MA, Reinoso E, Ordaz M, Huerta B, Silva R, Rodríguez ME, JC, (2016) A new approach to probabilistic earthquake-induced tsunami risk assessment. Ocean Coast Manag 119:68–75

Løvholt F, Griffin J, Salgado-Gálvez MA (2022) Tsunami hazard and risk assessment on the global scale. Encyclopedia of Complexity and Systems Science

Ordaz M (2000) Metodología para la evaluación del riesgo sísmico enfocada a la gerencia de seguros por terremoto. Universidad Nacional Autónoma de México, Mexico City, Mexico

Ordaz M (2015) A simple probabilistic model to combine losses arising from the simultaneous occurrence of several hazards. Nat Hazards 76(1):389–396

Ordaz M, Salgado-Gálvez MA (2020) R-CRISIS validation and verification document. ERN International, Mexico City, Mexico

Ordaz M, Salgado-Gálvez MA, Huerta B, Rodríguez JC Avelar C (2019) Considering the impacts of simultaneous perils. Disaster Prev Manag 28(6):823–837

Ordaz M, Salgado-Gálvez MA, Giraldo S (2021) R-CRISIS: 35 years of continuous developments and improvements for probabilistic seismic hazard analysis. Bull Earthq Eng 19:2797–2816

Salgado-Gálvez MA, Zuloaga-Romero D, Bernal GA, Mora MG, Cardona OD (2014) Fully probabilistic seismic risk assessment considering local site effects for the portfolio of buildings in Medellín Colombia. Bull Earthq Eng 12(2):671–695

Salgado-Gálvez MA, Ordaz M, Singh SK, Pérez-Campos X, Huerta B, Bazzurro P, Fagà E (2023) A caribbean and central america seismic hazard model for sovereign parametric insurance coverage. Bull Seismol Soc Am 113(1):1–22

Singh SK, Pérez-Campos X, Iglesias A, Melgar D (2012) A method for rapid estimation of moment magnitude for early tsunami warning based on coastal GPS networks. Seismol Res Lett 83(2):516–530

Suppasri A, Mas E, Charvet I, Gunasekera R, Imai K, Fukutani Y, Abe Y, Imamura F (2013) Building damage characteristics based on surveyed data and fragility curves of the 2011 Great East Japan tsunami. Nat Hazards 66:319–341

Chapter 24
R15—Multi-Hazard Risk Assessment for Earthquakes and Tsunamis

Katsuichiro Goda and Raffaele De Risi

24.1 Introduction

This recipe describes a computational framework for conducting a multi-hazard risk assessment of a building portfolio subjected to megathrust subduction earthquakes and tsunamis. The simulation-based framework implements time-dependent earthquake occurrence models and characterises uncertain earthquake rupture models with variable fault geometry and heterogeneous slip distribution, all of which are common processes for ground motions and tsunami inundations. Integrating earthquake-tsunami hazard models with an exposure model for the built environment and seismic and tsunami fragility functions enables quantitative risk assessment. The final outputs from the developed tool include single-hazard and multi-hazard loss exceedance probability curves and related risk metrics (e.g., average annual loss and value at risk). The developed multi-hazard risk model is illustrated by considering a coastal community of Iwanuma City, Miyagi Prefecture, Japan, devastated by the 2011 Tohoku earthquake and tsunami. Detailed descriptions of the method and model implementation can be found in Goda and De Risi (2017) and Goda (2020).

K. Goda (✉)
Earth Sciences, Western University, London, ON, Canada
e-mail: kgoda2@uwo.ca

R. De Risi
School of Civil, Aerospace and Design Engineering, University of Bristol, Bristol, UK
e-mail: raffaele.derisi@bristol.ac.uk

M. B. Sørensen et al. (eds.), *Probabilistic Tsunami Hazard and Risk Analysis*, Mathematics of Planet Earth 14, https://doi.org/10.1007/978-3-031-98115-9_24

24.2 Framework

An overview of the computational procedure for quantifying time-dependent earthquake and tsunami risks due to megathrust subduction events is illustrated in Fig. 24.1. The numerical evaluation is based on Monte Carlo simulations, which facilitate the flexible incorporation of time-dependent earthquake occurrence models and stochastic earthquake rupture models, both of which affect earthquake and tsunami risks. Therefore, this recipe follows FW1A (see Chap. 2.2).

The first step of the portfolio loss model is generating stochastic seismic events for a specified duration of interest from a set of earthquake occurrence models. The occurrence of major tsunamigenic seismic events is modelled based on renewal processes

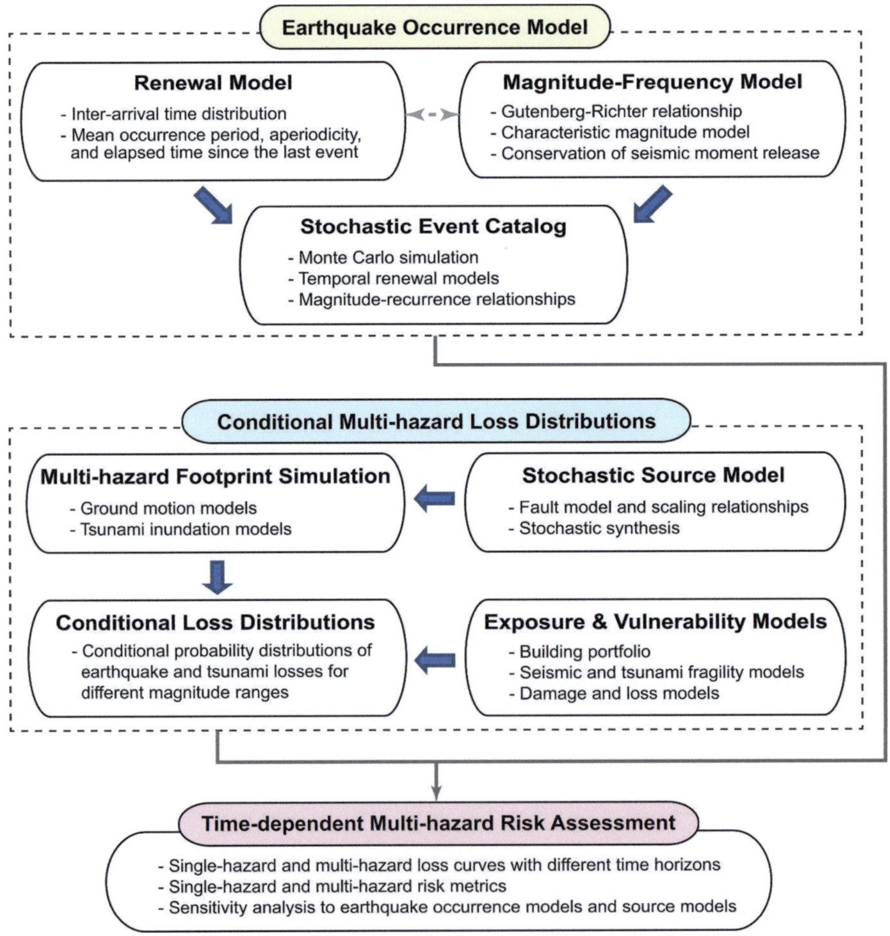

Fig. 24.1 Computational framework for multi-hazard risk assessment for earthquakes and Tsunamis

capturing quasi-periodic characteristics of major tsunamigenic earthquakes via non-exponential inter-arrival time distributions by keeping the memory of the last event occurrence. The inter-arrival time refers to the time differences between successive events. The magnitude of these major events is characterised by a set of magnitude recurrence models, such as truncated exponential model and characteristic model. The second step of the portfolio loss model is developing the conditional multi-hazard loss distribution. In deriving the conditional multi-hazard loss distributions for discretized magnitude ranges, numerous earthquake rupture models with variable geometry and location and heterogeneous earthquake slip distribution are generated. For each rupture scenario, the hazard footprints in terms of shaking intensity and tsunami inundation are computed at building locations of interest. Subsequently, seismic and tsunami fragility functions are applied to the building portfolio of interest. Also, ground-motion and tsunami damage severities are evaluated for individual buildings and aggregated portfolios to obtain the probability distribution functions of single-hazard and multi-hazard loss metrics for different magnitude ranges. When the multi-hazard loss is computed for each building, the larger loss between shaking and tsunami is adopted; this approach does not account for damage accumulation from cascading ground-motion and tsunami hazards. To combine the outputs from the earthquake occurrence model and the conditional multi-hazard loss distributions, for each event in the stochastic event catalogues, the single-hazard and multi-hazard loss values are sampled from the conditional loss distribution corresponding to the event magnitude. By repeating this loss sampling, the stochastic event catalogue is expanded to include information on the single-hazard and multi-hazard building portfolio loss and is used to derive the loss exceedance curves and related risk metrics for the building portfolio.

Regarding the consideration of multiple alternative models for earthquake occurrence (e.g., different inter-arrival time distributions and their parameters), the developed method is computationally efficient because of the decoupling of the earthquake occurrence model and the conditional multi-hazard loss distributions. Therefore, the process can be implemented to create a full logic tree to capture the epistemic uncertainty related to earthquake occurrence modelling. On the other hand, the development of conditional multi-hazard loss distributions involves significant computations (mainly for high-resolution tsunami inundation simulations). The computational effort can be reduced by fitting an analytical loss distribution (e.g., Pareto distribution) to a smaller sample size of simulated conditional loss data.

24.3 Source Identification and Characterisation

The Tohoku region of Japan is considered for the case study. The earthquake occurrence for large earthquakes is characterised by renewal models whose inter-arrival time distribution is specified by either exponential distribution, lognormal distribution, Weibull distribution, or Brownian Passage Time distribution (Matthews et al. 2002). The minimum and maximum earthquake magnitudes for tsunami generation are 7.5 and 9.1, respectively. The annual occurrence rate is evaluated using a seismic catalogue of the Tohoku region. Due to the lack of historical data other than the 2011 $M9.0$ event, the magnitude-frequency distribution is assumed to follow either the truncated exponential model with the mean recurrence period of 12.5 years for $M > 7.5$ events and b-value $= 0.9$ or the characteristic uniform magnitude models with different upper and lower magnitude bounds (Youngs and Coppersmith 1985). The moment release rate is conserved for different magnitude models. The mean recurrence time of the inter-arrival time is determined based on the magnitude-frequency analysis, whereas multiple plausible values are selected for the coefficient of variation of the inter-arrival time based on the literature.

24.4 Source Modelling

The fault model for the Tohoku region covers an offshore area of 650 km by 250 km. For each magnitude value, eight source parameters, i.e., fault width, fault length, mean slip, maximum slip, Box-Cox power parameter, correlation length along dip, correlation length along strike, and Hurst number, are generated using empirical prediction equations derived from many finite-fault models (Goda et al. 2016). Once the geometry and position of a stochastic source model are determined, a random heterogeneous slip distribution is generated using a Fourier integral method, where the amplitude spectrum is represented by von Kármán spectrum and the phase is random (Mai and Beroza 2002). To generate a slip distribution with realistic right-heavy tail features, the synthesized slip distribution is converted via Box-Cox power transformation. The transformed slip distribution is then adjusted to achieve suitable slip characteristics, such as mean slip and maximum slip. 500 stochastic source models are generated for eight magnitude ranges with 0.2 bin width spanning from $M7.5$ to $M9.1$. The synthesised earthquake source models reflect the possible variability of tsunamigenic earthquakes in geometry, fault location, and slip distribution.

24.5 Multi-Hazard Earthquake-Tsunami Modelling

For a given earthquake source model, shaking and inundation hazard intensities at building locations are evaluated using a ground motion model and solving nonlinear shallow water equations for initial boundary conditions of the sea surface caused by an earthquake rupture, respectively. An empirical model for the peak ground velocity (Morikawa et al. 2013) is adopted for ground motion intensities by considering local site conditions from the J-SHIS database (https://www.j-shis.bosai.go.jp/en/) and an intra-event spatial correlation model (Goda and Atkinson 2010). For tsunami inundation, nonlinear shallow water equations are solved using TUNAMI computer code (Goto et al. 1997) and considering the nested grid system of 1350-m, 450-m, 150-m, and 50-m resolutions. Tsunami simulations are conducted for all 4,000 stochastic sources for a duration of 2 h.

24.6 Exposure Modelling

Using a building dataset compiled by the Japanese Ministry of Land Infrastructure and Transportation (MLIT) for the post-2011-Tohoku tsunami damage assessment, an exposure model is developed for Iwanuma City. Building types are mainly low-rise wooden structures, for which well-calibrated seismic fragility and tsunami fragility models are available. The exposure model is composed of 6,152 wooden structures. To obtain estimates of the building costs for the wooden buildings in Iwanuma, the Japanese building cost information handbook and the MLIT building stock database are utilised. The mean and coefficient of variation of the unit replacement cost for wooden buildings are obtained as 1,600 US$/m^2 and 0.33, respectively; the mean and coefficient of variation of typical floor areas for wooden houses are determined as 130 m^2 and 0.33, respectively. Both unit cost and floor area are modelled by the lognormal distribution. Based on the above building cost information, the expected total cost of the 6,152 buildings in Iwanuma is 1.28 billion US$.

24.7 Vulnerability Modelling

Damage ratios for shaking and tsunami are estimated using seismic and tsunami fragility functions. The seismic fragility models are based on three empirical functions for low-rise wooden buildings in Japan (Yamaguchi and Yamazaki 2001; Midorikawa et al. 2011; Wu et al. 2016), whereas the tsunami fragility model is based on the tsunami damage data from the 2011 Tohoku earthquake and tsunami (De Risi et al. 2017). The damage states for shaking are defined as: partial damage, half collapse, and complete collapse, and the corresponding damage ratios are assigned as 0.03–0.2, 0.2–0.5, and 0.5–1.0, respectively (Kusaka et al. 2015). For tsunami damage, the following five damage states are considered: minor, moderate, extensive, complete, and collapse, together with the damage ratios of 0.03–0.1, 0.1–0.3, 0.3–0.5, 0.5–1.0, and 1.0, respectively. Subsequently, for each building, the final damage ratio is determined by taking the maximum of the estimated shaking and tsunami damage ratios. A multi-hazard loss value is calculated by sampling the value of the building replacement cost from the lognormal distribution and multiplying it by the final damage ratio.

24.8 Risk Communication

The outputs from the developed multi-hazard risk model for earthquakes and tsunamis are represented by an exceedance probability curve for a building portfolio loss and derived risk metrics to facilitate risk management decisions (Mitchell-Wallace et al. 2017). The developed framework is versatile and incorporates various uncertainties associated with earthquake occurrence modelling and earthquake rupture characterisation. Thereby, sensitivity analyses are enabled by considering different model components and parameters, including the mean recurrence period, coefficient of variation, elapsed time since the last major event, time window, and inter-arrival time distribution. The sensitivity analysis results highlight not only the significant influences of individual model components but also the impact of their interaction. They also indicate that the time-dependent earthquake occurrence model should be specified carefully and account for a range of parameter combinations in a logic tree (Fukutani et al. 2015). Another important observation from the numerical results is that the degree of epistemic uncertainty associated with temporal earthquake occurrence changes with time.

24.9 Discussion

The multi-hazard portfolio loss estimation framework for time-dependent earthquake and tsunami hazards has limitations. Some of the limitations of the developed multi-hazard risk model are listed below, which are worthy of future investigations.

- A simple renewal process is adopted in modelling interaction between spatiotemporal earthquake occurrence and magnitude. Other applicable models, such as time-predictable and slip-predictable models (Shimazaki and Nakata 1980; Kiremidjian and Anagnos 1984), can capture causal relationships between the inter-arrival time and magnitude.
- The framework only considers a single source region. Since other sources could cause major destruction to the building portfolio, they can be incorporated. Such additional sources include crustal and in-slab seismic sources for shaking damage and far-field tsunami sources for tsunami damage.
- Ground motion simulation is based on an empirical statistical model of ground motion intensity. When a high-resolution 3D crustal velocity model is available and more computational resources are dispensable, hybrid ground motion simulations can be implemented to incorporate more physical features in evaluating shaking intensity (Frankel et al. 2018).
- In assessing the damage extent of a building subjected to a sequence of shaking and tsunami hazards, the damage accumulation of the cascading hazards becomes an essential factor (Park et al. 2012). Such a damage accumulation model can be developed based on reliable numerical models of structures by subjecting these to a sequence of earthquake-tsunami loads.

References

De Risi R, Goda K, Yasuda T, Mori N (2017) Is flow velocity important in tsunami empirical fragility modeling? Earth-Sci Rev 166:64–82

Frankel A, Wirth E, Marafi N, Vidale J, Stephenson W (2018) Broadband synthetic seismograms for magnitude 9 earthquakes on the Cascadia megathrust based on 3D simulations and stochastic synthetics, Part 1: methodology and overall results. Bull Seismol Soc Am 108:2347–2369

Fukutani Y, Suppasri A, Imamura F (2015) Stochastic analysis and uncertainty assessment of tsunami wave height using a random source parameter model that targets a Tohoku-type earthquake fault. Stoch Env Res Risk Assess 29:1763–1779

Goda K, Atkinson GM (2010) Intraevent spatial correlation of ground-motion parameters using SK-net data. Bull Seismol Soc Am 100:3055–3067

Goda K, Yasuda T, Mori N, Maruyama T (2016) New scaling relationships of earthquake source parameters for stochastic tsunami simulation. Coast Eng J 58:1650010

Goda K, De Risi R (2017) Probabilistic tsunami loss estimation: stochastic earthquake scenario approach. Earthq Spectra 33(4):1301–1323. https://doi.org/10.1193/012617EQS019M

Goda K (2020) Multi-hazard portfolio loss estimation for time-dependent shaking and tsunami hazards. Frontiers Earth Sci 8:592444. https://doi.org/10.3389/feart.2020.592444

Goto C, Ogawa Y, Shuto N, Imamura F (1997) Numerical method of tsunami simulation with the leap-frog scheme, IOC Manual, UNESCO, No. 35, Paris, France

Kiremidjian AS, Anagnos T (1984) Stochastic slip-predictable model for earthquake occurrences. Bull Seismol Soc Am 74:739–755

Kusaka A, Ishida H, Torisawa K, Doi H, Yamada K (2015) Vulnerability functions in terms of ground motion characteristics for wooden houses evaluated by use of earthquake insurance experience. Arch Inst Jpn J Technol Des 21:527–532

Matthews MV, Ellsworth WL, Reasenberg PA (2002) A brownian model for recurrent earthquakes. Bull Seismol Soc Am 92:2233–2250

Midorikawa S, Ito Y, Miura H (2011) Vulnerability functions of buildings based on damage survey data of earthquakes after the 1995 Kobe earthquake. J Jpn Assoc Earthq Eng 11:34–47

Mitchell-Wallace K, Jones M, Hillier J, Foote M (2017) Natural catastrophe risk management and modelling: a practitioner's guide. Wiley-Blackwell, p 536

Morikawa N, Fujiwara H (2013) A new ground motion prediction equation for Japan applicable up to M9 mega-earthquake. J Disaster Res 8:878–888

Park P, van de Lindt JW, Cox D, Gupta R, Aguiniga F (2012) Successive earthquake-tsunami analysis to develop collapse fragilities. J Earthquake Eng 16:851–863

Shimazaki K, Nakata T (1980) Time-predictable recurrence model for large earthquakes. Geophys Res Lett 7:279–282

Wu H, Masaki K, Irikura K, Kurahashi S (2016) Empirical fragility curves of buildings in northern Miyagi prefecture during the 2011 off the Pacific coast of Tohoku earthquake. J Disaster Res 11:1253–1270

Yamaguchi N, Yamazaki F (2001) Estimation of strong motion distribution in the 1995 Kobe earthquake based on building damage data. Earthquake Eng Struct Dynam 30:787–801

Youngs RR, Coppersmith KJ (1985) Implications of fault slip rates and earthquake recurrence models to probabilistic seismic hazard estimates. Bull Seismol Soc Am 75:939–964

Mai PM and Beroza GC (2002) A spatial random field model to characterize complexity in earthquake slip. J Geophys Res 107(B11):2308. https://doi.org/10.1029/2001JB000588

Chapter 25
R16—Bayesian Empirical Fragility Assessment for Hierarchical Damage Levels: 2018 Sulawesi-Palu Tsunami

Fatemeh Jalayer and Hossein Ebrahimian

25.1 Introduction

This recipe describes a procedure for empirical tsunami fragility assessment, for hierarchical damage levels, using generalized linear regression based on Bayesian updating and model class selection (BMCS). Although developed and described for tsunamis, the recipe is applicable to other types of hazards as long as they can be described by a given intensity measure. This procedure is primarily used for fragility assessment; nevertheless, it can be used also for vulnerability assessment for hierarchical fatality and loss levels. This demonstrates the versatility and utility of this recipe beyond the specific use case described herein for empirical fragility assessment, based on observed damage data from Sulawesi-Palu 2018 Tsunami. For definitions of fragility curves and vulnerability curves and the distinction between them, the reader is referred to Chap. 7 (Vulnerability Modelling).

Hierarchical damage levels can be ordered across a scale in a monotonically increasing manner (see Fig. 25.1). The main feature of the fragility curves for hierarchical damage levels is that they should never cross. The basic empirical fragility assessment procedure (denoted herein as *Basic*) assesses the fragility curves for one damage level at a time. This means that there is no specific guarantee that the curves will not intersect. This recipe provides a rigorous Bayesian workflow (denoted herein

F. Jalayer (✉)
Department of Disaster Risk Reduction, University College London, London, UK
e-mail: f.jalayer@ucl.ac.uk

H. Ebrahimian
Department of Structures for Engineering and Architecture, Università di Napoli Federico II, Naples, Italy
e-mail: hossein.ebrahimianchelehkhaneh@unina.it

M. B. Sørensen et al. (eds.), *Probabilistic Tsunami Hazard and Risk Analysis*, Mathematics of Planet Earth 14, https://doi.org/10.1007/978-3-031-98115-9_25

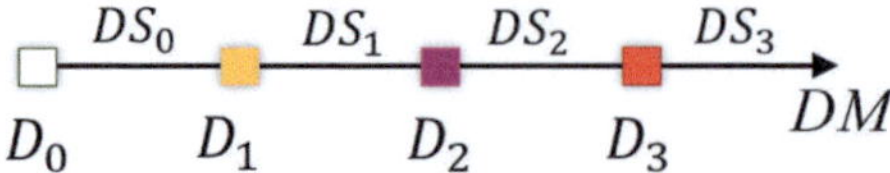

Fig. 25.1 Graphical representation of hierarchical damage levels (the damage scale used in Paulik et al. 2019)

as *BMCS*) for, (1) obtaining fragility curves for the ensemble of damage levels (D), ensuring that they do not intersect; (2) obtaining confidence intervals for the resulting fragility curves to reflect the epistemic uncertainty due to the quantity of empirical data available and, (3) finding the best generalized regression model that fits the data among a predefined set of regression model alternatives.

Figure 25.2 shows the empirical fragility curves obtained based on observed damage to buildings after the Sulawesi-Palu 2018 Tsunami for the "*non engineered light timber*" building class, based on survey data from N = 14 buildings for two damage levels D2 (unrepairable, magenta) and D3 (complete, red). It is noteworthy that the observed damage to buildings encompasses also the possible damage caused by the preceding earthquake and it should be viewed as a "cumulative damage". However, damage reports and satellite images after the tsunami indicated that most impacts occurred within 300 m of the coastline (Paulik et al. 2019). Therefore, it can be deducted that, while many buildings could have endured some minor damage (D1) due to the earthquake, the tsunami has been responsible for damaging the building beyond repair (D2) or washing them totally away (D3).

The vertical axis is the "fragility" and shows the probability of exceeding the damage level as a function of the hazard intensity measure (IM). For example, in this specific case, the fragility for D2 depicts the probability of being in damage state DS2 or being in damage state DS3, whereas the fragility for D3 depicts the probability

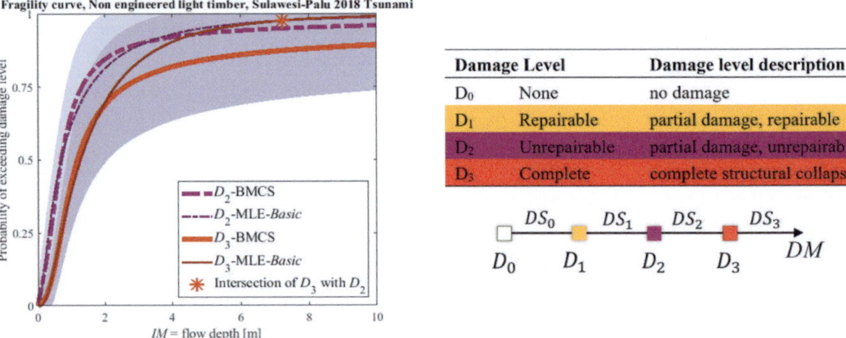

Damage Level		Damage level description
D_0	None	no damage
D_1	Repairable	partial damage, repairable
D_2	Unrepairable	partial damage, unrepairable
D_3	Complete	complete structural collapse

Fig. 25.2 (a, left) The fragility curves for D2 (magenta) and D3 (red) obtained based on Basic (dashed line) and BMCS (solid line) methods; (b, right) the description of the damage scale used

of being in DS3 (see Fig. 25.1). The horizontal axis shows the IM which in this case corresponds to the tsunami flow depth. Using the *Basic* method, the curves for D2 and D3 will cross at a flow depth of around 7 m. However, using *BMCS*, laminar fragility curves for D2 and D3 are obtained, and each fragility curve is accompanied by a confidence interval, and the best regression model that fits the data is selected.

This recipe connects to the following ingredients chapters:

- PTRA: Framework FW1B (Probabilistic Tsunami Risk Assessment; Sect. 2.2)
- Tsunami inundation propagation: Tsunami modelling—Chap. 5
- Physical exposure: building taxonomy—Chap. 6
- Vulnerability: empirical fragility—Chap. 7

Before describing the type of data needed for this recipe, it is useful to go over some of the basic requirements and definitions: (1) The first requirement for using this recipe is to define a **damage scale.** This means defining hierarchical (i.e., monotonically increasing) levels of damage that cover the entire range of possibilities. The graphic representation of such a scale is shown in Fig. 25.1; (2) The second requirement is to define the type of **physical asset** (e.g., a specific type of building, road, pipeline, bridge) for which the fragility curve is going to be developed; (3) The final requirement is to define the baseline set of physical assets or spatial extent covered by the data. For example, this can be defined as a set of physical assets within the tsunami inundation extent, or the administrative boundaries of the coastal city affected by the tsunami inundation.

Two types of raw data can be used for this recipe:

1. **Point-wise**: For this category, the raw data are in the form of a set of N (number of data points) pairs of IM values (e.g., flow depth, momentum flux) either calculated through numerical simulation of tsunami inundation or through observed tsunami heights at the position of the surveyed physical assets, and the damage states observed for each physical asset. The damage states can be acquired through field surveys, remote-sensing, and machine detection (artificial intelligence).
2. **Aggregated**: For this category, the raw data are in the form of a set of N (number of data points) pairs of IM values (e.g., flow depth, momentum flux) either calculated through numerical simulation of tsunami inundation or through observed tsunami heights at the centroid of spatial or administrative units and number/percentage of buildings in specific damage states within the spatial/administrative units.

This recipe is accompanied by standalone open-source software available on the GitHub repository of the European Tsunami Risk Service (ETRiS). The workflow and background theory for this recipe is fully described in Jalayer et al. (2023). The recipe uses open access raw data from Paulik et al. (2019). The resulting empirical fragility curves are available on the EPOS Data Portal through the GitHub repository of ETRiS. There is a Jupyter Notebook available for visualizing the fragility curves based on data in *.csv format available on the EPOS portal. The links to these materials are listed in the resource's sections at the end of this recipe.

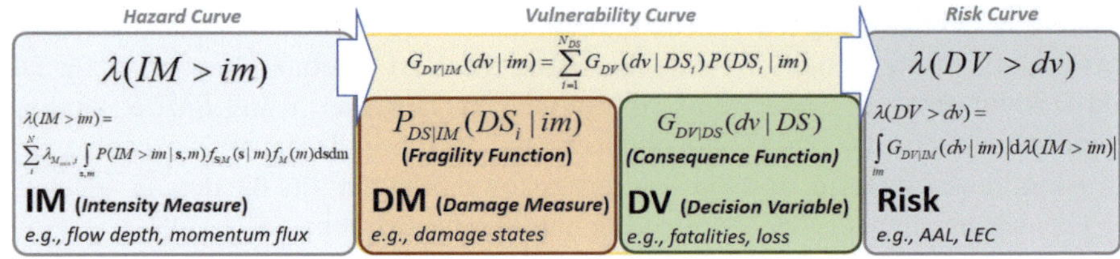

Fig. 25.3 PEER-type (modular) framework for PTRA (Behrens et al. 2021)

25.2 Framework Used in Recipe

The fragility curves are best defined within a PEER-type framework (the orange box Fig. 25.3, see Behrens et al. (2021) for more details); that is, a modular framework that moves from a hazard curve to a risk curve via convolution of these curves.

Among the frameworks discussed in Chap. 2, fragility curves are relevant tools for risk assessment in Framework 1 (i.e., all those related to quantitative risk and impact assessments). Framework 1B which describes the fully probabilistic approach (many scenarios) relies on the fragility curves by direct integration of the hazard curves (obtained based on many scenarios) and the fragility curves. Alternatively, the fragility curves, which are probability distributions, can be directly implemented in a fully simulation-based approach (Framework 1A). Framework 1D can find perfect use of fragility curves in a scenario-based approach. That is, the IM values (e.g., flow depth) propagated for a given (or a few) scenarios can be used as input to the fragility curves to obtain probabilities of exceeding different damage levels. This leads to the generation of scenario-based damage grading maps.

25.3 Tsunami Modelling

For this specific use case, raw data for the intensity measures (i.e., flow depth) at the position of each building were obtained based on field survey (observing the water marks some weeks after the event, Paulik et al. 2019). However, it is possible to use numerical tsunami inundation simulation to generate the IM values at the position of the buildings of interest.

25.4 Exposure Modelling

Exposure modelling is closely related to fragility assessment. This recipe regards fragility assessment for buildings and therefore it will rely on exposure models for the built environment (see Chap. 6.2). More specifically, each fragility curve is associated with a specific class of physical assets (e.g., buildings, bridges, roads). The physical asset class or typology is described by a string known as taxonomy. The taxonomy used in this recipe for labeling physical assets exposed to risk is GED4ALL, which is an extended version of the Global Exposure Database by the Global Earthquake Model (Silva et al. 2018; Murnane et al. 2019).

25.5 Vulnerability Modelling

The fragility curves derived according to this recipe are classified as empirical fragility curves, which is one of the approaches for deriving physical vulnerability models. The raw data used for fragility assessment are point-wise, meaning that they contain N pairs of IM and DS assigned to each surveyed building. Both the flow depth (IM) values and the damage levels were acquired through field surveys (Paulik et al. 2019). A four-tier damage scale containing the following damage levels is used: None (D0), repairable (D1), non-repairable (D2), complete (D3). Generalized regression with three link functions, namely, logit, probit, and cumulative log-log, has been used for deriving the fragility curves. The fragility curves are obtained through Bayesian model class selection. In simple terms, the generalized regression model (i.e., link function), that fits the data best with a smaller number of parameters is chosen. A hierarchical algorithm is used for deriving the fragility curves for the ensemble of the damage levels through a recursive formula. This recursive formula calculates the probability of being in each damage state i as the product of the probability exceeding the damage level i and the probability of not exceeding damage level $i + 1$ conditioned that damage level i is exceeded. This is to ensure that the probability of being in two distinct damage states is zero (no intersection of the curves). The epistemic uncertainty in fragility estimation is propagated by considering the uncertainty in the fragility model parameters and propagating it through Bayesian model updating. This epistemic uncertainty mainly presents the confidence interval for the estimated fragility curves due to the limited sample size of the raw data.

25.6 Discussion

The approach presented in this recipe is very versatile and can be applied to other hazards if a meaningful IM can be used for their characterization. Moreover, it can be used for vulnerability curve derivation providing the probability of exceeding a certain number of fatalities, or a certain level of economic losses as a function of the selected IM. This approach can be applied to hierarchical scales of damage, fatalities, and loss. This is hardly a limitation in the case of fragility assessment since the hierarchical damage levels are necessary conditions for integration of the fragility curves with the consequence functions for different DS to obtain vulnerability functions (see Fig. 25.3). There are some pros of using this method: (1) it is accompanied by a standalone open source software and complete documentation; (2) it provides fragility curves that are ready for integration with the consequence function and do not need any adjustment to fix the issues of intersecting curves; (3) it quantifies the uncertainty in the resulting fragility curves; (4) it identifies the best regression model for the data.

An alternative to this method is to use the hierarchical fragility derivation described above outside of the Bayesian inference and rely on maximum likelihood estimation (MLE). The advantage of this alternative is that toolboxes within Matlab, R, and Python can be used for fragility derivation. Deriving the confidence bands and ranking the regression models is a bit more complicated with the MLE with respect to BMCS. Another alternative is to use the MLE for the generalized regression and apply it to each damage level separately. The main problem with this approach is that the fragility curves could sometimes intersect, creating difficulties in the integration with the consequence function and risk assessment.

25.7 Links to Data, Models and Other Resources

- **EPOS DATA PORTAL**: The empirical fragility curves derived based on this recipe can be accessed via EPOS Data Portal (EPOS-ICS-C, https://www.ics-c.epos-eu.org/). These fragility curves are featured in two layers called Empirical Tsunami Risk Products Dataset ETRIS V0 (OGC WFS and OGC WMS).
- **European Tsunami Risk Service ETRiS**: The empirical fragility curves can be also accessed from the site of Tsunami Risk Service: https://github.com/eurotsunamirisk/etris_data_and_data_products/tree/main/etris_data_products/Fragility_Curves
- **Open-source Software**: ComputeFrag, Version 2, GitHub [code], https://github.com/eurotsunamirisk/computeFrag/tree/main/Code_ver2
- **Jupyter notebook**: https://eurotsunamirisk.org/tsunamirisktoolkit/
- **Complete documentation**: Jalayer F, Ebrahimian H, Trevlopoulos K and Bradley B (2023) Empirical tsunami fragility modelling for hierarchical damage levels. Natural Hazards and Earth System Sciences, 23(2), pp. 909–931, https://doi.org/10.5194/nhess-23-909-2023

References

Behrens J, Løvholt F, Jalayer F, Lorito S, Salgado-Gálvez MA, Sørensen M, Abadie S, Aguirre-Ayerbe I, Aniel-Quiroga I, Babeyko A, Baiguera M (2021) Probabilistic tsunami hazard and risk analysis: a review of research gaps. Front Earth Sci 9:628772

Ebrahimian H and Jalayer F (2022) ComputeFrag, Version 2, GitHub [code]. https://github.com/eurotsunamirisk/computeFrag/tree/main/Code_ver2. Access 30 Nov 2022

https://www.globalquakemodel.org/gempublications/Global-Exposure-Database-for-Multi-Hazard-Risk-Analysis-Multi-hazard-Exposure-Taxonomy

Jalayer F, Ebrahimian H, Trevlopoulos K, Bradley B (2023) Empirical tsunami fragility modelling for hierarchical damage levels. Nat Hazard 23(2):909–931. https://doi.org/10.5194/nhess-23-909-2023

Jalayer F, De Risi R, Kyessi A, Mbuya E and Yonas N (2015) Vulnerability of built environment to flooding in African cities. Urban Vulnerability and Climate Change in Africa: A Multidisciplinary Approach, pp77–106

Murnane RJ, Allegri G, Bushi A, Dabbeek J, de Moel H, Duncan M, Fraser S, Galasso C, Giovando C, Henshaw P, Horsburgh K (2019) Data schemas for multiple hazards, exposure and vulnerability. Disaster Prev Manag: Int J 28(6):752–763

Paulik R, Gusman A, Williams JH, Pratama GM, Lin SL, Prawirabhakti A, Sulendra K, Zavhari MY, Dwi Fortuna ZE, Pare Lakuy NB, Suwarni NWI (2019) Tsunami hazard and built environment damage observations from Palu city after the September 28 2018 Sulawesi earthquake and tsunami. Pure Appl Geophys 176(8):3305–3321

Silva V, Yepes-Estrada C, Dabbeek J, Martins L. and Brzev S (2018) Global Exposure Database for Multi-Hazard Risk Analysis. D2 – Multi-hazard Exposure Taxonomy.

Chapter 26
R17—Nonlinear Pushover Analysis for Tsunami

Marco Baiguera, Marta Del Zoppo, and Tiziana Rossetto

26.1 Introduction

The use of structural analysis methodologies is fundamental to assess the vulnerability of buildings for tsunami loading due to the scarcity of empirical data. This recipe presents a novel engineering method, namely the Variable Depth Pushover (VDPO) analysis. This approach is used to generate tsunami analytical fragility functions (*tsunami fragility assessment*), as well as to support tsunami performance-based design methodologies (*tsunami design*). This tool requires a combination of the following ingredients:

- Any framework (Chap. 2), source selection (Chap. 3), source modelling (Chap. 4) and tsunami modelling (Chap. 5) can be chosen to provide with an estimate of one or more tsunami intensity measures.
- Exposure data—Chap. 6.
- Analytical vulnerability—Chap. 7.3.

The recipe can be used in local decision-making processes. VDPO has been used to support tsunami risk studies by engineering companies (Arup) and insurance companies (Wills Re/Gallagher Re). VDPO is also implemented in the commercial finite element software SeismoStruct 2023 (SeismoStruct 2022) as the main procedure for

M. Baiguera (✉)
Risk Team, Global Earthquake Model (GEM) Foundation, Pavia, Italy
e-mail: marco.baiguera@globalquakemodel.org

M. Del Zoppo
Università di Napoli Federico II, Naples, Italy
e-mail: marta.delzoppo@unina.it

T. Rossetto
Faculty of Civil Engineering and Geosciences (CEG), Delft University of Technology (TU Delft), Delft, The Netherlands
e-mail: T.Rossetto-1@tudelft.nl

M. B. Sørensen et al. (eds.), *Probabilistic Tsunami Hazard and Risk Analysis*, Mathematics of Planet Earth 14, https://doi.org/10.1007/978-3-031-98115-9_26

tsunami nonlinear analysis. Improved versions of the VDPO have been developed in recent years and are presented in the next sections. VDPO2 is referenced in ASCE7-22 (ASCE 2022) as one of the recommended methods for the tsunami design of buildings.

26.2 Frameworks, Sources and Modelling

This recipe relies on hazard data from probabilistic or scenario-based studies as input. The choice of source characterization, source modelling and tsunami modelling are free, as long as the process results with an estimate of tsunami inundation depth and flow velocity. These parameters are fundamental for the estimation of the tsunami loading acting on the structure.

The advantage of using the probabilistic frameworks FW1A or FW1B (Sect. 2.2) is the use of hazard data that can help develop a design for a given return period. For instance, the ASCE7-22 Standard (ASCE 2022) provides probabilistic tsunami design zone maps for tsunami prone regions in the USA. Those areas are vulnerable to being inundated by the Maximum Considered Tsunami (MCT), defined as having 2% probability of being exceeded in a 50-year period. In this case, the results can be directly compared with other hazards (e.g., earthquakes). However, there are many tsunami-prone regions where no site-specific hazard data exist. Hence, scenario-based data may be a valid alternative, particularly for the fragility assessment of buildings.

26.3 Exposure Modelling

Exposure data related to the geometry, structural details, non-structural arrangements and material mechanical properties are needed. Such information represents the input data required to construct the finite element model (FEM) of any building. These can be collected using available design structural drawings, supplemented by in-situ surveys, when possible. The latter surveys can only be carried out for existing buildings and are particularly useful in collecting as-built information, particularly on non-structural elements and openings, as well as on quality of construction and condition. Where structural drawings do not exist, information on section detailing (e.g. reinforcement arrangement and size), may need to be assumed based on standard local construction practice. Baiguera et al. (2021) and Cels et al. (2023) propose survey forms that are specifically developed for collecting of field data on community facilities such as hospitals and schools that are exposed to tsunami hazard. Such tools can be used to generate exposure databases relevant to both tsunami vulnerability and more general hazard loss modelling, and are useful in identifying a number of archetype buildings, which can be analysed using the VDPO tool.

26.4 Vulnerability Modelling

26.4.1 Tsunami Loading

Tsunami induces lateral and vertical (uplift) hydrostatic and hydrodynamic pressures on the structure. Such flow can be considered as quasi-steady according to Qi et al. (2014), this is because of the typical extremely long wavelengths associated with tsunami, which result in long-duration inundation where the temporal variation of the flow is small with respect to the length scale of a building. Tsunami-flow induced lateral pressure can be computed following the empirical equations presented by Foster et al. (2017) or based on design codes, such as ASCE7-22. Recent studies indicate that the pressure distribution due to tsunami flow is linear, leading to either triangular or trapezoidal load patterns, when the flows affect the front surface, or both front and back surfaces, of a structural component, respectively. A conservative approach is adopted by ASCE7-22, for which the load pattern is assumed to be uniform on all elements. There are no empirical models for induced uplift loads on structures, hence the equations provided by the ASCE7-22 can be adopted to compute such loading.

26.4.2 Tsunami Nonlinear Static Analysis

The VDPO is a nonlinear static analysis to be carried out using finite element (FEM) software packages. Examples of commercial FEM packages that have been used for tsunami structural response simulations include OpenSees (McKenna et al. 2010) and SeismoStruct (SeismoStruct 2022). As tsunami forces are typically of long duration, the static nature of the analysis is able to appropriately represent the actual hydrodynamic drag on the overall structure and individual components.

The starting point is the development of a structural model, which is then subjected to the VDPO. The tsunami force applied to the structure is incrementally increased by monotonically increasing the tsunami inundation depth. At each depth value, the corresponding flow velocity is calculated assuming a constant Froude number (Fr). The VDPO as originally conceived in Petrone et al. (2017) is a load-control analysis, and is not capable of capturing the post-peak degrading behaviour of the structure. This limitation is overcome by the VDPO2 approach, developed by Baiguera et al (2022). VDPO2 consists of a two-phase nonlinear static analysis (Fig. 26.1): in the first phase a load-control pushover analysis is conducted assuming that the inundation depth and flow velocity increase incrementally; in the second phase, the analysis switches to response-control pushover analysis, where the displacement is increased incrementally and the corresponding tsunami force is calculated, assuming that there is no further increase in inundation depth. The switch from the first phase to the second phase occurs either when a predefined load level is reached or when the analysis encounters a numerical convergence issue, whichever occurs first.

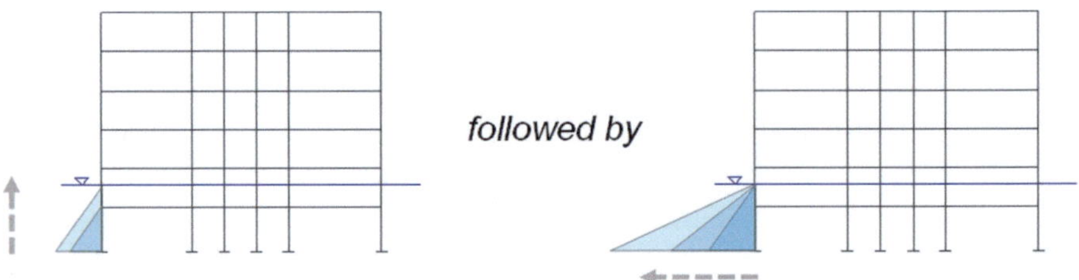

Fig. 26.1 VDPO2 analysis for a watertight structure: (left) load-controlled phase 1 with incremental increase in inundation depth and flow velocity; and (right) response-controlled phase 2 with no further increase in inundation depth

The VDPO method has recently been modified for the analysis of reinforced concrete frames with breakaway infill walls. Del Zoppo et al. (2020) developed a bespoke method, referred to as VDPO-BI, in which the tsunami-induced lateral and vertical loads are monotonically applied to structural and non-structural components (i.e., breakaway cladding) as a function of the tsunami inundation depth up to the structural collapse (Del Zoppo et al. 2021, 2023).

26.4.3 Loading Discretisation

There are two main approaches for discretising and applying the tsunami force along the building:

- Storey load discretisation (Fig. 26.2a), in which nodal loads are applied to each storey. The tsunami forces are calculated using a simple influence area approach.
- Distributed load discretisation (Fig. 26.2b), in which the same hydrodynamic forces are discretised into point loads distributed along the building height. It is recommended that for each column, the tsunami loads are applied at least at 5 load application points along the height of vertical structural elements (e.g., columns)

Storey load discretisation is useful when estimating the overall systemic response of the lateral load resisting system. However, the distributed load discretisation is recommended, since it allows to simulate the local drag effects on single components (e.g., columns).

26.4.4 Structural Assessment

The assessment evaluates the ability of the structure to sustain tsunami loads. This requires the definition of the following items:

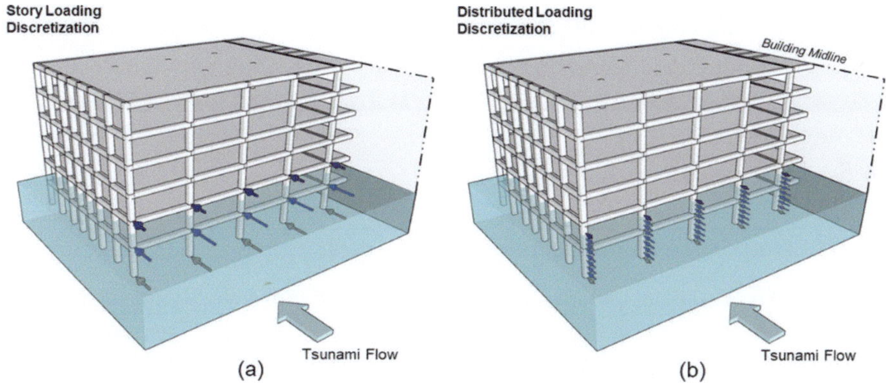

Fig. 26.2 Tsunami loading discretisation approaches: **a** story loading; and **b** distributed loading

- Intensity Measure (IM). The tsunami force F_T is an efficient IM since the structural response is typically controlled by the actual hydrodynamic drag on the overall structure. Other IMs are the tsunami inundation depth (h) or the momentum flux (hu^2).
- Damage states (DS). The definition of DSs should be based on good engineering understanding of the particular structure being assessed. Damage mechanisms typically form at section, member, storey and global structural level. These damage mechanisms should be represented by threshold values of Engineering Demand Parameters (EDPs) that define, unambiguously, the progression between one damage state and the next one; with the occurrence of the first of these indicating initiation of the damage state. Examples of commonly-used EDPs are the interstorey drift (IDR) and column shear force.
- The structural performance is evaluated at an applied tsunami force corresponding to the design (or assessment) tsunami force. In the context of the ASCE7-22, there are two main types of structural checks:
- A systemic check. The structural lateral-force-resisting system (LFRS) is checked for its adequacy in resisting the prescribed tsunami loads. The maximum strength of the lateral force resisting system (LFRS) of a building is compared to the design basis event, the so-called Load Case 2 (LC2). This occurs when the velocity reaches its peak and the inundation depth is two-thirds of the maximum inundation depth at the site. If the tsunami pushover curve shows the building to have a strength larger than the applied loading at LC2, then the systemic check is deemed to have been passed.
- A component-based check. The design strength of each individual structural component is checked for adequacy against an enhanced tsunami loading that accounts for local drag effects. Component failure is indicated when either the component axial, moment or shear capacity is reached.

This VDPO-based approach has the potential for providing a more economical design as compared to the prescriptive ASCE 7–22 approach, which promotes an enhanced seismic design of the structure to meet the systemic acceptance criteria of the standard.

26.5 Discussion

The VDPO method requires the following input data:

- Tsunami hazard data: tsunami inundation depth and flow velocity for the building site.
- Exposure data on the structure: particularly structural and non-structural details, geometric and mechanical properties and opening arrangements.
- The development of a structural model using a finite element method (FEM).
- For fragility assessment, the definition of a bespoke damage scale.

The following limitations are noted:

- The VDPO analysis is implemented in few software packages only and requires detailed exposure data at structure level.
- Other tsunami-induced effects, such as scour or debris impact, are not considered in the VDPO, and separate analyses should be conducted.

References

ASCE (2022) Minimum design loads and associated criteria for buildings and other structures. ASCE/SEI 7–22. Reston, VA, USA

Baiguera M, Rossetto T, Robertson IN, Petrone C (2022) A procedure for performing nonlinear pushover analysis for tsunami loading to ASCE 7. ASCE J Struct Eng 148(2). https://doi.org/10.1061/(ASCE)ST.1943-541X.0003256

Baiguera M, Rossetto T, Palomino J, Dias P, Lopez-Querol S, Siriwardana C, Hasalanka H, Ioannou I, Robinson D (2021) A new relative risk index for hospitals exposed to tsunami. Front Earth Sci 9:626809. https://doi.org/10.3389/feart.2021.626809

Cels J, Rossetto T, Dias P, Thamboo J, Wijesundara K, Baiguera M, Del Zoppo M (2023) Engineering surveys of Sri Lankan schools exposed to tsunami. Front Earth Sci 11:1075290. https://doi.org/10.3389/feart.2023.1075290

Del Zoppo M, Di Ludovico M, Prota A (2020) Methodology for assessing the performance of RC structures with breakaway infill walls under tsunami inundation. J Struct Eng 147(2):04020330. https://doi.org/10.1061/(ASCE)ST.1943-541X.0002900

Del Zoppo M, Wijesundara K, Rossetto T, Dias P, Baiguera M, Di Ludovico M, Thamboo J, Prota A (2021) Influence of exterior infill walls on the performance of RC frames under tsunami loads: case study of school buildings in Sri Lanka. Eng Struct 234:111920. https://doi.org/10.1016/j.engstruct.2021.111920

Del Zoppo M, Rossetto T, Di Ludovico M, Prota A (2023) Effect of buoyancy loads on the tsunami fragility of existing reinforced concrete frames including consideration of blow-out slabs. Sci Rep 13(1):9015. https://doi.org/10.1038/s41598-023-36237-7

Foster ASJ, Rossetto T, Allsop W (2017) An experimentally validated approach for evaluating tsunami inundation forces on rectangular buildings. Coast Eng J 128:44–57

McKenna F, Scott MH, Fenves GL (2010) Nonlinear finite-element analysis software architecture using object composition. J Comput Civ Eng 24(1):95–107

Petrone C, Rossetto T, De la Barra C, Baiguera M, Ioannou I (2020) Fragility of a reinforced concrete structure subjected to earthquake and tsunami in sequence. Eng Struct 205:1102120. https://doi.org/10.1016/j.engstruct.2019.110120

Petrone C, Rossetto T, Goda K (2017) Fragility assessment of a RC structure under tsunami actions via nonlinear static and dynamic analyses. Eng Struct 136:36–53

Qi ZX, Eames I, Johnson ER (2014) Force acting on a square cylinder fixed in a free-surface channel flow. J Fluid Mech 756:716–727

Rossetto T, De la Barra C, Petrone C, De la Llera JC, Vasquez J, Baiguera M (2019) Comparison of nonlinear static and dynamic analysis methods for assessing structural response under earthquake and tsunami in sequence. Earthquake Eng Struct Dynam 48:867–887. https://doi.org/10.1002/eqe.3167

SeismoStruct (2022) A computer program for static and dynamic nonlinear analysis of framed structures. https://seismosoft.com/

Tagle SJ, Jünemann R, Vásquez J, De la Llera JC, Baiguera M (2021) Performance of a RC wall building subjected to earthquake and tsunami loads in sequence. Eng Struct 238:111995. https://doi.org/10.1016/j.engstruct.2021.111995

Chapter 27
R18—Analytical Vulnerability Assessment for Italian Buildings Portfolio

Marta Del Zoppo, Marco Di Ludovico, and Andrea Prota

27.1 Introduction

The purpose of this recipe is to provide a classification of existing buildings in Italy based on their physical vulnerability to tsunami loads. The development of methodologies for large-scale (i.e., urban or regional scale) tsunami vulnerability assessment is an essential tool for tsunami loss analysis, risk management and for prioritizing interventions to protect the built environment. The analytical vulnerability assessment procedure includes the following ingredients:

- Tsunami modelling (Chap. 5)
- Tsunami exposure—(Chap. 6)
- Vulnerability modelling—(Chap. 7)

Physical vulnerability is represented by the mean damage ratios of buildings for increasing levels of a tsunami intensity measure. The recipe is suitable for both single buildings and building portfolios and allows to derive vulnerability curves correlating the mean damage index of a building/building class with a tsunami intensity measure. The mean damage index for a building portfolio is calculated as the weighted average of the frequencies of each damage level in a defined tsunami damage scale. Such curves can be converted into loss curves introducing appropriate consequence functions.

M. Del Zoppo (✉) · A. Prota
Università di Napoli Federico II, Naples, Italy
e-mail: marta.delzoppo@unina.it

A. Prota
e-mail: aprota@unina.it

M. Di Ludovico
Department of Structures for Engineering and Architecture, Università di Napoli Federico II, Naples, Italy
e-mail: marco.diludovico@unina.it

M. B. Sørensen et al. (eds.), *Probabilistic Tsunami Hazard and Risk Analysis*, Mathematics of Planet Earth 14, https://doi.org/10.1007/978-3-031-98115-9_27

27.2 Framework

To perform the analytical vulnerability assessment, a probabilistic analysis is conducted to compute the probability of damage of buildings during a tsunami event. The probability of buildings damage is computed performing a structural analysis where tsunami loads are applied on buildings to simulate a tsunami event (see the recipe R17). To characterise the tsunami loads and their uncertainties, both fully probabilistic (FW1A) and scenario-based (FW1D) tsunami hazard simulations can be adopted based on the available data for the study region. For the case of Italy, scenario-based hazard analyses have been conducted so far to characterise tsunami loads on buildings.

The Intensity Measure (IM) adopted for the hazard analysis influences the selection of the one adopted for the vulnerability analysis. For the Italian case, the tsunami inundation depth (h) is selected as IM given the high uncertainties in simulating velocities. Uncertain parameters for the tsunami loading characterization are the Froude number and the flow velocity. In the absence of such data, records from past tsunami events can help in the characterization of uncertainties. The lack of a proper uncertainty estimation in the hazard simulation may reduce the accuracy level of the analytical vulnerability model.

27.3 Exposure

27.3.1 Definition of Typological Classes and Uncertainties

When dealing with large-scale vulnerability assessment, the existing building portfolio in the study region is divided into typological classes and analytical fragility or vulnerability curves are derived for each typological class. The definition of classes depends on the available Exposure data for the physical assets in the study region. The attributes for the Exposure characterization for tsunami can follow the GED4ALL recommendations (Silva et al. 2022).

For the Italian case, open-source national repositories are available with information about the existing residential buildings portfolio such as the age of construction, the construction material and the number of storeys. The availability of local repositories allows a refined definition of classes for the study area.

Notwithstanding the general validity of the methodology presented in the recipe for the analytical vulnerability assessment, the case of Italian reinforced concrete (RC) buildings is specifically addressed in the following. The main attributes selected are the height of the building (i.e., number of storeys) and the age of construction. Two classes are defined based on the number of storeys: low-rise buildings (with less than 4 storeys) and mid-rise buildings (with 4 to 6 storeys). For the age of construction, the definition of classes can be correlated to the evolution of seismic design codes. Indeed, the design method (i.e., for gravity loads or for seismic excitation) significantly affects the lateral response of a structure under tsunami loading. The selection of the number of classes is a balance between the desired accuracy level of the analytical vulnerability model and the computational effort. Based on the age of construction three classes are defined for buildings designed for gravity loads and following obsolete or current seismic design method; thus for the Italian RC buildings, six typological classes are defined (Del Zoppo et al. 2022a).

Data repositories may provide useful information for modelling uncertainties related to the geometric and mechanical characteristic of physical assets for each typological class. Uncertainties are usually described through statistical distributions. In the literature, few historical data related to the evolution of mechanical properties of concrete and steel in the study region can be found. A limitation in the analytical vulnerability assessment process can be represented by the lack of information regarding the uncertainties related to the knowledge of the existing assets (Subsect. 7.3.3). To perform the probabilistic damage assessment, Monte Carlo simulation is adopted to generate different tsunami loading conditions to account for the uncertainties.

27.4 Vulnerability

27.4.1 Simulation of a Buildings Portfolio

When dealing with a single building or with a small building portfolio, we often have a good knowledge of the assets in terms of geometry, mechanical properties of materials and structural details. This kind of information is needed to perform a detailed structural analysis and predict the damage mechanisms under tsunami loads.

For a large-scale vulnerability assessment encompassing many buildings, it is not feasible to have a detailed knowledge of each building in the portfolio. Hence, a buildings portfolio is simulated via Monte Carlo simulations or other similar methods, in order to generate building properties statistically representative of existing assets based on the available uncertainties characterization.

Missing data regarding the structural details (i.e., cross-section dimensions, longitudinal and transverse reinforcement details) needed for the numerical modelling and the structural analysis of buildings can be completed through a simulated design procedure. The simulated design procedure consists in simulating the structural design of a building following the design code and construction practices in force at the time of construction to obtain the missing data needed to perform the structural analysis. For the Italian case, the simulated design can refer to the design codes RD (1937), RD (1939), DM (1975) and IBC (2018). Information about historical construction practices in the study region are of paramount importance for a correct simulation of a building portfolio.

27.4.2 Damage Scale

The definition of a damage scale for tsunami loads is required for the fragility and vulnerability analysis. Given the lack of damage data from past tsunami events in Italy, the definition of a damage scale for Italian RC buildings is based on damage observations from past tsunamis on similar structural types (from e.g. Japan, US, see below). Empirical damage caused by a tsunami inundation on RC buildings is commonly classified following the Ministry of Land, Infrastructure and Transportation of Japan (MLIT 2011) guidelines. According to the MLIT, six damage levels are identified ranging from minor damage to washed away. Even though the qualitative nature of this classification, it is helpful to understand the progression of damage. Damage states DS1-3 (i.e., minor to major) are mainly related to non-structural components. Conversely, only damage states DS4-5 (i.e., complete and collapsed) focus on structural damage. Damage state DS6 (i.e., washed away) refers to the observed complete destruction of a building.

The HAZUS Tsunami Model Technical Guidance (FEMA 2017) provides a quantitative estimation of damage levels for risk and loss analysis purposes. This classification is proposed for RC buildings. In detail, four structural damage levels are considered, from Slight to Complete. Following the HAZUS approach, structural damage associated to a global ductile failure mechanism is considered for RC buildings. Indeed, the HAZUS classification neglects the occurrence of local failure mechanism of structural components and, hence, does not provide any information on how this kind of damage mechanism should be classified for a vulnerability analysis. However, previous numerical studies have found that the brittle failure of RC columns due to shear can be a damage mechanism for such structures.

Based on MLIT and HAZUS damage classifications, mechanical damage scales for RC buildings are proposed in the recent literature by Petrone et al 2020 and Del Zoppo et al. (2022b), considering both member-level and global-level structural damage mechanisms. In Del Zoppo et al. (2022b) a damage classification for non-structural components (i.e., masonry infill walls) is also proposed. For this recipe, the following damage scale based on the HAZUS model is adopted for Italian RC buildings (Table 27.1):

The reaching of a certain damage level during the structural analysis is computed by defining threshold values for significant Engineering Demand Parameters (EDPs). EDPs are defined based on the mechanical damage mechanism reported in Table R18.1, and are monitored during the structural analysis to assess the progressive reaching of damage levels. For this recipe, EDPs are the longitudinal steel strain, bending moments and shear forces in columns.

Table 27.1 Damage scale for Italian RC buildings (Del Zoppo et al. 2023)

	Slight	Moderate	Extensive	Complete
Structural damage definition	First reaching in any vertical member of concrete cracking	First reaching in any vertical member of ½ steel yield strain in the longitudinal steel Or First reaching of slab flexural capacity	First reaching in any vertical member of steel yield strain in the longitudinal steel Or First reaching in any vertical member of shear capacity	Peak base shear Or Reaching of the shear capacity in two consecutive vertical members

27.4.3 Tsunami Structural Analysis and Damage Assessment

To derive the probability of occurrence of a certain damage level for a building or a buildings portfolio subjected to tsunami loads, it is necessary to:

- Build a numerical model for each building in the portfolio (i.e., finite element modelling, see Subsection 7.3.1).
- Perform the structural analysis applying tsunami loads on the building increasing the inundation depth up to the structural collapse (i.e., by using the Variable Depth Pushover, see recipe R17).
- Assess during the structural analysis the reaching of damage levels in relation to the tsunami IM (i.e., inundation depth) by monitoring the selected EDPs.
- Store the values of IM (i.e., inundation depth) associated with the reaching of each damage level for each building.

Damage data derived from the analytical procedure described in the bullet list above are treated as the empirical damage data derived after a tsunami event. Damage data are collected in the form of inundation depth values corresponding to the reaching of each damage level for all the buildings in the portfolio or for the single building. The derivation of fragility and vulnerability curves from damage data follow the same approaches adopted for the empirical case. Probabilistic models able to describe the distribution of damage data as a function of the IM are adopted as for fragility curves. Lognormal or exponential functions are usually adopted for fragility curves. Several fitting methods can be used, such as the Least Square Estimation method or the Maximum Likelihood method, among others.

The limitation of the analytical damage assessment procedure for large-scale vulnerability analysis is related to the high computational effort needed to model all buildings in the portfolio and perform on each a tsunami structural analysis accounting for the uncertainties. One solution to reduce the computational effort relies on the limitation of the uncertainties considered during the analysis, allowing to reduce the number of simulations. A second solution consists in the adoption of simplified models able to simulate the structural response of buildings avoiding refined numerical simulations. A mechanic-based method, called SAFETI (Structural Assessment to Flows and Extreme Tsunami Inundation) has been developed by Del Zoppo et al. (2022b) for RC buildings and implemented in a Matlab tool for the derivation of fragility curves. The method implements the Variable Depth Pushover for Breakaway Infilled frames (VDPO-BI, Del Zoppo et al. 2021).

27.4.4 *Mean Damage Index*

A mean damage index (DI) is computed to easily compute the probability of damage of a building portfolio as a function of the selected IM, and to compare different building portfolios. The DI is computed as the weighted average of the frequencies of occurrence of each damage level in the tsunami damage scale, as follows:

$$DI = \frac{\sum f_i}{n}$$

where f_i is the frequency of occurrence of a given damage level for a certain value of IM (i = number of each damage level) and n is the total number of damage levels. The DI ranges between 0 and 1, where 0 indicates the absence of damage and 1 corresponds to a complete damage. The distribution of DI derived for the portfolio is fitted by a lognormal function to provide a continuous description of DI as a function of the IM. Vulnerability curves for Italian RC buildings are reported in Del Zoppo et al. (2022b).

27.5 Discussion

The tsunami analytical vulnerability analysis requires:

1. A good characterization of the exposed physical assets, as well as of the uncertainties related to the geometric and mechanical properties and structural details.
2. A good characterization of the expected tsunami loads on structures and their uncertainties.
3. A deep knowledge of the evolution of design codes and construction practice in the study region to perform a simulated design, if structural details are missing.
4. A representative tsunami damage scale for the building typology under investigation.
5. A tool to perform tsunami structural analysis and damage assessment of buildings.

The following limitations of the methodology are noted:

- Only damage related to hydrostatic and hydrodynamic tsunami-induced loads are considered in the recipe, and other effects such as scour or debris impact are neglected.
- The lack of data to properly model the uncertainties related to the tsunami loads and to the buildings characteristics may limit the validity of the obtained vulnerability models.

References

Del Zoppo M, Di Ludovico M, Prota A (2021) Methodology for assessing the performance of RC structures with breakaway infill walls under tsunami inundation. J Struct Eng 147(2):04020330. https://doi.org/10.1061/(ASCE)ST.1943-541X.0002900

Del Zoppo M, Belliazzi S, Lignola GP, Di Ludovico M, Prota A (2022a) Analytical structural vulnerability and classification of Italian residential building stock to tsunami hazard. Bull Geophys Ocean 63(4):555–574. https://doi.org/10.4430/bgo00398

Del Zoppo M, Di Ludovico M, Prota A (2022b) A mechanics-based method towards risk assessment of RC buildings under tsunami and flow-type hazards. Eng Struct 264:114452. https://doi.org/10.1016/j.engstruct.2022.114452

Del Zoppo M, Rossetto T, Di Ludovico M, Prota A (2023) Effect of buoyancy loads on the tsunami fragility of existing reinforced concrete frames including consideration of blow-out slabs. Sci Rep 13(1):9015. https://doi.org/10.1038/s41598-023-36237-7

DM (1975) Decreto Ministeriale 3 marzo 1975. Approvazione delle norme tecniche per le costruzioni in zone sismiche. Gazzetta Ufficiale 8 April 1975, n.93 (in Italian)

FEMA (2017) Hazus tsunami model technical guidance

IBC (2018) Norme Tecniche per le Costruzioni. Ministerial Decree, Rome, Italy. in Italian

MLIT (Ministry of Land, Infrastructure, Transport and Tourism) (2011) The interim guideline for the structural conditions of tsunami evacuation buildings etc, based on the building damage from the tsunami in the Tohoku earthquake. Rep. No. 2570. Chiyoda, Japan: MLIT

Petrone C, Rossetto T, De la Barra C, Baiguera M, Ioannou I (2020) Fragility of a reinforced concrete structure subjected to earthquake and tsunami in sequence. Eng Struct 205:1102120. https://doi.org/10.1016/j.engstruct.2019.110120

RD (1939) Regio Decreto Legge 16 novembre 1939 n. 2229, che riguarda le Norme per la esecuzione delle opere in conglomerato cementizio semplice od armato. Gazzetta Ufficiale 18 aprile 1940 n.92 (in Italian)

RD (1937) Regio Decreto Legge 22 novembre 1937 n. 2015, che riguarda le norme tecniche ed igieniche da osservarsi per i lavori edilizi nelle località colpite dal terremoto del 13 gennaio 1915. Gazzetta Ufficiale 27 dicembre 1937 n.298 (in Italian)

Silva V, Brzev S, Scawthorn C, Yepes C, Dabbeek J, Crowley H (2022) A building classification system for multi-hazard risk assessment. Int J Disaster Risk Sci 13(2):161–177. https://doi.org/10.1007/s13753-022-00400-x

Chapter 28
R19—An Indicator-Based Method for the Assessment of the Physical Vulnerability of Buildings to Tsunami

Maria Papathoma-Köhle

28.1 Introduction

The recipe described herein shows an indicator-based method with a clear focus on the vulnerability of buildings to tsunami impact. The ingredients considered in the recipe are: (a) an inundation zone identification based on the potential wave height of a specific scenario, and (b) an indicator-based methodology for the assessment of the physical vulnerability of buildings. A case study in Italy is presented. In more detail, the vulnerability of buildings located in a coastal segment in Apulia (San Isidoro and Torre Squilace), Italy, is assessed (Papathoma-Köhle et al. 2019). The recipe is suitable for the local scale in areas with or without a tsunami record and is based on a method (Papathoma Tsunami Vulnerability Assessment—PTVA) that has been developed by using data from Peloponnese and Crete, Greece (Papathoma et al. 2003; Papathoma and Dominey-Howes 2003), and then used for tsunami vulnerability assessments in many areas of the world, such as the Maldives (Dominey-Howes and Papathoma 2007), Australia (Dall'Osso et al. 2009, 2016), Morocco, Chile, Italy, and Iran, among others.

This recipe connects to the following ingredients chapters:

- Framework FW2 (Tsunami risk assessment with special focus on indicator-based vulnerability assessment; Sect. 2.2)
- Vulnerability Sect. 7.4 Indicator-based physical-vulnerability

M. Papathoma-Köhle (✉)
Institute for Mountain Risk Engineering (IAN), BOKU University, Vienna, Austria
e-mail: maria.papathoma-koehle@boku.ac.at

M. B. Sørensen et al. (eds.), *Probabilistic Tsunami Hazard and Risk Analysis*,
Mathematics of Planet Earth 14, https://doi.org/10.1007/978-3-031-98115-9_28

28.2 Framework used in the Recipe

This recipe follows the framework FW2 (Tsunami risk assessment with a special focus on indicator-based vulnerability assessment; Sect. 2.2). The method makes use of deterministic tsunami scenarios and focuses on the built environment within the specified inundation zone at the local scale. The results are qualitative (i.e., high–medium–low vulnerability) and can be visualised through maps that can form the basis for decision-making and planning by several end-users (e.g., authorities, emergency services, insurance companies, and homeowners). Indicator-based methods may be reproduced in areas with similar building characteristics, or they can be adapted for areas with different building typologies.

28.3 Source Identification and Characterisation Used in the Recipe

The method used is designed for all tsunamigenic sources (e.g., seismic, volcanic, or landslide-related sources) and is based on a specific tsunami scenario. Information regarding the tsunami event and tsunami characteristics may be acquired from tsunami catalogues, photographs, historical documents, newspapers, etc. In the recipe presented here, the tsunami of 1743 (Tinti et al. 2004) is used as the worst-case scenario for the study area. The tsunami catalogue of the area comprises four events (occurred in 1627, 1731, 1743, and 1889) all of which were triggered by earthquakes. Given the incompleteness of the catalogue, it is difficult to determine a return period for tsunamis of this size.

28.4 Tsunami Modelling

The only information that is needed regarding the tsunami hazard is the wave height, which can be acquired through historical documentation, photos, or modelling. In this recipe, the wave height of the tsunami of 1743 was estimated through modelling. In more detail, Mastronuzzi et al. (2004) carried out a study focusing on large boulders that were found in Apulia, at the coast of Torre Squilace. The authors suggested that the boulders' origin is either the tsunami of 1667 (originated in Croatia) or the tsunami of 1743, that affected Apulia. Pignatelli et al. (2009) produced a flooding map of the area based on the size, weight, and rock density of the boulders, as well as their original position, and concluding that the tsunami had a minimum height of 4.94 m. This height was taken into account for the identification of the potential inundation zone, which was defined as the area between the coastline and the 5-m contour. The advantage of this approach is that when we have limited information about an event, we can still get a picture of its potential impact under the current circumstances.

Nevertheless, there is significant uncertainty in the identification of the potential inundation zone, since many sources of information including bathymetry, vegetation in the inundation area, and the role of existing structures in the propagation of the tsunami on land, are not considered.

28.5 Exposure Modelling

The approach focuses entirely on buildings that are located between the coastline and the contour of the height of the wave. The buildings can be digitised from an orthophoto if they are not provided by the authorities as a shapefile. As in some places, the available orthophotos/shapefiles may be old, the dataset can be updated through a field trip in the study area. The buildings considered herein are mainly residential, although commercial and hotel units are included too.

28.6 Vulnerability Modelling

The method for the assessment of vulnerability is chosen to satisfy the following needs:

- To focus on buildings at a local scale, as buildings are critical for the safety of the population.
- To enable decision-makers, governments, and planners to focus their resources on "vulnerability hotspots" rather than spreading the efforts across large areas.

The indicator identification is based on existing descriptions of past events. The selection, scoring, normalisation, and weighting procedures follow the guidelines of the Handbook on Constructing Composite Indicators (OECD 2008) and are based on two versions of the PTVA method (PTVA 3 and PTVA 4) described in the studies of Dall'Osso (2009, 2016) respectively. One of the most important parts of the development of an index is the weighting of the individual indicators since different weighting methods can lead to different vulnerability indices per building, which may affect decision-making and plans for the reduction of tsunami vulnerability and risk (Fig. 28.1). We use here a participatory method that involves expert judgement (Analytic Hierarchy Process—AHP). In fact, in this recipe we use two different variants: AHP based on the expert judgement of the authors (Dall'Osso et al. 2009) and AHP based on the judgements of experts (Dall'Osso et al. 2016). Apart from the weighting of the indicators, the classification of the vulnerability categories is crucial and depends on the aim of the study. The indicators are weighted and aggregated into an index assigned to each building. The indicators and resulting index can form a GIS database and can be mapped (see Fig. 28.1).

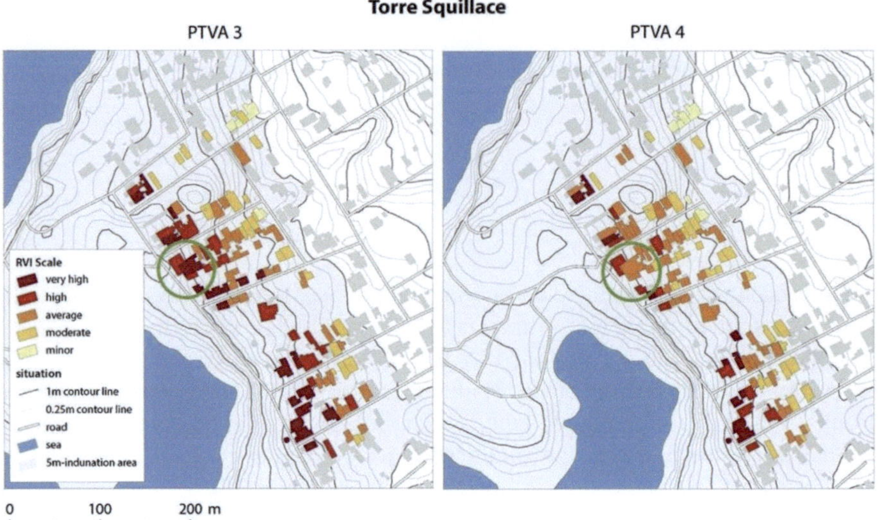

Fig. 28.1 Vulnerability maps using two different weighting methods in Torre Squillace (Apulia, Italy). The green circle shows one of the areas where the differences in the results of the vulnerability assessment are evident (Papathoma-Köhle et al. 2019)

28.7 Risk Communication

The mapping of the results is a valuable and practical tool that can be used for the communication of the results to different stakeholders and the public, although different issues must be considered, such as the choice of the classification method for the different vulnerability classes (e.g., natural breaks, equal intervals, quantiles, or standard deviation) that will be shown on the map, which in turn depends on the aim of the study.

The map indicates the potential inundation zone and the spatial pattern of the physical vulnerability of the buildings. It can be used by a variety of end-users in different ways (e.g., spatial and land-use planning, emergency planning). A disadvantage of the method as a tool for risk communication is the fact that, in such a scale, personal or sensitive data may be revealed. Moreover, although the uncertainties associated with the resulting vulnerability map or vulnerability scoring of buildings must be clearly communicated to the users (e.g., uncertainties concerning the inundation zone, the return period of the potential event, etc.), their quantification has not been done in this recipe.

28.8 Discussion

The geodatabase containing the indicators for each building and the overall vulnerability index is a dynamic tool that can be easily updated. The visualisation of the new spatial distribution of the results can be used to evaluate the BBB (Build Back Better) or as a basis for plans for the reinforcement of buildings or the development of local adaptation measures. Advantages associated with the method include the consideration of building characteristics, the support of decision-making (vulnerability hotspots), and the predictive power of indicators for areas without empirical data. Nevertheless, the following challenges are to be considered:

- The results are qualitative and cannot be directly translated into economic costs.
- The results of the assessment depend on the classification decided by the user.
- There is a high demand for data that may be obtained through time-consuming fieldwork.
- The weighting of the indicators is identified through expert judgement.
- The method may be transferable to other locations with similar types of houses. Further research is required to investigate the transferability of the index to other areas.
- The intensity of the tsunami is considered in the method based only on one parameter (i.e., wave height). Other properties of the process such as the direction of the flow, its velocity, and the duration of impact are not considered.

Alternative methods for the assessment of the vulnerability of buildings may include fragility curves and vulnerability functions. Fragility curves describe the probability of reaching a damage state given an intensity measure, whereas vulnerability functions express the relationship between the wave height and the degree of economic loss. These curves were developed after the two most recent large tsunami events (the 2004 Indian Ocean tsunami and the 2011 tsunami in Japan) because empirical data were available (Tarbotton et al. 2015). The advantage of the curves is that the method is quantitative and can "translate" the damages and impacts of specific events into monetary losses. It demonstrates, however, a number of disadvantages including the difficulty of transferring the function to different geographical areas and the requirement of a relatively large amount of empirical data.

28.9 Links to Data, Models and Other Resources

Supplementary material with the publication of Dall'Osso et al. (2016): https://link.springer.com/article/10.1007/s11069-016-2387-9#Sec18

References

Dall'Osso F, Dominey-Howes D, Tarbotton C, Summerhayes S, Withycombe G (2016) Revision and improvement of the PTVA-3 model for assessing tsunami building vulnerability using 'international expert judgement': introducing the PTVA-4 model. Nat Hazard 83:1229–1256

Dall'Osso F, Gonella M, Gabbianelli G, Withycombe G, Dominey-Howes D (2009) A revised (PTVA) model for assessing the vulnerability of buildings to tsunami. Nat Hazards Earth Syst Sci 9:1557–1565. https://doi.org/10.5194/nhess-9-1557-2009

Dominey-Howes D, Papathoma M (2007) Validating the papathoma tsunami vulnerability assessment model (PTVAM) using field data from the 2004 Indian Ocean tsunami. Nat Hazards 53:43–61

Mastronuzzi G, Pignatelli C, and Sanso P (2004) Assessment of catastrophic wave impact in Apulia region (Southern Italy). Risk Analysis 4:681–689

OECD (2008) Handbook on Constructing Composite Indicators: Methodology and User Guide. OECD Publications

Papathoma M, Dominey-Howes D (2003) Tsunami vulnerability assessment and its implications for coastal hazard analysis and disaster management. Nat Hazard 6:733–744

Papathoma M, Dominey-Howes D, Zong Y, Smith D (2003) Assessing tsunami vulnerability, an example from Herakleio, Crete. Nat Hazard 3:377–389. https://doi.org/10.5194/nhess-3-377-2003

Papathoma-Köhle M, Cristofari G, Wenk M, Fuchs S (2019) The importance of indicator weights for vulnerability indices and implications for decision making in disaster management. Int J Disaster Risk Reduct 36:1–12. https://doi.org/10.1016/j.ijdrr.2019.101103

Pignatelli C, Sanso P, Mastronuzzi G (2009) Evaluation of tsunami flooding using geomorphological evidence. Mar Geol 260:6–18

Tarbotton C, Dall'Osso F, Dominey-Howes D, Goff J (2015) The use of empirical vulnerability functions to assess the response of buildings to tsunami impact: comparative review and summary of best practice. Earth-Sci Rev 142:120–134. https://doi.org/10.1016/j.earscirev.2015.01.002

Tinti S, Maramai A, Graziani L (2004) The new catalogue of Italian Tsunamis. Nat Hazards 33(3):439–465. https://doi.org/10.1023/B:NHAZ.0000048469.51059.65

Chapter 29
R20—Monte Carlo Approach for Tsunami Economic Impact Estimation in a Pseudo-Probabilistic Framework

Alejandro González del Pino, Jorge Macías, Miguel Llorente Isidro, and Carlos Paredes Bartolomé

29.1 Introduction

The purpose of this recipe is to provide an estimation of the economic impact probability distribution that a random seismic-triggered tsunami may produce using Monte Carlo methods for seismic-source selection. The main necessity to accomplish this task is to provide a way of linking thousands of full-fledged tsunami inundation simulations with an estimation of direct economic loss for each one. The procedure uses the following main ingredients:

- Seismic fault catalogue with well-defined geometric parameters and standard errors of some of the parameters (QAFI database)—Sect. 3.2

A. González del Pino · J. Macías (✉)
Departamento de Análisis Matemático, Estadística e Investigación Operativa y Mathemática Aplicada, University of Málaga, Málaga, Spain
e-mail: jmacias@uma.es

A. González del Pino
e-mail: alexgp@uma.es

M. Llorente Isidro
Geological Risks and Climate Change, IGME (CSIC), Madrid, Spain
e-mail: m.llorente@igme.es

C. Paredes Bartolomé
Geological and Mining Engineering, Escuela de Minas y Energía, Universidad Politécnica de Madrid, Madrid, Spain
e-mail: carlos.paredes@upm.es

M. B. Sørensen et al. (eds.), *Probabilistic Tsunami Hazard and Risk Analysis*, Mathematics of Planet Earth 14, https://doi.org/10.1007/978-3-031-98115-9_29

- Parametric sea floor dislocation model for seismic fault sources—Sect. 4.2
- Tsunami modelling—Chap. 5
- High-resolution topobathymetric data—Sect. 5.2
- Spatially distributed exposure database—Chap. 6
- Vulnerability model and an algorithm relating tsunami simulation outputs to a spatially distributed economic database—Chap. 7

The recipe is suitable for studies aiming to assess direct economic losses due to tsunami impact for regional coastal zones. Given that it is based on maximum plausible earthquake approach, this recipe does not fully account for the aleatory variability of the source, as prescribed in standard PTHA. Thus, it may be assimilated to the framework FW1D (Working on a limited number of scenarios). In particular, it is suitable where economic and infrastructural information is available, either actual or hypothetical. An efficient tsunami simulation software, as well as high-resolution emerged and submerged terrain elevation data are mandatory in order to carry out the study. This type of study is not intended to be used for global analysis, as the high level of detail required for the tsunami simulations and impact estimation makes it unfeasible in such cases. The pseudo-probabilistic approach, which can be described as a Monte Carlo method, to generate the artificial fault source database, together with a general description of the economic estimation, is described in González et al. (2022).

29.2 Framework Used in Recipe

This recipe uses a pseudo-probabilistic framework for the generation of an artificially crafted seismic catalogue. Although likelihood of occurrence is associated with the maximum plausible earthquake (in terms of seismic magnitude), hence an estimation of annual exceedance chances may be derived, probabilities have not been specifically addressed in the simulation process. Moreover, no other probabilities have been considered other than the maximum plausible earthquake, therefore this recipe should not be considered as PTHA. Geometric fault known uncertainties are modelled and sampled using a quasi-random sequence to provide a wide coverage of possible scenarios. Then each simulated scenario is considered as input for the economic estimation algorithm. The main final output of this recipe is the economic impact frequency distribution. Additionally, relative flood uncertainty maps are produced as derived products.

The advantage of using the framework is, on the one hand, the generation of an artificially crafted seismic catalogue that covers very well the chosen input parametric space (space of variables with uncertainty considered for the work). On the other hand, it provides a very good approach to a probability density function out of the economic impact frequencies by using information on fault seismotectonic parameters (geometry and seismicity) and its uncertainties via Monte Carlo simulation. It provides an easy-to-implement economic estimation algorithm from the results of the numerical simulations as well. Additionally, the high-resolution nature of the topobathymetric data used in the simulations takes advantage of building-to-building data, making this approach very robust from a statistical perspective. The limitations of the framework are both the simplistic approximation of the source definition and the overlook of the epistemic uncertainties. Furthermore, the uncertainty is only accounted for three of the fault parameters and the vulnerability functions are quite elementary, not capable of distinguishing between different ground, type of buildings or type of objects within each building.

29.3 Source Selection and Activity Rate Characterization Used

This recipe represents the Spanish tsunami hazard owing to earthquake sources described in Sect. 3.2, using the following source selection models:

- Fault sources are used. Potential planar fault sources capable of producing seafloor deformation in SW offshore Iberia and nearshore and offshore western Mediterranean (the Ibero-Magreb-Algerian arch) were retrieved from the latest version of the QAFI database (García-Mayordomo 2022). The QAFI database compiles information on Quaternary-active faults in the Iberia region both onshore and offshore. The database provides basic geometric parameters (e.g., length, strike, dip), as well as a summary of the available evidence of Quaternary activity for each fault. A total amount of 62 faults were selected to be used in this study after careful review and update of available published information and considering the opinion of a number of experts gathered in a workshop devoted to this task held in 2017 (Expert Workshop 2017).
- The probabilistic treatment and parameters sampling were implemented in Python. The fault-source modelling assumes geometric values for which maximum seismic rupture is plausible, hence parameters such as dip angle, fault length and width, remain fixed in order to illustrate maximum potential values. With respect to the remaining parameters, some tests have been carried out to assess which of them may be described as the driving factors. It is clear that some parameters involve more uncertainty than others, such as strike, rake and slip. These three parameters are chosen to reflect the uncertainty. Normal distribution is considered for each parameter. However, given strike and rake are degree-angular magnitudes, they will be assumed to follow a Von-Mises distribution (also known

as a circular normal distribution), which is a close approximation to the normal distribution wrapped over a circle. Once the probability distributions are properly defined for strike, rake and slip parameters of each one of the considered fault sources, they are sampled using the Sobol sequences technique (Sobol 1976), which are a type of quasi-random sampling sequences, able to adequately cover the input parameter space with little amount of samples. For each one of the 62 well-known selected faults from the QAFI, 64 variants are sampled, leading to a total of 3,968 different seismic events for the artificial catalogue to be simulated.

The advantage of this source representation is, among others, the simplicity of its definition, allowing a straightforward determination of the probabilistic distribution for some of the geometric parameters. A crucial limitation is the lack of inclusion of heterogeneous slip that oversimplifies the complex rupture mechanism. Also, not including smaller magnitude earthquakes or other sources than earthquakes limit the analysis.

29.4 Source Mechanics Modelling Used in Recipe

An Okada type elastic analytical model (see Seafloor Deformation in Sect. 4.2) is used to simulate the co-seismic seabed displacement response for each fault source. All faults are assumed to be composed of one single segment, in addition to a homogeneous slip model. This is the simplest type of seafloor deformation model among those in Sect. 4.2. The strengths and weaknesses of this method are shared with those described for the source selection procedure above.

29.5 Tsunami Modelling

The model uses the following methods and steps as described in Chap. 5 following all the different steps from data preparation, tsunami generation, tsunami modelling, and inundation modelling.

- For the first step on data preparation (see Sect. 5.2), a heavy workload is concentrated on generating the emerged and submerged elevation data. Nearly 450 high-resolution meshes (five-metres pixel resolution) are employed in simulating inundation along all Spanish SW Atlantic and western Mediterranean coastlines (including Balearic Islands). Moreover, each computed simulation consists of a four-level nested grid system with resolutions 640 m - > 160 m - > 40 m - > 5 m. A total amount of 565 computational grids were generated using both Python and ESRI ArcGIS Pro©. Raw datasets used to construct the final topo-bathymetric meshes were retrieved from different sources: GEBCO, EMODnet, Ministerio para la Transición Ecológica y Reto Demográfico (MITECO), Instituto

Geológico y Minero de España (IGME), Instituto Geográfico Nacional (IGN) and Red de Información Ambiental de Andalucía (REDIAM).

- For the second step (see Sect. 5.3, interface model between the earthquake unit source and tsunami model), the co-seismic seabed displacement obtained from the Okada model is directly transmitted to the free-surface (no Kajiura filter is applied), providing the initial condition for tsunami modelling.
- The third and fourth steps (tsunami propagation and inundation) are simulated within a single code. The Nonlinear Shallow Water (NLSW) equations are considered and the GPU-based numerical code Tsunami-HySEA is used (Macías et al. 2017 and references therein). This model solves the conservative form of the NLSW equations using finite volume techniques. One key feature implemented in this numerical model is the possibility of using two-way nested meshing for high-resolution simulations. The nested mesh system approach allows computing of open ocean and offshore wave propagation using meshes of lower pixel resolution. Near the coast, however, higher pixel resolution meshes are used, both to reproduce wave shape and to capture complex inundation features.
- A total amount of 3,968 tsunamis were simulated. Each one of them inundated every portion of the considered coastlines. The large extension of the coastlines to be inundated forced each source simulation to be partitioned into 37 groups (divided by provinces). In the end, 109,312 high-resolution tsunami inundations were computed.

Regarding model validation (Sect. 5.6), the Tsunami-HySEA code has been widely tested (Macías et al. 2017 and references therein) and has also been validated and verified following the standards of the National Tsunami Hazard and Mitigation Program (NTHMP) of the US (Macías et al. 2017, 2020a, b).

The advantage of this simulation procedure is that the fault rupture process, the corresponding initial water displacement and the calculation of its propagation and inundation along the selected nested grid system are all incorporated into a single code. Information such as inundation depth, maximum current speed, maximum mass flow, etc., is produced for each five-metre pixel resolution cell. The disadvantage of this simulation procedure is that not homogeneous highly detailed bathymetric data is available, generating all computational grids is a heavy time-consuming task. The model does not include building information, thus adding another layer of inaccuracy. The large amount of simulations computed requires large computational resources, making it mandatory to apply for extensive High Performance Computer (HPC) projects.

29.6 Exposure Modelling

Exposure data is incorporated into the study through an exhaustive analysis of the cadastral database. Cadastral values are assigned by Spanish administrations based on data about the property itself (true ground and building) and neighbouring effects (proximity to other services or infrastructures) and it is used for tax purposes. Hence, it can be used as a very good proxy of the value of a property. The cadastral value of a building is usually much lower than the market value of the property due to several factors (it does not integrate economic benefits, true ground value nor market interest fluctuations). However, the cadastral value is homogeneously estimated over the whole country and it is readily available.

Exposure data come in the form of vector data files of polygon type, which represent cadastral parcels that may integrate different buildings and properties. Each cadastral parcel used for the study only contains data about building value, as it is assumed that ground cannot be lost, hence its value is retained. Each cadastral parcel value is increased by 30% to account for items within properties (furniture, appliances). Further work was undertaken in order to assign people and vehicle distribution to each parcel, according to the information gathered from the Instituto Nacional de Estadística (INE) and parcel value.

The hypothesis for the exposure modelling used for this study state that whole cadastral constructed value is transferred to the base floor, which is assumed to have an extension of three metres height (from ground height to three metres above that); and the value of the objects within a cadastral parcel amounts to a 30% of the total constructed value.

29.7 Vulnerability Modelling

Vulnerability functions adopted for this work relate tsunami hazard and exposure in such a way that enables the estimation of direct building and contained-objects losses. Following the empirical vulnerability description from Sect. 7.2, two vulnerability functions correlating the Explanatory Variable "maximum water height" to the Response Variable "damage percentage" are established. One function accounts for building damage, and the other one for the contained objects. Each one is applied to individual flooded cells in the economic estimation algorithm. The former consists of a piecewise function of value zero if maximum water column height is less than two metres, and value 1 otherwise. The latter is a linear continuous function that takes value 0 for maximum water column height of 0.25 m, linearly increasing till value 1 for maximum water column height of 3 m.

Two additional vulnerability functions are prescribed to estimate the number of people and vehicles affected, although these are not included in the economic frequency distribution approximation. For vehicles (resp. people), the function is piecewise constant of value zero if maximum water column height is less than 0.2 m (resp. 0.5 m) and value 1 otherwise.

To match the exposure modelling with the inundation results, each polygon representing a cadastral parcel is transformed into a raster-type set of five-metre pixel resolution, where each cell in the raster set accounts for twenty-five squared metres constructed value of the parcel. The values for the estimation of the final frequency distribution were calculated by assigning an economic value to each flooded cell, taking into consideration the maximum water height, the cadastral value registered on that cell and the corresponding vulnerability function. Then, all economic cell values are added together, producing a final total economic value. This iterative procedure was implemented in Python, which, in the end, allows relating a direct economic loss in a scenario-by-scenario manner.

The advantage of this procedure is that it provides highly detailed economic impact data for each simulated seismic source, enabling exhaustive analysis at various scales. It also allows the development of an economic frequency distribution. The main disadvantage is that the data used in this study requires substantial anonymization given its very high detail.

A shortcoming of the study is the lack of distinction between building types. Apart from the fact that the prescribed vulnerability functions are too simplistic, there are no site-specific models and the uncertainties associated with the vulnerability functions are not taken into account in the estimation of losses.

29.8 Risk Communication

The aim of this recipe is to ensure preparedness in dealing with the economic cost of a tsunami in a particular area, within a region or in a whole country or several countries. Regarding risk communication, a key question is what to communicate (Chap. 8). In this particular case, the message that should be conveyed may include authority readiness and public awareness through comprehensive risk maps and risk diagrams as well as tsunami and seismic hazard maps, faults and earthquake cartographic charts and complementary explanatory material.

29.9 Discussion

The tsunami risk model regarding economic impact naturally demanded a highly elaborate study, which required:

1. Preliminary analysis of the potential faults affecting the coastal areas of interest.
2. Integration of loss metrics to produce a probability function for the risk.
3. Estimation of inundation-prone areas, exposure analysis, and risk aggregation.
4. Use of country-scale source elevation data to fulfill the desired goals.

The selected approach to proceed with the study has some advantages:

- High level of detail obtained in the simulations and the economic impact aspects.
- Economic losses data produced for the main populated areas along the Ibero-Magreb-Algerian arch coasts.
- Data produced throughout the work can also be used to understand nearshore and onshore flood distribution from an arbitrary tsunami of Atlantic or Mediterranean

origin, facilitating the production of stochastic inundations maps and evacuation routes for people living near the coast. Not to mention the usefulness of data regarding maximum velocity and maximum mass flow in approaching the evaluation of particular structural damage.

Due to the high-resolution nature of the study, in terms of pixel count per portion of terrain covered by inundations, and considering the scale of it (there are no genuinely cross-national studies of this nature), the following limitations arise:

- Actual state of the study does not allow generation of hazard curves, thus not being able to be fully incorporated in a PTHA framework.
- Fault models are highly simplified, only accounting for homogeneous slip models and one-segment faults.
- The great amount of full-fledged simulations required HPC resources. To be precise, this work demanded about 950,000 GPU hours (Nvidia V100) to be fully completed.
- The algorithm for economic impact computation is rather simple, in which vulnerability functions do not consider different types of buildings and objects contained within cadastral parcels. Besides, only direct economic losses are accounted for, excluding topological issues (such as interruption of services) and derived effects (such as loss of opportunities or subsequently triggered effects).
- Uncertainty is only considered for three of the Okada parameters that define a fault, thus damage spectra might present some bias from this perspective.

29.10 Links to Data, Models and Other Resources

- Tsunami-HySEA code: https://zenodo.org/record/6400815#.ZB8f8OzMJQJ
- QAFI database v.4: http://info.igme.es/qafi/
- GEBCO https://www.gebco.net
- EMODnet: https://emodnet.ec.europa.eu/en
- Cadastral database: https://sedecatastro.gob.es/

References

Expert Workshop (2017) Expert Meeting on Tsunamigenic Sources with Potential Impact in the Iberian Coast, Balearic and Canary Islands, Málaga, Spain, 6–7 Nov 2017. https://shorturl.at/qUW78

González A, Fernández M, Llorente M, Macías J, Sánchez-Linares C, García-Mayordomo J, Paredes C (2022) Pseudo-probabilistic design for high-resolution tsunami simulations in the Southwestern Spanish Coast. GeoHazards 3:294–322. https://doi.org/10.3390/geohazards3020016

García-Mayordomo J, Martín-Banda R (2022) Guide for the use of QAFI v.4. In: http://info.igme.es/qafi/-Instituto Geológico y Minero de España (IGME), CSIC, Madrid, Spain. https://doi.org/10.13140/RG.2.2.12873.01121

Macías J, Castro MJ, Ortega S, Escalante C, González-Vida JM (2017) Performance benchmarking of Tsunami-HySEA model for NTHMP's inundation mapping activities. Pure Appl Geophys 174:3147–3183. https://doi.org/10.1007/s00024-017-1583-1

Macías J, Castro MJ, Escalante C (2020a) Performance assessment of Tsunami-HySEA model for NTHMP tsunami currents benchmarking. Laboratory Data Coast Eng 158:103667. https://doi.org/10.1016/j.coastaleng.2020.103667

Macías J, Castro MJ, Ortega S, González-Vida JM (2020b) Performance assessment of Tsunami-HySEA model for NTHMP tsunami currents benchmarking. Field Cases Ocean Model 152:101645. https://doi.org/10.1016/j.ocemod.2020.101645

Sobol IM (1976) Uniformly distributed sequences with an additional property of uniformity. Zh Vychisl Mat Mat Fiz 16(5):1332–1337; U.S.S.R. Comput Math Math Phys 16(5):236–242. https://doi.org/10.1016/0041-5553(76)90154-3

Chapter 30
R21—GITEWS End-To-End Early Warning

Jörn Lauterjung, Harald Spahn, and Jörn Behrens

30.1 Introduction

The German-Indonesian Tsunami Early Warning System (GITEWS) project was implemented after the devastating Tsunami in the Indian Ocean on December 26, 2004 over a period of 6 years and was followed by the PROTECTS project (3 years) to accompany the operational Indonesian Tsunami Early Warning System by extensive training and education for the operation centre and for the communities at risk during the first operational years of the system. The project follows an "end-to-end" approach to cover the complete warning chain from rapid hazard detection over decision support to capacity development of communities at risk and the implementation of disaster reduction measures. A tsunami scenario database including propagation and coastal inundation serves as potential input for probabilistic hazard and risk analysis. The corresponding hazard and risk maps on the other hand, serve the purpose to derive warning products in line with consistent response procedures. Consequently, the warning levels in an event are compatible with corresponding response activities (see Fig. 30.1 for examples of such maps).

J. Lauterjung
Geophysics, Helmholtz Centre Potsdam GFZ German Research Centre for Geociences, Potsdam, Germany
e-mail: joern.lauterjung@gfz.de

H. Spahn
Independent Consultant, Apen, Germany
e-mail: harald.spahn@web.de

J. Behrens (✉)
Department of Mathematics, Universität Hamburg, Hamburg, Germany
e-mail: joern.behrens@uni-hamburg.de

M. B. Sørensen et al. (eds.), *Probabilistic Tsunami Hazard and Risk Analysis*, Mathematics of Planet Earth 14, https://doi.org/10.1007/978-3-031-98115-9_30

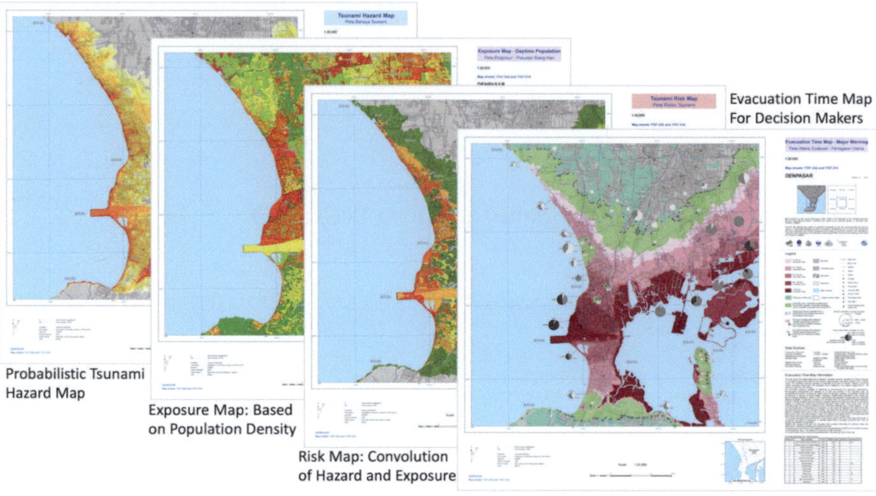

Fig. 30.1 A sequence of map products for Denpasar, Bali, from the GITEWS end-to-end warning system: From a scenario database, a probabilistic tsunami hazard map (indicating the probability of inundation) is generated. This together with a population-density-based exposure map serves as input to a risk map. Taking population density and composition, hazard level and geographical features (elevation, vertical evacuation shelters) into account, an evacuation time map serves decision makers. These Maps were produced by the German Aerospace Center (DLR) in the framework of the German-Indonesian Tsunami Early Warning System (GITEWS, see Strunz et al. 2011)

Indonesia is faced with all "types" of tsunamis: near-field and far-field tsunamis generated by earthquakes, landslides and volcanic activities, and originating from the Indian Ocean, the Pacific, and sources in connected seas in eastern Indonesia. GITEWS concentrated on earthquake induced near-field tsunamis originating from the Sunda trench (typically with travel times from the source to the coastline between 20–40 min) so that the system was technically designed with a focus on high speed, incorporating early input parameters with high uncertainties. Near-field tsunami forecasting is particularly challenging due to the necessity for precise characterisation of the earthquake rupture including details of the slip distribution in order to avoid excessive epistemic uncertainties in the forecast. In order to provide an early warning, this has to be achieved as quickly as possible. From the highest political level, it was demanded that a warning should be disseminated within 5 min after an earthquake. Seismological observations can only provide the primary earthquake parameters such as location, depth and magnitude within 2–4 min. Thus, the epicentre and magnitude are poorly constrained immediately after the earthquake, and errors in both depth and location of up to 50 km can occur. Therefore, an assessment of tsunami potential and—if positive—selection of propagation models from a tsunami scenario database has to be made on the basis of highly uncertain parameters.

Additionally, in order to be an effective tool for mitigation, the hazard assessment needs to be complemented with disaster management strategies that are viable on a local level and compatible with the magnitude of the potential hazard. Even risk assessment may be incorporated into the decision-making process. These aspects were considered in designing the end-to-end system, as described below.

30.2 Framework Used in Recipe

The GITEWS approach to end-to-end early warning follows FW1A, i.e. is based on a large number of scenarios and assesses hazard in a probabilistic way to derive hazard maps. Since the assessment focuses on assisting the early warning system, and the hazard and risk analysis serves the purpose of enabling consistent mitigation measures in case of an event, return period analysis is not considered for the early warning purpose. The system uses a large number of seismic source events along the Sunda Arc and associates approximate probabilities to the events (Strunz et al. 2011). Each scenario is used for propagation and inundation modelling, giving rise to flow depths at the whole Indian Ocean coast of Indonesia with high-resolution simulations in three pilot areas (Padang/Western Sumatra, Cilacap/Java, Denpasar/Bali). From the occurrences of inundation in certain locations, inundation likelihoods are derived, yielding probabilistic tsunami hazard data. Finally, simple vulnerability indices (day/night population density, horizontal evacuation times) serve as a preliminary risk assessment (simplified variant of FW2). A consultation process involving national and local stakeholders (science, politics, civil society) and decision makers is employed to derive the vulnerability indices and determine acceptable risk levels for eventually deriving evacuation zones corresponding to warning levels offered by the national monitoring and warning centre (InaTEWS).

30.3 Source Identification and Characterisation Used in Recipe

Main hazard source for tsunami in the Indian Ocean for Indonesia is the Sunda Arc subduction zone, reaching from the northern edge of Sumatra to Papua New Guinea, almost parallel to the coastline of the major islands Sumatra, Java, and Flores Islands facing the Indian Ocean. Hence, we are dealing over a distance of ~5.000 km with a near-field tsunami situation. The system uses seismic sources according to Sect. 3.2. To handle this situation the subduction zone is divided evenly into hypothetical earthquake hypocentres. To constrain many seismic parameters (such as epicentre location, depth, focal mechanism) for fast early warning purposes, pre-existing information such as fault geometry (based on seismicity) and scaling laws are employed. A gridded patch-based micro plate discretization leads to approximately 3750 potential locations. Considering scaling laws, patches are combined to form earthquake scenarios of moment magnitudes (M_w) between 7.5 and 9.0 (Babeyko et al. 2010).

The tsunami propagation model is calculated for earthquakes in several depths (0–60 km) and magnitudes in the range $M_w = 7.5–9$ with 0.3 magnitude steps. Strike and dip are oriented according to the Benioff zone. Tsunami models are precalculated for every hypothetical hypocentre with the complete range of magnitudes and put in a database.

30.4 Source Mechanics Modelling Used in Recipe

The seafloor displacement due to earthquake (see Sect. 4.2) sources are modelled based on the following characteristics (Babeyko et al. 2010):

- Width and length of source areas follow scaling laws (Wells Coppersmith, 1994),
- Epicentre equals the geometrical centre of rupture area,
- Slip distribution has a symmetrical Gaussian shape with maximum at epicentre location,
- Coseismic slip distribution is computed from a scaling law (see Babeyko et al. 2010).

A sensitivity analysis showed that such simplified rupture models were able to represent the coastal hazard of complex slip distributions in the early warning context. However, some caution needs to be taken, since according to Davies and Griffin (2020) sensitivity to complex slip distributions may be large.

The described modelling approach is implemented in the software tool RuptGen, specifically developed in the GITEWS project. It provides output on a given unstructured mesh as defined in the tsunami modelling tool TsunAWI (Behrens 2008; Rakowsky et al. 2013).

30.5 Tsunami Modelling

A new finite element nonlinear shallow water tsunami model with non-uniform locally refined mesh for the Indian Ocean was developed (TsunAWI, Behrens 2008, Rakowsky et al. 2013). This model allows seamless propagation and inundation modelling. Approx. 4000 scenarios for the Sunda-Arc were computed, based on seismic sources from RuptGen. Each scenario contains seismic information, such as hypocenter location, and earth crust dislocation vectors corresponding to measurable GNSS dislocations, as well as wave height, arrival time and inundation information. With respect to the modelling Chap. 5, this is an example of a shallow-water based, non-uniform modelling approach, including inundation, where the interface to the source is handled by a time-dependent ocean bottom uplift.

For tsunami early warning a novel approach to diminish uncertainty early after the earthquake was developed, combining different sensor systems in a multi-sensor uncertainty reducing approach (see Behrens et al. 2010). The tsunami model computes wave heights, arrival times, as well as inundation depths in each grid cell of a variable resolution mesh of approx. 10 million cells covering the entire Indian Ocean. In pilot areas the mesh resolution is as high as approximately 50 m whereas on average the coastal area is resolved with approximately 200 m mesh resolution. While the latter is not sufficient for detailed inundation mapping, it suffices for a preliminary hazard assessment and classification of warning levels.

Additional high-resolution local regular mesh model simulations of mesh resolutions of 10's of metres were conducted in pilot areas. These simulations complemented TsunAWI unstructured propagation and inundation modelling results for detailed hazard mapping.

30.6 Exposure Modelling

Based on the tsunami simulation database a probabilistic hazard analysis was performed with the aim to develop products useful for the early warning process. On the one hand these have been "normal" hazard and risk maps (with the restrictions mentioned in the previous paragraph) for Indonesia for disaster risk reduction (DRR) purposes, but also products reflecting and supporting the warning process (see Strunz et al. 2011).

The warning system in Indonesia delivers warnings on the geographical scale of district administrative units along the coastline. Depending on the source region of a tsunami, each district receives a warning message, either no tsunami, tsunami advisory (height < 0.5 m), tsunami warning (0,5 m < height < 3 m) and major tsunami warning (height > 3 m). For these wave heights inundation lines have been calculated and respective evacuation maps or time-for-evacuation maps have been developed. The evacuation strategies were derived based on the previously developed hazard and risk maps, considering exposure and vulnerability assessments.

Beyond hazard and exposure modelling, risk assessment was only performed in the three pilot regions, for the remaining (major) part of the Indonesian coastline it was not performed (for resource reasons, personal cost, computing time (for high resolution inundation modelling), data availability).

30.7 Risk Communication

The results from the tsunami and exposure modelling were fed into processes for a range of practical applications involving different stakeholders. As a consequence, different methods of risk communication were used, each of which was agreed with the partners involved.

The modelling results played a fundamental role in the design of the tsunami warning processes in Indonesia, particularly with regard to the development of geographic differentiated warnings and the definition of warning levels to be communicated to the users. This work process required close bilateral collaboration with the warning provider and gathering feedback from other related stakeholders and the user perspective. For this purpose, small group work and workshops were the most important methods applied.

Furthermore, the provision of local hazard and inundation maps derived from the modelling results played a central role, as those were considered as the prerequisite for adequate local preparedness and mitigation efforts. A number of challenges needed to be addressed here. In absence of a guiding standard, a variety of tsunami hazard maps developed by various national and international scientific institutions using a range of different approaches, methodologies and data were circulating in the country at that time. In addition, each map is always subject to uncertainties. These must be understood and taken into account when decisions are made about the designation of safe zones, which in the end is a political decision. To enable local governments to make the necessary decisions, the dialogue format 'Science meets Politics' was promoted, allowing scientists and local decision makers to work hand in hand. The initiative was implemented in selected pilot areas through a series of facilitated workshops and aimed at an agreement on reference scenario(s) and zoning criteria for an officially approved local tsunami hazard map and the provision of recommendations for official policies as a framework for tsunami early warning, evacuation planning and preparedness activities. An important part of this process has been to compare different scientific approaches and perspectives that underlie the different hazard maps that were already available and to develop aligned recommendations for local policy.

The developed hazard and inundation maps were furthermore used in a structured capacity development process to enable local communities to better prepare for and respond to a tsunami emergency. The maps played an important role in raising awareness and improving knowledge regarding the local tsunami threat, but there was always a need for a detailed explanation of the underlying natural tsunami generation processes, the scenario approach, and how the results were presented on the map. Such explanations were provided mostly in risk communication formats like meetings and workshops with local stakeholders and usually required some expert involvement. It was avoided to present and discuss these maps with the general public as the risk of misinterpretation or misunderstanding was considered too high. Instead, it was recommended to share evacuation maps to the public as those provide concrete guidance on how to respond in an emergency.

The approaches described above, the methods and tools used, and the main results were systematically documented throughout the intervention and made available on a website (https://www.gitews.de/tsunami-kit/index_en.html), which remains relevant even 10 years after its launch.

30.8 Discussion

30.8.1 Practical Application in the Warning Process

The decision support system developed during the GITEWS project allows the display of different situation pictures. The "standard" view is a map of the tsunami region with propagating wave front in real time display. The administrative districts along the affected coastline are coloured according to their warning status (white, yellow, orange, red). The warning status results from the calculated wave height at the coastline. Instead of this "standard" view also a risk view (depending on the warning level) can be displayed, which was precalculated from inundation (of the selected scenario), exposure information and vulnerability data.

30.8.2 Practical Application for Local Disaster Risk Management

Most important for local disaster risk management was the provision of zoned hazard maps. These maps followed a zoning concept that distinguished between high and low to medium probability tsunami hazard zones. These two zones were directly related to the warning levels used by the National Tsunami Warning Center (NTWC). The high probability hazard zone included all tsunamis within the lower warning level (tsunami wave at coastlines < 3 m) and was represented by a uniform red colour, while the low and medium probability zone covered all tsunamis > 3 m corresponding to the high warning level. In this second zone the probability was visualised by different colouring.

These hazard maps played an important role in the evacuation planning processes, as they allowed local stakeholders to assess the local tsunami risk in a more differentiated way and helped to delineate evacuation zones, especially in areas that were considered difficult to evacuate. Initially, the warning level zoning concept of the hazard maps was also used for evacuation planning, but practical experience showed that the designation of two evacuation zones according to the NTWC warning levels is not easy to explain to the general public and even more difficult to communicate during an emergency. In addition, evacuation time maps were developed for some densely populated areas, which supported the discussion on appropriate evacuation strategies, especially regarding vertical evacuation options.

30.8.3 Conclusions

From 2005 to 2014, the GITEWS (2005–2011) and Protects (2011–2014) projects ran as part of the German response to the Boxing Day tsunami of 26 December 2004.

These projects (total funding €65 million) developed the core elements of a national end-to-end tsunami early warning system in Indonesia from scratch. The system, known as InaTEWS since 2011, is operated entirely and independently by Indonesian institutions and was accredited by UNESCO in 2014 as one of three regional centres in the Indian Ocean. The system was initially designed for seismically induced tsunamis along the Sunda Trench, but was later extended to the entire country, and a recent project is now investigating how non-seismically induced tsunamis can also be taken into account. Indicators of the success of the projects and the implementation of an operational early warning system were as follows:

- Sufficient funding and adequate time to implement and support a pre-operational and initial operational phase.
- Political commitment and backing, a formal cooperation agreement and a regular steering mechanism among German and Indonesian partners. Establishment of trusting cooperation and communication at eye level.
- Sufficient time and resources to implement a comprehensive training and capacity development program.
- Integration of all international donors who delivered instruments and subsystems.
- Integration of all activities into international coordination activities of UNESCO.
- Permanent presence of German staff in Indonesia to collaborate with the Indonesian key actors providing direct support and thereby contributing to build confidence as well as to avoid misunderstandings and misinterpretations.
- Integration of operational processes and developed knowledge products into existing structures with an end-to-end perspective. This also supports the sustainability of the system and the investment in future developments and improvements.

Main obstacles and difficulties encountered related principally to a rather difficult and hesitant cooperation between institutions for monitoring relevant parameters, especially when they belong to different ministries, and the provision of data (e.g. shallow water bathymetry, open access to monitoring data, demographic data).

30.9 Links to Data, Models and Other Resources

- Link to special Issue with scientific articles describing the components of GITEWS: https://nhess.copernicus.org/articles/special_issue100.html
- Link to TsunAWI: https://tsunami.awi.de/index.html
- Lint to InaTEWS: https://inatews.bmkg.go.id/
- Link to GITEWS Project: https://www.gitews.org/
- Tsunami Kit: https://www.gitews.org/tsunami-kit/index_en.html

References

Babeyko A, Hoechner A, Sobolev SV (2010) Source modeling and inversion with near real-time GPS: a GITEWS perspective for Indonesia. Nat Hazards Earth Syst Sci 10:1617–1627. https://doi.org/10.5194/nhess-10-1617-2010

Behrens J (2008) Unstrukturierte finite Elemente Simulationen für die Tsunami Frühwarnung in Indonesien, NAFEMS Magazin 10:41–51. https://hdl.handle.net/10013/epic.32423.d001

Behrens J, Androsov A, Babeyko A, Harig S, Klaschka F, Mentrup L (2010) A new multi-sensor approach to simulation assisted tsunami early warning. Nat Hazards Earth Syst Sci 10:1085–1100. https://doi.org/10.5194/nhess-10-1085-2010

Davies G, Griffin J (2020) Sensitivity of probabilistic tsunami hazard assessment to far-field earthquake slip complexity and rigidity depth-dependence: case study of Australia. Pure Appl Geophys 177:1521–1548. https://doi.org/10.1007/s00024-019-02299-w

Rakowsky N, Androsov A, Fuchs A, Harig S, Immerz A, Danilov S, Hiller W, Schröter J (2013) Operational tsunami modelling with TsunAWI-recent developments and applications. Nat Hazards Earth Syst Sci 13:1629–1642. https://doi.org/10.5194/nhess-13-1629-2013

Rudloff A, Lauterjung J, Münch U (eds) (2011) The GITEWS project (German-Indonesian Tsunami Early Warning System). https://nhess.copernicus.org/articles/special_issue100.html. Accessed 30 Mar 2023

Spahn H (2008) Our experience-Science meets Politics, Fact Sheet available at https://www.gitews.de/tsunami-kit/en/E1/our_experience/Science%20meets%20Politics.pdf

Spahn H, Hoppe M, Kodijat A, Rafliana I, Usdianto B, Dwi Vidiarina H (2013) Walking the last mile: contributions to the development of an end-to-end tsunami early warning system in Indonesia. In Friedemann and Zschau (eds) Early Warning for Geological Disasters, Scientific Methods and Current Practice, Springer Berlin, Heidelberg. https://doi.org/10.1007/978-3-642-12233-0

Strunz G, Post J, Zosseder K et al (2011) Tsunami risk assessment in Indonesia. Nat Hazards Earth Syst Sci 11:67–82. https://doi.org/10.5194/nhess-11-67-2011

Tsunami-Kit. A collection of guidelines, material on early warning, with checklists and tools for practical implementation, portrays experiences and best practices, presents a compilation of various outreach materials and offers further resources. https://www.gitews.org/tsunami-kit/index_en.html. Accessed 30 Mar 2023

TsunAWI Web. https://tsunami.awi.de/index.html. Accessed 30 Mar 2023

Wells DL, Coppersmith KJ (1994) New empirical relationships among magnitude, rupture length, rupture width, rupture area, and surface displacement. Bull Seism Soc Am 84:974–1002

Chapter 31
R22—Probabilistic Tsunami Forecasting for Tsunami Warning

Stefano Lorito and Jacopo Selva

31.1 Introduction

This recipe illustrates the Probabilistic Tsunami Forecasting (PTF), an application of the ingredients more commonly used for long-term PTHA to the case of tsunami warning. This PTF recipe regards tsunamis generated by *Earthquake Sources* (Sect. 3. 1), that is, tsunamis generated directly by the co-seismic displacement of the sea floor.

To identify and characterise the sources both *Fault and Earthquake Catalogues* are used. The sources are modelled both as *Fault and Gridded Sources*. The *Tsunami Source Modelling* (Sect. 4.1) is performed with a two-step procedure. First, the *Slip Distribution* is obtained for each individual earthquake scenario included in the earthquake source model. Then, the *Seafloor Deformation* is calculated for each slip distribution.

The *Tsunami Modelling* (Sect. 31.5) is performed on a realistic *bathymetric model* (Sect. 5.2), applying a *Filter* to transform the seafloor displacement into a tsunami initial condition (Sect. 5.3). The *Propagation* (Sect. 5.4) of the tsunami is performed from the offshore up to a given point of interest in front of the coast with a numerical simulation, using linear combinations of pre-calculated scenarios, to which the Green's law amplification is applied.

The source probabilities are finally combined with the output of the numerical simulations. The PTF output is the probability distribution of the tsunami intensity at given forecast points. The PTF can be seen as a PTHA performed in real time for the ongoing potentially tsunamigenic event.

S. Lorito (✉)
Istituto Nazionale di Geofisica e Vulcanologia, Roma, Italy
e-mail: stefano.lorito@ingv.it

J. Selva
Dipartimento di Scienze della Terra, dell' Ambiente e delle Risorse (DiSTAR) , Università di Napoli Federico II, Naples, Italy
e-mail: jacopo.selva@unina.it

M. B. Sørensen et al. (eds.), *Probabilistic Tsunami Hazard and Risk Analysis*, Mathematics of Planet Earth 14, https://doi.org/10.1007/978-3-031-98115-9_31

This is a good recipe for tsunami warning centres to manage the large uncertainty characterising the earthquake source parameter estimates, and to combine the uncertainty regarding the source with that associated with numerical tsunami modelling, as described in more detail hereafter.

A pre-calculated ensemble of tsunami scenarios, with their long-term rates must be prepared in advance. Applying the recipe in real-time is then fast and computationally cheap, as it becomes essentially a query from a database, followed by simple transformations to prepare the probability of the tsunami at each forecast point. The recipe also allows to convert the PTF output into alert levels, based on a pre-defined conversion rule.

This workflow has been published as Selva et al. (2021).

31.2 Framework Used in Recipe

This recipe uses Framework 1A (FW1A, Sect. 2.2), the scenario-based with many scenarios and uncertainty treatment. FW1A is selected because tsunami warning is an extremely critical activity principally aimed at saving lives. It then requires the best possible approach to uncertainty quantification. Since the uncertainty regarding the source parameters is large in the first minutes after an earthquake occurrence, it is required to explore a large number of potential scenarios. Using pre-calculated scenarios makes this task feasible in a very short amount of time.

The goal of PTF is to provide the hazard curve H(x) for the tsunami intensity metric x, given that a potentially tsunamigenic earthquake event EV has occurred:

$$H(x) = P(x > x_{th}|EV, FP) \tag{31.1}$$

$H(x)$ is the probability of exceedance of different thresholds x_{th}, conditional upon the occurrence of the seismic event. FP is a generic forecast point. The tsunami intensity metric x can be, for example, the maximum run-up, or the tsunami wave height in front of the coast. The uncertainty in EV, in turn, can be expressed by the probabilistic estimates of the earthquake parameters, such as an empirical earthquake magnitude probability density function—a histogram—, and a covariance matrix, or a likelihood function, for the earthquake hypocentre location. The pre-defined scenario database is described by the earthquake source parameter list (σ_i: strike, dip, rake, rupture area and slip distribution) with their time-independent probability estimated prior to the specific earthquake occurrence. To quantify $H(x)$ we apply the total probability theorem:

$$H(x) = P(x > x_{th}|EV, FP) = \sum_i \mathrm{P}(x > x_{th}|\sigma_i, FP)\mathrm{P}(\sigma_i|EV) \tag{31.2}$$

In practice, to obtain $P(\sigma_i|EV)$, we extract a subset of $\{\sigma_i\}$ consistent with EV by truncation of the summation according to a given level of accuracy, and we combine the long-term probabilities with the uncertainty on the source inversion from the monitoring. $P(x > x_{th}|\sigma_i, FP)$ is instead the pre-calculated tsunami scenario probability, obtained from numerical simulations of individual scenarios combined with a log-normal distribution that describes modelling uncertainties.

The procedure can be repeated if new information regarding EV becomes available. EV, in principle, may also contain measurements of the ongoing tsunami, and/or be based on tsunami data assimilation.

The main advantage of PTF is that the uncertainty is formally quantified. The rule to convert the PTF into an alert level can then be decided in advance, and automatically applied in real time. This reduces the need to make difficult decisions in a regime of large uncertainty.

The main limitation is that the PTF is as good as the models and assumptions used are. Despite the efforts to take everything into account, the sparseness of the tsunami phenomenon makes calibration and validation a challenging task, and moreover there may always be "hidden" uncertainties (sometimes called *unknown unknowns*).

31.3 Source Identification and Characterisation Used in Recipe

The real-time estimates of magnitude and location (epicentre and focal depth) feature some uncertainty, which depends on the density and azimuthal coverage of the seismic stations used, and on the accuracy of the seismic velocity model used to invert for these parameters. The uncertainties affecting magnitude and location tend to decrease with time as the seismic signal continues to be acquired and as more stations are reached by the seismic waves and used for the estimation. Several earthquake parameters estimates are not robust enough or can be totally unconstrained in the first minutes after the earthquake origin time. This regards the faulting orientation (strike, dip) and style (rake); and the source complexity, as slip can depart in different directions from the earthquake location and create very heterogeneous spatial distributions (for example, as far as the amount of potentially less seismic shallow slip is concerned).

As a consequence, a mixed approach is required. As described in the previous section, the source uncertainties are formally propagated. The probabilities of magnitude and location can be taken from the real-time monitoring systems. The probability regarding the faulting mechanism and slip distribution are taken using the long-term seismotectonic behavior of the source region derived from NEAMTHM18 (Basili et al. 2021, Recipe R1), since we present here an application for tsunami forecasting in the Mediterranean Sea.

The NEAMTHM18 approach is used for the overall discretization and characterization of the seismic sources used by the PTF. The reason for using this model is that a long-term hazard model should in principle contain a complete description of the potential earthquake sources. NEAMTHM18 uses both faults and gridded sources. Faults are used for the three subduction zones in the Mediterranean Sea: the Calabrian, Hellenic, and Cyprus Arcs. For them, the geometrical and faulting parameters are bound to happen with a purely reverse mechanism on 3D subduction interfaces. Several alternative models are used to deal with the possibility of shallow slip amplification in the subduction zones. The crustal earthquakes feature a larger uncertainty and are modelled as planar faults with variable orientation. The role of crustal fault and earthquake catalogues is to constrain the priors for the faulting mechanism and scaling, but also to quantify the probability that an earthquake occurred on a crustal fault or on a nearby subduction zone. Different alternative scaling relations are used to determine fault length and width and slip. We refer the reader to the dedicated recipe for details (Recipe R1).

The PTF is as good as the model for source identification and characterization that lies at its back-end. For example, this approach makes it possible to assign a relatively larger or smaller uncertainty to different earthquake parameters, depending on the distance from a well-characterized subduction zone. This model could be improved if more local fault and seismicity data would be available to better constrain the probability of the source parameters for specific fault systems.

In the first minutes after the earthquake occurrence, which are the most relevant for the coastlines in the near-field of the seismic event, the uncertainty associated with the source characterization remains rather significant. Nevertheless, the PTF framework is totally compatible with real time tsunami data assimilation, if useful observations are available early enough, to reduce the uncertainty.

31.4 Source Mechanics Modelling Used in Recipe

It is important that the scenarios explore sufficient source variability. If a large source area is covered, a very large number of potential tsunamis must be explored. Even if the numerical simulations of the individual tsunami scenarios are pre-calculated, some simplifications regarding the numerical modelling are required to make it computationally cheaper, both for modelling the tsunami generation by an earthquake, and for the hydrodynamic modelling (dealt with in the next section).

A dislocation with uniform instantaneous slip on a planar fault embedded in a homogenous, isotropic, elastic Poisson's solid infinite half-space is used to calculate the co-seismic sea floor displacement. For crustal earthquakes, rectangular faults are used (Okada 1992). For subduction earthquakes, linear combinations of the displacement caused by triangular elements ("subfaults") are used (Meade 2007); the triangular elements belong to a mesh which approximates the 3D subduction interface. The slip distributions are stochastic, and for some alternative models are modulated according to generic depth-dependent rigidity and coupling (Scala et al. 2020).

These simplifications introduce uncertainties that are taken into account in the PTF by means of a lognormal distribution superimposed to the results of the deterministic modelling of the tsunami wave height.

The advantage is that this simplified approach to source mechanics modelling is widely used and well understood. It is considered a good compromise because more sophisticated ones, such as dynamic rupture models, are generally less established and require local information on the fault system structure, friction and pre-stress that are not generally well constrained. However, the uncertainties associated with this type of approach needs further evaluation (Davies 2019). One further limitation may be the absence of specific earthquake sources in the PTHA source model, such as multiple ruptures on complicated fault systems which have not been yet addressed in the present PTF formulation.

31.5 Tsunami Modelling

In the spirit of computational cost containment, some (common) simplifications are employed also for tsunami modelling.

A Kajiura filter is applied to transform the initial sea floor displacement into the initial static sea level anomaly, as a simplification of a non-hydrostatic effect during the generation phase (Kajiura 1963). Elementary Gaussian-shaped sea level elevations are propagated with the Tsunami-HySEA shallow water numerical model up to the coastal forecast points in proximity of the 50 m isobath and extrapolated to it using the Green's law (dela Asunción et al. 2013; Molinari et al. 2016). The coefficients that allow to approximate the initial displacement with a linear superposition of Gaussians are used to superpose the marigrams produced by the elementary sources. The Green's law is used to approximate the shoaling and inundation phases.

These simplifications introduce further uncertainty in addition to those propagating from the previous step. They are taken into account as well in the PTF by means of the lognormal distribution superimposed to the results of the deterministic modelling results.

This approach makes it feasible to consider a large variability to span the uncertainty regarding the source parameters and to consider several alternative models. The limited resolution of the bathymetry used for the tsunami simulations may lead to underestimating local smaller scale variability and resonances. The calibration of the PTF with past tsunami observations and higher resolution simulations, built ad-hoc, is then necessary.

31.6 Risk Communication

The PTF makes it possible to distinguish, and eventually separate, the scientific from the decision-making aspects. It provides a probability distribution based on which decisions can be made with consideration of the existing (and quantified) uncertainties.

For tsunami warning, the probability distribution must be converted into an alert level (evacuation yes/no). The conversion has to happen in real-time based on a pre-defined conversion rule, and, if possible, according to the local standard operating procedures, also automatically, to maximize the lead time for evacuation and other countermeasures.

It may be difficult to change the common practice in which the upstream scientific component of the tsunami warning systems is expected to provide an alert level right away, thus intrinsically playing the role of the decision-maker. Nonetheless, if PTF is adopted, the transparency of its approach may allow to improve both the upstream and downstream sides of the alerting chains.

A close interaction is needed between the scientific and political components of the warning system, to ensure a proper communication of the meaning of the probabilistic estimates, on the one hand, and of the needs of the decision-makers, on the other, to find optimal solutions.

31.7 Discussion

To manage the seismic risk, scientists determine the probability of different shaking intensities in the target area in a given time window (e.g., 50 years). Decision-makers, with the support of engineers and other risk analysts, then define a design exceedance probability. This approach is becoming a standard also for long-term coastal planning against tsunamis and tsunami building code definition in several countries (ref. MCDEM 2016; Tonini et al. 2021; Chock et al. 2016). For tsunami evacuation planning, this also clarifies the separation between scientific components (uncertain tsunami forecast through hazard curves) and political duties (acceptable risk definition for risk mitigation). Applying similar principles to the tsunami warning systems for the conversion of the forecast into alert level, in combination with the definition of corresponding evacuation maps would go in a similar direction.

The uncertainties associated with PTF based on seismic detection may be large and difficult to manage. In principle, given a certain conversion rule, it should be possible to estimate in advance the expected relative amount of missed, false, and correct alarms.

The PTF is conceived in the attempt of quantifying, and consequently reducing the uncertainties when possible. This could lead to reducing the rate of false alarms inherent with the use of traditional decision matrices. Bearing in mind that the uncertainty is large, and there are human lives at stake, it is completely licit that authorities decide to take the upper envelope to the PTF. This is currently done both in Chile and in New Zealand, where, even if less formally, uncertainty is treated by considering numerous scenarios compatible with the seismic observations, and taking the maximum impacts predicted by the ensemble of scenarios. Defining the alert levels from the PTF based on different probability thresholds corresponds to applying different levels of conservatism.

The PTF is expected to be soon introduced operationally at CAT-INGV which is a Tsunami Service Provider operating in the Mediterranean Sea for the NEAMTWS (http://www.ioc-tsunami.org/index.php?option=com_content&view=article&id=10&Itemid=14&lang=en). Testing, calibration, and validation of the PTF against past tsunami events worldwide, in hind-casting mode, and against synthetic tsunami catalogues, will be first performed to ensure a more robust prediction, and also to stress-test the alerting chain from both the technical and the human-preparation angles.

The only alternatives remain the assimilation of tsunami observations, if time allows. And actually, this is not an alternative, as data assimilation can be used as a complementary approach to reduce the uncertainty of the PTF. Self-evacuation, based on preparation of the population and of the different actors involved, remains a fundamental asset, when and where natural signs (long and strong shaking, sea level anomalies, noise, etc.) permit to anticipate the impending tsunami (see the discussion in Chap. 8).

31.8 Links to Data, Models and Other Resources

The PTF recipe is available through a Matlab code available on Github at https://github.com/INGV/matPTF.

An urgent computing version of the PTF, for rapid post-event assessment (e.g., ARISTOTLE http://aristotle.ingv.it/tiki-index.php), or future inclusion in Destination Earth (https://digital-strategy.ec.europa.eu/en/policies/destination-earth), is being developed throughout several European projects (https://cheese-coe.eu/; https://cheese2.eu/; https://eflows4hpc.eu/; https://dtgeo.eu/). This development is bringing towards an open-source collaborative development, whose results are being progressively made available from different repositories, such as https://zenodo.org/record/6376520#.ZDwbvexBz0o.

The distribution will also continue via EPOS-ERIC (https://www.epos-eu.org/tcs/tsunami) Thematic Core Service Tsunami portal https://tsunamidata.org/.

References

Basili R, Brizuela B, Herrero A, Iqbal S, Lorito S, Maesano FE, Murphy S, Perfetti P, Romano F, Scala A, Selva J, Taroni M, Tiberti MM, Thio HK, Tonini R, Volpe M, Glimsdal S, Harbitz CB, Løvholt F, Baptista MA, Carrilho F, Matias LM, Omira R, Babeyko A, Hoechner A, Gürbüz M, Pekcan O, Yalçıner A, Canals M, Lastras G, Agalos A, Papadopoulos G, Triantafyllou I, Benchekroun S, Agrebi Jaouadi H, Ben Abdallah S, Bouallegue A, Hamdi H, Oueslati F, Amato A, Armigliato A, Behrens J, Davies G, Di Bucci D, Dolce M, Geist E, Gonzalez Vida JM, González M, Macías Sánchez J, Meletti C, Ozer Sozdinler C, Pagani M, Parsons T, Polet J, Power W, Sørensen M, Zaytsev A (2021) The making of the NEAM tsunami hazard model 2018 (NEAMTHM18). Front Earth Sci 8:616594. https://doi.org/10.3389/feart.2020.616594

Chock G, Yu G, Thio HK, Lynett PJ (2016) Target structural reliability analysis for tsunami hydrodynamic loads of the ASCE 7 standard. J Struct Eng 142(11):04016092. https://doi.org/10.1061/(ASCE)ST.1943-541X.0001499

Davies G (2019) Tsunami variability from uncalibrated stochastic earthquake models: tests against deep ocean observations 2006–2016. Geophys J Int 218(3):1939–1960. https://doi.org/10.1093/gji/ggz260

de la Asunción M, Castro MJ, Fernández-Nieto ED, Mantas JM, Acosta SO, González-Vida JM (2013) Efficient GPU implementation of a two waves TVD-WAF method for the two-dimensional one layer shallow water system on structured meshes. Comput Fluids 80:441–452. https://doi.org/10.1016/j.compfluid.2012.01.012

Kajiura K (1963) Leading wave of a tsunami. Bull Earthq Reserch Institue 41:535–571

MCDEM (2016) Tsunami evacuation zones—director's guideline for civil defence emergency management groups. New Zealand: Ministry of civil defence and emergency management

Meade BJ (2007) Algorithms for the calculation of exact displacements, strains, and stresses for triangular dislocation elements in a uniform elastic half space. Comput Geosci 33(8):1064–1075. https://doi.org/10.1016/j.cageo.2006.12.003

Molinari I, Tonini R, Lorito S, Piatanesi A, Romano F, Melini D, Hoechner A, Gonzàlez Vida JM, Maciás J, Castro MJ, de la Asunción M (2016) Fast evaluation of tsunami scenarios: uncertainty assessment for a Mediterranean sea database. Nat Hazard 16(12):2593–2602. https://doi.org/10.5194/nhess-16-2593-2016

Okada Y (1992) Internal deformation due to shear and tensile faults in a half-space. Bull Seismol Soc Am 82(2):1018–1040

Scala A, Lorito S, Romano F, Murphy S, Selva J, Basili R, Babeyko A, Herrero A, Hoechner A, Løvholt F, Maesano FE (2020) Effect of shallow slip amplification uncertainty on probabilistic tsunami hazard analysis in subduction zones: use of long-term balanced stochastic slip models. Pure Appl Geophys 177(3):1497–1520. https://doi.org/10.1007/s00024-019-02260-x

Selva J, Lorito S, Volpe M, Romano F, Tonini R, Perfetti P, Bernardi F, Taroni M, Scala A, Babeyko A, Løvholt F (2021) Probabilistic tsunami forecasting for early warning. Nat Commun 12:5677. https://doi.org/10.1038/s41467-021-25815-w

Tonini R, Di Manna P, Lorito S, Selva J, Volpe M, Romano F, Basili R, Brizuela B, Castro MJ, de la Asunción M, Di Bucci D, Dolce M, Garcia A, Gibbons SJ, Glimsdal S, González-Vida JM, Løvholt F, Macías J, Piatanesi A, Pizzimenti L, Sánchez-Linares C, Vittori E (2021) Testing tsunami inundation maps for evacuation planning in Italy. Front Earth Sci 9:628061. https://doi.org/10.3389/feart.2021.628061

Chapter 32
R23—Use of Neural Networks for Tsunami Arrival Time and Maximum Height Prediction

Juan Francisco Rodríguez Gálvez and Jorge Macías

32.1 Introduction

The present recipe aims at proposing a prototype modelling tool to infer tsunami arrival times and maximum wave heights at selected points of interest (PoI) for an operational tsunami early warning system (TEWS). The operational tool is based on the use of machine learning (ML) techniques, more specifically, on neural network models (NN). The application of such methodology is performed for a prototype for the coasts of Huelva and Cádiz, in the Atlantic coast in southern Spain, the most affected Spanish coastal area from tsunamis generated in the NE Atlantic and Gulf of Cadiz and, in particular, the 1755 event. Besides, this is part of a joint work between the University of Malaga and the Instituto Geográfico Nacional, IGN (the body responsible for the Spanish TEWS) to implement such a system in the Spanish TEWS. This recipe does not implement a hazard or risk analysis, but enables the evaluation of the hazard based on many scenarios, to implement the hazard quantification of FW1A (Fully simulation-based risk assessment) or FW1B (Simulation-based hazard convoluted with vulnerability and loss using numerical integration).

The methodology proposed here can be applied to any other prone-to-tsunami coastal area. To do so, in the present recipe, we describe the way to construct the building blocks of such a system, how to implement them and we present results that serve to show the accuracy of the method for one particular component of such a system: the NN models for earthquake-generated tsunami produced by a particular fault. Each of these building blocks consists of a NN model to forecast arrival times and maximum water height at selected PoIs for tsunamis generated by a given fault.

J. F. Rodríguez Gálvez (✉) · J. Macías
Departamento de Análisis Matemático, Estadística e Investigación Operativa y Matemática Aplicada, Universidad de Málaga, Málaga, Spain
e-mail: juanrg@uma.es

J. Macías
e-mail: jmacias@uma.es

M. B. Sørensen et al. (eds.), *Probabilistic Tsunami Hazard and Risk Analysis*, Mathematics of Planet Earth 14, https://doi.org/10.1007/978-3-031-98115-9_32

To complete the system, analogous NN models must be trained for all the faults in the catalogue (Chap. 3) in a particular marine region, in the case presented here, for the NE Atlantic and Gulf of Cadiz. Here, in order to test several approaches, we have developed many single neural networks and also some ensemble models (i.e., a combination of several single models in order to produce improved results), these ones aimed at improving single models' performance. Much faster predictions are provided by NN models compared to traditional forecasting models (shallow water equations, for example, see Chap. 5) and, at the same time, the possibility of including uncertainty without an increase in computing time.

The use of ML techniques requires a large amount of data as a key ingredient. Regarding tsunamis, which are rare events, the amount of data required to train NN models does not exist. This seems to prevent the use of ML techniques for the study of tsunamis. But an alternative approach to using observed data is to generate the huge amount of data needed to train the NN model using traditional tsunami numerical models. This requires the use of large HPC (High Performance Computing) resources and a robust and efficient numerical model. Besides this, tsunamigenic faults or source areas in the selected region must be identified and the description and variability in the parameter space in order to produce a full description of the source geometry and kinematics, to enable tsunami modelling (Chap. 4) is necessary.

A great advantage of this methodological approach is that, once all the NN models are trained (one for each active fault considered) the computational resources required for the inference step (prediction) are really basic (a simple laptop would suffice) and the time required for producing such response is below one second, for a single prediction but also if some information including variability/uncertainty is required, that information is also obtained within very few seconds.

The present recipe includes the following ingredients:

- Scenario-based with large number of scenarios and uncertainty treatment (FW1A (Fully simulation-based risk assessment) or FW1B (Simulation-based hazard convoluted with vulnerability and loss using numerical integration)—Sect. 2.2
- Source selection for earthquakes sources—Sect. 3.2
- Source modelling for earthquakes sources—Sect. 4.2
- Tsunami modelling—Chap. 5.

32.2 Framework Used in Recipe

Traditionally, TEWS in the NEAM (North Eastern Atlantic and Mediterranean) region have been based on the implementation of decision matrices. In many cases, the very short arrival times of the tsunami waves from generation to impact in this region make it not possible to use real-time on-the-fly simulations to produce more accurate alert levels. In these cases, when time restriction is so demanding, an alternative to the use of decision matrices is the use of datasets of precomputed tsunami scenarios. The present recipe presents the use of neural networks to predict the tsunami maximum height and arrival time in the context of TEWS and, more specifically, multilayer perceptron NN models (details can be seen in Rodríguez et al. 2022).

In recent years, deep learning has become very popular in all fields, not only of science but also in social studies, marketing, image and speech recognition or self-driving, among many other applications. This is because of its efficiency and accuracy. Tsunami studies have not been immune to this unstoppable trend, and many new works applying ML to tsunamis are continuously appearing (Romano et al. 2009; Delashmith et al. 2005; Fauzi et al. 2020; Giles et al. 2021; Makinoshima et al. 2021). As already mentioned, to implement ML algorithms, a large amount of data is required. When it comes to tsunamis this data is not available, thus we need to generate a database of numerical simulations of tsunamis, extracting from it the data required to train the NN model. In particular, in the present recipe, these data are maximum wave heights and arrival times at selected locations. The simplified sea-floor deformation model used is the Okada model (see Seafloor Deformation in Sect. 4.2), therefore the input parameters for the numerical modelling are those required for the Okada model, in particular, hypocenter location (lon, lat and depth), fault length (L) and width (W) (or magnitude and empirical relationships for L and W), strike, dip, rake and slip. The numerical code used is the Tsunami-HySEA code (Macías et al. 2017) and the variability in the Okada parameters used is generated by means of a Sobol sequence (Bratley et al. 1988). Sobol sequences are one of the most popular quasi-random sequences due to their simplicity and efficiency in implementation. Sobol numbers offer a lower discrepancy resulting in a faster convergence. Once this is done, we need to generate different neural networks for both problems (maximum height and arrival time). Once these networks are developed, they can be combined using ensemble techniques (different types of the networks combinations) to improve single model results. When the final model is selected (single or ensemble), a simple workflow must be implemented to be able to easily apply the model.

This technique has several advantages:

- Improves the current TEWS in several aspects.
- Allows us to predict in a few seconds the outputs required for multiple scenarios, or including variability associated to uncertainties and all this in very few seconds.
- It is easy to use, the inputs and their ranges can be adjusted, i.e., they do not have to be necessarily the nine Okada parameters, we can include magnitude or other parameters.
- The maximum absolute error is small (compared to the reference results provided by the numerical model).
- If more PoIs need to be included because more forecast locations are required, this number can be increased, maintaining good results. Two options are presented in this case: training a new model with every PoI or we can train a different model (or several models) for the new PoIs.
- The PoIs can be located as close to the coast as we want, avoiding the use of Green's functions.
- Include the variability associated with the uncertainty in the seismic source.

- Variability in epicentre/hypocentre location allows it to cover a large area around the fault, avoiding unexpected events for the system.

Regarding the limitations, we see that the models are developed for a specific range of Okada parameters. In case the parameters are not in this range the results will not be reliable. This implies we need to define ranges of the Okada parameters very rigorously. Besides, it is limited to a coast zone, to create a TEWS we need to create lots of neural networks. However, the number of points to be predicted by our networks can be increased, thus reducing the number of networks needed to cover the coast. Very large computational resources are required to fully train the complete system.

32.3 Source Identification and Characterisation Used in Recipe

In order to construct a complete TEWS based on NN models for tsunamis generated in a particular area, we need to have identified and characterised all the potential tsunamigenic faults in that area. This means at least to have access to a complete and homogeneous catalogue. In the present recipe, we use the QAFI v.4 catalogue (Sect. 3.2). In Rodríguez et al. (2022), several of these NN models are implemented for tsunamis generated in the Horseshoe fault because it is supposed to be the source for the major event in the region, the Lisbon 1755 earthquake and tsunami. The same approach must be applied to all the rest of faults in the area to complete the TEWS and explore also the possibility for other source areas not covered by the known faults.

32.4 Source Mechanics Modelling Used in Recipe

An Okada model (see Sect. 4.2) is used. The Okada model is a simple model for seafloor deformation widely used in tsunami simulation studies, as it can be solved rapidly and permits easily generating new synthetic events by varying some values of the Okada parameters and, thus, easily treating uncertainty in the source. However, it also has limitations as complex ruptures cannot be represented.

32.5 Tsunami Modelling

Regarding tsunami modelling (Chap. 5), the code Tsunami-HySEA has been used to perform the large number of simulations necessary to train, validate and test the different NN models proposed. In particular, in Rodríguez et al. (2022) a total of

16,000 numerical simulations are produced for every single fault. The Tsunami-HySEA model (Macías et al. 2017) implements in the same code the three phases involved in the simulation of an earthquake generated tsunami: seismic source generation, propagation, and coastal inundation. In the generation stage, several initial conditions are available in the Tsunami-HySEA code. For the present recipe, the basic one-segment Okada initial deformation is used, that is transmitted instantaneously to the sea surface generating the tsunami wave. Regarding the propagation and inundation steps, the Tsunami-HySEA model implements the NLSW equations (Sects. 5.4 and 5.5) and nested grids for high-resolution inundation. Besides, the Tsunami-HySEA code has been widely benchmarked (Sect. 5.6).

In the context of TEWS, when FTRT (Faster Than Real Time) numerical simulations are used, no inundation is computed, and alert levels are given based on the maximum height at virtual gauges located at some distance from the coast (for example at the 50 m isobath). Therefore, the numerical resolution used is coarse, for example, currently in the Spanish TEWS the standard resolution used is 30 arc-sec (approx. 900 m). The methodology proposed here, although aimed for TEWS, as numerical simulations are not performed on-the-fly, permits using finer resolution, nested grids (computationally much more expensive) and to locate the PoIs (the virtual gauges) as close to land as model resolution permits. In Rodríguez et al. (2022), the best available topo-bathymetric data (Chap. 5.2) was used and a three level nested grid (320 m-160 m-40 m) spatial discretization was designed to fit this purpose. Finally, large HPC resources were needed to run all the source simulations and aggregate the results from the different scenarios. A total of 16,000 simulations were developed.

32.6 Machine Learning Modelling

Neural networks are a fundamental concept in the field of artificial intelligence and machine learning. They have gained widespread popularity in various fields due to their ability to approximate complex functions. In this work we have made use of Multilayer Perceptrons (MLPs) which are a specific type of artificial neural network architecture that has a significant prominence in the field of deep learning.

The fundamental components of a MLP include an input layer that receives data, one or more hidden layers for performing complex computations, and an output layer responsible for generating predictions. Regarding our cases, the input layer will be the Okada parameters and the output layer will be the maximum height or arrival time in the PoIs.

These types of networks require a previous configuration. It is required to configure the number of hidden layers and neurons in each layer, among other important parameters (learning rate, batch size, error function, etc.), to tailor the network's performance to specific tasks. MLPs have demonstrated remarkable success in many different tasks and, as it has been demonstrated, we can take advantage of them to predict some key parameters in tsunami early warning systems such as maximum height and arrival time.

After several configurations for the neural networks developed for the two problems we have addressed, we have found that good results were obtained. This makes its use in TEWS possible, since these networks can be executed in a matter of seconds without the need for large computational resources, allowing, at the same time, to assess the impact on tsunami simulations and associated predicted hazard of uncertainty in seismic source parameters.

32.7 Discussion

Deep learning, specifically neural networks, is a relatively new approach for tsunami scenarios. Despite this novelty, there are already many works using deep learning techniques for tsunami modelling and results are promising. It is a field still to be explored in depth.

Some of the pros are:

- Time-to-solution is reduced compared to FTRT on-the-fly simulations, while accuracy is maintained.
- Even more, uncertainty can also be included, still in extremely short response times.
- They can help to improve current TEWS as closer to coast predictions are available.
- Easy to use and just basic computational resources are required for the inference phase (prediction).
- More local and precise predictions are available, as more PoIs closer to the coast can be considered.
- Small errors are obtained compared with reference numerical simulations.
- If necessary, more tsunami simulations can be subsequently included and train again the models.

It also has some limitations:

- We cannot trust in results if the inputs are out of the input range defined.
- It is limited to a coastal zone. We need to generate a neural network for each coastal zone.
- Large computational resources and a huge computational effort are required.

Nowadays, neural networks can offer a complementary methodology to be used in TEWS. Decision matrices can provide a quick and safe response, but being a rudimentary approach, that do not consider local effects, nonlinearities of the phenomena or any complexity associated with bathymetry or coastal shape. In order to include a numerical approach in these systems, databases of precomputed scenarios have been generated and used. The system searches for a best-fit scenario in the database to be used as a model for the actual event. The main disadvantage of such a system relies on unexpected events not included in the database. Along with these techniques, Probabilistic Tsunami Forecast (PTF) (Selva et al. 2021) and FTRT on-the-fly simulations (Gaite et al. 2022) have been possible since recently and, if more computational HPC resources available, these simulations can also include some variability in tsunami source parameters.

32.8 Links to Data, Models and Other Resources

- Tsunami-HySEA code: https://zenodo.org/record/6400815#.ZB8f8OzMJQJ

References

Bratley P, Fox BL (1988) Algorithm 659: implementing Sobol's quasirandom sequence generator. ACM Trans Math Softw (TOMS) 14(1):88–100

Delashmit WH, Missiles LM, Manry MT (2005) Recent developments in multilayer perceptron neural networks. In: Proceedings of the 7th annual memphis area engineering and science conference MAESC, Memphis, TN, USA, 11 May 2005

Fauzi A, Mizutani N (2020) Machine learning algorithms for real-time tsunami inundation forecasting: a case study in Nankai region. Pure Appl Geophys 177:1437–1450

Gaite B, Macías J, Cantavella JV, Sánchez-Linares C, González C, Puertas LC (2022) Analysis of faster-than-real-time (FTRT) tsunami simulations for the Spanish tsunami warning system for the Atlantic. GeoHazards 3:371–394. https://doi.org/10.3390/geohazards3030019

Giles D, Gopinathan D, Guillas S, Dias F (2021) Faster than real time tsunami warning with associated Hazard uncertainties. Front Earth Sci 8:560. https://doi.org/10.3389/feart.2020.597865

Macías J, Castro MJ, Ortega S, Escalante C, González-Vida JM (2017) Performance benchmarking of tsunami-HySEA model for NTHMP's inundation mapping activities. Pure Appl Geophys 174:3147–3183. https://doi.org/10.1007/s00024-017-1583-1

Makinoshima F, Oishi Y, Yamazaki T (2021) Early forecasting of tsunami inundation from tsunami and geodetic observation data with convolutional neural networks. Nat Commun 12:2253. https://doi.org/10.1038/s41467-021-22348-0

Rodríguez JF, Macías J, Castro MJ, de la Asunción M, Sánchez-Linares C (2022) Use of neural networks for tsunami maximum height and arrival time predictions. GeoHazards 3:323–344. https://doi.org/10.3390/geohazards3020017

Romano M, Liong SY, Vu MT, Zemskyy P, Doan CD, Dao MH, Tkalich P (2009) Artificial neural network for tsunami forecasting. J Asian Earth Sci 36:29–37

Selva J, Lorito S, Volpe M, Romano F, Tonini R, Perfetti P, Bernardi F, Taroni M, Scala A, Babeyko A, Løvholt F, Gibbons S, Macías J, Castro M, González Vida J, Sánchez-Linares C, Bayraktar B, Basili R, Maesano F, Amato A (2021) Probabilistic tsunami forecasting for early warning. Nat Comm 12:5677. https://doi.org/10.1038/s41467-021-25815-w

BY

Chapter 33
R24—PTHA-Based Tsunami Inundation Mapping in Italy

Roberto Tonini, Daniela Di Bucci, and Pio Di Manna

33.1 Introduction

The purpose of this recipe is to present the methodology used in Italy to quickly design tsunami inundation maps for civil protection planning (DPCM 2017; Dipartimento della Protezione Civile 2018) and provide a first attempt to explore the uncertainties associated with these kinds of applications.

Ingredients for this recipe:

- This recipe follows framework FW1A (Fully simulation-based risk assessment; Sect. 2.2) but is concluded at the completion of the PTHA.
- Source selection and characterisation were based on geologic fault sources and seismicity from publicly available datasets (Sect. 3.2). These were used to model earthquake sources as a regularly spaced grid and individual three-dimensional faults.
- The process of tsunami source modelling (Sect. 4.2) involved two steps. The slip distribution for each earthquake scenario in the earthquake source model was determined in the first step. The seafloor deformation for each slip distribution was calculated in the second step.

R. Tonini (✉)
Istituto Nazionale di Geofisica e Vulcanologia, Roma, Italy
e-mail: roberto.tonini@ingv.it

D. Di Bucci
Italian Civil Protection Department, Presidency of the Council of Ministers, Roma, Italy
e-mail: daniela.dibucci@protezionecivile.it

P. Di Manna
ISPRA-Italian Institute for Environmental Protection and Research, Geological Survey of Italy, Roma, Italy
e-mail: pio.dimanna@isprambiente.it

M. B. Sørensen et al. (eds.), *Probabilistic Tsunami Hazard and Risk Analysis*, Mathematics of Planet Earth 14, https://doi.org/10.1007/978-3-031-98115-9_33

- The tsunami modelling (Chap. 5) for each slip distribution was carried out using a realistic bathymetric model (Sect. 5.2) and by applying a filter to transform the seafloor displacement into a tsunami initial condition (Sect. 5.3).

The recipe describes a solution to quickly derive reliable tsunami inundation maps for coastal areas at a regional scale (i.e., with an extension that is typical of national boundaries) with relatively cheap computational resources. The reported approach is the one followed by Italian authorities to define inundation maps as the basic tool for land use planning or evacuation strategies. Italian inundation maps are based on NEAMTHM18, a probabilistic regional hazard model for tsunamis generated by earthquakes (Basili et al. 2021; Recipe R1). The method consists of defining a design probability for the maximum—watch/red—alert level defined in the frame of the Italian tsunami alert system (SiAM, Sistema di Allertamento Maremoti; DPCM 2017; Dipartimento della Protezione Civile 2018) in which the Italian Tsunami Alert Centre—CAT operates (Amato et al. 2021). This design probability corresponds to the 2,500 years average return period (ARP) and the 84th percentile of the epistemic uncertainty. The tsunami intensity measure (Maximum Inundation Height, MIH) extracted from the regional hazard model is converted into a value of maximum runup R_d^* (see Fig. 33.1, panel A) and, in turn, in evacuation and civil protection planning area using a chain of simplified and conservative approximations. These include some safety factors that allow to account for uncertainties in the tsunami modelling during the inundation phase as well as in the digital terrain model (DTM) and the coastline. The performance of this approach has been evaluated by comparing these inundation maps with the ones derived from a site-specific Seismic Probabilistic Tsunami Hazard Analysis (SPTHA, Gibbons et al. 2020; Recipe R2), including massive high-resolution numerical inundation simulations at the same percentile and ARP.

It is demonstrated that the inundation maps proposed for civil protection planning can conservatively account for the unmodeled uncertainties in the source and tsunami numerical models. The method used to draw inundation maps in Italy and the associated analysis and evaluation are fully described in Tonini et al. (2021).

33.2 Framework Used in Recipe

This recipe makes use of the hazard component of Framework FW1A (Fully simulation-based risk assessment; Sect. 2.2), aiming to treat the propagation of aleatoric and epistemic uncertainties associated with the production of inundation maps.

This was realised with two different approaches:

1. Using the regional tsunami hazard model for the NEAM area (NEAMTHM18, Basili et al. 2021; Recipe R1) as the main direct input to draw the inundation maps proposed for civil protection planning.

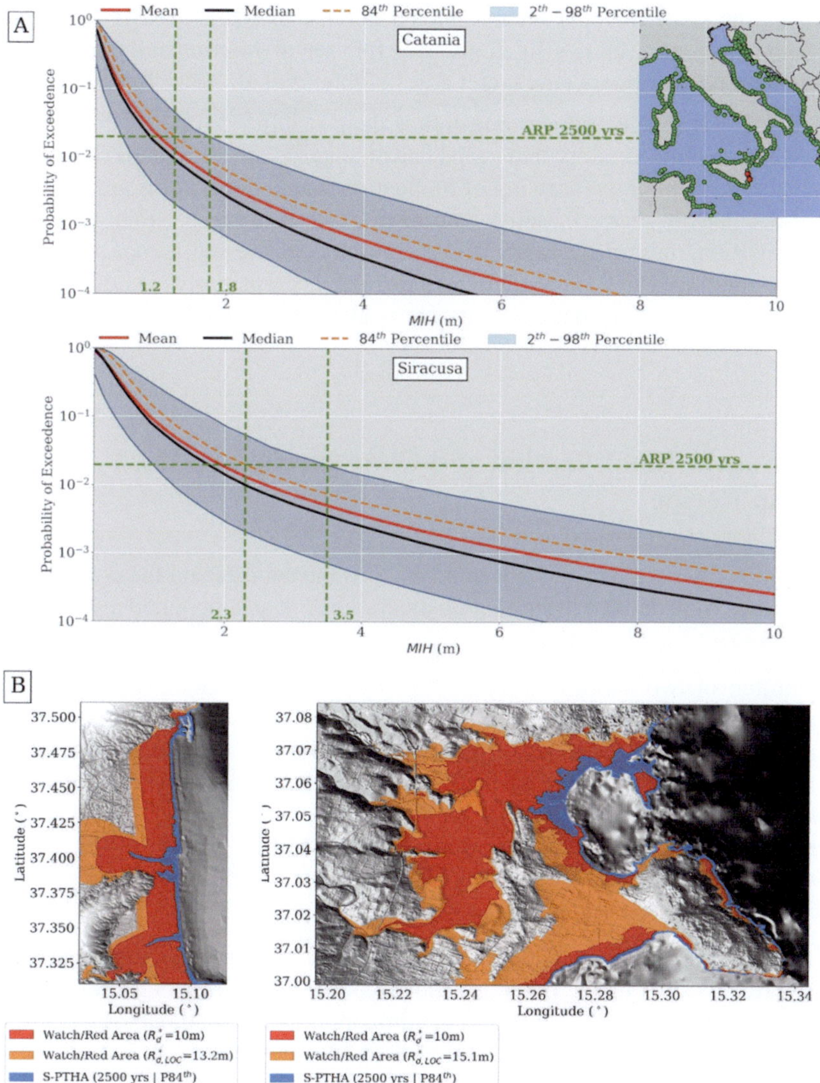

Fig. 33.1 **a** Hazard curves for Catania (upper) and Siracusa (lower) with different quantities: the mean (red line), the median (black line), the 84th percentile (orange line) and the interval 2nd–98th percentiles (blue area) and the selected ARP of 2,500 years (green lines). Inset: location of the points of interest (POIs, green dots) zoomed around Italy in the frame of the NEAMTHM18 tsunami hazard model; the two POIs in front of Catania and Siracusa are highlighted in red. The curves can be downloaded from the NEAMTHM18 model (Basili et al. 2018, clicking on the web mapper link and then using the interactive tsunami hazard tool to select the points corresponding to the geographical position of Catania and Siracusa, labelled with POI codes med9153 and med9174, respectively). **b** Comparison of inundation maps for both Catania (left) and Siracusa (right), designed using (i) the values of R_d^* obtained from the local SPTHA and increased with the uncertainty (orange areas), (ii) the R_d^* 10 m (red areas) and (iii) the local SPTHA inundation corresponding to 2,500 years ARP and the 84th percentile (blue areas)

2. Using the same regional seismic model defined in NEAMTHM18 as input to perform a local SPTHA by running a full set of high-resolution inundation simulations for two specific target areas (Gibbons et al. 2020; Recipe R2).

The main advantage of this framework is the possibility to quantify, in principle, all the sources of uncertainty that propagate among the different steps of the hazard calculation up to the final drawing of inundation maps. Limitations are mainly due to the fact that, in practice, some complexities of both the seismic source (e.g., slip distribution, rupture dynamics) and tsunami modelling (e.g., propagation errors, friction modelling) can be difficult to include. This is particularly true at local scales, where the availability of detailed datasets is also needed to reduce uncertainty as much as possible.

33.3 Source Identification and Characterisation Used in Recipe

This recipe only considers seismic sources. Seismic sources and the corresponding uncertainties were defined in a probabilistic framework and modelled using different approaches:

- definition of two classes of seismic sources: predominant seismicity (PS) and background seismicity (BS). PS corresponds to earthquakes generated by well-known and larger faults (i.e., subduction interfaces), whereas BS includes all the relatively smaller, crustal earthquakes that are analysed without a direct association with specific seismic sources;
- exploration of source parameters for both PS and BS;
- use of geologic knowledge, faults, and seismic catalogues to inform the parameter distribution and the seismicity rates.

This approach has the advantage of quantifying the uncertainties where the seismic sources are less known (i.e., offshore) and providing data-driven constraints in those seismic areas that are better known. One limitation of this methodology is the very large number of defined scenarios, which often require strategies to be reduced with the smallest possible loss in accuracy.

33.4 Source Mechanics Modelling Used in Recipe

The co-seismic displacement caused by the activation of seismic sources is modelled as follows:

- BS sources: the classical Okada (1992) model is used for the co-seismic displacement caused by the dislocation along a rectangular fault buried in an infinite elastic half-space.

- PS sources: these sources are modelled over a mesh of triangular elements, and the co-seismic displacement corresponding to each earthquake scenario is obtained by superposition of the contribution of these individual triangular patches as described by the used fault rupture model (e.g., Scala et al. 2020).

For BS sources, the modelling has the advantage of being very fast (analytical solutions) and can therefore be used to rapidly explore the main parameters of the earthquake sources (location, orientation, magnitude), but source complexities are neglected (such as not uniform slip distribution or rupture area).

For PS sources, uncertainties were reduced by the quality of the available geological and seismological data, which provided constraints on the geometry of the faults. Also, using dense triangular meshes allowed to include heterogeneous slip models along fault surfaces with complex geometry (i.e., variable strike and dip, and non-rectangular shape).

33.5 Tsunami Modelling

In this recipe, tsunami modelling was obtained through two alternative approaches:

- regional offshore modelling: offshore sea surface elevations are obtained by linear combinations of pre-computed elementary source scenarios, which are propagated with the 2D non-linear hydrostatic shallow-water and multi-GPU code Tsunami-HySEA (de la Asunción et al. 2013), up to water depths of 50 m, along a set of points of interest (POIs) spaced about 20 km from each other (Molinari et al. 2016). Offshore sea surface elevations are then transformed into maximum inundation heights (MIH) through amplification factors (AF), accounting for the amplitudes, periods, and polarities of the incident waves (Glimsdal et al. 2019). This is the input for the NEAMTHM18 tsunami hazard model;
- local inundation modelling: the tsunami scenarios are simulated with Tsunami-HySEA, exploiting a system of nested topo/bathymetric grids to inundate the target areas. Four-level nested grids were used (640, 160, 40, and 10 m). The 640 m grid was derived from the GEBCO model, while the local 10 m grid was produced by mosaicking several rasters with different resolutions (as described in Gibbons et al. 2020) and then resampling them to a 10 m horizontal cell. This final 10 m grid is the same DTM on which all kinds of inundation maps presented in Tonini et al. (2021) are calculated.

In both cases, in each simulation the seafloor displacement described in the previous section is converted into a tsunami initial condition using the attenuation filter described by Kajiura (1963), which mimics the short-wavelength attenuation effect due to the water column.

Uncertainties associated with tsunami modelling are accounted for in two different ways:

- for the regional modelling, the uncertainty associated with the projection from the 50 m isobath to inland, based on 1D assumptions and models and therefore neglecting any lateral effect, was accounted for by using a multiplicative factor (resulting in a value of 3) in the estimation of the maximum runup R_d^*. This value is derived from the 98th percentile of the uncertainty estimated for the 1D modelled amplification factors. The uncertainty on DTM is also considered and balanced by applying a safety factor of 1 m to the runup R_d^*;
- for the local modelling, since the variety of target areas can be very heterogeneous in terms of coastal morphology and DTM (e.g., flat versus steep coasts, uncovered terrain, and few buildings versus dense urban infrastructures), we defined an "error bar" where tsunami modelling without and with epistemic uncertainty represent the lower and upper limits of this uncertainty. A simplified way to include this uncertainty in this kind of inundation maps is derived by the work of Davies et al. (2017) for events at global scale, considering it a sort of maximum uncertainty with respect to local analyses.

The approach used for regional modelling has the advantage of performing millions of tsunami scenarios with relatively small computational resources but has the limitation of being less accurate at the local scale. For the local model, advantages and limitations are inverted.

33.6 Discussion

The methodology used in Italy to process the tsunami inundation maps provided at national scale for the civil protection plans resulted in a good degree of reliability, since it lies within the error bar defined by the inundation maps based on high-resolution inundation modelling, computed with and without the uncertainty treatment (Fig. 33.1, panel B). The main advantages of this approach are the low computational cost and the possibility of applying the method to extended coastal areas (national level).

Some limitations and concerns about this method have also been identified:

- the method depends on the quality (resolution and elevation accuracy) of the used DTM; in particular, the presence of secondary riverbeds seems to be relevant for the inundation areas. The impact of less accurate DTM models had, at least in our case-study, a relatively smaller impact on the definition of inundation zones. However, this cannot be generalised, since it could depend on the specific local morphology (e.g., the presence of steep slopes);
- there is a need to improve the accuracy of the regional hazard model (e.g., higher density of POIs, refinement of source parameters, simulations up to the 10 m isobath) to estimate the design runup R_d^*;
- for near-source coasts, the maximum inundation distance can also depend on the effect of the possible co-seismic displacement in altering the coastal elevation, which is not modelled in the regional SPTHA (Volpe et al. 2019).

In general, for tsunami inundation modelling, we also highlight two further important findings:

- ignoring the epistemic uncertainty of tsunami modelling may have a major impact on local hazard results, potentially producing severe underestimations;
- detailed analyses based on specific local data are needed to better calibrate the parametrization of the uncertainties.

In conclusion, it seems that in the absence of more detailed studies, the currently adopted choices for the national planning, which were the outcome of extensive discussions among the decision-makers and different scientific actors at the national level, remain fully justified.

Similar methods have been adopted to define inundation maps also in other countries, e.g., New Zealand (Leonard et al. 2008) and Samoa (Wright et al. 2011). As said, calculating inundation maps using a local SPTHA with high-resolution inundation simulations is likely the best alternative for specific areas, but this requires huge resources in terms of computational efforts and availability of detailed datasets. In this case, a further alternative consists of providing a worst-case scenario based on the maximum impact obtained from the envelope of all the scenarios defined in the SPTHA affecting the target area. This could be the best option for very critical infrastructures (Tonini et al. 2021).

Acknowledgements The contents of this recipe represent the authors' ideas and do not necessarily correspond to the official opinion and policies of the Italian Civil Protection Department.

Links to Data, Models and Other Resources

- NEAMTHM18 hazard model: https://doi.org/10.13127/tsunami/neamthm18
- Italian inundation map viewer: http://sgi2.isprambiente.it/tsunamimap

References

Amato A, Avallone A, Basili R, Bernardi F, Brizuela B, Graziani L, Herrero A, Lorenzino MC, Lorito S, Mele FM, Michelini A, Piatanesi A, Pintore S, Romano F, Selva J, Stramondo S, Tonini R, Volpe M (2021) From seismic monitoring to tsunami warning in the mediterranean sea. Seismol Res Lett 92:1796–1816. https://doi.org/10.1785/0220200437

Basili R, Brizuela B, Herrero A, Iqbal S, Lorito S, Maesano FE, Murphy S, Perfetti P, Romano F, Scala A, Selva J, Taroni M, Thio HK, Tiberti MM, Tonini R, Volpe M, Glimsdal S, Harbitz CB, Løvholt F, Baptista MA, Carrilho F, Matias LM, Omira R, Babeyko A, Hoechner A, Gurbuz M, Pekcan O, Yalçıner A, Canals M, Lastras G, Agalos A, Papadopoulos G, Triantafyllou I, Benchekroun S, Agrebi Jaouadi H, Attafi K, Ben Abdallah S, Bouallegue A, Hamdi H, Oueslati F (2018) NEAM tsunami hazard model 2018 (NEAMTHM18): online data of the probabilistic tsunami hazard model for the NEAM Region from the TSUMAPS-NEAM project. Istituto Nazionale di Geofisica e Vulcanologia (INGV). https://doi.org/10.13127/tsunami/neamthm18

Basili R, Brizuela B, Herrero A, Iqbal S, Lorito S, Maesano FE et al (2021) The making of the NEAM tsunami hazard model 2018 (NEAMTHM18). Front Earth Sci 8. https://doi.org/10.3389/feart.2020.616594

Davies G, Griffin J, Løvholt F, Glimsdal S, Harbitz C, Thio HK et al (2017) A global probabilistic tsunami hazard assessment from earthquake sources. Geol Soc Lond Spec Publ 456:219–244. https://doi.org/10.1144/SP456.5

de la Asunción M, Castro MJ, Fernández-Nieto ED, Mantas JM, Acosta SO, González-Vida JM (2013) Efficient GPU implementation of a two waves TVD-WAF method for the two-dimensional one layer shallow water system on structured meshes. Comput Fluids 80:441–452. https://doi.org/10.1016/j.compfluid.2012.01.012

Dipartimento della Protezione Civile (2018) Presidenza del Consiglio dei Ministri—Dipartimento della Protezione Civile—decreto 2 ottobre 2018. Indicazioni alle Componenti ed alle Strutture operative del Servizio nazionale di protezione civile per l'aggiornamento delle pianificazioni di protezione civile per il rischio maremoto, (18A07309) (GU Serie Generale n.266 del 15-11-2018). http://www.gazzettaufficiale.it/eli/gu/2018/11/15/266/sg/pdf. Accessed 29 July 2025

DPCM (2017) Direttiva Del Presidente Del Consiglio Dei Ministri 17 febbraio 2017 Istituzione del Sistema d'Allertamento nazionale per i Maremoti generati da sisma—SiAM. (17A03755) (GU Serie Generale n.128 del 05-06-2017). http://www.gazzettaufficiale.it/eli/gu/2017/06/05/128/sg/pdf. Accessed 29 July 2025

Gibbons SJ, Lorito S, Macías J, Løvholt F, Selva J, Volpe M et al (2020) Probabilistic tsunami hazard analysis: high performance computing for massive scale inundation simulations. Front Earth Sci 8:591549. https://doi.org/10.3389/feart.2020.591549

Glimsdal S, Løvholt F, Harbitz CB, Romano F, Lorito S, Orefice S et al (2019) A new approximate method for quantifying tsunami maximum inundation height probability. Pure Appl Geophys 176:3227–3246. https://doi.org/10.1007/s00024-019-02091-w

Kaijura K (1963) The leading wave of a tsunami. Bull Earthq Res Inst Univ Tokyo 41:535–571

Leonard GS, Power W, Lukovic B, Smith W et al (2008) Interim tsunami evacuation planning zone boundary mapping for the Wellington and Horizons regions defined by a GIS-calculated attenuation rule. GNS Science Report 2008/30

Molinari I, Tonini R, Lorito S, Piatanesi A, Romano F, Melini D et al (2016) Fast evaluation of tsunami scenarios: uncertainty assessment for a Mediterranean sea database. Nat Hazards Earth Syst Sci 16:2593–2602. https://doi.org/10.5194/nhess-16-2593-2016

Okada Y (1992) Internal deformation due to shear and tensile faults in a half-space. Bull Seismol Soc Am 82(2):1018–1040

Scala A, Lorito S, Romano F, Murphy S, Selva J, Basili R et al (2020) Effect of shallow slip amplification uncertainty on probabilistic tsunami hazard analysis in subduction zones: use of long-term balanced stochastic slip models. Pure Appl Geophys 177:1497–1520. https://doi.org/10.1007/s00024-019-02260-x

Tonini R, Di Manna P, Lorito S, Selva J, Volpe M, Romano F et al. (2021) Testing tsunami inundation maps for evacuation planning in Italy. Front Earth Sc 9. https://doi.org/10.3389/feart.2021.628061

Volpe M, Lorito S, Selva J, Tonini R, Romano F, Brizuela B (2019) From regional to local SPTHA: efficient computation of probabilistic tsunami inundation maps addressing near-field sources. Nat Hazards Earth Syst Sci 19:455–469. https://doi.org/10.5194/nhess-19-455-2019

Wright KC, Beavan RJ, Daly MC, Gale NH, Leonard GS, Lukovic B, Palmer NG et al (2011) Filling a critical gap in end-to-end tsunami warning in the Southwest Pacific: a pilot project in Samoa to create scientifically robust, community-based evacuation maps. GNS Science Report 2011/53. Lower Hutt, New Zealand, 140 p

Chapter 34
R25—Methodology for Tsunami Hazard Microzoning Applied to Evacuation Decision Making

Natalia Zamora, Patricio A. Catalan, and Alejandra Gubler

34.1 Introduction

The aim of this recipe is to develop a methodology for assessing the hazard that takes into account the arrival times and expected flow depths of a tsunami. It is hypothesized that this information would allow the definition or modification of evacuation routes and/or placement and design of vertical evacuation infrastructure. Moreover, data are grouped in terms of categories, yielding an easy to interpret map for decision makers, at the cost of losing direct use of quantitative hazard metrics. This recipe is not intended as a surrogate of PTHA, as the recurrence of events is discarded to assess the worst possible conditions, based on such selected scenarios.

The workflow described in this recipe was originally devised for Central Chile (Zamora et al. 2021). This area has been affected by large earthquakes, particularly in 1730 when a mega-earthquake (Mw 9.1–9.3) generated a widespread tsunami that affected several coastal regions in Chile and throughout the Pacific basin (Carvajal et al. 2017a). Considering such events and the long history of tsunamis along Chile (e.g. Valdivia Mw 9.5 earthquake), a national strategy is needed for establishing evacuation procedures to prepare for future events which has motivated the development of novel strategies for alerting people and to proceed with evacuations (Catalán et al. 2020; León et al. 2020).

N. Zamora (✉)
Wave Phenomena Group, Computer Applications in Science and Engineering, Barcelona Supercomputing Center, Barcelona, Spain
e-mail: natalia.zamora@bsc.es

P. A. Catalan
Departamento de Obras Civiles, Universidad Tecnica Federico Santa Maria, Valparaiso, Chile
e-mail: patricio.catalan@usm.cl

P. A. Catalan · A. Gubler
Research Center for Integrated Disaster Risk Management (CIGIDEN), Santiago, Chile
e-mail: alejandra.gubler@usm.cl

M. B. Sørensen et al. (eds.), *Probabilistic Tsunami Hazard and Risk Analysis*, Mathematics of Planet Earth 14, https://doi.org/10.1007/978-3-031-98115-9_34

The ingredients used only some elements of the framework FW1D (Working on a limited number of scenarios; Sect. 2.2), that is a scenario-based approach with thousands of scenarios. However, no probabilistic assessment is conducted in our study. While PTHA has been used for evacuation elsewhere, here we adopt the standpoint of a High Impact, Low Likelihood (HILL) event, whereby the probability of occurrence is discarded under the assumption that evacuation routes need to work under all possible circumstances, including the worst-case scenario. Hence statistical metrics are used to obtain the maximum envelope of flow depths and minimum time arrival because the focus is evacuation to minimize life loss. However, the approach used here can be introduced in a probabilistic framework.

The steps are based on source characterization, numerical modelling of inundation and aggregation of results with focus on cross-evaluation of flow depths and arrival time leading to a categorization of hazard levels. Other ingredients are used which relate to:

- Selection for earthquake sources along a seismotectonic segment in the central Chilean margin as described in Chap. 3.
- Stochastic source modelling for earthquake sources as defined in Chap. 4.
- Tsunami numerical modelling using high performance computing (HPC) resources, which is described in Chap. 5.

A range of scenarios was set to account for the intra-and-inter event variability modeled with Tsunami-HySEA and using HPC. Thus, in this study HPC was one key aspect. In terms of data availability, a key dataset is the high-resolution topo-bathymetry required for modeling inundation. The various datasets were merged to generate the nested grids used to increase progressively the spatial resolution from the generation to the inundation phase for the numerical tsunami simulations.

34.2 Framework Used in Recipe

This example fits in the framework FW1D (Working on a limited number of scenarios; Sect. 2.2). It is a scenario-based approach with thousands of earthquake scenarios (2800 with Mw range 8.6–9.2). This choice was based on the estimated range of magnitudes in this region, which implicitly limits the maximum return period. In other implementations, a wider magnitude range with associated probabilities can be considered to accommodate a fully probabilistic approach. However, only the slip distribution of the inter-event size (different Mw range) and the intra-event characterization (earthquakes with the same moment magnitude Mw but varied slip) have been addressed for the purposes of this study. This allows for taking into consideration large variability of tsunami events that are aggregated. Thus, the statistical treatment allows for considering variability for the hazard microzoning.

The approach used in this case is a hybrid-probabilistic, stochastic sampling of seismic sources without the use of any specific criteria, such as the frequency-magnitude distributions described in Chap. 3. The fundamental idea behind this study's setup focuses on the variety of tsunami scenarios and cross-evaluation of two parameters: tsunami intensity measures and tsunami arrival times.

Implementation steps:

1. Definition of the source generation (seismogenic) zone and its relevant parameter space
2. Single seismic scenario generation (stochastic)
   - Definition of rupture area
   - Slip distribution along the rupture area
3. Numerical modeling of tsunami propagation and inundation
4. Estimation of tsunami metrics: maximum flow depth and arrival time obtained from each scenario at every grid point
5. Integration of the information of all scenarios between Mw 8.6–9.2. Statistical analysis to yield fields of maximum flow depth envelope, 95 percentile of maximum flow depth, fraction of events that inundate each cell, minimum arrival time in each cell, among others
6. Hazard categorization
   - Definition of two thresholds for maximum flow depths and two for arrival time
   - Categorize the hazard at each grid point in accordance to flow depth and arrival time, independently. Flow depth levels are categorized 1–3 and arrival times are categorized A-C. Both cases consider increasing hazard in the nomenclature
   - Aggregation of hazard using a 9-level qualitative hazard scale, from A1 (least hazardous) to C3 (most hazardous)
   - Application of categorization thresholds to qualify hazard for a selected single scenario or set of scenarios

This approach allows for identifying high hazards (large flow depth and/or short arrival times). As it will be seen below, these areas can be close to lower hazard areas (smaller flow depth and/or long arrival times), which allows for the efficient design of alternative evacuation strategies. Although no other tsunami-triggering sources have been considered, they could be easily implemented. A disadvantage, instead, is that the selection of hazard levels is subjective or based on thresholds used by national civil protection agencies. This may require elicitation. Besides, another inherent disadvantage of thresholds is that they introduce a discontinuity hazard level from one category to another in correspondence with a slight flow depth or time variation around the threshold.

In this work, the uncertainty propagation is applied on the seismic source variability generating hazard mapping based on a large tsunami simulation database that considers random positions of the rupture extent and slip distributions for a range of moment magnitudes. Thus, it only applies to source characterization-related uncertainties. Epistemic uncertainties were addressed by a sensitivity analysis of the changes in the values for arrival time and wave height, with larger changes for scenarios at the furthest distances than for the scenarios near the coast. This sensitivity to the hazard level thresholds shows that according to particular applications, further studies may be needed to define such levels.

34.3 Source Identification and Characterisation Used in Recipe

Only seismic sources are considered, with a range of tsunamigenic magnitudes with sizes ranging from Mw 8.6–9.2. In this example, 400 stochastic slip distributions per magnitude were considered. Each of them was estimated using a predefined algorithm based on the Karhunen-Loève (KL) expansion to decompose a random process (Melgar et al. 2016). The method consists of defining a random location of the hypocenter within the seismogenic zone. The slip distribution is generated starting from this 'seed'. The total areal extent is bound by the length and width which are calculated based on the magnitude using the scaling relations of Blaser et al. (2010).

The definition of the geometry of the subfaults and the spatial extent of the maximum earthquake has been based on the 1730 Mw ~ 9.2 earthquake and tsunami, and also from paleoseismic data used to set up assumptions. Previous studies justify this seismotectonic zone to be suitable to generate stochastic ruptures (e.g. Carvajal et al. 2017a,b; Dura et al. 2015). This geometry that corresponds to the interplate seismogenic zone is divided in a 20 $\times$ 20 km grid and information (e.g., strike, dip, depth) of the Slab1 model (Hayes et al. 2012) is retrieved for the centre of each grid's cell.

The main advantage of implementing this methodology is that the inputs are relatively easy to define, and both slip variability and kinematic rupture can be modelled.

34.4 Source Mechanics Modelling Used in Recipe

Only interplate earthquake sources have been considered in this study. An extensive seismogenic subduction zone area of about 1200 km has been considered. Inside this domain earthquakes of magnitudes ranging from Mw 8.6 to 9.2 have been generated using a stochastic slip distribution approach (Melgar et al. 2016). Over this source zone, a total set of 2,800 rupture scenarios (400 per magnitude bin) with varying slip over the fault plane are generated, where the fault centre and rupture area for each event are first prescribed.

- The geometry of the subduction interface where synthetic seismicity is derived follows the extent of a seismotectonic region, in this case, the one 2 as in Poulos et al. (2019). Gridding within the segment and parameters characterizing the subfaults are taken from the on Slab1.0 (Hayes et al. 2012).
- Each scenario does not necessarily use the whole extent of this region (zone 2), and their individual sizing follows empirical earthquake scaling relations (e.g., Blaser et al. 2010).
- Slip distribution modelling was done based on the Karhunen–Loeve expansion approach implemented in MudPy (Melgar et al. 2016).
- The seismic velocity model required in MudPy is taken from Mendoza et al. (1994).

It is possible to include variability on source parameters, correlation lengths, and others, and finally, it is possible to include other seismogenic zones and/or sources to account for more tsunami scenario variability. These extensions are not tested herein.

The main limitations of the source modelling are the implementation of only static initial conditions, which means no rupture kinematics were implemented in the datasets in this study.

34.5 Tsunami Modelling

- Numerical modelling including inundation was performed using Tsunami-HySEA (Macías et al. 2017), a GPU-based numerical code, using high performance computing of the CTE-Power9 (Barcelona Supercomputing Center).
- The static Okada model (Okada 1985) was used to get initial conditions for tsunami modelling.
- Nesting topo-bathymetry grids: different sources of topo-bathymetry were interpolated to get a uniform computational grid, using triangular irregular networks, and Delaunay triangulation. Four nested regular grids were generated, with resolutions of 30, 7.5, 1.875 and 0.234 arc seconds.
- Topography and bathymetry datasets with high resolution were used. GEBCO (2019) provides free datasets used for the coarser grids, however, high resolution topo-bathymetry datasets are normally available from national or local organizations, with high levels of restriction. While this data is restricted to the public, for this study it was shared by SHOA under an agreement.
- No Manning friction coefficient spatial variability has been considered within the domain, hence a uniform Manning coefficient of n = 0.025 is used in this case. However, different n-values may be tested according to new findings (Cárdenas and Catalán 2022).

Table 34.1 Choices of the metrics' thresholds used in this study (Zamora et al. 2021)

Max. flow depth (*dm*)/ Arrival time (*at*)	Very short *at* < = 10 min	Short *10 min* < *at* < = 20 min	Delayed *at* > 20 min
Small: dm < = 0.3 [m]	C1	B1	A1
Medium: 0.3 [m] < dm < = 1.0 [m]	C2	B2	A2
Large: dm > 1.0 [m]	C3	B3	A3

34.6 Hazard Microzoning and Metrics

For each of the 2800 simulated scenarios two outputs were obtained at the finest nested grid (level 4): maximum flow depth and arrival time. These outputs are 2D variables (x,y). The thresholds for assessing the hazard based on maximum flow depths (*dm*) and arrival times (*at*) are defined in Table 34.1.

- The *dm* thresholds were defined according to the Chilean tsunami hazard levels used for evacuation purposes (see Catalán et al. 2020). Thresholds used: 0.3 [m] and 1.0 [m].
- The *at* thresholds were defined arbitrarily. The expected issuance of the first assessment by the Chilean Tsunami Warning System in Chile (rounded up to 10 min), and twice that time (20 min).
- Definition of 9-level qualitative hazard scale, from A1 (least hazardous) to C3 (most hazardous).

Application of categorization thresholds to qualify hazard. Intensity estimations can be applied to a single scenario, or a set of scenarios (e.g., all scenarios between Mw 8.6–9.2), but the categorization will be more appropriate if using a large set of events (Fig. 34.1).

34.7 Discussion

This method has been used to conduct microzoning of the tsunami hazard and can be used for evacuation maps and tsunami evacuation planning. For the present test case, the results show that the earliest tsunami arrival times have a significant spread among all scenarios. While most are relatively short, some events can cause inundation relatively late with respect to the onset of the earthquake. This method can help redefine tsunami watch periods.

One crucial application of such methods is the definition of evacuation routes, and the assessment of human behaviour to model their exposure. The latter will aid in decision making regarding, for example, the vertical evacuation procedures. Hence, these types of studies where a variation of tsunami scenarios are used to estimate the evacuation routes may benefit from this approach (e.g., León et al. 2022). In a

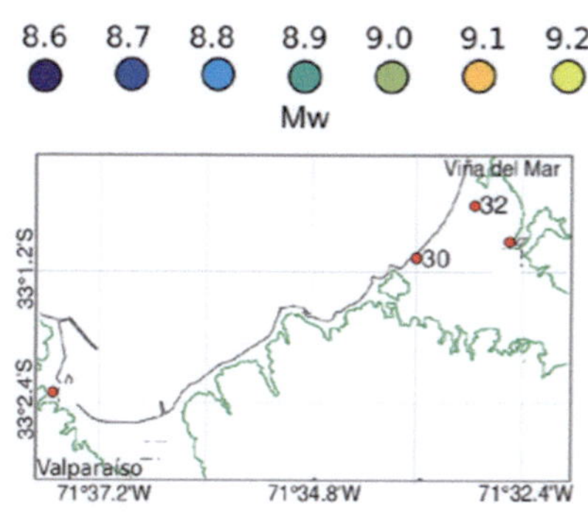

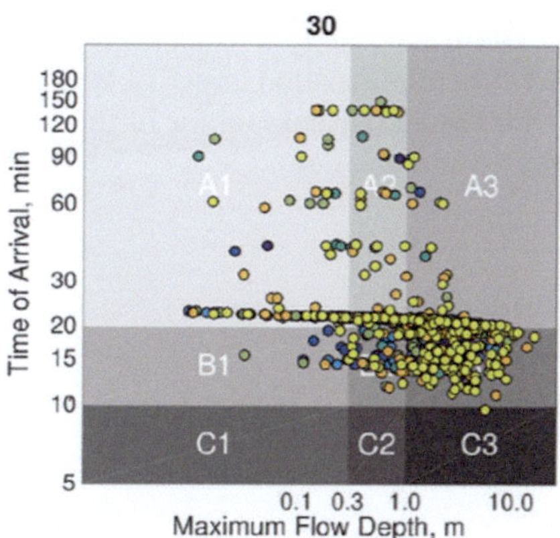

Fig. 34.1 Left: Analysis in one point (#30) in Viña del Mar. Right: Analysis Mw and arrival times against wave height. Letters and numbers stand for the categorization selected for this case. Source: Zamora et al. (2021), licensed under CC-BY 4.0

nation-wide strategy, this method is now tested in more cities in Chile, and more data can be introduced. In this regard, several real cases are used to test the method and drills (e.g., 2010 Maule, 2014 Iquique, 2015 Illapel).

Following these results, a clustering, ML-based approach (Nuñez et al. 2022) was developed as a proof of concept, where a few points were trained along the coast in central Chile and Coquimbo. Thus, further applications of this approach allow downstream coupling with agent-based evacuation modelling with actual tsunami data. Some studies have followed or benefited from this approach: Zamora et al. (2021); Nuñez et al. (2022); Zamora et al. (2023).

Finally, an aspect that was not considered in this recipe is the number of scenarios methods such as filtering/clustering (Lorito et al. 2015; Williamson et al. 2020) for reducing the number of scenarios that can be used without losing large variability of rupture characterizations. This will need further exploration.

34.8 Final Remarks

- This ‘recipe’ is used to generate a hazard mapping for decision making considering not only flow depths but minimum arrival times, providing an enhanced procedure for decision making for risk analysts or authorities.
- It has been observed that the large inundations seem to impose a high weight on the hazard zoning.
- In this method, a thorough treatment of uncertainties was not conducted only relying on the seismic source variabilities. Numerical simulations using the same Manning-value and similar setups in the numerical modelling neglect some effects in the inundation, but this is out of the scope of this recipe.

– The results related to this recipe can be found in Zamora et al. (2021), where further information is available regarding the rationale of the thresholds, the main results and discussion for further applications.

References

Blaser L, Kruger F, Ohrnberger M, Scherbaum F (2010) Scaling relations of earthquake source parameter estimates with special focus on subduction environment. Bull Seismol Soc Am 100:2914–2926. https://doi.org/10.1785/0120100111

Cárdenas G, Catalán PA (2022) Accelerating tsunami modeling for evacuation studies through modification of the manning roughness values. GeoHazards 3:492–507. https://doi.org/10.3390/geohazards3040025

Carvajal M, Cisternas M, Gubler A, Catalán P, Winckler P, Wesson R (2017b) Reexamination of the magnitudes for the 1906 and 1922 Chilean earthquakes using Japanese tsunami amplitudes: implications for source depth constraints. J Geophys Res 122:4–17. https://doi.org/10.1002/2016JB013269

Carvajal M, Cisternas M, Catalán PA (2017a) Source of the 1730 Chilean earthquake from historical records: implications for the future tsunami hazard on the coast of Metropolitan Chile. J Geophys Res Solid Earth 122:3648–3660. https://doi.org/10.1002/2017JB014063

Catalán PA, Gubler A, Cañas J, Zuñiga C, Zelaya C, Pizarro L et al (2020) Design and operational implementation of the integrated tsunami forecast and warning system in Chile (SIPAT). Coast Eng J 62:373–388. https://doi.org/10.1080/21664250.2020.1727402

Dura T, Cisternas M, Horton BP, Ely LL, Nelson AR, Wesson RL et al (2015) Coastal evidence for Holocene subduction-zone earthquakes and tsunamis in central Chile. Quat Sci Rev 113:93–111. https://doi.org/10.1016/j.quascirev.2014.10.015

Hayes GP, Wald DJ, Johnson RL (2012) Slab1.0: a three-dimensional model of global subduction zone geometries. J Geophys Res Solid Earth 117(B01). https://doi.org/10.1029/2011JB008524

León J, Gubler A, Ogueda A (2022) Modelling geographical and built-environment attributes as predictors of human vulnerability during tsunami evacuations: a multi-case-study and paths to improvement. Nat Hazards Earth Syst Sci 22(9):2857–2878. https://doi.org/10.5194/nhess-22-2857-2022

León J, Castro S, Mokrani C, Gubler A (2020) Tsunami evacuation analysis in the urban built environment: a multi-scale perspective through two modeling approaches in viña del mar, Chile. Coast Eng J 62:389–404. https://doi.org/10.1080/21664250.2020.1738073

Macías J, Castro MJ, Ortega S, Escalante C, González-Vida JM (2017) Performance benchmarking of tsunami-HySEA model for NTHMP's inundation mapping activities. Pure Appl Geophys 174:3147–3183. https://doi.org/10.1007/s00024-017-1583-1

Melgar D, LeVeque RJ, Dreger DS, Allen RM (2016) Kinematic rupture scenarios and synthetic displacement data: an example application to the Cascadia subduction zone. J Geophys Res Solid Earth 121:6658–6674. https://doi.org/10.1002/2016JB013314

Mendoza C, Hartzell S, Monfret T (1994) Wide-band analysis of the 3 March 1985 central Chile earthquake: overall source process and rupture history. Bull Seism Soc Amer 84(2):269–283. https://doi.org/10.1785/BSSA0840020269

Núñez J, Catalán PA, Valle C et al (2022) Discriminating the occurrence of inundation in tsunami early warning with one-dimensional convolutional neural networks. Sci Rep 12:10321. https://doi.org/10.1038/s41598-022-13788-9

Okada Y (1985) Surface deformation due to shear and tensile faults in a half-space. Bull Seismol Soc Am 75:1135–1154. https://doi.org/10.1785/BSSA0750041135

Poulos A, Monsalve M, Zamora N, de la Llera JC (2019) An updated recurrence model for Chilean subduction seismicity and statistical validation of its poisson nature. Bull Seismol Soc Am 109:66–74. https://doi.org/10.1785/0120170160

Zamora N, Catalán PA, Gubler A, Carvajal M (2021) Microzoning tsunami hazard by combining flow depths and arrival times. Front Earth Sci 8:591514. https://doi.org/10.3389/feart.2020.591514

Zamora N, León J, Gubler A, Catalán P (2023) Tsunami evacuation using an agent-based model in Chile. In: Galileo conference: solid earth and geohazards in the exascale Era, Barcelona, Spain, 23–26 May 2023, GC11-solidearth-29. https://doi.org/10.5194/egusphere-gc11-solidearth-29

MIX
Papier aus verantwortungsvollen Quellen
Paper from responsible sources
FSC® C105338

If you have any concerns about our products,
you can contact us on
ProductSafety@springernature.com

In case Publisher is established outside the EU,
the EU authorized representative is:
Springer Nature Customer Service Center GmbH
Europaplatz 3, 69115 Heidelberg, Germany

Printed by Libri Plureos GmbH
in Hamburg, Germany